INTERMEDIATE ALGEBRA WITH EARLY FUNCTIONS AND GRAPHING

SIXTH EDITION

LIAL • HORNSBY • MILLER

STUDENT'S SOLUTIONS MANUAL

INTERMEDIATE ALGEBRA
WITH
EARLY FUNCTIONS AND GRAPHING

SIXTH EDITION

LIAL • HORNSBY • MILLER

STUDENT'S SOLUTIONS MANUAL

PREPARED WITH THE ASSISTANCE OF

THERESA MCGINNIS

An imprint of Addison Wesley Longman, Inc.

Reading, Massachusetts • Menlo Park, California • New York • Harlow, England
Don Mills, Ontario • Sydney • Mexico City • Madrid • Amsterdam

Reproduced by Addison Wesley Educational Publishers from camera-ready copy supplied by the author.

ISBN 0-321-01319-0

2 3 4 5 6 7 8 9 10 VG 00 99 98

PREFACE

This book provides complete solutions for all the margin exercises, section exercises numbered 3, 7, 11, ..., and all the chapter review, chapter test, and cumulative review exercises in *Intermediate Algebra*, Sixth Edition, by Margaret L. Lial, John Hornsby, and Charles D. Miller. Some solutions are presented in more detail than others. Thus, you may need to refer to a similar exercise to find a solution that is presented in sufficient detail. As needed, artwork is provided to clarify and illustrate solutions.

The following people have made valuable contributions to the production of this *Student's Solutions Manual*: Theresa McGinnis, editor; Judy Martinez, typist; Darryl Nester, artist; and Laura Tanenbaum, proofreader.

CONTENTS

CHAPTER 4 SYSTEMS OF LINEAR EQUATIONS

CHAPTER 5 POLYNOMIALS

CHAPTER 6 RATIONAL EXPRESSIONS

CHAPTER 7 ROOTS AND RADICALS

CHAPTER 8 QUADRATIC EQUATIONS AND INEQUALITIES

CHAPTER 9 GRAPHS OF NONLINEAR FUNCTIONS AND CONIC SECTIONS

CHAPTER 10 EXPONENTIAL AND LOGARITHMIC FUNCTIONS

Chapter 1

THE REAL NUMBERS

1.1 Basic Terms

1.1 Margin Exercises

1. $\{10, \frac{3}{10}, 52, 98.6\}$

   The natural numbers are $\{1, 2, 3, 4, 5, 6, ...\}$. In the given set, the elements 10 and 52 are natural numbers.

2. **(a)** $\{x \mid x \text{ is a natural number less than } 5\}$

   The natural numbers are $\{1, 2, 3, 4, 5, 6, ...\}$. The ones less than 5 would be the set $\{1, 2, 3, 4\}$.

   (b) $\{13, 14, 15, ...\}$ is the set of natural numbers greater than 12. In set-builder notation, this set would be written as $\{x \mid x \text{ is a natural number greater than } 12\}$.

3. See the answer graphs for the margin exercises in the textbook.

   (a) $\{-4, -2, 0, 2, 4, 6\}$

   These are the even integers between and including -4 and 6. Place dots for $-4, -2, 0, 2, 4,$ and 6 on a number line.

   (b) $\{-1, 0, \frac{2}{3}, \frac{5}{2}\}$

   Place dots at -1 and 0 on a number line. To graph $\frac{2}{3}$, divide the interval between 0 and 1 into three equal parts. $\frac{2}{3}$ is located two-thirds of the distance from 0 to 1. To graph $\frac{5}{2}$, first change the fraction to a mixed number.

   $$\frac{5}{2} = 2\frac{1}{2}$$

   Divide the interval between 2 and 3 into two equal parts. $\frac{5}{2}$ is located halfway between 2 and 3.

   (c) $\{5, \frac{16}{3}, 6, \frac{13}{2}, 7, \frac{29}{4}\}$

   Place dots at 5, 6, and 7 on a number line. To locate the fractions, first change each to a mixed number and place dots at $\frac{16}{3} = 5\frac{1}{3}$, $\frac{13}{2} = 6\frac{1}{2}$, and $\frac{29}{4} = 7\frac{1}{4}$.

4. **(a)** -6 is a rational number because

   $$-6 = -\frac{6}{1}.$$

   -6 is also a real number.

   (b) 12 is a whole number, a rational number $\left(12 = \frac{12}{1}\right)$, and a real number.

   (c) $.\bar{3}$, a repeating decimal, is a rational number and a real number.

   (d) $-\sqrt{15}$ is an irrational number and a real number.

   (e) $-\frac{6}{0}$ is undefined.

   (f) π, a nonrepeating, nonterminating decimal, is an irrational number and a real number.

   (g) $\frac{22}{7}$ is a rational number and a real number.

   (h) 3.14, a terminating decimal, is a rational number and a real number.

5. **(a)** The statement "All whole numbers are integers" is true. As shown in Figure 3 in the textbook, the set of integers includes the set of whole numbers.

   (b) The statement "Some integers are whole numbers" is true since 0 and the positive integers are whole numbers. The negative integers are not whole numbers.

   (c) The statement "Every real number is irrational" is false. Some real numbers such as $\frac{4}{9}$, $.\bar{3}$, and $\sqrt{16}$ are rational. Others such as $\sqrt{2}$ and π are irrational.

6.

	Number	*Additive Inverse*
(a)	9	-9
(b)	-12	12
(c)	$-\frac{6}{5}$	$\frac{6}{5}$
(d)	0	0

7. **(a)** $|6| = 6$

   (b) $|-3| = 3$

(c) $-|5|$

Find the absolute value first, then find the additive inverse.

$$-|5| = -(5) = -5$$

(d) $-|-2| = -(2) = -2$

(e) $-(-7) = 7$

(f) $|-6| + |-3|$

Evaluate each absolute value first, then add.

$$|-6| + |-3| = 6 + 3 = 9$$

(g) $|-9| - |-4| = 9 - 4 = 5$

(h) $-|9-4| = -|5| = -5$

1.1 Section Exercises

3. The statement "Every number has an additive inverse" is true. All real numbers have additive inverses.

7. Graph $\{-3, -1, 0, 4, 6\}$.

   Place a dot for $-3, -1, 0, 4$, and 6 on a number line. See the answer graph in the back of the textbook.

11. The graph of a number is the point on the number line. The coordinate of a point is the number that corresponds to the point.

15. $\{z \mid z$ is an integer greater than $4\}$ is the set $\{5, 6, 7, 8, ...\}$.

19. $\{x \mid x$ is an irrational number that is also rational$\}$ is the empty set $\emptyset$. Irrational numbers cannot also be rational numbers.

23. $\{z \mid z$ is a whole number and a multiple of $3\}$

    The multiples of 3 are $3 \cdot 0, 3 \cdot 1, 3 \cdot 2, ...$ The set of whole numbers that are multiples of 3 is $\{0, 3, 6, 9, ...\}$.

27. $\{4, 8, 12, 16, ...\}$

    One way to describe this set is $\{x \mid x$ is a multiple of 4 greater than $0\}$.

31. $\{-8, -\sqrt{5}, -.6, 0, \frac{1}{0}, \frac{3}{4}, \sqrt{3}, 4, 5, \frac{13}{2}, 17, \frac{40}{2}\}$

    (a) The elements 4, 5, 17, and $\frac{40}{2}$ (or 20) are natural numbers.

    (b) The elements 0, 4, 5, 17, and $\frac{40}{2}$ (or 20) are whole numbers.

    (c) The elements -8, 0, 4, 5, 17, and $\frac{40}{2}$ (or 20) are integers.

    (d) The elements $-8, -.6, 0, \frac{3}{4}, 4, 5, \frac{13}{2}$, 17, and $\frac{40}{2}$ (or 20) are rational numbers.

    (e) The elements $-\sqrt{5}$ and $\sqrt{3}$ are irrational numbers.

    (f) All the elements are real numbers except $\frac{1}{0}$.

    (g) The element $\frac{1}{0}$ is undefined.

35. **(a)** The additive inverse of -12 is 12.

    (b) $|-12| = 12$

39. **(a)** The additive inverse of $\frac{6}{5}$ is $-\frac{6}{5}$.

    (b) $\left|\frac{6}{5}\right| = \frac{6}{5}$

43. $-|5| = -(5) = -5$

47. $-|4.5| = -(4.5) = -4.5$

51. $|-9| - |-3| = 9 - 3 = 6$

55. $|-1| + |-2| - |-3| = 1 + 2 - 3 = 3 - 3 = 0$

59. The statement "Every irrational number is an integer" is false. Irrational numbers cannot also be rational numbers. Therefore, every irrational number is not an integer.

63. The statement "Some rational numbers are irrational" is false. Rational numbers cannot also be irrational numbers.

67. The statement "The absolute value of any number is the same as the absolute value of its additive inverse" is true. The absolute value of any number represents its distance from zero on a number line. Since additive inverses are the same distance from zero, the absolute values of additive inverses must be the same.

71. The absolute value of the depth of the Pacific Ocean,

    $$|-12,925| = 12,925,$$

    is greater than the absolute value of the depth of the Indian Ocean,

    $$|-12,598| = 12,598.$$

    The statement is true.

75. $|-x| = x$

In words, the absolute value of the additive inverse of a number is the same as the number. This can only be true if the number is positive or equal to zero. The statement is true for all real numbers greater than or equal to 0.

1.2 Inequality

1.2 Margin Exercises

1. **(a)** $3 < 7$ since 3 is farther to the left on a number line than 7.

   (b) $9 > 2$ since 9 is farther to the right than 2.

   (c) $-4 > -8$ since -4 is farther to the right than -8.

   (d) $-2 < -1$ since -2 is farther to the left than -1.

   (e) $0 > -3$ since 0 is farther to the right than -3.

2. **(a)** $-2 \leq -3$ is false.

   Since -2 is farther to the right on a number line, it must be larger.

   (b) $8 \leq 8$ is true.

   The statement is "8 is less than or equal to 8." Since 8 is equal to 8, the statement is true.

   (c) $-9 \geq -1$ reads "-9 is greater than or equal to -1." This is false, since -9 is to the left of -1 on a number line.

   (d) $5 \cdot 8 \leq 7 \cdot 7$

   Since $5 \cdot 8 = 40$ and $7 \cdot 7 = 49$, then $40 \leq 49$. The statement is true.

   (e) $3(4) > (2)6$

   Since $3(4) = 12$ and $2(6) = 12$, the statement becomes $12 > 12$, which is false.

For Exercises 3-5, see the answer graphs for the margin exercises in the textbook.

3. **(a)** In interval notation, $\{x \mid x < -1\}$ is written as $(-\infty, -1)$. The right parenthesis indicates that -1 is not included. The graph extends from -1 to the left on a number line.

   (b) In interval notation, $\{x \mid x > 0\}$ is written as $(0, \infty)$. The left parenthesis indicates that 0 is not included. The graph extends from 0 to the right on a number line.

4. **(a)** In interval notation, $\{x \mid x \geq -3\}$ is written as $[-3, \infty)$. The bracket at -3 indicates that -3 is included. The graph extends from -3 to the right on a number line.

   (b) In interval notation, $\{x \mid x \leq 5\}$ is written as $(-\infty, 5]$. The bracket at 5 indicates that 5 is included. The graph extends from 5 to the left on a number line.

5. **(a)** In interval notation, $\{x \mid -1 \leq x \leq 2\}$ is written as $[-1, 2]$. The brackets indicate that -1 and 2 are both included. The graph goes from -1 to 2 on a number line.

   (b) In interval notation, $\{x \mid 8 < x < 12\}$ is written as $(8, 12)$. The parentheses indicate that neither 8 nor 12 is included. The graph goes from 8 to 12 on a number line.

   (c) In interval notation, $\{x \mid -4 \leq x < 2\}$ is written as $[-4, 2)$. The bracket at -4 indicates that -4 is included, while the parenthesis at 2 indicates that 2 is not included. The graph goes from -4 to 2 on a number line.

1.2 Section Exercises

3. $-4 > -3$ is false.

   Since -4 is farther to the left on a number line than -3, it must be smaller.

7. $-3 \geq -3$ is true.

   The statement is "-3 is greater than or equal to -3." Since -3 is equal to -3, the statement is true.

11. In $x < 5$, x represents any real number less than (but not equal to) 5. In $x \leq 5$, x represents any real number less than or equal to 5.

15. $3t - 4 \leq 10$

    The symbol for "is less than or equal to" is $\leq$.

19. $-3 < t < 5$

    Since t is between -3 and 5, it must be greater than -3 and less than 5.

23. $5x + 3 \neq 0$

    The symbol for "is not equal to" is $\neq$.

27. $2 \cdot 5 \geq 4 + 6$

    Since $2 \cdot 5 = 10$ and $4 + 6 = 10$, the statement becomes $10 \geq 10$, which is true.

31. $3 \not< 2$ reads "3 is not less than 2." The equivalent statement is "3 is greater than or equal to 2," which is $3 \geq 2$.

35. $5 \geq 3$ reads "5 is greater than or equal to 3." The equivalent statement is "5 is not less than 3," which is $5 \not< 3$.

For Exercises 39 and 43, see the answer graphs in the back of the textbook.

39. In interval notation, $\{x \mid x \leq 6\}$ is written as $(-\infty, 6]$. The bracket at 6 indicates that 6 is included. The graph extends from 6 to the left on a number line.

43. In interval notation, $\{x \mid 2 \leq x \leq 7\}$ is written as $[2, 7]$. The brackets indicate that 2 and 7 are both included. The graph goes from 2 to 7 on a number line.

47. From the graph, the number of Oregon permits was greater than 20,000 in the years 1989, 1990, 1993, 1994, and 1995.

1.3 Operations on Real Numbers

1.3 Margin Exercises

1. Add the absolute values of the numbers. The answer is negative since two negative numbers are being added.

(a) $-2+(-7)=-(2+7)=-9$

(b) $-15+(-6)=-(15+6)=-21$

(c) $-1.1+(-1.2)=-(1.1+1.2)=-2.3$

(d)
$$\begin{aligned}-\frac{3}{4}+\left(-\frac{1}{2}\right)&=-\left(\frac{3}{4}+\frac{1}{2}\right)\\&=-\left(\frac{3}{4}+\frac{2}{4}\right)\\&=-\frac{5}{4}\end{aligned}$$

2. **(a)** $12+(-1)=11$

Subtract the smaller absolute value from the larger absolute value $(12-1)$, and take the sign of the number with the larger absolute value.

(b) $3+(-7)=-4$

Subtract the smaller absolute value from the larger absolute value $(7-3)$, and take the sign of the larger.

(c) $-17+5=-12$

(d) $-\frac{3}{4}+\frac{1}{2}=-\frac{3}{4}+\frac{2}{4}=-\frac{1}{4}$

3. **(a)** $9-12$

Change the sign and add.

$$9-12=9+(-12)=-3$$

(b) $-7-2=-7+(-2)=-9$

(c) $-8-(-2)$

Change the sign of -2, then add.

$$-8-(-2)=-8+2=-6$$

(d) $-6.3-(-11.5)=-6.3+11.5=5.2$

(e) $12-(-5)=12+5=17$

4. **(a)** $-6+9-2$

Add and subtract in order from left to right.

$$\begin{aligned}-6+9-2&=(-6+9)-2\\&=3-2\\&=1\end{aligned}$$

(b)
$$\begin{aligned}12-(-4)+8&=(12+4)+8\\&=16+8\\&=24\end{aligned}$$

(c)
$$\begin{aligned}-6-(-2)-8-1&=-6+2-8-1\\&=-4+(-8)-1\\&=-12-1\\&=-13\end{aligned}$$

(d)
$$\begin{aligned}-3-[(-7)+15]+6&=-3-8+6\\&=-11+6=-5\end{aligned}$$

5. **(a)** $(-7)(-5)=7\cdot 5=35$

The numbers have the same sign, so the product is positive.

(b) $(-.9)(-15)=13.5$

(c) $\left(-\frac{4}{7}\right)\left(-\frac{14}{3}\right)=\frac{56}{21}=\frac{7\cdot 8}{7\cdot 3}=\frac{8}{3}$

(d) $7(-2)=-14$

The numbers have different signs, so the product is negative.

(e) $(-.8)(.006) = -.0048$

(f) $\frac{5}{8}(-16) = \frac{-80}{8} = -10$

(g) $\left(-\frac{2}{3}\right)(12) = \frac{-24}{3} = -8$

6.

	Number	*Reciprocal*
(a)	15	$\frac{1}{15}$
(b)	-7	$-\frac{1}{7}$
(c)	$\frac{8}{9}$	$\frac{9}{8}$
(d)	$.125 = \frac{1}{8}$	8
(e)	$.0\bar{5} = \frac{1}{18}$	18

7. **(a)** $\frac{-16}{4} = -16\left(\frac{1}{4}\right) = -4$

The numbers have opposite signs, so the quotient is negative.

(b) $\frac{8}{-2} = 8\left(\frac{1}{-2}\right) = -4$

(c) $\frac{-.15}{-.3} = -.15\left(-\frac{1}{.3}\right) = .5$

(d) $\frac{\frac{3}{8}}{\frac{11}{16}} = \frac{3}{8} \div \frac{11}{16} = \frac{3}{8} \cdot \frac{16}{11} = \frac{6}{11}$

1.3 Section Exercises

3. The sum of two negative numbers is a *negative* number. For example,

$$-7 + (-21) = -28.$$

7. The difference between two negative numbers is negative if *the one with smaller absolute value is subtracted from the one with larger absolute value.* For example,

$$-15 - (-3) = -12.$$

11. $13 + (-4) = 9$

15. $-\frac{7}{3} + \frac{3}{4}$

Write each number with a common denominator.

$$-\frac{7}{3} = -\frac{7 \cdot 4}{3 \cdot 4} = -\frac{28}{12}$$
$$\frac{3}{4} = \frac{3 \cdot 3}{4 \cdot 3} = \frac{9}{12}$$
$$-\frac{7}{3} + \frac{3}{4} = -\frac{28}{12} + \frac{9}{12} = -\frac{19}{12}$$

19. $-6 - 5 = -6 + (-5) = -11$

23. $-16 - (-3) = -16 + 3 = -13$

27. $\frac{9}{10} - \left(-\frac{4}{3}\right) = \frac{9}{10} + \frac{4}{3} = \frac{27}{30} + \frac{40}{30} = \frac{67}{30}$

31. Answers will vary. One example is

$$-4 - (-9) = -4 + 9 = 5.$$

In Exercises 35-47, the product of two numbers with the same sign is positive. The product of two numbers with different signs is negative.

35. $5(-7) = -35$

39. $-10\left(-\frac{1}{5}\right) = \frac{10}{5} = 2$

43. $-\frac{5}{2}\left(-\frac{12}{25}\right) = \frac{60}{50} = \frac{6}{5}$

47. $-2.4(-2.45) = 5.88$

In Exercises 51-59, a number and its reciprocal always have the same sign.

51. The reciprocal of 6 is $\frac{1}{6}$ since

$$6 \cdot \frac{1}{6} = 1.$$

55. The reciprocal of $-\frac{2}{3}$ is $-\frac{3}{2}$ since

$$-\frac{2}{3}\left(-\frac{3}{2}\right) = 1.$$

59. $-.001 = -\frac{1}{1000}$

The reciprocal of $-.001$ is -1000.

In Exercises 63-75, the quotient of two nonzero real numbers with the same sign is positive. The quotient of two nonzero real numbers with different signs is negative.

63. $\frac{-14}{2} = -14\left(\frac{1}{2}\right) = -7$

67. $\frac{100}{-25} = 100\left(\frac{1}{-25}\right) = -4$

71. $-\frac{10}{17} \div \left(-\frac{12}{5}\right) = -\frac{10}{17} \cdot \left(-\frac{5}{12}\right)$
$= \frac{50}{204} = \frac{25}{102}$

75. $-\dfrac{27.72}{13.2} = -27.72\left(\dfrac{1}{13.2}\right) = -2.1$

79. $-7+5-9 = (-7+5)-9 = -2-9 = -11$

83. $$\begin{aligned} -9-4-(-3)+6 &= (-9-4)-(-3)+6 \\ &= -13-(-3)+6 \\ &= (-13+3)+6 \\ &= -10+6 \\ &= -4 \end{aligned}$$

87. To find the difference between these two temperatures, subtract the low temperature from the high temperature.

$$90^\circ - (-22^\circ) = 90^\circ + 22^\circ = 112^\circ\text{F}$$

91. Locate the percents for 1993 in the graph. To find the difference between the percent change for the U.S., 2.0, and that for California, $-.9$, subtract.

$$2.0 - (-.9) = 2.0 + .9 = 2.9$$

1.4 Exponents and Roots; Order of Operations

1.4 Margin Exercises

1. **(a)** $5^3 = 5\cdot5\cdot5 = 25\cdot5 = 125$

   (b) $3^4 = 3\cdot3\cdot3\cdot3 = 81$

   (c) $$\begin{aligned} (-4)^5 &= (-4)(-4)(-4)(-4)(-4) \\ &= 16(-4)(-4)(-4) \\ &= -64(-4)(-4) \\ &= 256(-4) \\ &= -1024 \end{aligned}$$

   (d) $(-3)^4 = (-3)(-3)(-3)(-3) = 81$

   (e) $(.75)^3 = (.75)(.75)(.75) = (.5625)(.75) = .421875$

   (f) $\left(\dfrac{2}{5}\right)^4 = \dfrac{2}{5}\cdot\dfrac{2}{5}\cdot\dfrac{2}{5}\cdot\dfrac{2}{5} = \dfrac{16}{625}$

2.

		Exponent	*Base*
(a)	7^5	5	7
(b)	m^3	3	m
(c)	$(-5)^7$	7	-5
(d)	-12^4	4	12
(e)	$-(.9)^4$	4	.9

3. **(a)** $\sqrt{9} = 3$ since 3 is positive and $3^2 = 9$.

   (b) $\sqrt{49} = 7$ since 7 is positive and $7^2 = 49$.

   (c) $-\sqrt{81} = -9$ since the negative sign is outside the radical sign.

   (d) $\sqrt{\dfrac{121}{81}} = \dfrac{11}{9}$ since $\left(\dfrac{11}{9}\right)^2 = \dfrac{121}{81}$.

   (e) $\sqrt{.25} = .5$ since $(.5)^2 = .25$.

   (f) $\sqrt{-9}$ is not a real number.

   (g) $-\sqrt{-169}$ is not a real number.

4. **(a)** $\sqrt[3]{64} = 4$ since $4^3 = 64$.

   (b) $\sqrt[3]{-\dfrac{27}{8}} = -\dfrac{3}{2}$ since $\left(-\dfrac{3}{2}\right)^3 = -\dfrac{27}{8}$.

   (c) $\sqrt[3]{0} = 0$ since $0^3 = 0$.

   (d) $\sqrt[4]{625} = 5$ since $5^4 = 625$.

   (e) $\sqrt[4]{\dfrac{16}{625}} = \dfrac{2}{5}$ since $\left(\dfrac{2}{5}\right)^4 = \dfrac{16}{625}$.

   (f) $\sqrt[4]{-16}$ is not a real number since no real number raised to the fourth power is -16.

5. $5\cdot9+2\cdot4$

   Multiply first, then add.

   $5\cdot9+2\cdot4 = 45+8 = 53$

6. $(4+2)-3^2-(8-3)$

   Evaluate the exponent first.

   $= (4+2)-9-(8-3)$

   Add and subtract inside parentheses.

   $= 6-9-5$

   Subtract from left to right.

   $= -3-5$
   $= -8$

7. $6+\dfrac{2}{3}(-9)-\dfrac{5}{8}\cdot16$

   Do the multiplications first.

   $= 6+(-6)-10$
   $= 0-10$
   $= -10$

8. $\dfrac{\frac{1}{2}\cdot10-6+\sqrt{9}}{\frac{5}{6}\cdot12-3(2)^2}$

   Evaluate powers and roots.

   $= \dfrac{\frac{1}{2}\cdot10-6+3}{\frac{5}{6}\cdot12-3(4)}$

Multiply in the numerator and denominator.

$$= \frac{5-6+3}{10-12}$$

Add and subtract in the numerator and denominator.

$$= \frac{2}{-2} = -1$$

9. Let $x = -12$, $y = 64$, and $z = -3$. Substitute the appropriate values in each expression.

(a) $$\begin{aligned} 5x - 2 \cdot \sqrt{y} &= 5(-12) - 2\sqrt{64} \\ &= 5(-12) - 2(8) \\ &= -60 - 16 \\ &= -76 \end{aligned}$$

(b) $$\begin{aligned} -6(x - \sqrt[3]{y}) &= -6[(-12) - \sqrt[3]{64}] \\ &= -6[(-12) - 4] \\ &= -6(-16) \\ &= 96 \end{aligned}$$

(c) $$\begin{aligned} \frac{5x - 3 \cdot \sqrt{y}}{x-1} &= \frac{5(-12) - 3\sqrt{64}}{-12-1} \\ &= \frac{5(-12) - 3(8)}{-12-1} \\ &= \frac{-60-24}{-12-1} \\ &= \frac{-84}{-13} \\ &= \frac{84}{13} \end{aligned}$$

(d) $$\begin{aligned} x^2 + 2z^3 &= (-12)^2 + 2(-3)^3 \\ &= 144 + 2(-27) \\ &= 144 + (-54) \\ &= 90 \end{aligned}$$

1.4 Section Exercises

3. The statement "$\sqrt{16}$ is a positive number" is true. The symbol $\sqrt{\ }$ always gives a positive square root.

7. The statement "The product of 8 positive factors and 8 negative factors is positive" is true. The product of an even number of negative factors is positive.

11. $4^2 = 4 \cdot 4 = 16$

15. $\left(\frac{1}{5}\right)^3 = \frac{1}{5} \cdot \frac{1}{5} \cdot \frac{1}{5} = \frac{1}{125}$

19. $(-5)^3 = (-5)(-5)(-5) = -125$

23. $-3^6 = -(3 \cdot 3 \cdot 3 \cdot 3 \cdot 3 \cdot 3) = -729$

27. $(-a)^n = -a^n$ if a is a real number and n is an odd natural number.

31. In -4.1^7, the exponent is 7 and the base is 4.1.

35. $\sqrt{169} = 13$ since $13^2 = 169$.

39. $\sqrt{\frac{100}{121}} = \frac{10}{11}$ since $\left(\frac{10}{11}\right)^2 = \frac{100}{121}$.

43. $\sqrt{-25}$ is not a real number since no real number squared equals -25.

47. If a is a positive number, then $-\sqrt{-a}$ would be the negative square root of a negative number which is not a real number.

51. $\sqrt[4]{625} = 5$ since $5^4 = 625$.

55. $-\sqrt[6]{729} = -3$ since $3^6 = 729$. (Don't forget the negative sign from outside the radical sign.)

59. $-12\left(-\frac{3}{4}\right) - (-5) = 9 - (-5) = 9 + 5 = 14$

63. $(-7)(\sqrt{36}) - (-2)(-3)$

Find the square root first.

$$\begin{aligned} &= (-7)(6) - (-2)(-3) \\ &= -42 - 6 \\ &= -48 \end{aligned}$$

67. $\dfrac{(-6+3)(-2^2)}{-5-1}$

Evaluate the power.

$$= \frac{(-6+3)(-4)}{-5-1}$$

Add within the parentheses.

$$= \frac{(-3)(-4)}{-5-1}$$

$$= \frac{12}{-5-1}$$

$$= \frac{12}{-6}$$

$$= -2$$

71. $\dfrac{\frac{1}{4} \cdot 16 + 3}{\frac{1}{2} \cdot 12 - 1} = \dfrac{4+3}{6-1} = \dfrac{7}{5}$

In Exercises 75 and 79, let $a = -3$, $b = 64$, and $c = 6$. Substitute the appropriate values in each expression.

75. $3a + \sqrt{b} = 3(-3) + \sqrt{64}$
$= 3(-3) + 8$
$= -9 + 8$
$= -1$

79. $4a^3 + 2c = 4(-3)^3 + 2(6)$
$= 4(-27) + 2(6)$
$= -108 + 12$
$= -96$

83. $(-9-5)(-2^3+5) = (-9-5)(-8+5)$
$= (-14)(-3)$
$= 42$

1.5 Properties of Real Numbers

1.5 Margin Exercises

1. **(a)** Let $a = 8$, $b = m$, and $c = n$ in the statement of the distributive property. Then,

$$8(m+n) = 8m + 8n.$$

(b) $-4(p-5) = -4(p + (-5))$
$= -4(p) + (-4)(-5)$
$= -4p + 20$

(c) Use the second form of the distributive property.

$$3k + 6k = (3+6)k = 9k$$

(d) Use the second form of the distributive property.

$$-6m + 2m = (-6+2)m$$
$$= -4m$$

(e) $2r + 3s$

Since there is no common number or variable here, the distributive property cannot be used.

2. **(a)** $14 \cdot 5 + 14 \cdot 85 = 14(5 + 85)$
$= 14(90)$
$= 1260$

(b) $78 \cdot 33 + 22 \cdot 33 = (78 + 22)33$
$= 100(33)$
$= 3300$

3. **(a)** The number that must be added to 4 to get 0 is -4.

(b) The number that must be added to -7.1 to get 0 is 7.1.

(c) The sum of -9 and 9 is 0.

(d) The number that must be multiplied by 5 to get 1 is the reciprocal of 5, which is $\frac{1}{5}$.

(e) The number that must be multiplied by $-\frac{3}{4}$ to get 1 is the reciprocal of $-\frac{3}{4}$, which is $-\frac{4}{3}$.

(f) The product of 7 and $\frac{1}{7}$ is 1.

4. **(a)** $p - 3p = 1p - 3p$
Identity property
$= (1-3)p$
Distributive property
$= -2p$

(b) $r + r + r = 1r + 1r + 1r$
Identity property
$= (1+1+1)r$
Distributive property
$= 3r$

(c) $-(3 + 4p) = -1(3 + 4p)$
Identity property
$= -1(3) + (-1)(4p)$
Distributive property
$= -3 + (-4p)$
$= -3 - 4p$

(d) $-(k-2) = -1(k-2)$
Identity property
$= -1(k) + (-1)(-2)$
Distributive property
$= -k + 2$

5. **(a)** $12b - 9b + 4b - 7b + b$
$= (12 - 9 + 4 - 7 + 1)b$
Distributive property
$= b$

(b) $-3w + 7 - 8w - 2$
$= -3w - 8w + 7 - 2$
Commutative and associative properties
$= -11w + 5$
Combine like terms.

(c) $-3(6 + 2t) = -18 - 6t$
Distributive property

(d) $9 - 2(a - 3) + 4 - a$
$= 9 - 2a + 6 + 4 - a$
Distributive property
$= -2a - a + 9 + 6 + 4$
Commutative and associative properties
$= -3a + 19$ or $19 - 3a$
Combine like terms.

(e) $(4m)(2n) = (4m)(2)(n)$
Order of operations
$= (4)(m \cdot 2)(n)$
Associative property
$= (4)(2 \cdot m)(n)$
Commutative property
$= (4 \cdot 2)(m \cdot n)$
Associative property
$= 8mn$

6. **(a)** Using the multiplication property of zero,

$$197 \cdot 0 = 0.$$

(b) Using the multiplication property of zero,

$$(0)\left(-\frac{8}{9}\right) = 0.$$

(c) $0 \cdot \_\_\_ = 0$

Any real number will be a solution, since zero times any number will equal zero.

1.5 Section Exercises

3. $58 \cdot \frac{3}{2} - 8 \cdot \frac{3}{2} = (58 - 8)\frac{3}{2} = 50\left(\frac{3}{2}\right) = 75$

7. Like terms are terms with exactly the same variables raised to exactly the same powers.

11. Using the second form of the distributive property,

$$5k + 3k = (5 + 3)k$$
$$= 8k.$$

15. $-8z + 4w$

This expression cannot be simplified. Since there is no common number or variable, the distributive property cannot be used.

19. Using the distributive property,

$$2(m + p) = 2m + 2p.$$

23. $-12y + 4y + 3 + 2y = -12y + 4y + 2y + 3$
$= (-12 + 4 + 2)y + 3$
$= -6y + 3$

In Exercises 27-35, use the distributive property to remove parentheses and then combine terms.

27. $3(k + 2) - 5k + 6 + 3$
$= 3k + 6 - 5k + 6 + 3$
$= 3k - 5k + 6 + 6 + 3$
$= (3 - 5)k + 6 + 6 + 3$
$= -2k + 15$

31. $.25(8 + 4p) - .5(6 + 2p)$
$= 2 + p - 3 - p$
$= p - p + 2 - 3$
$= (1 - 1)p + 2 - 3$
$= 0p - 1$
$= -1$

35. $2 + 3(2z - 5) - 3(4z + 6) - 8$
$= 2 + 6z - 15 - 12z - 18 - 8$
$= 6z - 12z + 2 - 15 - 18 - 8$
$= (6 - 12)z + 2 - 15 - 18 - 8$
$= -6z - 39$

39. $5(9r) = (5 \cdot 9)r = 45r$ *Associative property*

43. $1 \cdot 7 = 7$ *Identity property*

47. Answers will vary. One example of a commutative operation is the activities of washing your face and brushing your teeth. The activities can be carried out in either order.

51. $3x + 4 + 2x + 7$
$= (3x + 4) + (2x + 7)$
Associative property of addition

52. $= 3x + (4 + 2x) + 7$
Associative property of addition

53. $= 3x + (2x + 4) + 7$
Commutative property of addition

54. $= (3x + 2x) + (4 + 7)$
Associative property of addition

55. $= (3 + 2)x + (4 + 7)$
Distributive property

56. $= 5x + 11$
Arithmetic facts

Chapter 1 Review Exercises

For Exercises 1 and 2, see the answer graphs in the back of the textbook.

1. $\{-4, -1, 2, \frac{9}{4}, 4\}$

   Place dots at $-4, -1, 2$, and 4 on a number line. To locate $\frac{9}{4}$, change the fraction to a mixed number. Place a dot at $2\frac{1}{4}$.

2. $\{-5, -\frac{11}{4}, -.5, 0, 3, \frac{13}{3}\}$

   Place dots at $-5, -\frac{11}{4} = -2\frac{3}{4}, -.5$ (or $-\frac{1}{2}$), 0, 3, and $\frac{13}{3} = 4\frac{1}{3}$ on a number line.

3. $|-16| = 16$

4. $|23| = 23$

5. $-|-4| = -(4) = -4$

6. $-|-8| + |-3| = -8 + 3 = -5$

In Exercises 7-10,

$$S = \left\{-9, -\frac{4}{3}, -\sqrt{4}, -.25, 0, .\overline{35}, \frac{5}{3}, \sqrt{7}, \sqrt{-9}, \frac{12}{3}\right\}.$$

7. The elements 0 and $\frac{12}{3}$ (or 4) are whole numbers.

8. The elements $-9, -\sqrt{4}$ (or -2), 0, and $\frac{12}{3}$ (or 4) are integers.

9. The elements $-9, -\sqrt{4}$ (or -2), $-\frac{4}{3}$, $-.25$, 0, $.\overline{35}$, $\frac{5}{3}$, and $\frac{12}{3}$ (or 4) are rational numbers. (Remember that terminating and repeating decimals are rational numbers.)

10. All the elements in the set are real numbers except $\sqrt{-9}$.

11. $\{x \mid x$ is a natural number between 3 and 9$\}$

    The natural numbers between 3 and 9 are 4, 5, 6, 7, and 8. Therefore, the set is $\{4, 5, 6, 7, 8\}$.

12. $\{y \mid y$ is a whole number less than 4$\}$ is the set $\{0, 1, 2, 3\}$.

13. $4 \cdot 2 \le |12 - 4|$

    The statement is true.
    Since $4 \cdot 2 = 8$, and $|12 - 4| = 8$, the statement becomes $8 \le 8$, which is true.

14. $2 + |-2| > 4$

    The statement is false.
    Since $2+|-2| = 2+2 = 4$, the statement becomes $4 > 4$, which is false.

15. $4(3 + 7) > -|40|$

    The statement is true.
    Since $4(3 + 7) = 4(10) = 40$, and $-|40| = -40$, the statement becomes $40 > -40$, which is true.

For Exercises 16 and 17, see the answer graphs in the back of the textbook.

16. $\{x \mid x < -5\}$

    In interval notation, $x < -5$ is written as $(-\infty, -5)$. The parenthesis at -5 indicates that -5 is not included. The graph extends from -5 to the left on a number line.

17. $\{x \mid -2 < x \le 3\}$

    In interval notation, $-2 < x \le 3$ is written as $(-2, 3]$. The parenthesis indicates that -2 is not included, while the bracket indicates that 3 is included. The graph goes from -2 to 3 on a number line.

18. $$-\frac{5}{8} - \left(-\frac{7}{3}\right) = -\frac{5}{8} + \frac{7}{3} = -\frac{15}{24} + \frac{56}{24} = \frac{41}{24}$$

19. $$-\frac{4}{5} - \left(-\frac{3}{10}\right) = -\frac{4}{5} + \frac{3}{10} = -\frac{8}{10} + \frac{3}{10} = -\frac{5}{10} = -\frac{1}{2}$$

20. $-5 + (-11) + 20 - 7 = -16 + 20 - 7 = 4 - 7 = -3$

21. $-9.42 + 1.83 - 7.6 - 1.9$
    $= -7.59 - 7.6 - 1.9$
    $= -15.19 - 1.9$
    $= -17.09$

22. $-15 + (-13) + (-11) = -28 + (-11) = -39$

23. $-1-3-(-10)+(-7)$
$= -4-(-10)+(-7)$
$= -4+10+(-7)$
$= 6+(-7)$
$= -1$

24. $\frac{3}{4}-\left(\frac{1}{2}-\frac{9}{10}\right)=\frac{3}{4}-\left(\frac{5}{10}-\frac{9}{10}\right)$
$= \frac{3}{4}-\left(-\frac{4}{10}\right)$
$= \frac{3}{4}+\frac{4}{10}$
$= \frac{15}{20}+\frac{8}{20}$
$= \frac{23}{20}$

25. $-\frac{2}{3}-\left(\frac{1}{6}-\frac{5}{9}\right)=-\frac{2}{3}-\left(\frac{3}{18}-\frac{10}{18}\right)$
$= -\frac{2}{3}-\left(-\frac{7}{18}\right)$
$= -\frac{2}{3}+\frac{7}{18}$
$= -\frac{12}{18}+\frac{7}{18}$
$= -\frac{5}{18}$

26. $-|-12|-|-9|+(-4)-|10|$
$= -12-9+(-4)-10$
$= -21+(-4)-10$
$= -25-10$
$= -35$

27. If the signs of the numbers being added are the same, then the sign of the sum matches the sign of the numbers. If the signs are different, then the sign of the sum matches the sign of the number with the larger absolute value.

28. To subtract $a-b$, write as an addition problem, $a+(-b)$, and add.

29. $2(-5)(-3)(-3)=(-10)(-3)(-3)$
$= (30)(-3)$
$= -90$

30. $-\frac{3}{7}\left(-\frac{14}{9}\right)=\frac{42}{63}=\frac{2}{3}$

31. $-4.6(2.48)=-11.408$

32. $\frac{-38}{-19}=-38\left(\frac{1}{-19}\right)=2$

33. $\frac{75}{-5}=75\left(\frac{1}{-5}\right)=-15$

34. $\frac{\frac{2}{3}}{-\frac{1}{6}}=\frac{2}{3}\cdot\frac{1}{-\frac{1}{6}}=\frac{2}{3}\cdot\frac{-6}{1}=-4$

35. $\frac{-2.3754}{-.74}=3.21$

36. $\frac{5}{7-7}=\frac{5}{0}$, which is undefined.
$\frac{7-7}{5}=\frac{0}{5}=0$

37. $10^4=10\cdot10\cdot10\cdot10=10,000$

38. $\left(\frac{3}{7}\right)^3=\frac{3}{7}\cdot\frac{3}{7}\cdot\frac{3}{7}=\frac{27}{343}$

39. $(-5)^3=(-5)(-5)(-5)=-125$

40. $-5^3=-(5\cdot5\cdot5)=-125$

41. $(1.7)^2=(1.7)(1.7)=2.89$

42. $\sqrt{400}=20$ since $20^2=400$.

43. $\sqrt[3]{27}=3$ since $3^3=27$.

44. $\sqrt[3]{-343}=-7$ since $(-7)^3=-343$.

45. $\sqrt[4]{81}=3$ since $3^4=81$.

46. $\sqrt[6]{-64}$ is not a real number.

47. $-14\left(\frac{3}{7}\right)+6\div3$

Multiply and divide in order first.

$= -6+2$
$= -4$

48. $-\frac{2}{3}[5(-2)+8-4^3]$

Evaluate the power.

$= -\frac{2}{3}[5(-2)+8-64]$

$= -\frac{2}{3}(-10+8-64)$

$= -\frac{2}{3}(-66)$

$= 44$

49. $\frac{-4(\sqrt{25})-(-3)(-5)}{3+(-6)\sqrt{9})}$

Evaluate the roots.

$= \frac{-4(5)-(-3)(-5)}{3+(-6)(3)}$

Multiply in the numerator and denominator.

$= \frac{-20-15}{3+(-18)}$

Add and subtract in the numerator and denominator.

$= \frac{-35}{-15}$

Simplify.

$= \frac{7}{3}$

50. $\frac{-5(3^2)+9(\sqrt{4})-5}{6-5(\sqrt[3]{-8})}$

Evaluate the power and roots.

$= \frac{-5(9)+9(2)-5}{6-5(-2)}$

Multiply in the numerator and denominator.

$= \frac{-45+18-5}{6+10}$

Add and subtract in the numerator and denominator.

$= \frac{-32}{16}$

Simplify.

$= -2$

In Exercises 51-54, let $k = -4$, $m = 2$, and $n = 16$. Substitute the appropriate values in each expression.

51. $4k - 7m = 4(-4) - 7(2)$
$= -16 - 14$
$= -30$

52. $-3(\sqrt{n}) + m + 5k$
$= -3(\sqrt{16}) + 2 + 5(-4)$
$= -3(4) + 2 + 5(-4)$
$= -12 + 2 - 20$
$= -30$

53. $-2(3k^2 + 5m) = -2[3(-4)^2 + 5(2)]$
$= -2[3(16) + 5(2)]$
$= -2[48 + 10]$
$= -2[58]$
$= -116$

54. $\frac{4m^3 - 3n}{7k^2 - 10} = \frac{4(2)^3 - 3(16)}{7(-4)^2 - 10}$
$= \frac{4(8) - 3(16)}{7(16) - 10}$
$= \frac{32 - 48}{112 - 10}$
$= \frac{-16}{102}$
$= -\frac{8}{51}$

55. In order to evaluate $(3 + 2)^2$, you should work within the parentheses first.

56. Let $a = 4$ and $b = 6$.

$(a+b)^2 = (4+6)^2 = 10^2 = 100$
$a^2 + b^2 = 4^2 + 6^2 = 16 + 36 = 52$

Since $100 \neq 52$, this shows that $(a+b)^2 \neq a^2 + b^2$.

57. $2q + 19q$
$= (2 + 19)q$ *Distributive property*
$= 21q$

58. $13z - 17z$
$= (13 - 17)z$ *Distributive property*
$= -4z$

59. $-m + 6m$
$= -1m + 6m$ *Identity property*
$= (-1 + 6)m$ *Distributive property*
$= 5m$

60. $5p - p$
$= 5p + (-1)p$ *Identity property*
$= (5 + (-1))p$ *Distributive property*
$= 4p$

61. $-2(k + 3)$
$= -2(k) + (-2)(3)$ *Distributive property*
$= -2k - 6$

62. $6(r + 3)$
$= 6(r) + 6(3)$ *Distributive property*
$= 6r + 18$

63. $9(2m + 3n)$
$= 9(2m) + 9(3n)$ *Distributive property*
$= 18m + 27n$

64. $-(3k - 4h)$
$= -1(3k - 4h)$ *Identity property*
$= -1(3k) + (-1)(-4h)$ *Distributive property*
$= -3k + 4h$

65. $-(-p+6q)-(2p-3q)$
$= -1(-p+6q)+(-1)(2p-3q)$
$= -1(-p)+(-1)(6q)+(-1)(2p)+(-1)(-3q)$
$= p-6q-2p+3q$
$= p-2p-6q+3q$
$= -p-3q$

66. $-2x+5-4x+1 = -2x-4x+5+1 = -6x+6$

67. $-3y+6-5+4y = -3y+4y+6-5 = y+1$

68. $2a+3-a-1-a-2$
$= 2a-a-a+3-1-2$
$= 0$

69. $-2(k-1)+3k-k = -2k+2+3k-k$
$= -2k+3k-k+2$
$= 2$

70. $-3(4m-2)+2(3m-1)-4(3m+1)$
$= -12m+6+6m-2-12m-4$
$= -12m+6m-12m+6-2-4$
$= -18m$

71. $2x+3x = (2+3)x = 5x$ *Distributive property*

72. $-4\cdot 1 = -4$ *Identity property*

73. $2(4x) = (2\cdot 4)x = 8x$ *Associative property*

74. $-3+13 = 13+(-3) = 10$ *Commutative property*

75. $-3+3 = 0$ *Inverse property*

76. $5(x+z) = 5x+5z$ *Distributive property*

77. $0+7 = 7$ *Identity property*

78. $8\cdot\frac{1}{8} = 1$ *Inverse property*

79. $3a+5a+6a = (3+5+6)a = 14a$
Distributive property

80. $\frac{9}{28}\cdot 0 = 0$ *Multiplication property of 0*

81. $\left(-\frac{4}{5}\right)^4 = \left(-\frac{4}{5}\right)\left(-\frac{4}{5}\right)\left(-\frac{4}{5}\right)\left(-\frac{4}{5}\right) = \frac{256}{625}$

82. $-\frac{5}{8}(-40) = -\frac{5}{8}\cdot\frac{-40}{1} = 25$

83. $-25\left(-\frac{4}{5}\right)+3^3-32\div\sqrt{4}$
$= -25\left(-\frac{4}{5}\right)+27-32\div 2$
$= 20+27-16$
$= 31$

84. $-8+|-14|+|-3| = -8+14+3 = 9$

85. $\frac{6\cdot\sqrt{4}-3\cdot\sqrt{16}}{-2\cdot 5+7(-3)-10} = \frac{6\cdot 2-3\cdot 4}{-2\cdot 5+7(-3)-10}$
$= \frac{12-12}{-10-21-10}$
$= \frac{0}{-41}$
$= 0$

86. $-\sqrt[5]{32} = -2$ since $2^5 = 32$. (Don't forget the negative sign from outside the radical.)

87. $-\frac{10}{21}\div\left(-\frac{5}{14}\right) = -\frac{10}{21}\left(-\frac{14}{5}\right)$
$= \frac{140}{105}$
$= \frac{4}{3}$

88. $.8-4.9-3.2+1.14 = -4.1-3.2+1.14$
$= -7.3+1.14$
$= -6.16$

89. $-3^2 = -(3\cdot 3) = -9$

Chapter 1 Test

1. $\{-3, .75, \frac{5}{3}, 5, 6.3\}$

To locate the fraction, change it to a mixed number. See the answer graph in the back of the textbook.

In Exercises 2-5,

$A = \left\{-\sqrt{6}, -1, -.5, 0, 3, \sqrt{25}, 7.5, \frac{24}{2}, \sqrt{-4}\right\}.$

2. The elements 0, 3, $\sqrt{25}$ (or 5), and $\frac{24}{2}$ (or 12) are whole numbers.

3. The elements $-1, 0, 3$, $\sqrt{25}$ (or 5), and $\frac{24}{2}$ (or 12) are integers.

4. The elements $-1, -.5, 0, 3$, $\sqrt{25}$ (or 5), 7.5, and $\frac{24}{2}$ (or 12) are rational numbers.

5. All the elements in the set are real numbers except $\sqrt{-4}$.

For Exercises 6 and 7, see the answer graphs in the back of the textbook.

6. $\{x \mid x < -3\}$

In interval notation, $x < -3$ is written as $(-\infty, -3)$. The parenthesis at -3 indicates that -3 is not included. The graph extends from -3 to the left on a number line.

7. $\{y \mid -4 < y \le 2\}$

In interval notation, $-4 < y \le 2$ is written as $(-4, 2]$. The parenthesis indicates that -4 is not included, while the bracket indicates that 2 is included. The graph goes from -4 to 2 on a number line.

8. $-6 + 14 + (-11) - (-3)$
$= 8 + (-11) - (-3)$
$= -3 - (-3)$
$= -3 + 3$
$= 0$

9. $10 - 4 \cdot 3 + 6(-4) = 10 - 12 + (-24)$
$= -2 + (-24)$
$= -26$

10. $7 - 4^2 + 2(6) + (-4)^2$
$= 7 - 16 + 2(6) + 16$
$= 7 - 16 + 12 + 16$
$= 19$

11. $$\frac{10 - 24 + (-6)}{\sqrt{16}(-5)} = \frac{10 - 24 + (-6)}{4(-5)} = \frac{10 - 24 + (-6)}{-20} = \frac{-20}{-20} = 1$$

12. $$\frac{-2[3 - (-1 - 2) + 2]}{\sqrt{9}(-3) - (-2)} = \frac{-2[3 - (-1 - 2) + 2]}{3(-3) - (-2)} = \frac{-2[3 - (-3) + 2]}{3(-3) - (-2)} = \frac{-2(8)}{3(-3) - (-2)} = \frac{-16}{-9 - (-2)} = \frac{-16}{-7} = \frac{16}{7}$$

13. $$\frac{8 \cdot 4 - 3^2 \cdot 5 - 2(-1)}{-3 \cdot 2^3 + 1} = \frac{8 \cdot 4 - 9 \cdot 5 - 2(-1)}{-3 \cdot 8 + 1} = \frac{32 - 45 + 2}{-24 + 1} = \frac{-11}{-23} = \frac{11}{23}$$

14. $\sqrt{196} = 14$ since $14^2 = 196$.

15. $-\sqrt{225} = -15$ since $15^2 = 225$ and we want the negative square root.

16. $\sqrt[3]{-27} = -3$ since $(-3)^3 = -27$.

17. $\sqrt[4]{-16}$ is not a real number.

18. If a is negative, then $\sqrt[n]{a}$ will represent a real number if n is odd.

In Exercises 19 and 20, let $k = -3$, $m = -3$, and $r = 25$. Substitute the appropriate values in each expression.

19. $\sqrt{r} + 2k - m = \sqrt{25} + 2(-3) - (-3)$
$= 5 + 2(-3) - (-3)$
$= 5 - 6 + 3$
$= 2$

20. $$\frac{8k + 2m^2}{r - 2} = \frac{8(-3) + 2(-3)^2}{25 - 2} = \frac{8(-3) + 2(9)}{25 - 2} = \frac{-24 + 18}{25 - 2} = \frac{-6}{23} \text{ or } -\frac{6}{23}$$

21. $-3(2k - 4) + 4(3k - 5) - 2 + 4k$
$= -3(2k) + (-3)(-4) + 4(3k) + 4(-5) - 2 + 4k$
$= -6k + 12 + 12k - 20 - 2 + 4k$
$= -6k + 12k + 4k + 12 - 20 - 2$
$= -10k - 10$

22. When simplifying

$$(3r + 8) - (-4r + 6),$$

the subtraction sign in front of $(-4r + 6)$ changes the signs of the terms $-4r$ and 6.

$$\begin{aligned}(3r+8)-(-4r+6) &= 3r+8-(-4r)-6\\ &= 3r+8+4r-6\\ &= 3r+4r+8-6\\ &= 7r+2\end{aligned}$$

23. $6+(-6)=0$

    The answer is B. Inverse Property. The sum of 6 and its inverse, -6, equals zero.

24. $4+5=5+4$

    The answer is E. Commutative Property. The order of the terms is reversed.

25. $-2+(3+6)=(-2+3)+6$

    The answer is D. Associative Property. The order of the terms is the same, but the grouping has changed.

26. $5x+15x=(5+15)x$

    The answer is A. Distributive Property. This is the second form of the distributive property.

27. $13\cdot 0=0$

    The answer is F. Multiplication Property of Zero. Multiplication by 0 always equals 0.

28. $-9+0=-9$

    The answer is C. Identity Property. The addition of 0 to any number does not change the number.

29. $4\cdot 1=4$

    The answer is C. Identity Property. Multiplication of any number by 1 does not change the number.

30. $(a+b)+c=(b+a)+c$

    The answer is E. Commutative Property. The order of the terms a and b is reversed.

Chapter 2

LINEAR EQUATIONS AND INEQUALITIES

2.1 Linear Equations in One Variable

2.1 Margin Exercises

1. To decide if a given number is a solution, substitute that number in the equation to see if the result is true or false.

(a) $3k = 15$; 5

The number 5 is a solution since $3 \cdot 5 = 15$.

(b) $r + 5 = 4$; 1

The number 1 is not a solution since $1+5 = 6 \neq 4$.

(c) $-8m = 12$; $\frac{3}{2}$
The number $\frac{3}{2}$ is not a solution since $-8\left(\frac{3}{2}\right) = -12 \neq 12$.

2. **(a)**

$$\begin{aligned} 3p + 2p + 1 &= -24 \\ 5p + 1 &= -24 \\ &\quad \textit{Combine terms.} \\ 5p + 1 - 1 &= -24 - 1 \\ &\quad \textit{Subtract 1.} \\ 5p &= -25 \\ \frac{5p}{5} &= \frac{-25}{5} \\ &\quad \textit{Divide by 5.} \\ p &= -5 \end{aligned}$$

To check, substitute -5 for p in the original equation.

$$\begin{aligned} 3p + 2p + 1 &= 24 \\ 3(-5) + 2(-5) + 1 &= -24 \quad ? \quad \textit{Let } p = -5. \\ -15 - 10 + 1 &= -24 \quad ? \\ -24 &= -24 \quad \textit{True} \end{aligned}$$

Solution set: $\{-5\}$

(b)

$$\begin{aligned} 3p &= 2p + 4p + 5 \\ 3p &= 6p + 5 \\ &\quad \textit{Combine terms.} \\ 3p - 6p &= 6p + 5 - 6p \\ &\quad \textit{Subtract } 6p. \\ -3p &= 5 \\ \frac{-3p}{-3} &= \frac{5}{-3} \\ &\quad \textit{Divide by } -3. \\ p &= -\frac{5}{3} \end{aligned}$$

Substitute to check that $-\frac{5}{3}$ is the solution.

$$\begin{aligned} 3p &= 2p + 4p + 5 \\ 3\left(-\frac{5}{3}\right) &= 2\left(-\frac{5}{3}\right) + 4\left(-\frac{5}{3}\right) + 5 \quad ? \\ -5 &= -\frac{10}{3} - \frac{20}{3} + 5 \quad ? \\ -5 &= -\frac{30}{3} + 5 \quad ? \\ -5 &= -10 + 5 \quad ? \\ -5 &= -5 \quad \textit{True} \end{aligned}$$

Solution set: $\{-\frac{5}{3}\}$

(c)

$$\begin{aligned} 4a + 8a &= 17a - 9 - 1 \\ 12a &= 17a - 10 \\ &\quad \textit{Combine terms.} \\ 12a - 17a &= 17a - 10 - 17a \\ &\quad \textit{Subtract } 17a. \\ -5a &= -10 \\ \frac{-5a}{-5} &= \frac{-10}{-5} \\ &\quad \textit{Divide by } -5. \\ a &= 2 \end{aligned}$$

Check. Let $a = 2$.

$$\begin{aligned} 4a + 8a &= 17a - 9 - 1 \\ 4(2) + 8(2) &= 17(2) - 9 - 1 \quad ? \\ 8 + 16 &= 34 - 10 \quad ? \\ 24 &= 24 \quad \textit{True} \end{aligned}$$

Solution set: $\{2\}$

(d)

$$\begin{aligned} -7+3y-9y &= 12y-5 \\ -7-6y &= 12y-5 \\ -7-6y+6y+5 &= 12y-5+6y+5 \\ &\quad \textit{Add } 6y\textit{; add } 5. \\ -2 &= 18y \\ \frac{-2}{18} &= \frac{18y}{18} \\ -\frac{1}{9} &= y \end{aligned}$$

Check. Let $y = -\frac{1}{9}$.

$$\begin{aligned} 7+3y-9y &= 12y-5 \\ -7+3\left(-\frac{1}{9}\right)-9\left(-\frac{1}{9}\right) &= 12\left(-\frac{1}{9}\right)-5 \quad ? \\ -7-\frac{1}{3}+1 &= -\frac{4}{3}-5 \quad ? \\ \frac{-21-1+3}{3} &= \frac{-4-15}{3} \quad ? \\ -\frac{19}{3} &= -\frac{19}{3} \quad \textit{True} \end{aligned}$$

Solution set: $\{-\frac{1}{9}\}$

3. **(a)**

$$\begin{aligned} 5p+4(3-2p) &= 2+p-10 \\ 5p+12-8p &= p-8 \\ &\quad \textit{Clear parentheses.} \\ 12-3p &= p-8 \\ &\quad \textit{Combine terms.} \\ 12-3p+3p+8 &= p-8+3p+8 \\ &\quad \textit{Add } 3p\textit{; add } 8. \\ 20 &= 4p \\ \frac{20}{4} &= \frac{4p}{4} \\ 5 &= p \end{aligned}$$

To check, substitute 5 for p in the original equation.

$$\begin{aligned} 5p+4(3-2p) &= 2+p-10 \\ 5(5)+4[3-2(5)] &= 2+5-10 \quad ? \\ 25+4(3-10) &= -3 \quad ? \\ 25+4(-7) &= -3 \quad ? \\ 25-28 &= -3 \quad ? \\ -3 &= -3 \quad \textit{True} \end{aligned}$$

Solution set: $\{5\}$

(b)

$$\begin{aligned} 3(z-2)+5z &= 2 \\ 3z-6+5z &= 2 \quad \textit{Clear parentheses.} \\ 8z-6 &= 2 \\ 8z-6+6 &= 2+6 \\ 8z &= 8 \\ \frac{8z}{8} &= \frac{8}{8} \\ z &= 1 \end{aligned}$$

Substitute to check that 1 is the solution.

$$\begin{aligned} 3(z-2)+5z &= 2 \\ 3(1-2)+5(1) &= 2 \quad ? \\ 3(-1)+5(1) &= 2 \quad ? \\ -3+5 &= 2 \quad ? \\ 2 &= 2 \quad \textit{True} \end{aligned}$$

Solution set: $\{1\}$

(c)

$$\begin{aligned} -2+3(y+4) &= 8y \\ -2+3y+12 &= 8y \\ 3y+10 &= 8y \\ 3y+10-3y &= 8y-3y \\ 10 &= 5y \\ \frac{10}{5} &= \frac{5y}{5} \\ 2 &= y \end{aligned}$$

Check. Let $y = 2$.

$$\begin{aligned} -2+3(y+4) &= 8y \\ -2+3(2+4) &= 8(2) \quad ? \\ -2+3(6) &= 8(2) \quad ? \\ -2+18 &= 16 \quad ? \\ 16 &= 16 \quad \textit{True} \end{aligned}$$

Solution set: $\{2\}$

(d)

$$\begin{aligned} 6-(4+m) &= 8m-2(3m+5) \\ 6-4-m &= 8m-6m-10 \\ 2-m &= 2m-10 \\ 2-m+m+10 &= 2m-10+m+10 \\ &\quad \textit{Add } m\textit{; add } 10. \\ 12 &= 3m \\ \frac{12}{3} &= \frac{3m}{3} \\ 4 &= m \end{aligned}$$

Check. Let $m = 4$.

$$\begin{aligned} 6-(4+m) &= 8m-2(3m+5) \\ 6-(4+4) &= 8(4)-2[3(4)+5] \quad ? \\ 6-8 &= 32-2(17) \quad ? \\ -2 &= 32-34 \quad ? \\ -2 &= -2 \quad \textit{True} \end{aligned}$$

Solution set: $\{4\}$

4. **(a)** $\frac{2p}{7} - \frac{p}{2} = -3$

Multiply both sides by the LCD, 14.

$$\begin{aligned} 14\left(\frac{2p}{7}\right) - 14\left(\frac{p}{2}\right) &= 14(-3) \\ 4p - 7p &= -42 \\ -3p &= -42 \\ \frac{-3p}{-3} &= \frac{-42}{-3} \\ p &= 14 \end{aligned}$$

Check. Let $p = 14$.

$$\begin{aligned} \frac{2p}{7} - \frac{p}{2} &= -3 \\ \frac{2(14)}{7} - \frac{14}{2} &= -3 \quad ? \\ \frac{28}{7} - \frac{14}{2} &= -3 \quad ? \\ 4 - 7 &= -3 \quad ? \\ -3 &= -3 \quad \textit{True} \end{aligned}$$

Solution set: $\{14\}$

(b) $\frac{k+1}{2} + \frac{k+3}{4} = \frac{1}{2}$

Multiply both sides by the LCD, 4.

$$\begin{aligned} 4\left(\frac{k+1}{2}\right) + 4\left(\frac{k+3}{4}\right) &= 4\left(\frac{1}{2}\right) \\ 2(k+1) + 1(k+3) &= 2 \\ 2k + 2 + k + 3 &= 2 \\ 3k + 5 &= 2 \\ 3k + 5 - 5 &= 2 - 5 \\ 3k &= -3 \\ \frac{3k}{3} &= \frac{-3}{3} \\ k &= -1 \end{aligned}$$

Check. Let $k = -1$.

$$\begin{aligned} \frac{k+1}{2} + \frac{k+3}{4} &= \frac{1}{2} \\ \frac{-1+1}{2} + \frac{-1+3}{4} &= \frac{1}{2} \quad ? \\ 0 + \frac{2}{4} &= \frac{1}{2} \quad ? \\ \frac{1}{2} &= \frac{1}{2} \quad \textit{True} \end{aligned}$$

Solution set: $\{-1\}$

5. **(a)** $.04x + .06(20 - x) = .05(50)$

Multiply both sides by 100.

$$\begin{aligned} 4x + 6(20 - x) &= 5(50) \\ 4x + 120 - 6x &= 250 \\ -2x + 120 &= 250 \\ -2x + 120 - 120 &= 250 - 120 \\ -2x &= 130 \\ \frac{-2x}{-2} &= \frac{130}{-2} \\ x &= -65 \end{aligned}$$

Check. Let $x = -65$.

$$\begin{aligned} .04x + .06(20 - x) &= .05(50) \\ .04(-65) + .06[20 - (-65)] &= .05(50) \quad ? \\ -2.6 + .06(85) &= 2.5 \quad ? \\ -2.6 + 5.1 &= 2.5 \quad ? \\ 2.5 &= 2.5 \quad \textit{True} \end{aligned}$$

Solution set: $\{-65\}$

(b) $.10(x - 6) + .05x = .06(50)$

Multiply both sides by 100.

$$\begin{aligned} 10(x - 6) + 5x &= 6(50) \\ 10x - 60 + 5x &= 300 \\ 15x - 60 &= 300 \\ 15x - 60 + 60 &= 300 + 60 \\ 15x &= 360 \\ \frac{15x}{15} &= \frac{360}{15} \\ x &= 24 \end{aligned}$$

Check. Let $x = 24$.

$$\begin{aligned} .10(x - 6) + .05x &= .06(50) \\ .10(24 - 6) + .05(24) &= .06(50) \quad ? \\ .10(18) + 1.2 &= 3 \quad ? \\ 1.8 + 1.2 &= 3 \quad ? \\ 3 &= 3 \quad \textit{True} \end{aligned}$$

Solution set: $\{24\}$

6. **(a)** $5(x + 2) - 2(x + 1) = 3x + 1$

$$\begin{aligned} 5x + 10 - 2x - 2 &= 3x + 1 \\ 3x + 8 &= 3x + 1 \\ 3x + 8 - 3x - 1 &= 3x + 1 - 3x - 1 \\ & \textit{Subtract } 3x\textit{; subtract } 1. \\ 7 &= 0 \quad \textit{False} \end{aligned}$$

Since the result, $7 = 0$, is false, the equation is a contradiction. There is no solution.

Solution set: $\emptyset$

(b) $\frac{x+1}{3}+\frac{2x}{3}=x+\frac{1}{3}$

Multiply both sides by 3.

$$3\left(\frac{x+1}{3}\right)+3\left(\frac{2x}{3}\right)=3\left(x+\frac{1}{3}\right)$$
$$x+1+2x=3x+1$$
$$3x+1=3x+1$$

This is an identity. Any real number will make the equation true.

Solution set: {all real numbers}

(c)
$$5(3x+1)=x+5$$
$$15x+5=x+5$$
$$15x+5-x-5=x+5-x-5$$
Subtract x; subtract 5.
$$14x=0$$
$$\frac{14x}{14}=\frac{0}{14}$$
$$x=0$$

This is a conditional equation.

Solution set: {0}

2.1 Section Exercises

3. For any number, we will have a solution since the equation is an identity. For example, if your last name is Lincoln, which has seven letters, we have $x = 7$.

$$4[x+(2-3x)]=2(4-4x)$$
$$4[7+(2-3(7))]=2[4-4(7)] \quad ?$$
$$4[7+(-19)]=2(4-28) \quad ?$$
$$4(-12)=2(-24) \quad ?$$
$$-48=-48 \quad \textit{True}$$

7.
$$8-8x=-16$$
$$8-8x-8=-16-8 \quad \textit{Subtract 8.}$$
$$-8x=-24$$
$$\frac{-8x}{-8}=\frac{-24}{-8} \quad \textit{Divide by } -8.$$
$$x=3$$

To check, substitute 3 for x in the original equation.

$$8-8x=-16$$
$$8-8(3)=-16 \quad ?$$
$$8-24=-16 \quad ?$$
$$-16=-16 \quad \textit{True}$$

Solution set: {3}

11.
$$12w+15w-9+5=-3w+5-9$$
$$27w-4=-3w-4$$
Combine terms.
$$27w-4+3w+4=-3w-4+3w+4$$
Add $3w$; add 4.
$$30w=0$$
$$\frac{30w}{30}=\frac{0}{30}$$
$$w=0$$

Check. Let $w = 0$.

$$12w+15w-9+5=-3w+5-9$$
$$12(0)+15(0)-9+5=-3(0)+5-9 \quad ?$$
$$-4=-4 \quad \textit{True}$$

Solution set: {0}

15.
$$3(2w+1)-2(w-2)=5$$
$$6w+3-2w+4=5 \quad \textit{Clear parentheses.}$$
$$4w+7=5$$
$$4w+7-7=5-7$$
$$4w=-2$$
$$\frac{4w}{4}=\frac{-2}{4}$$
$$w=-\frac{1}{2}$$

Check. Let $w=-\frac{1}{2}$.

$$3(2w+1)-2(w-2)=5$$
$$3\left[2\left(-\frac{1}{2}\right)+1\right]-2\left(-\frac{1}{2}-2\right)=5 \quad ?$$
$$3(-1+1)-2\left(-\frac{5}{2}\right)=5 \quad ?$$
$$0+5=5 \quad ?$$
$$5=5 \quad \textit{True}$$

Solution set: $\{-\frac{1}{2}\}$

In Exercises 19 and 23, we do not show the checks of the solutions. To be sure that your solutions are correct, check them by substituting in the original equations.

19.
$$6p-4(3-2p)=5(p-4)-10$$
$$6p-12+8p=5p-20-10$$
$$14p-12=5p-30$$
$$14p-12-5p+12=5p-30-5p+12$$
$$9p=-18$$
$$\frac{9p}{9}=\frac{-18}{9}$$
$$p=-2$$

Solution set: {−2}

23. $$\begin{aligned} -(9-3a)-(4+2a)-3 &= -(2-5a)+(-a)+1 \\ -9+3a-4-2a-3 &= -2+5a-a+1 \\ a-16 &= 4a-1 \\ a-16-a+1 &= 4a-1-a+1 \\ -15 &= 3a \\ \frac{-15}{3} &= \frac{3a}{3} \\ -5 &= a \end{aligned}$$

Solution set: $\{-5\}$

27. Yes, you should get the correct solution. The coefficients will be larger, but in the end the solution will be the same.

31. $\dfrac{x-8}{5}+\dfrac{8}{5}=-\dfrac{x}{3}$

Multiply both sides by the LCD, 15.

$$\begin{aligned} 15\left(\frac{x-8}{5}\right)+15\left(\frac{8}{5}\right) &= 15\left(-\frac{x}{3}\right) \\ 3(x-8)+3(8) &= -5x \\ 3x-24+24 &= -5x \\ 8x &= 0 \\ \frac{8x}{8} &= \frac{0}{8} \\ x &= 0 \end{aligned}$$

Check. Let $x=0$.

$$\begin{aligned} \frac{x-8}{5}+\frac{8}{5} &= -\frac{x}{3} \\ \frac{0-8}{5}+\frac{8}{5} &= -\frac{0}{3} \quad ? \\ -\frac{8}{5}+\frac{8}{5} &= 0 \quad ? \\ 0 &= 0 \quad \textit{True} \end{aligned}$$

Solution set: $\{0\}$

35. $.05y+.12(y+5000)=940$

Multiply both sides by 100.

$$\begin{aligned} 5y+12(y+5000) &= 100(940) \\ 5y+12y+60,000 &= 94,000 \\ 17y+60,000 &= 94,000 \\ 17y+60,000-60,000 &= 94,000-60,000 \\ 17y &= 34,000 \\ \frac{17y}{17} &= \frac{34,000}{17} \\ y &= 2000 \end{aligned}$$

Check. Let $y=2000$.

$$\begin{aligned} .05y+.12(y+5000) &= 940 \\ .05(2000)+.12(2000+5000) &= 940 \quad ? \\ 100+.12(7000) &= 940 \quad ? \\ 100+840 &= 940 \quad ? \\ 940 &= 940 \quad \textit{True} \end{aligned}$$

Solution set: $\{2000\}$

39. $.05x+.10(200-x)=.45x$

Multiply both sides by 100.

$$\begin{aligned} 5x+10(200-x) &= 45x \\ 5x+2000-10x &= 45x \\ 2000-5x &= 45x \\ 2000-5x+5x &= 45x+5x \\ 2000 &= 50x \\ \frac{2000}{50} &= \frac{50x}{50} \\ 40 &= x \end{aligned}$$

Check by substituting 40 for x in the original equation.

Solution set: $\{40\}$

43. $$\begin{aligned} -2p+5p-9 &= 3(p-4)-5 \\ 3p-9 &= 3p-12-5 \\ 3p-9 &= 3p-17 \\ 3p-9-3p+9 &= 3p-17-3p+9 \\ 0 &= -8 \quad \textit{False} \end{aligned}$$

The equation is a contradiction.

Solution set: $\emptyset$

47. $$\begin{aligned} 7[2-(3+4r)]-2r &= -9+2(1-15r) \\ 7[2-3-4r]-2r &= -9+2(1-15r) \\ 7[-1-4r]-2r &= -9+2(1-15r) \\ -7-28r-2r &= -9+2-30r \\ -7-30r &= -7-30r \end{aligned}$$

Since both sides are exactly the same, any real number would make the equation true. The equation is an identity.

Solution set: {all real numbers}

49. $$\begin{aligned} 2[3x+(x-2)] &= 9x+4 \\ 2(4x-2) &= 9x+4 \\ 8x-4 &= 9x+4 \\ 8x-4-8x-4 &= 9x+4-8x-4 \\ -8 &= x \end{aligned}$$

Solution set: $\{-8\}$

50. An equivalent equation will have the same solution set, that is, $\{-8\}$.

51. $-4(x+2)-3(x+5)=k$

Let $x=-8$.

$$\begin{aligned}-4(-8+2)-3(-8+5)&=k\\-4(-6)-3(-3)&=k\\24+9&=k\\33&=k\end{aligned}$$

52. If $k=33$, then the equation in Exercise 51 will be equivalent to the one in Exercise 49. The solution set for both will be $\{-8\}$, as shown in Exercise 53.

53. $$\begin{aligned}-4(x+2)-3(x+5)&=33\\-4x-8-3x-15&=33\\-7x-23&=33\\-7x-23+23&=33+23\\-7x&=56\\\frac{-7x}{-7}&=\frac{56}{-7}\\x&=-8\end{aligned}$$

Solution set: $\{-8\}$

54. To find the value of k that will make

$$3x+k=11 \text{ and } 5x-8=22$$

equivalent, first solve the second equation (like in Exercise 49).

$$\begin{aligned}5x-8&=22\\5x-8+8&=22+8\\5x&=30\\\frac{5x}{5}&=\frac{30}{5}\\x&=6\end{aligned}$$

Solution set: $\{6\}$

Now replace x with 6 in the first equation (like in Exercise 51).

$$\begin{aligned}3x+k&=11\\3(6)+k&=11\\18+k&=11\\k&=-7\end{aligned}$$

If $k=-7$ in the first equation, then the two equations should both have solution set $\{6\}$ and be equivalent. To see if this is the case, let $k=-7$ in the first equation and solve (like in Exercise 53).

$$\begin{aligned}3x+k&=11\\3x-7&=11\\3x-7+7&=11+7\\3x&=18\\\frac{3x}{3}&=\frac{18}{3}\\x&=6\end{aligned}$$

Solution set: $\{6\}$

Therefore, if $k=-7$, then the equations are equivalent.

55. If two equations are equivalent, they have the same *solution set*.

59. $k=4$

The solution set is $\{4\}$.

$k^2=16$

Since $(-4)^2=16$ and $4^2=16$, the solution set is $\{-4,4\}$.

The solution sets are not the same. Therefore, the equations are not equivalent.

63. **(a)** According to the bar graph, the charges in 1989 were about \$230 billion.

(b) In 1990, the charges were about \$275 billion.

(c) In 1991, the charges were about \$300 billion.

(d) In 1992, the charges were about \$335 billion.

2.2 Formulas

2.2 Margin Exercises

The strategy is to isolate the given variable on one side of the equals sign by treating every other variable as a constant using the steps for solving a linear equation in one variable.

1. $I=prt$

(a) To solve $I=prt$, notice that rt multiplies the variable p. To isolate p, divide by rt.

$$\frac{I}{rt}=\frac{p(rt)}{rt}$$

$$\frac{I}{rt}=p \text{ or } p=\frac{I}{rt}$$

(b) To solve $I = prt$ for r, divide by pt.

$$I = prt$$
$$\frac{I}{pt} = \frac{r(pt)}{pt}$$
$$\frac{I}{pt} = r \text{ or } r = \frac{I}{pt}$$

2. $m = 2k + 3b$

(a) Solve for k.

$$m = 2k + 3b$$
$$m - 3b = 2k + 3b - 3b$$
$$m - 3b = 2k$$
$$\frac{m - 3b}{2} = \frac{2k}{2}$$
$$\frac{m - 3b}{2} = k$$

(b) Solve for b.

$$m = 2k + 3b$$
$$m - 2k = 3b \quad \textit{Subtract } 2k.$$
$$\frac{m - 2k}{3} = b \quad \textit{Divide by } 3.$$

3. $P = a + b + c$

(a) Solve for b.

$$P = a + b + c$$
$$P = b + (a + c)$$
$$P - (a + c) = b + (a + c) - (a + c)$$
Subtract $(a + c)$.
$$P - (a + c) = b$$
$$\text{or } P - a - c = b$$

(b) Solve for c.

$$P = (a + b) + c$$
$$P - (a + b) = (a + b) + c - (a + b)$$
Subtract $(a + b)$.
$$P - (a + b) = c$$
$$\text{or } P - a - b = c$$

4. **(a)** Solve for x.

$$2x + ky = p - qx$$
$$2x + ky + qx = p - qx + qx \quad \textit{Add } qx.$$
$$2x + ky + qx = p$$
$$2x + ky + qx - ky = p - ky \quad \textit{Subtract } ky.$$
$$2x + qx = p - ky$$
$$x(2 + q) = p - ky \quad \textit{Factor.}$$
$$x = \frac{p - ky}{2 + q} \quad \textit{Divide by } 2 + q.$$

(b) Solve for x.

$$m = \frac{2x - z}{x}$$
$$x(m) = x\left(\frac{2x - z}{x}\right)$$
Multiply by x.
$$xm = 2x - z$$
$$mx + z = 2x - z + z \quad \textit{Add } z.$$
$$mx + z = 2x$$
$$mx + z - mx = 2x - mx$$
Subtract mx.
$$z = 2x - mx$$
$$z = (2 - m)x \quad \textit{Factor.}$$
$$\frac{z}{2 - m} = x$$
Divide by $2 - m$.
$$\text{or } \frac{-z}{m - 2} = x$$

5. **(a)** Use the formula for the area of a triangle. Solve for h. Multiply by $\frac{2}{b}$.

$$A = \frac{1}{2}bh$$
$$A\left(\frac{2}{b}\right) = \frac{1}{2}bh\left(\frac{2}{b}\right)$$
$$\frac{2A}{b} = h \text{ or } h = \frac{2A}{b}$$

Substitute $A = 36$ and $b = 12$.

$$h = \frac{2(36)}{12} = 6$$

The height is 6 in.

(b) Use $d = rt$. Solve for t.

$$\frac{d}{r} = \frac{rt}{r}$$
$$\frac{d}{r} = t \text{ or } t = \frac{d}{r}$$

Substitute $d = 500$ and $r = 153.6$.

$$t = \frac{500}{153.6} \approx 3.255$$

His time is about 3.255 hr.

6. **(a)** The given amount of mixture is 20 oz. The part that is oil is 1 oz. Thus, the percent of oil is

$$\frac{1}{20} = \frac{5}{100} = 5\%.$$

(b) Here we find the commission by multiplying the commission percent by the amount sold.

$$8\% \text{ of } 12{,}000 = .08(12{,}000) = 960$$

The salesman earns \$960.

7. More than 5 times means 6 or more times. According to the graph, 13% of the population uses their cards 6-9 times and 20% uses their cards 10 or more times. We must find 13% + 20% = 33% of the 4000 users. Multiplying by .33 gives

$$.33(4000) = 1320.$$

Therefore, 1320 people in the group of 4000 use their cards more than 5 times.

2.2 Section Exercises

1. **(a)** $x = \dfrac{5x+8}{3}$

$$3x = 3\left(\frac{5x+8}{3}\right)$$
$$3x = 5x + 8$$

(b) $t = \dfrac{bt+k}{c}\ (c \neq 0)$

$$ct = c\left(\frac{bt+k}{c}\right)$$
$$ct = bt + k$$

2. **(a)** $3x - 5x = 5x + 8 - 5x$
$3x - 5x = 8$

(b) $ct - bt = bt + k - bt$
$ct - bt = k$

3. Use the distributive property in each case.

(a) $-2x = 8$ **(b)** $(c-b)t = k$

4. **(a)** $\dfrac{-2x}{-2} = \dfrac{8}{-2}$ **(b)** $\dfrac{(c-b)t}{c-b} = \dfrac{k}{c-b}$

$x = -4$ $t = \dfrac{k}{c-b}$

5. The restriction $b \neq c$ must be applied. If $b = c$, the denominator becomes zero and division by zero is undefined.

6. The process for solving an equation of letters (or a formula) for a specified variable is the same as solving an equation. Treat all the variables except the one being solved for as constants. Additional restrictions may have to be applied, however, to insure that denominators do not become zero.

7. $P = 2L + 2W$ is equivalent to $P = 2(L + W)$ by the distributive property.

11. Solve $P = 2L + 2W$ for L.
Subtract $2L$; then divide by 2.

$$P - 2W = 2L + 2W - 2W$$
$$P - 2W = 2L$$
$$\frac{P-2W}{2} = \frac{2L}{2}$$
$$\frac{P-2W}{2} = L$$

15. Solve $C = 2\pi r$ for r.
Divide by 2π.

$$\frac{C}{2\pi} = \frac{2\pi r}{2\pi}$$
$$\frac{C}{2\pi} = r$$

19. Solve $\mathrm{F} = \frac{9}{5}\mathrm{C} + 32$ for C.
Subtract 32; then multiply by $\frac{5}{9}$.

$$\mathrm{F} - 32 = \frac{9}{5}\mathrm{C} + 32 - 32$$
$$\mathrm{F} - 32 = \frac{9}{5}\mathrm{C}$$
$$\frac{5}{9}(\mathrm{F} - 32) = \frac{5}{9}\left(\frac{9}{5}\mathrm{C}\right)$$
$$\frac{5}{9}(\mathrm{F} - 32) = \mathrm{C}$$

23. Solve $w = \dfrac{3y - x}{y}$ for y.

Multiply by y.

$$y(w) = y\left(\frac{3y-x}{y}\right)$$
$$wy = 3y - x$$
$$wy - 3y = 3y - x - 3y$$
$$wy - 3y = -x$$
$$(w-3)y = -x \quad \textit{Factor.}$$
$$y = \frac{-x}{w-3} \quad \textit{Divide by } w - 3.$$
$$\text{or} \quad y = \frac{x}{3-w}$$

27. Use $d = rt$, solving for t.

$$t = \frac{d}{r}$$

Substitute $d = 500$ and $r = 160.9$.

$$t = \frac{500}{160.9} \approx 3.108$$

His time is about 3.108 hr.

31. Use $P = 4s$, solving for s.

$$s = \frac{P}{4}$$

Substitute $P = 920$.

$$s = \frac{920}{4} = 230$$

The length of each side is 230 m.

35. Use $V = LWH$, solving for L.

$$L = \frac{V}{WH}$$

Substitute $V = 128$, $W = 4$, and $H = 4$.

$$L = \frac{128}{4(4)} = 8$$

The stack is 8 ft long.

39. **(a)** From the pie chart, the 24-to-34 age group spent 23%. Multiply .23 times the total sales to get

$$.23(\$2,295,000,000) = \$527,850,000.$$

(b) The 35-to-44 age group spent 26%. Multiply by .26 to get

$$.26(\$2,295,000,000) = \$596,700,000.$$

(c) The 65-and-older age group spent 7%. Multiply by .07 to get

$$.07(\$2,295,000,000) = \$160,650,000.$$

43. Since 24% did not read the Sunday paper, multiply .24 times the number of people surveyed to get

$$.24(263,000) = 63,120.$$

Therefore, 63,120 people did not read the Sunday paper.

47. $A = \dfrac{24f}{b(p+1)}$

Let $p = 48$, $f = 400$, and $b = 3840$.

$$A = \frac{24(400)}{3840(48+1)} = \frac{9600}{3840(49)} \approx .051$$

The approximate annual interest rate is 5.1%.

2.3 Applications of Linear Equations

2.3 Margin Exercises

1. **(a)** 9 added to a number is expressed $9 + x$, or $x + 9$.

(b) To express the difference between 7 and a number, subtract x from 7, that is, $7 - x$.

(c) Four times a number is expressed $4 \cdot x$, or $4x$.

(d) To express the product of -6 and a number, multiply -6 and x, that is, $-6x$.

(e) To express the quotient of 7 and a nonzero number, divide 7 by x, that is, $\frac{7}{x}$.

2. **(a)**

The sum of a number and 6	is	28.
↓	↓	↓
$x + 6$	$=$	28,

or $x + 6 = 28$.

(b)

If twice a number	is decreased by 3,	the result is	17.
↓	↓	↓	↓
$2x$	$-\ 3$	$=$	17,

or $2x - 3 = 17$.

(c)

The product of a number and 7	is	twice the number	plus	12.
↓	↓	↓	↓	↓
$7x$	$=$	$2x$	$+$	12,

or $7x = 2x + 12$.

(d)

The quotient of a number and 6	added to	twice the number	is	7.
↓	↓	↓	↓	↓
$\frac{x}{6}$	$+$	$2x$	$=$	7,

or $\dfrac{x}{6} + 2x = 7$.

(e)

The quotient of a number and 3	is	8.
↓	↓	↓
$\frac{x}{3}$	$=$	8,

or $\dfrac{x}{3} = 8$.

(f) The sum of a number and twice the number is 3.

$$x + 2x \quad = \quad 3,$$

or $x + 2x = 3$.

3. **(a)** $5x - 3(x + 2) = 7$ is an equation because it has an equals sign.

(b) $5x - 3(x + 2)$ is an expression because there is no equals sign.

4. **(a)** First, the length and perimeter are given in terms of the width. The length L is 5 cm more than the width W. Thus,

$$L = W + 5.$$

Perimeter P is 5 times the width, so

$$P = 5W.$$

Use the formula for perimeter of a rectangle.

$$P = 2L + 2W$$

Substitute $5W$ for P and $W + 5$ for L.

$$\begin{aligned} 5W &= 2(W + 5) + 2W \\ 5W &= 2W + 10 + 2W \\ 5W &= 4W + 10 \\ 5W - 4W &= 4W + 10 - 4W \\ W &= 10 \\ L &= W + 5 = 10 + 5 = 15 \end{aligned}$$

The rectangle is 10 cm by 15 cm.

(b) Let $x =$ the age of Cindy's son, Cody;
$3x + 5 =$ Cindy's age.

The sum of their ages is 45, so the equation is

$$\begin{aligned} x + (3x + 5) &= 45 \\ 4x + 5 &= 45 \\ 4x + 5 - 5 &= 45 - 5 \\ 4x &= 40 \\ \frac{4x}{4} &= \frac{40}{4} \\ x &= 10 \\ 3x + 5 &= 3(10) + 5 = 35 \end{aligned}$$

Cody is 10 yr old, and Cindy is 35 yr old.

5. **(a)** Let x be the number, which is increased by 15% of x, or $.15x$. Then,

$$\begin{aligned} x + .15x &= 287.5 \\ 1.15x &= 287.5 \\ \frac{1.15x}{1.15} &= \frac{287.5}{1.15} \\ x &= 250. \end{aligned}$$

The number is 250.

(b) Let x be the amount she earned before deductions. Then 10% of x, or $.10x$, is the amount of her deductions.

$$\begin{aligned} x - .10x &= 162 \\ .90x &= 162 \\ \frac{.90x}{.90} &= \frac{162}{.90} \\ x &= 180 \end{aligned}$$

Michelle earned $180.

6. **(a)** Let $x =$ the amount invested at 5%;
$72,000 - x =$ the amount invested at 3%.

Use $I = prt$ with $t = 1$.
Make a table to organize the information.

% as a decimal	Amount invested	Interest in 1 year
.05	x	$.05x$
.03	$72,000 - x$	$.03(72,000 - x)$
		3160

The last column gives the equation.

$$\begin{aligned} .05x + .03(72,000 - x) &= 3160 \\ .05x + 2160 - .03x &= 3160 \quad \textit{Clear parentheses.} \\ .02x + 2160 &= 3160 \\ .02x + 2160 - 2160 &= 3160 - 2160 \\ .02x &= 1000 \\ \frac{.02x}{.02} &= \frac{1000}{.02} \\ x &= 50,000 \\ 72,000 - x &= 72,000 - 50,000 \\ &= 22,000 \end{aligned}$$

The woman should invest $50,000 at 5% and $22,000 at 3%.

(b) Let $x =$ the amount invested at 5%;
$34,000 - x =$ the amount invested at 4%.

Use $I = prt$ with $t = 1$.
Make a table.

% as a decimal	Amount invested	Interest in 1 year
.05	x	$.05x$
.04	$34,000 - x$	$.04(34,000 - x)$
		1545

The last column gives the equation.

$$.05x + .04(34,000 - x) = 1545$$
$$.05x + 1360 - .04x = 1545 \quad \textit{Clear parentheses.}$$
$$.01x + 1360 = 1545$$
$$.01x + 1360 - 1360 = 1545 - 1360$$
$$.01x = 185$$
$$\frac{.01x}{.01} = \frac{185}{.01}$$
$$x = 18,500$$
$$34,000 - x = 34,000 - 18,500$$
$$= 15,500$$

The man should invest \$18,500 at 5% and \$15,500 at 4%.

7. **(a)** Let $x =$ the number of liters of the 10% solution;
$x + 60 =$ the number of liters of the 15% solution.

Make a table.

Strength	Liters of solution	Liters of pure solution
10%	x	$.10x$
25%	60	$.25(60)$
15%	$x + 60$	$.15(x + 60)$

The last column gives the equation.

$$.10x + .25(60) = .15(x + 60)$$
$$.10x + 15 = .15x + 9$$
$$6 = .05x$$

Subtract $.10x$; *subtract* 9.

$$\frac{6}{.05} = x$$

Divide by .05.

$$120 = x$$

120 L of 10% solution should be used.

(b) Let $x =$ the amount of \$8 candy;
$x + 100 =$ the amount of \$7 candy.

Make a table.

Price per pound	Number of pounds	Value
\$8	x	$8x$
\$4	100	400
\$7	$x + 100$	$7(x + 100)$

The last column gives the equation.

$$8x + 400 = 7(x + 100)$$
$$8x + 400 = 7x + 700$$

Subtract $7x$; subtract 400.

$$x = 300$$

300 lb of candy worth \$8 per lb should be used.

2.3 Section Exercises

3. A number decreased by 13 is translated $x - 13$.

7. The product of 8 and 12 more than a number is translated $8(x + 12)$.

Let x represent the unknown in Exercises 11 and 15.

11. The sum of a number and 6 is -31.
Solve the equation.

$$x + 6 = -31$$
$$x + 6 - 6 = -31 - 6$$
$$x = -37$$

The number is -37.

15. Subtract $\frac{2}{3}$ of a number from 12 to get a result of 10 is translated

$$12 - \frac{2}{3}x = 10.$$
$$3(12) - 3\left(\frac{2}{3}x\right) = 3(10) \quad \textit{Multiply by 3.}$$
$$36 - 2x = 30$$
$$36 - 2x - 36 = 30 - 36$$
$$-2x = -6$$
$$x = \frac{-6}{-2}$$
$$x = 3$$

The number is 3.

19. $5(x+3) - 8(2x-6)$ is an expression because there is no equals sign.

23. $\frac{r}{2} - \frac{r+9}{6} - 8$ is an expression because there is no equals sign.

27. Let $x =$ the length of the middle side.

The shortest side is 75 mi less than the middle side; therefore, the shortest side has length $x - 75$. The longest side is 375 mi more than the middle side; therefore, the longest side has length $x + 375$. The perimeter is 3075 mi. Add the three lengths.

$$\begin{aligned} x + (x - 75) + (x + 375) &= 3075 \\ 3x + 300 &= 3075 \\ 3x + 300 - 300 &= 3075 - 300 \\ 3x &= 2775 \\ x &= \frac{2775}{3} \\ x &= 925 \\ x - 75 = 925 - 75 &= 850 \\ x + 375 = 925 + 375 &= 1300 \end{aligned}$$

The sides have length of 850 mi, 925 mi, and 1300 mi.

31. Let $x =$ General Motors' sales in billions of dollars;
$x - 28.7 =$ Ford's sales in billions of dollars.

Total sales for both companies were 213.5 billion dollars.

$$\begin{aligned} x + (x - 28.7) &= 213.5 \\ 2x - 28.7 &= 213.5 \\ 2x - 28.7 + 28.7 &= 213.5 + 28.7 \\ 2x &= 242.2 \\ x &= \frac{242.2}{2} \\ x &= 121.1 \\ x - 28.7 &= 121.1 - 28.7 \\ &= 92.4 \end{aligned}$$

General Motors' sales were 121.1 billion dollars; Ford Motor's sales were 92.4 billion dollars.

35. Let $x =$ the cost of tuition in 1990.

$$\begin{aligned} x &= 840 + 139\%(840) \\ x &= 840 + 1.39(840) \\ x &= 2007.6 \end{aligned}$$

The cost of tuition was approximately $2008.

39. Let $x =$ the amount invested at 3%;
$12{,}000 - x =$ the amount invested at 4%.

Make a table to organize the information.

% as a decimal	Amount invested	Interest in 1 year
.03	x	$.03x$
.04	$12{,}000 - x$	$.04(12{,}000 - x)$
		440

The last column gives the equation.

$$\begin{aligned} .03x + .04(12{,}000 - x) &= 440 \\ .03x + 480 - .04x &= 440 \end{aligned}$$

Clear the parentheses.

$$\begin{aligned} -.01 + 480 &= 440 \\ -.01x + 480 - 480 &= 440 - 480 \\ -.01x &= -40 \\ x &= \frac{-40}{-.01} \\ x &= 4000 \\ 12{,}000 - x &= 12{,}000 - 4000 \\ &= 8000 \end{aligned}$$

George should invest $4000 at 3% and $8000 at 4%.

43. Let $x =$ the amount invested at 2%;
$x + 29{,}000 =$ the amount invested at 3%.

Make a table.

% as a decimal	Amount invested	Interest in 1 year
.05	29,000	$.05(29{,}000)$
.02	x	$.02x$
.03	$x + 29{,}000$	$.03(x + 29{,}000)$

The last column gives the equation.

$$.05(29{,}000) + .02x = .03(x + 29{,}000)$$

$$1450 + .02x = .03x + 870$$

Clear parentheses.

$$580 = .01x$$

Subtract $.02x$; *subtract* 870.

$$\frac{580}{.01} = x$$

$$58{,}000 = x$$

Ed should invest $58,000 at 2%.

47. Let $x =$ the number of liters of the 20% alcohol solution;
$x + 12 =$ the number of liters of the 14% alcohol solution.

Make a table.

Strength	Liters of solution	Liters of pure alcohol
12%	12	$.12(12)$
20%	x	$.20x$
14%	$x + 12$	$.14(x + 12)$

The last column gives the equation.

$$.12(12) + .20x = .14(x + 12)$$
$$1.44 + .20x = .14x + 1.68$$
$$.06x = .24$$

Subtract $.14x$; *subtract* 1.44.

$$x = \frac{.24}{.06}$$
$$x = 4$$

4 L of the 20% alcohol solution are needed.

51. Let $x =$ the amount of \$6 nuts.

Make a table.

Cost per lb	lb of nuts	Total cost
\$2	50	$2(50) = 100$
\$6	x	$6x$
\$5	$x + 50$	$5(x + 50)$

The last column gives the equation.

$$100 + 6x = 5(x + 50)$$
$$100 + 6x = 5x + 250$$
$$x = 150$$

Subtract $5x$; *subtract* 100.

He should use 150 lb of \$6 nuts.

55. **(a)** Let $x =$ the amount invested at 5%;
$800 - x =$ the amount invested at 10%.

(b) Let $y =$ the amount of 5% acid used;
$800 - y =$ the amount of 10% acid used.

56. Organize the information in a table.

(a)

% as a decimal	Amount invested	Interest in 1 year
.05	x	$.05x$
.10	$800 - x$	$.10(800 - x)$
.0875	800	$.0875(800)$

The amount of interest earned at 5% and 10% is found in the last column of the table, $.05x$ and $.10(800 - x)$.

(b)

Strength	Liters of solution	Liters of pure acid
.05	y	$.05y$
.10	$800 - y$	$.10(800 - y)$
.0875	800	$.0875(800)$

The amount of pure acid in the 5% and 10% mixtures is found in the last column of the table, $.05y$ and $.10(800 - y)$.

57. Refer to the tables for Exercise 56. In each case, the last column gives the equation.

(a) $.05x + .10(800 - x) = .0875(800)$

(b) $.05y + .10(800 - y) = .0875(800)$

58. In both cases, multiply by 10,000 to clear the decimals.

(a)
$$.05x + .10(800 - x) = .0875(800)$$
$$500x + 1000(800 - x) = 875(800)$$
$$500x + 800,000 - 1000x = 700,000$$
$$800,000 - 500x = 700,000$$
$$-500x = -100,000$$
$$x = 200$$

Then

$$800 - x = 800 - 200 = 600.$$

Jack invested \$200 at 5% and \$600 at 10%.

(b)
$$.05y + 10(800 - y) = .0875(800)$$
$$500y + 1000(800 - y) = 875(800)$$
$$500y + 800,000 - 1000y = 700,000$$
$$800,000 - 500y = 700,000$$
$$-500y = -100,000$$
$$y = 200$$

Then

$$800 - y = 800 - 200 = 600.$$

Jill used 200 L of 5% acid solution and 600 L of 10% acid solution.

59. The processes used to solve Problems A and B were virtually the same. Aside from the variables chosen, the problem information was organized in similar tables and the equations solved were the same. In both problems, one amount was added to a second amount to arrive at a total amount.

2.4 More Applications of Linear Equations

2.4 Margin Exercises

1. Let $x =$ the number of dimes;
$26 - x =$ the number of half-dollars.

Denomination	Number of coins	Value
$.10	x	$.10x$
$.50	$26 - x$	$.50(26 - x)$
		8.60

Multiply the number of coins by the denominations, and add the results to get 8.60.

$$.10x + .50(26 - x) = 8.60$$

Multiply by 100.

$$10x + 50(26 - x) = 860$$
$$10x + 1300 - 50x = 860$$
$$-40x = -440$$
$$x = 11$$
$$26 - x = 26 - 11 = 15$$

The cashier has 11 dimes and 15 half-dollars.

2. Let $x =$ the time until the cars will be 420 mi apart.

Use the formula $d = rt$. Make a table.

	r	t	d
Northbound car	60	x	$60x$
Southbound car	45	x	$45x$
			420

The total distance traveled is the sum of the distances traveled by each car, since they are traveling in opposite directions. This total is 420 mi. Therefore,

$$60x + 45x = 420$$
$$105x = 420$$
$$x = \frac{420}{105} = 4.$$

The cars will be 420 mi apart in 4 hr.

3. Let $x =$ the time it takes Clay to catch up to Elayn;
then $x + \frac{1}{2} =$ Elayn's time.

Use the formula $d = rt$. Make a table.

	r	t	d
Elayn	3	$x + \frac{1}{2}$	$3\left(x + \frac{1}{2}\right)$
Clay	5	x	$5x$

Since they travel the same distance, the equation will be

$$3\left(x + \frac{1}{2}\right) = 5x$$
$$3x + \frac{3}{2} = 5x.$$

Multiply by 2.

$$6x + 3 = 10x$$
$$3 = 4x$$
$$\frac{3}{4} = x$$

It takes Clay $\frac{3}{4}$ hr or 45 min to catch up to Elayn.

4. Let $x =$ the measure of the second angle;
$x + 15 =$ the measure of the first angle;
$2x + 25 =$ the measure of the third angle.

The sum of the three angles must equal 180°.

$$x + (x + 15) + (2x + 25) = 180$$
$$4x + 40 = 180$$
$$4x = 140$$
$$x = 35$$
$$x + 15 = 35 + 15 = 50$$
$$2x + 25 = 2(35) + 25 = 95$$

The angles measure 35°, 50°, and 95°.

2.4 Section Exercises

3. To calculate Anh's rate, use the formula $d = rt$ or $r = \frac{d}{t}$ with $d = 520$ and $t = 10$.

$$r = \frac{520}{10} = 52$$

Anh's rate is 52 mph.

7. Let $x =$ the number of quarters;
$36 - x =$ the number of half-dollars.

Denomination	Number of coins	Value
\$.25	x	$.25x$
\$.50	$36 - x$	$.50(36 - x)$
		14.75

Multiply the number of coins by the denominations, and add the results to get 14.75.

$$.25x + .50(36 - x) = 14.75$$

Multiply by 100.

$$\begin{aligned} 25x + 50(36 - x) &= 1475 \\ 25x + 1800 - 50x &= 1475 \\ -25x + 1800 &= 1475 \\ -25x &= -325 \\ x &= 13 \\ 36 - x = 36 - 13 &= 23 \end{aligned}$$

Lisa has 13 quarters and 23 half-dollars.

11. Let $x =$ the number of two-cent pieces;
then $3x =$ the number of three-cent pieces.

Denomination	Number of coins	Value
\$.02	x	$.02x$
\$.03	$3x$	$.03(3x)$
		1.21

Multiply the number of coins by the denominations, and add the results to get 1.21.

$$.02x + .03(3x) = 1.21$$

Multiply by 100.

$$\begin{aligned} 2x + 3(3x) &= 121 \\ 2x + 9x &= 121 \\ 11x &= 121 \\ x &= 11 \\ 3x = 3(11) &= 33 \end{aligned}$$

Frances has 11 two-cent pieces and 33 three-cent pieces.

15. Use the formula $d = rt$ or $r = \frac{d}{t}$. Let $d = 200$ and $t = 19.32$.

$$\begin{aligned} r &= \frac{d}{t} \\ r &= \frac{200}{19.32} \\ r &\approx 10.35 \end{aligned}$$

Michael Johnson's rate was 10.35 m/sec.

19. Let $x =$ the time it takes for the trains to be 315 km apart.

Use the formula $d = rt$.

	r	t	d
Northbound	85	x	$85x$
Southbound	95	x	$95x$
			315

The total distance traveled is the sum of the distances traveled by each train, since they are traveling in opposite directions. This total is 315 km. Therefore,

$$\begin{aligned} 85x + 95x &= 315 \\ 180x &= 315 \\ x &= \frac{315}{180} = 1\frac{3}{4}. \end{aligned}$$

It will take the trains $1\frac{3}{4}$ hr before they are 315 km apart.

23. Let $x =$ Tri's speed when he drives his car.

Use the formula $d = rt$.

	r	t	d
Travel by car	x	$\frac{1}{2}$	$\frac{1}{2}x$
Travel by bus	$x - 12$	$\frac{3}{4}$	$\frac{3}{4}(x - 12)$

The distance by car and the distance by bus are the same. Therefore,

$$\frac{1}{2}x = \frac{3}{4}(x - 12).$$

Multiply by the LCD, 4.

$$\begin{aligned} 2x &= 3(x - 12) \\ 2x &= 3x - 36 \\ -x &= -36 \\ x &= 36 \end{aligned}$$

Tri's speed by car is 36 mph. The distance he travels to work is

$$\frac{1}{2}(36) = 18 \text{ mi}.$$

27. The sum of the angles must be 180°.

$$\begin{aligned}(x+15)+(x+5)+(10x-20)&=180\\12x&=180\\x&=15\end{aligned}$$

The angle measures are

$$\begin{aligned}(x+5)^\circ&=(15+5)^\circ=20^\circ,\\(x+15)^\circ&=(15+15)^\circ=30^\circ,\\(10x-20)&=(10\cdot 15-20)^\circ=130^\circ.\end{aligned}$$

29. The sum of the measures of the angles of a triangle is 180°, so the equation is

$$\begin{aligned}x+2x+60&=180\\3x+60&=180\\3x&=120\\x&=40.\\2x&=80\end{aligned}$$

The measures of the unknown angles are 40° and 80°.

30. Supplementary angles sum to 180°, so $180°-60° = 120°$. The measure of the unknown angle is 120°.

31. The sum of the measures of the unknown angles in Exercise 29 is $40° + 80° = 120°$. This is equal to the measure of the angle in Exercise 30.

32. The sum of the measures of angles 1 and 2 is equal to the measure of angle 3.

35. Since the angles are supplementary, their sum must be 180°.

$$\begin{aligned}(3x+5)+(5x+15)&=180\\8x+20&=180\\x&=160\\&=20\end{aligned}$$

The angle measures are

$$(3x+5)^\circ=(3\cdot 20+5)^\circ=65^\circ$$

and

$$(5x+15)^\circ=(5\cdot 20+15)^\circ=115^\circ.$$

39. Let x = the page number on one page;
$x+1$ = the page number on the next page.

Since the sum of the page numbers is 153, the equation is

$$\begin{aligned}x+(x+1)&=153\\2x+1&=153\\2x&=152\\x&=76.\\x+1&=77\end{aligned}$$

The page numbers are 76 and 77.

2.5 Linear Inequalities in One Variable

2.5 Margin Exercises

1. See the answer graphs for the margin exercises in the textbook.

(a)
$$\begin{aligned}p+6&\le 8\\p+6-6&\le 8-6 \quad \textit{Subtract 6.}\\p&\le 2\end{aligned}$$

Solution set: $(-\infty, 2]$

(b)
$$\begin{aligned}k-5&\ge 1\\k-5+5&\ge 1+5 \quad \textit{Add 5.}\\k&\ge 6\end{aligned}$$

Solution set: $[6, \infty)$

(c)
$$\begin{aligned}8y&<7y-6\\8y-7y&<7y-6-7y \quad \textit{Subtract 7y.}\\y&<-6\end{aligned}$$

Solution set: $(-\infty, -6)$

2. This exercise is to reinforce the idea that multiplying both sides of an inequality by a negative number reverses the direction of the inequality symbol.

(a) $7 < 8$

Multiply by -5.

$$\begin{aligned}7(-5)&=-35\\8(-5)&=-40\end{aligned}$$

Therefore, $-35 > -40$.

(b) $-1 > -4$

Multiply by -5.

$$\begin{aligned}-1(-5)&=5\\-4(-5)&=20\end{aligned}$$

Therefore, $5 < 20$.

For Exercises 3-5, see the answer graphs for the margin exercises in the textbook.

3. **(a)** $2y < -10$

$$\frac{2y}{2} < \frac{-10}{2} \quad \textit{Divide by 2.}$$

$$y < -5$$

Solution set: $(-\infty, -5)$

(b) $-7k \geq 8$

Divide by -7 and reverse the inequality sign.

$$\frac{-7k}{-7} \leq \frac{8}{-7}$$

$$k \leq -\frac{8}{7}$$

Solution set: $(-\infty, -\frac{8}{7}]$

(c) $-9m < -81$

Divide by -9; reverse the inequality sign.

$$\frac{-9m}{-9} > \frac{-81}{-9}$$

$$m > 9$$

Solution set: $(9, \infty)$

4. **(a)** $-y \leq 2$

Multiply by -1 and reverse the inequality sign.

$$-1(-y) \geq -1(2)$$

$$y \geq -2$$

Solution set: $[-2, \infty)$

(b) $-z > -11$

Multiply by -1; reverse the inequality sign.

$$-1(-z) < -1(-11)$$

$$z < 11$$

Solution set: $(-\infty, 11)$

(c)

$$4(x+2) \geq 6x - 8$$

$$4x + 8 \geq 6x - 8 \quad \textit{Clear parentheses.}$$

$$4x + 8 - 8 - 6x \geq 6x - 8 - 8 - 6x$$

Subtract 8; *subtract* $6x$.

$$-2x \geq -16$$

Divide by -2; reverse the inequality sign.

$$\frac{-2x}{-2} \leq \frac{-16}{-2}$$

$$x \leq 8$$

Solution set: $(-\infty, 8]$

(d) $5 - 3(m-1) \leq 2(m+3) + 1$

$$5 - 3m + 3 \leq 2m + 6 + 1$$

Clear parentheses.

$$8 - 3m \leq 2m + 7$$

$$-5m \leq -1$$

Subtract $2m$; *subtract* 8.

Divide by -5; reverse the inequality sign.

$$\frac{-5m}{-5} \geq \frac{-1}{-5}$$

$$m \geq \frac{1}{5}$$

Solution set: $[\frac{1}{5}, \infty)$

5. **(a)** $-3 \leq x - 1 \leq 7$

Add 1 to each part.

$$-3 + 1 \leq x - 1 + 1 \leq 7 + 1$$

$$-2 \leq x \leq 8$$

Solution set: $[-2, 8]$

(b) $5 < 3x - 4 < 9$

Add 4 to each part; then divide each part by 3.

$$5 + 4 < 3x - 4 + 4 < 9 + 4$$

$$9 < 3x < 13$$

$$\frac{9}{3} < \frac{3x}{3} < \frac{13}{3}$$

$$3 < x < \frac{13}{3}$$

Solution set: $(3, \frac{13}{3})$

6. Let $x =$ the number of miles that results in the same rental cost.

The cost of renting from Ames is 48(3) plus the mileage cost of $.10x$. The cost of renting from Hughes is 40(3) plus the mileage cost of .15x. Write an inequality saying that the Hughes cost is at most as much as, that is, is less than or equal to, the Ames cost.

$$40(3) + .15x \leq 48(3) + .10x$$

$$4000(3) + 15x \leq 4800(3) + 10x$$

Multiply by 100.

$$12{,}000 + 15x \leq 14{,}400 + 10x$$

$$5x \leq 2400$$

Subtract $10x$; *subtract* 12,000.

$$x \leq 480$$

Divide by 5.

It would take a mileage of 480 mi or less to make Hughes cost at most as much as Ames.

7. Let $x =$ the grade the student must make on the fourth test.

To find the average of the four tests, add them and divide by 4. This average must be at least 90, that is, greater than or equal to 90.

$$\frac{92+90+84+x}{4} \geq 90$$
$$\frac{266+x}{4} \geq 90$$
$$266 + x \geq 360 \quad \textit{Multiply by 4.}$$
$$x \geq 94 \quad \textit{Subtract 266.}$$

The student must score at least 94 on the fourth test.

2.5 Section Exercises

For Exercises 3-27, see the answer graphs in the back of the textbook.

3. $x < 3$

In interval notation, this inequality is written $(-\infty, 3)$. The parenthesis indicates that 3 is not included. The graph of this inequality is shown in choice B.

7. Use a parenthesis when an endpoint is not included; use a bracket when it is included.

11. $4x + 1 \geq 21$
$$4x + 1 - 1 \geq 21 - 1$$
$$4x \geq 20$$
$$\frac{4x}{4} \geq \frac{20}{4}$$
$$x \geq 5$$

Solution set: $[5, \infty)$

15. $-4x < 16$

Divide by -4 and reverse the inequality sign.

$$\frac{-4x}{-4} > \frac{16}{-4}$$
$$x > -4$$

Solution set: $(-4, \infty)$

19. $-\frac{3}{2}y \leq -\frac{9}{2}$

Multiply by $-\frac{2}{3}$ and reverse the inequality sign.

$$-\frac{2}{3}\left(-\frac{3}{2}y\right) \geq -\frac{2}{3}\left(-\frac{9}{2}\right)$$
$$y \geq 3$$

Solution set: $[3, \infty)$

23. $\frac{2k-5}{-4} > 5$

Multiply by -4; reverse the inequality sign.

$$-4\left(\frac{2k-5}{-4}\right) < -4(5)$$
$$2k - 5 < -20$$
$$2k - 5 + 5 < -20 + 5$$
$$2k < -15$$
$$\frac{2k}{2} < \frac{-15}{2}$$
$$k < -\frac{15}{2}$$

Solution set: $(-\infty, -\frac{15}{2})$

27. $-(4 + r) + 2 - 3r < -14$
$$-4 - r + 2 - 3r < -14$$
$$-4r - 2 < -14$$
$$-4r < -12$$

Divide by -4; reverse the inequality sign.

$$\frac{-4r}{-4} > \frac{-12}{-4}$$
$$r > 3$$

Solution set: $(3, \infty)$

For Exercises 31-34, see the answer graphs in the back of the textbook.

31. $5(x + 3) - 2(x - 4) = 2(x + 7)$
$$5x + 15 - 2x + 8 = 2x + 14$$
$$3x + 23 = 2x + 14$$
$$x = -9$$

Solution set: $\{-9\}$

The graph is the point -9 on a number line.

32. $5(x + 3) - 2(x - 4) > 2(x + 7)$
$$5x + 15 - 2x + 8 > 2x + 14$$
$$3x + 23 > 2x + 14$$
$$x > -9$$

Solution set: $(-9, \infty)$

The graph extends from -9 to the right on the number line; -9 is not included in the graph.

33. $5(x + 3) - 2(x - 4) < 2(x + 7)$
$$5x + 15 - 2x + 8 < 2x + 14$$
$$3x + 23 < 2x + 14$$
$$x < -9$$

Solution set: $(-\infty, -9)$

The graph extends from -9 to the left on a number line; -9 is not included in the graph.

34. If we graph all the solution sets from Exercises 31-33, that is, $\{-9\}$, $(-9, \infty)$ and $(-\infty, -9)$, on the same number line, we will have graphed the set of all real numbers.

35. The solution set of the given equation is the point -3 on a number line. The solution set of the first inequality extends from -3 to the right (toward ∞) on the same number line. Based on Exercises 31-33, the solution set of the second inequality should then extend from -3 to the left (toward $-\infty$) on the number line. Complete the statement as follows:

    The solution set of $-3(x+2) = 3x + 12$ is $\{-3\}$, and the solution set of $-3(x+2) < 3x + 12$ is $(-3, \infty)$. Therefore, the solution set of $-3(x+2) > 3x + 12$ is $(-\infty, -3)$.

36. The statement "Equality is the boundary between less than and greater than" summarizes our findings in Exercises 31-35. There, the solution set of the equation was the point not included in the graphs of the inequalities.

For Exercises 39-47, see the answer graphs in the back of the textbook.

39. $-9 \le k + 5 \le 15$

    Subtract 5 from each part.

$$-9 - 5 \le k + 5 - 5 \le 15 - 5$$
$$-14 \le k \le 10$$

    Solution set: $[-14, 10]$

43. $-19 \le 3x - 5 \le 1$

    Add 5 to each part; then divide by 3.

$$-19 + 5 \le 3x - 5 + 5 \le 1 + 5$$
$$-14 \le 3x \le 6$$
$$\frac{-14}{3} \le \frac{3x}{3} \le \frac{6}{3}$$
$$-\frac{14}{3} \le x \le 2$$

    Solution set: $[-\frac{14}{3}, 2]$

47. $4 \le 5 - 9x < 8$

    Subtract 5 from each part.

$$4 - 5 \le 5 - 9x - 5 < 8 - 5$$
$$-1 \le -9x < 3$$

    Divide each part by -9; reverse the inequality signs.

$$\frac{-1}{-9} \ge \frac{-9x}{-9} > \frac{3}{-9}$$
$$\frac{1}{9} \ge x > -\frac{1}{3}$$
$$\text{or } -\frac{1}{3} < x \le \frac{1}{9}$$

    Solution set: $(-\frac{1}{3}, \frac{1}{9}]$

51. A total of 17,252 tornadoes were reported. To find the months in which fewer than (or less than) 1500 were reported, find what percent 1500 is of 17,252.

$$\frac{1500}{17,252} \approx .087 = 8.7\%$$

    The months when less than 8.7% of tornadoes were reported were January, February, March, August, September, October, November, and December.

55. Let $x =$ the number of miles that results in the same rental cost.

    The cost of renting from Ford is 15 plus the mileage cost of $.14x$. The cost of renting from Chevrolet is 14 plus the mileage cost of $.16x$. We must write an inequality saying that the cost to rent the Chevrolet exceeds, that is, is greater than, the cost to rent the Ford.

$$14 + .16x > 15 + .14x$$

    Multiply by 100.

$$1400 + 16x > 1500 + 14x$$

    Subtract $14x$; subtract 1400.

$$2x > 100$$

    Divide by 2.

$$x > 50$$

    After 50 mi the price to rent the Chevrolet would exceed the price to rent the Ford.

2.6 Set Operations and Compound Inequalities

2.6 Margin Exercises

1. Let $A = \{3, 4, 5, 6\}$ and $B = \{5, 6, 7\}$.

   The set $A \cap B$ is made up of those elements that belong to both A and B at the same time, in other words, the numbers 5 and 6. Therefore,

$$A \cap B = \{5, 6\}.$$

For Exercises 2 and 3, see the answer graphs for the margin exercises in the textbook

2. **(a)** $x < 10$ and $x > 2$
   Graph each inequality.

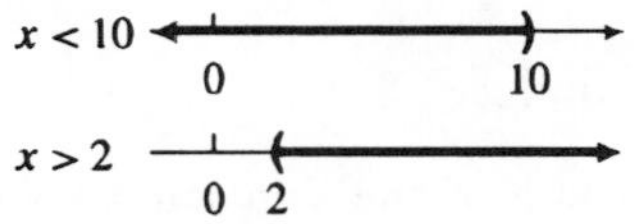

   The word "and" means to take the values that satisfy both inequalities, in other words, the numbers between 2 and 10, not including 2 or 10.

   Solution set: (2, 10)

   (b) $x + 3 < 1$ and $x - 4 > -12$

   Solve each inequality.

$$\begin{aligned} x + 3 &< 1 & \text{and} \quad x - 4 &> -12 \\ x + 3 - 3 &< 1 - 3 & x - 4 + 4 &> -12 + 4 \\ x &< -2 & \text{and} \quad x &> -8 \end{aligned}$$

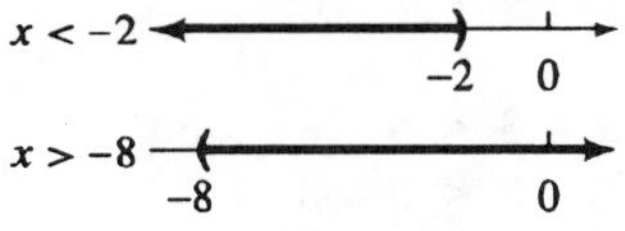

   The values that satisfy both inequalities are the numbers between -8 and -2, not including -8 or -2.

   Solution set: $(-8, -2)$

3. $$\begin{aligned} 2x &\geq x - 1 & \text{and} \quad 3x &\geq 3 + 2x \\ 2x - x &\geq x - 1 - x & 3x - 2x &\geq 3 + 2x - 2x \\ x &\geq -1 & \text{and} \quad x &\geq 3 \end{aligned}$$

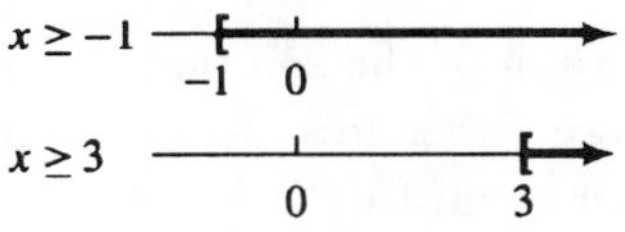

   The overlap of the two graphs consists of the numbers that are greater than or equal to -1 and also greater than or equal to 3, that is, the numbers greater than or equal to 3.

   Solution set: $[3, \infty)$

4. $$\begin{aligned} x + 2 &> 3 & \text{and} \quad 2x + 1 &< -3 \\ x + 2 - 2 &> 3 - 2 & 2x + 1 - 1 &< -3 - 1 \\ & & 2x &< -4 \\ & & \frac{2x}{2} &< \frac{-4}{2} \\ x &> 1 & \text{and} \quad x &< -2 \end{aligned}$$

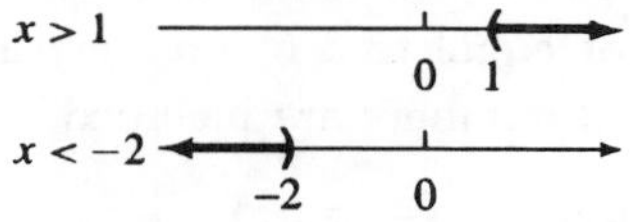

   The numbers that satisfy both inequalities must be greater than 1 and less than -2. There are no such numbers.

   Solution set: $\emptyset$

5. Let $A = \{3, 4, 5, 6\}$ and $B = \{5, 6, 7\}$.

   The set $A \cup B$ consists of all elements in either A or B (or both). Start by listing the elements of set A: 3, 4, 5, 6. Then list any additional elements from set B. In this case, the elements 5 and 6 are already listed, so the only additional element is 7. Therefore,

$$A \cup B = \{3, 4, 5, 6, 7\}.$$

6. See the answer graphs for the margin exercises in the textbook.

   (a) $$\begin{aligned} x + 2 > 3 \quad &\text{or} \quad 2x + 1 < -3 \\ & \quad\quad 2x < -4 \\ x > 1 \quad &\text{or} \quad x < -2 \end{aligned}$$

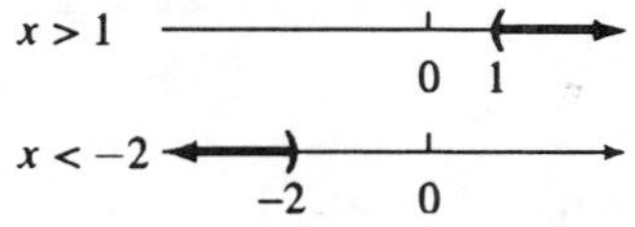

   The graph of the solution set will be all numbers greater than 1 or less than -2.

   Solution set: $(-\infty, -2) \cup (1, \infty)$

   (b) $$\begin{aligned} y - 1 > 2 \quad &\text{or} \quad 3y + 5 < 2y + 6 \\ y > 3 \quad &\text{or} \quad y < 1 \end{aligned}$$

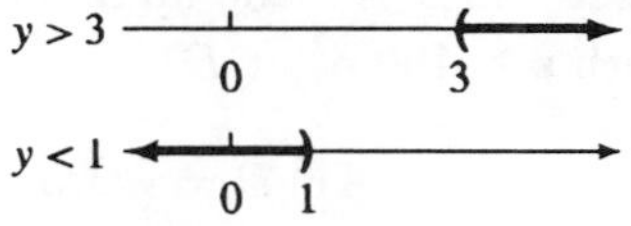

The graph of the solution set will be all numbers greater than 3 or less than 1.

Solution set: $(-\infty, 1) \cup (3, \infty)$

7. $3x - 2 \leq 13$ or $x + 5 \geq 7$
$3x \leq 15$
$x \leq 5$ or $x \geq 2$

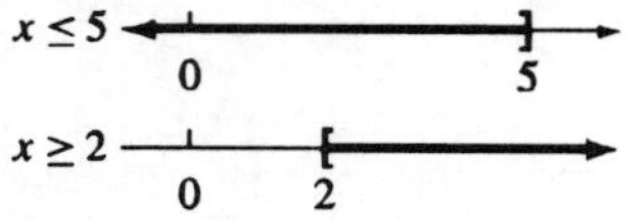

The solution set is all numbers that are either less than or equal to 5 or greater than or equal to 2. All real numbers are included.

Solution set: $(-\infty, \infty)$

2.6 Section Exercises

3. This statement is true.

$$2x + 1 = 3$$
$$2x + 1 - 1 = 3 - 1$$
$$2x = 2$$
$$\frac{2x}{2} = \frac{2}{2}$$
$$x = 1$$

Solution set: $\{1\}$

$$2x + 1 > 3$$
$$2x > 2$$
$$x > 1$$

Solution set: $(1, \infty)$

$$2x + 1 < 3$$
$$2x < 2$$
$$x < 1$$

Solution set: $(-\infty, 1)$

The union of the solution sets $\{1\}$, $(1, \infty)$, and $(-\infty, 1)$ is the set of $\{$all real numbers$\}$.

In Exercises 7-15, $A = \{1, 2, 3, 4, 5, 6\}$, $B = \{1, 3, 5\}$, $C = \{1, 6\}$, and $D = \{4\}$.

7. The set $A \cap D$ is made up of all numbers that are in both set A and set D.

$$A \cap D = \{4\} \text{ or } D$$

11. The set $A \cup B$ is made up of all numbers that are either in set A or in set B or both. Since all numbers in set B are also in set A, the set $A \cup B$ will be the same as set A.

$$A \cup B = \{1, 2, 3, 4, 5, 6\} \text{ or } A$$

15. The set $C \cup D$ is made up of all numbers that are either in set C or in set D or in both.

$$C \cup D = \{1, 4, 6\}$$

19. Answers will vary. For example, the intersection of two streets is the region common to *both* streets.

For Exercises 23-55, see the answer graphs in the back of the textbook.

23. $x \leq 2$ and $x \leq 5$

The graph of the solution set will be all numbers that are both less than or equal to 2 and less than or equal to 5. The overlap is the numbers less than or equal to 2.

Solution set: $(-\infty, 2]$

27. $x \geq -1$ and $x \geq 4$

The graph of the solution set will be all numbers that are both greater than or equal to -1 and greater than or equal to 4. The overlap is the numbers greater than or equal to 4.

Solution set: $[4, \infty)$

31. $x - 3 \leq 6$ and $x + 2 \geq 7$
$x - 3 + 3 \leq 6 + 3$ $\quad$ $x + 2 - 2 \geq 7 - 2$
$x \leq 9$ and $x \geq 5$

The graph of the solution set will be all numbers that are both less than or equal to 9 and greater than or equal to 5. The overlap is the numbers between 5 and 9, including the endpoints.

Solution set: $[5, 9]$

35. $x \leq 2$ or $x \geq 4$

The graph of the solution set will be all numbers that are either less than or equal to 2 or greater than or equal to 4.

Solution set: $(-\infty, 2] \cup [4, \infty)$

39. $x \le 1$ or $x \ge 10$

The graph of the solution set will be all numbers that are either less than or equal to 1 or greater than or equal to 10.

Solution set: $(-\infty, 1] \cup [10, \infty)$

43. $x \ge -2$ or $x \le 4$

The graph of the solution set will be all numbers that are either greater than or equal to -2 or less than or equal to 4. This is the set of all real numbers.

Solution set: $(-\infty, \infty)$

47.
$$\begin{array}{ccc} 4x - 8 > 0 & \text{or} & 4x - 1 < 7 \\ 4x - 8 + 8 > 0 + 8 & & 4x - 1 + 1 < 7 + 1 \\ 4x > 8 & & 4x < 8 \\ \frac{4x}{4} > \frac{8}{4} & & \frac{4x}{4} < \frac{8}{4} \\ x > 2 & \text{or} & x < 2 \end{array}$$

The word *or* means to take the union of both sets. The graph of the solution set will be all numbers that are either greater than 2 or less than 2. This is all real numbers except 2.

Solution set: $(-\infty, 2) \cup (2, \infty)$

51. $x < 4$ or $x < -2$

The word *or* means to take the union of both sets. The graph of the solution set will be all numbers that are either less than 4 or less than -2. This will be the same as all numbers less than 4.

Solution set: $(-\infty, 4)$

55.
$$\begin{array}{ccc} -3x \le -6 & \text{or} & -3x \ge 0 \\ \frac{-3x}{-3} \ge \frac{-6}{-3} & & \frac{-3x}{-3} \le \frac{0}{-3} \\ x \ge 2 & \text{or} & x \le 0 \end{array}$$

The word *or* means to take the union of both sets. The graph of the solution set will be all numbers that are either greater than or equal to 2 or less than or equal to 0.

Solution set: $(-\infty, 0] \cup [2, \infty)$

For Exercises 57-60, find the area and perimeter of each of the given yards.

For Luigi's, Mario's, and Than's yards, use the formulas $A = LW$ and $P = 2L + 2W$.

Luigi's yard

$$A = 50(30) = 1500 \text{ ft}^2$$
$$P = 2(50) + 2(30) = 160 \text{ ft}$$

Mario's yard

$$A = 40(35) = 1400 \text{ ft}^2$$
$$P = 2(40) + 2(35) = 150 \text{ ft}$$

Than's yard

$$A = 60(50) = 3000 \text{ ft}^2$$
$$P = 2(60) + 2(50) = 220 \text{ ft}$$

For Joe's yard, use the formulas $A = \frac{1}{2}bh$ and $P = a + b + c$.

Joe's yard

$$A = \frac{1}{2}(40)(30) = 600 \text{ ft}^2$$
$$P = 30 + 40 + 50 = 120 \text{ ft}$$

To be fenced, a yard must have a perimeter $P \le 150$ ft. To be sodded, a yard must have an area $A \le 1400$ ft^2.

57. Find "the yard can be fenced *and* the yard can be sodded."
A yard that can be fenced has $P \le 150$. Mario and Joe qualify.
A yard that can be sodded has $A \le 1400$. Again, Mario and Joe qualify.
Find the intersection. Mario's and Joe's yards are common to both sets, so Mario and Joe can have their yards both fenced and sodded.

58. Find "the yard can be fenced *and* the yard cannot be sodded."
A yard that can be fenced has $P \le 150$. From Exercise 57, Mario and Joe qualify.
A yard that cannot be sodded has $A > 1400$. This includes Luigi and Than.
Find the intersection. There are no yards common to both sets, so none of them qualify.

59. Find "the yard cannot be fenced *and* the yard can be sodded."
A yard that cannot be fenced has $P > 150$. This includes Luigi and Than.
A yard that can be sodded has $A \le 1400$. This includes Mario and Joe.
Find the intersection. There are no yards common to both sets, so none of them qualify.

60. Find "the yard cannot be fenced *and* the yard cannot be sodded."
A yard that cannot be fenced has $P > 150$. Luigi and Than qualify.
A yard that cannot be sodded has $A > 1400$. Again, Luigi and Than quality.
Find the intersection. Luigi's and Than's yards are common to both sets, so Luigi and Than qualify.

2.7 Absolute Value Equations and Inequalities

2.7 Margin Exercises

For Exercises 1-5, see the answer graphs for the margin exercises in the textbook.

1. **(a)** $|x| = 3$

The graph of the solution set will be the numbers that are 3 units from 0, in other words, the numbers 3 and -3.

Solution set: $\{-3, 3\}$

(b) $|x| > 3$

The graph of the solution set will be the numbers that are more than 3 units from 0. This will be all numbers greater than 3 or all numbers less than -3.

Solution set: $(-\infty, -3) \cup (3, \infty)$

(c) $|x| < 3$

The graph of the solution set will be the numbers that are less than 3 units from 0. This will be all numbers between -3 and 3.

Solution set: $(-3, 3)$

2. **(a)** $|x + 2| = 3$

$$x + 2 = 3 \quad \text{or} \quad x + 2 = -3$$
$$x = 1 \quad \text{or} \quad x = -5$$

Solution set: $\{-5, 1\}$

(b) $|3x - 4| = 11$

$$3x - 4 = 11 \quad \text{or} \quad 3x - 4 = -11$$
$$3x = 15 \qquad\qquad 3x = -7$$
$$x = 5 \quad \text{or} \quad x = -\frac{7}{3}$$

Solution set: $\{-\frac{7}{3}, 5\}$

3. **(a)** $|x + 2| > 3$

$$x + 2 > 3 \quad \text{or} \quad x + 2 < -3$$
$$x > 1 \quad \text{or} \quad x < -5$$

Solution set: $(-\infty, -5) \cup (1, \infty)$

(b) $|3x - 4| \geq 11$

$$3x - 4 \geq 11 \quad \text{or} \quad 3x - 4 \leq -11$$
$$3x \geq 15 \qquad\qquad 3x \leq -7$$
$$x \geq 5 \quad \text{or} \quad x \leq -\frac{7}{3}$$

Solution set: $(-\infty, -\frac{7}{3}] \cup [5, \infty)$

4. **(a)** $|x + 2| < 3$

$$-3 < x + 2 < 3$$
$$-5 < x < 1$$

Solution set: $(-5, 1)$

(b) $|3x - 4| \leq 11$

$$-11 \leq 3x - 4 \leq 11$$
$$-7 \leq 3x \leq 15$$
$$-\frac{7}{3} \leq x \leq 5$$

Solution set: $[-\frac{7}{3}, 5]$

5. **(a)** $|5a + 2| - 9 = -7$
$|5a + 2| = 2$

$$5a + 2 = 2 \quad \text{or} \quad 5a + 2 = -2$$
$$5a = 0 \qquad\qquad 5a = -4$$
$$a = 0 \quad \text{or} \quad a = -\frac{4}{5}$$

Solution set: $\{-\frac{4}{5}, 0\}$

(b) $|m + 2| - 3 > 2$
$|m + 2| > 5$

$$m + 2 > 5 \quad \text{or} \quad m + 2 < -5$$
$$m > 3 \quad \text{or} \quad m < -7$$

Solution set: $(-\infty, -7) \cup (3, \infty)$

(c) $|3a + 2| + 4 \leq 15$
$|3a + 2| \leq 11$

$$-11 \leq 3a + 2 \leq 11$$
$$-13 \leq 3a \leq 9$$
$$-\frac{13}{3} \leq a \leq 3$$

Solution set: $[-\frac{13}{3}, 3]$

6. **(a)** $|k-1|=|5k+7|$

$$k-1=5k+7$$
$$-8=4k$$
$$-2=k$$

or

$$k-1=-(5k+7)$$
$$k-1=-5k-7$$
$$6k=-6$$
$$k=-1$$

Solution set: $\{-2,-1\}$

(b) $|4r-1|=|3r+5|$

$$4r-1=3r+5$$
$$r=6$$

or

$$4r-1=-(3r+5)$$
$$4r-1=-3r-5$$
$$7r=-4$$
$$r=-\frac{4}{7}$$

Solution set: $\left\{-\frac{4}{7},6\right\}$

7. **(a)** $|6x+7|=-5$

Since an absolute value of an expression can never be negative, there is no solution to this equation.

Solution set: $\emptyset$

(b) $\left|\frac{1}{4}x-3\right|=0$

$$\frac{1}{4}x-3=0$$
$$x-12=0 \quad \textit{Multiply by } 4.$$
$$x=12$$

Solution set: $\{12\}$

8. **(a)** $|x|>-1$

The absolute value of a number is never negative. Therefore, the inequality is true for all real numbers.

Solution set: $(-\infty,\infty)$

(b) $|y|<-5$

There are no numbers whose absolute value is less than a negative number, so this inequality has no solution.

Solution set: $\emptyset$

(c) $|k+2|\le 0$

The value of $|k+2|$ will never be less than 0. The statement is true only when

$$k+2=0$$
$$k=-2.$$

Solution set: $\{-2\}$

2.7 Section Exercises

3. When solving an absolute value equation or inequality of the form

$$|ax+b|=k$$
$$|ax+b|<k, \text{ or}$$
$$|ax+b|>k$$

where k is a positive number, use *and* for the $<$ case and use *or* for both the $=$ case and the $>$ case.

7. $|4x|=20$

$$4x=20 \quad \text{or} \quad 4x=-20$$
$$x=5 \quad \text{or} \quad x=-5$$

Solution set: $\{-5,5\}$

11. $|2x+1|=7$

$$2x+1=7 \quad \text{or} \quad 2x+1=-7$$
$$2x=6 \qquad 2x=-8$$
$$x=3 \quad \text{or} \quad x=-4$$

Solution set: $\{-4,3\}$

15. $|2y+5|=14$

$$2y+5=14 \quad \text{or} \quad 2y+5=-14$$
$$2y=9 \qquad 2y=-19$$
$$y=\frac{9}{2} \quad \text{or} \quad y=-\frac{19}{2}$$

Solution set: $\left\{-\frac{19}{2},\frac{9}{2}\right\}$

19. $\left|1-\frac{3}{4}k\right|=7$

$$1-\frac{3}{4}k=7 \quad \text{or} \quad 1-\frac{3}{4}k=-7$$

Multiply all sides by 4.

$$4-3k=28 \quad \text{or} \quad 4-3k=-28$$
$$-3k=24 \qquad -3k=-32$$
$$k=-8 \quad \text{or} \quad k=\frac{32}{3}$$

Solution set: $\left\{-8,\frac{32}{3}\right\}$

For Exercises 23-47, see the answer graphs in the back of the textbook.

23. $|k| \geq 4$

$k \geq 4$ or $k \leq -4$

Solution set: $(-\infty, -4] \cup [4, \infty)$

27. $|3x - 1| \geq 8$

$$3x - 1 \geq 8 \quad \text{or} \quad 3x - 1 \leq -8$$
$$3x \geq 9 \qquad 3x \leq -7$$
$$x \geq 3 \quad \text{or} \quad x \leq -\frac{7}{3}$$

Solution set: $(-\infty, -\frac{7}{3}] \cup [3, \infty)$

31. **(a)** $|2x + 1| < 9$

The graph of the solution set will be all numbers between -5 and 4, since the absolute value is less than 9.

(b) $|2x + 1| > 9$

The graph of the solution set will be all numbers less than -5 or greater than 4, since the absolute value is greater than 9.

35. $|k| < 4$

$k < 4$ or $k > -4$

This is all numbers between -4 and 4.

Solution set: $(-4, 4)$

39. $|3x - 1| < 8$

$$-8 < 3x - 1 < 8$$
$$-7 < 3x < 9$$
$$-\frac{7}{3} < x < 3$$

Solution set: $(-\frac{7}{3}, 3)$

43. $|-4 + k| > 9$

$$-4 + k > 9 \quad \text{or} \quad -4 + k < -9$$
$$k > 13 \quad \text{or} \quad k < -5$$

Solution set: $(-\infty, -5) \cup (13, \infty)$

47. $|3r - 1| \leq 11$

$$-11 \leq 3r - 1 \leq 11$$
$$-10 \leq 3r \leq 12$$
$$-\frac{10}{3} \leq r \leq 4$$

Solution set: $[-\frac{10}{3}, 4]$

51. $|x| - 1 = 4$
$|x| = 5$

$$x = 5 \quad \text{or} \quad x = -5$$

Solution set: $\{-5, 5\}$

55. $|2x + 1| + 3 > 8$
$|2x + 1| > 5$

$$2x + 1 > 5 \quad \text{or} \quad 2x + 1 < -5$$
$$2x > 4 \qquad 2x < -6$$
$$x > 2 \quad \text{or} \quad x < -3$$

Solution set: $(-\infty, -3) \cup (2, \infty)$

59. $|3x + 1| = |2x + 4|$

$$3x + 1 = 2x + 4$$
$$x = 3$$

or

$$3x + 1 = -(2x + 4)$$
$$3x + 1 = -2x - 4$$
$$5x = -5$$
$$x = -1$$

Solution set: $\{-1, 3\}$

63. $|6x| = |9x + 1|$

$$6x = 9x + 1$$
$$-3x = 1$$
$$x = -\frac{1}{3}$$

or

$$6x = -(9x + 1)$$
$$6x = -9x - 1$$
$$15x = -1$$
$$x = -\frac{1}{15}$$

Solution set: $\{-\frac{1}{3}, -\frac{1}{15}\}$

67. $|12t - 3| = -8$

Since the absolute value of an expression can never be negative, there is no solution to this equation.

Solution set: $\emptyset$

71. $|2q - 1| < -6$

There are no numbers whose absolute value is less than a negative number, so this inequality has no solution.

Solution set: $\emptyset$

75. $|7x+3| \leq 0$

The absolute value of an expression can never be negative. This statement is true only when

$$\begin{aligned} 7x+3 &= 0 \\ 7x &= -3 \\ x &= -\frac{3}{7}. \end{aligned}$$

Solution set: $\{-\frac{3}{7}\}$

79. $|10z+7| > 0$

$$\begin{aligned} 10z+7 &> 0 \quad \text{or} \quad & 10z+7 &< 0 \\ 10z &> -7 & 10z &< -7 \\ z &> -\frac{7}{10} \quad \text{or} & z &< -\frac{7}{10} \end{aligned}$$

Solution set: $(-\infty, -\frac{7}{10}) \cup (-\frac{7}{10}, \infty)$

Alternatively, remember that an absolute value expression is always nonnegative. Therefore, there is only one possible value of z that makes this statement false, that is, when $10z+7=0$.

$$\begin{aligned} 10z+7 &= 0 \\ 10z &= -7 \\ z &= -\frac{7}{10} \end{aligned}$$

The solution set of the inequality is then

$$(-\infty, -\tfrac{7}{10}) \cup (-\tfrac{7}{10}, \infty),$$

which agrees with our first method of solution.

81. Add the given heights with a calculator to get 4602. There are 10 numbers, so divide the sum by 10.

$$\frac{4602}{10} = 460.2$$

The average height is 460.2 ft.

82. $|x-k| < t$

Substitute 460.2 for k and 50 for t. Solve the inequality.

$$\begin{aligned} |x-460.2| &< 50 \\ -50 < x - 460.2 &< 50 \\ 410.2 < x &< 510.2 \end{aligned}$$

The buildings with heights between 410.2 ft and 510.2 ft are the Federal Office Building, City Hall, Kansas City Power and Light, and the Hyatt Regency.

83. $|x-k| < t$

Substitute 460.2 for k and 75 for t. Solve the inequality.

$$\begin{aligned} |x-460.2| &< 75 \\ -75 < x - 460.2 &< 75 \\ 385.2 < x &< 535.2 \end{aligned}$$

The buildings with heights between 385.2 ft and 535.2 ft are Southwest Bell Telephone, City Center Square, Commerce Tower, the Federal Office Building, City Hall, Kansas City Power and Light, and the Hyatt Regency.

84. **(a)** This would be the opposite of the inequality in Exercise 83, that is,

$$|x-460.2| \geq 75.$$

(b) $|x-460.2| \geq 75$

$$\begin{aligned} x-460.2 &\geq 75 \quad \text{or} \quad & x-460.2 &\leq -75 \\ x &\geq 535.2 \quad \text{or} & x &\leq 385.2 \end{aligned}$$

(c) The buildings that are not within 75 ft of the average have height less than or equal to 385.2 ft or greater than or equal to 535.2 ft. This would include Pershing Road Associates, AT&T Town Pavilion, and One Kansas City Place.

(d) The answer makes sense because it includes all the buildings not listed earlier which had heights within 75 ft of the average.

Summary Exercises on Solving Linear and Absolute Value Equations and Inequalities

3. $6q-9 = 12+3q$

$$\begin{aligned} 3q &= 21 \quad \textit{Subtract } 3q\textit{; add } 9 \\ q &= 7 \end{aligned}$$

Solution set: $\{7\}$

7. $8r + 2 \geq 5r$
$3r \geq -2$
$r \geq -\frac{2}{3}$

Solution set: $[-\frac{2}{3}, \infty)$

11. $6z - 5 \leq 3z + 10$
$3z \leq 15$ *Subtract* $3z$; *add* 5.
$z \leq 5$

Solution set: $(-\infty, 5]$

15. $9y - 5 \geq 9y + 3$
$0 \geq 8$ *False*

The inequality is a contradiction.

Solution set: $\emptyset$

19. $\frac{2}{3}y + 8 = \frac{1}{4}y$

Multiply by the LCD, 12.

$$12\left(\frac{2}{3}y + 8\right) = 12\left(\frac{1}{4}y\right)$$
$$8y + 96 = 3y$$
$$5y = -96$$
$$y = -\frac{96}{5}$$

Solution set: $\{-\frac{96}{5}\}$

23. $\frac{3}{5}q - \frac{1}{10} = 2$

Multiply by the LCD, 10.

$$10\left(\frac{3}{5}q - \frac{1}{10}\right) = 10(2)$$
$$6q - 1 = 20$$
$$6q = 21$$
$$q = \frac{21}{6} \text{ or } \frac{7}{2}$$

Solution set: $\{\frac{7}{2}\}$

27. $|2p - 3| > 11$

$2p - 3 > 11$ or $2p - 3 < -11$
$2p > 14$ $\quad 2p < -8$
$p > 7$ or $p < -4$

Solution set: $(-\infty, -4) \cup (7, \infty)$

31. $-2 \leq 3x - 1 \leq 8$
$-1 \leq 3x \leq 9$
$-\frac{1}{3} \leq x \leq 3$

Solution set: $[-\frac{1}{3}, 3]$

35. $|1 - 3x| \geq 4$
$1 - 3x \geq 4$ or $1 - 3x \leq -4$
$-3x \geq 3$ or $-3x \leq -5$

Divide all sides by -3; reverse the direction of the inequality signs.

$$x \leq -1 \quad \text{or} \quad x \geq \frac{5}{3}$$

Solution set: $(-\infty, -1] \cup [\frac{5}{3}, \infty)$

39. $-6 \leq \frac{3}{2} - x \leq 6$

Multiply all parts by 2.

$$-12 \leq 3 - 2x \leq 12$$
$$-15 \leq -2x \leq 9$$

Divide all parts by -2; reverse the inequality signs.

$$\frac{15}{2} \geq x \geq -\frac{9}{2}$$
$$-\frac{9}{2} \leq x \leq \frac{15}{2}$$

Solution set: $[-\frac{9}{2}, \frac{15}{2}]$

43. $8q - (1 - q) = 3(1 + 3q) - 4$
$8q - 1 + q = 3 + 9q - 4$
$9q - 1 = 9q - 1$ *True*

The equation is true for every value of q. This is an identity.

Solution set: $(-\infty, \infty)$

47. $2x + 1 > 5$ or $3x + 4 < 1$
$2x > 4$ $\quad 3x < -3$
$x > 2$ or $x < -1$

Solution set: $(-\infty, -1) \cup (2, \infty)$

Chapter 2 Review Exercises

1. $-(8 + 3y) + 5 = 2y + 6$
$-8 - 3y + 5 = 2y + 6$
$-3y - 3 = 2y + 6$
$-5y = 9$
$y = -\frac{9}{5}$

Solution set: $\{-\frac{9}{5}\}$

2. $-(r + 5) - (2 + 7r) + 8r = 3r - 8$
$-r - 5 - 2 - 7r + 8r = 3r - 8$
$-7 = 3r - 8$
$1 = 3r$
$\frac{1}{3} = r$

Solution set: $\{\frac{1}{3}\}$

3. $\dfrac{m-2}{4}+\dfrac{m+2}{2}=8$

Multiply by the LCD, 4.

$$\begin{aligned} 4\left(\frac{m-2}{4}+\frac{m+2}{2}\right) &= 4(8) \\ (m-2)+2(m+2) &= 32 \\ m-2+2m+4 &= 32 \\ 3m+2 &= 32 \\ 3m &= 30 \\ m &= 10 \end{aligned}$$

Solution set: $\{10\}$

4. $\dfrac{2q+1}{3}-\dfrac{q-1}{4}=0$

Multiply by the LCD, 12.

$$\begin{aligned} 12\left(\frac{2q+1}{3}-\frac{q-1}{4}\right) &= 12(0) \\ 4(2q+1)-3(q-1) &= 0 \\ 8q+4-3q+3 &= 0 \\ 5q+7 &= 0 \\ 5q &= -7 \\ q &= -\frac{7}{5} \end{aligned}$$

Solution set: $\{-\frac{7}{5}\}$

5. $$\begin{aligned} 5(2x-3) &= 6(x-1)+4x \\ 10x-15 &= 6x-6+4x \\ 10x-15 &= 10x-6 \\ 0 &= 9 \quad \textit{False} \end{aligned}$$

The equation is a contradiction.

Solution set: $\emptyset$

6. $$\begin{aligned} -3x+2(4x+5) &= 10 \\ -3x+8x+10 &= 10 \\ 5x+10 &= 10 \\ 5x &= 0 \\ x &= 0 \end{aligned}$$

Solution set: $\{0\}$

7. $-\dfrac{3}{4}x=-12$

Multiply by $-\frac{4}{3}$.

$$\begin{aligned} x &= \frac{-48}{-3} \\ x &= 16 \end{aligned}$$

Solution set: $\{16\}$

8. $.05x+.03(1200-x)=42$

Multiply by 100.

$$\begin{aligned} 5x+3(1200-x) &= 4200 \\ 5x+3600-3x &= 4200 \\ 2x+3600 &= 4200 \\ 2x &= 600 \\ x &= 300 \end{aligned}$$

Solution set: $\{300\}$

9. Solve each equation.

(a) $$\begin{aligned} x-5 &= 5 \\ x &= 10 \end{aligned}$$

Solution set: $\{10\}$

(b) $$\begin{aligned} 4x &= 5x \\ 4x-4x &= 5x-4x \\ 0 &= x \end{aligned}$$

Solution set: $\{0\}$

(c) $$\begin{aligned} x+3 &= -3 \\ x &= -6 \end{aligned}$$

Solution set: $\{-6\}$

(d) $$\begin{aligned} 6x-6 &= 6 \\ 6x &= 12 \\ x &= 2 \end{aligned}$$

Solution set: $\{2\}$

Equation (b) has $\{0\}$ as its solution set.

10. Solve $-2x+5=7$.

Subtract 5 from both sides of the equation, then divide both sides by -2.

$$\begin{aligned} -2x &= 2 \\ x &= -1 \end{aligned}$$

Write the solution set.

Solution set: $\{-1\}$

11. $$\begin{aligned} 7r-3(2r-5)+5+3r &= 4r+20 \\ 7r-6r+15+5+3r &= 4r+20 \\ 4r+20 &= 4r+20 \quad \textit{True} \end{aligned}$$

This equation is an identity.

Solution set: $(-\infty,\infty)$

12. $$\begin{aligned} 8p - 4p - (p-7) + 9p + 6 &= 12p - 7 \\ 8p - 4p - p + 7 + 9p + 6 &= 12p - 7 \\ 12p + 13 &= 12p - 7 \\ 13 &= -7 \\ &\textit{False} \end{aligned}$$

This equation is a contradiction.

Solution set: $\emptyset$

13. $$\begin{aligned} -2r + 6(r-1) + 3r - (4-r) &= -(r+5) - 5 \\ -2r + 6r - 6 + 3r - 4 + r &= -r - 5 - 5 \\ 8r - 10 &= -r - 10 \\ 9r &= 0 \\ r &= 0 \end{aligned}$$

This equation is a conditional equation.

Solution set: $\{0\}$

14. Solve $V = LWH$ for H.
Divide by LW.

$$\frac{V}{LW} = \frac{LWH}{LW}$$
$$\frac{V}{LW} = H$$

15. Solve $A = \frac{1}{2}h(B+b)$ for h.

Multiply by 2; then divide by $B + b$.

$$2A = h(B+b)$$
$$\frac{2A}{B+b} = \frac{h(B+b)}{B+b}$$
$$\frac{2A}{B+b} = h$$

16. Solve $C = \pi d$ for d.
Divide by π.

$$\frac{C}{\pi} = \frac{\pi d}{\pi}$$
$$\frac{C}{\pi} = d$$

17. Use the formula $V = LWH$, and solve for H.

$$H = \frac{V}{LW}$$

Substitute $V = 180$, $L = 6$, and $W = 5$.

$$H = \frac{180}{6(5)} = 6$$

The height of the incinerator is 6 ft.

18. Divide the expected change by the original amount.

$$\frac{17.76 - 16.18}{16.18} = \frac{1.58}{16.18}$$
$$\approx .098 \text{ or } 9.8\%$$

The percent increase is about 9.8%.

19. Use the formula $I = prt$, and solve for r.

$$\frac{I}{pt} = \frac{prt}{pt}$$
$$\frac{I}{pt} = r$$

Substitute 30,000 for p, 7800 for I, and 4 for t.

$$r = \frac{7800}{30,000(4)} = \frac{7800}{120,000} = .065$$

The rate is 6.5%.

20. Use the formula $C = \frac{5}{9}(F - 32)$.

Substitute 68 for F.

$$C = \frac{5}{9}(68 - 32)$$
$$C = \frac{5}{9}(36)$$
$$C = 20$$

The Celsius temperature is 20°.

21. To find the percent of the total savings represented by staff cuts, divide the \$40.4 billion savings in staff cuts by the total savings of \$108 billion.

$$\frac{40.4}{108} \approx .374 \text{ or } 37.4\%$$

Staff cuts represent 37.4% of the total savings.

22. $$C = 2\pi r$$

Substitute 200π for C.

$$200\pi = 2\pi r$$
$$\frac{200\pi}{2\pi} = r$$
$$100 = r$$

The radius is 100 mm.

23. One-third of a number, subtracted from 9 is translated

$$9 - \frac{1}{3}x.$$

24. The product of 4 and a number, written $4x$, divided by 9 more than the number, written $x + 9$, is translated

$$\frac{4x}{x+9}.$$

25. Let $x =$ the width of the rectangle;
$2x - 3 =$ the length.

Twice the length plus twice the width equals the perimeter of the rectangle, 42 m.

$$\begin{aligned} 2x + 2(2x - 3) &= 42 \\ 2x + 4x - 6 &= 42 \\ 6x &= 48 \\ x &= 8 \\ 2x - 3 = 2(8) - 3 &= 13 \end{aligned}$$

The width is 8 m, and the length is 13 m.

26. Let $x =$ the length of each equal side;
$2x - 15 =$ the length of the third side.

The perimeter is 53 in. Add the lengths of the three sides.

$$\begin{aligned} x + x + (2x - 15) &= 53 \\ 4x - 15 &= 53 \\ 4x &= 68 \\ x &= 17 \\ 2x - 15 = 2(17) - 15 &= 19 \end{aligned}$$

The length of each equal side is 17 in, and the length of the third side is 19 in.

27. Let $x =$ the amount of peanut clusters;
$3x =$ the amount of chocolate creams.

The total weight is 48 kg.

$$\begin{aligned} x + 3x &= 48 \\ 4x &= 48 \\ x &= 12 \end{aligned}$$

The clerk has 12 kg of peanut clusters.

28. Let $x =$ the number of liters of the 20% chemical solution;
$x + 15 =$ the number of liters of the 30% chemical solution.

Make a table.

Strength	Liters of solution	Liters of pure chemical
20%	x	$.20x$
50%	15	$.50(15)$
30%	$x + 15$	$.30(x + 15)$

The last column gives the equation.

$$.20x + .50(15) = .30(x + 15)$$

Multiply by 10.

$$\begin{aligned} 2x + 5(15) &= 3(x + 15) \\ 2x + 75 &= 3x + 45 \\ 30 &= x \end{aligned}$$

30 L of the 20% solution should be mixed.

29. Let $x =$ the amount invested at 6%;
$x - 4000 =$ the amount invested at 4%.

% as a decimal	Amount invested	Interest in 1 year
.06	x	$.06x$
.04	$x - 4000$	$.04(x - 4000)$
		840

The last column gives the equation.

$$.06x + .04(x - 4000) = 840$$

Multiply by 100.

$$\begin{aligned} 6x + 4(x - 4000) &= 84,000 \\ 6x + 4x - 16,000 &= 84,000 \\ 10x &= 100,000 \\ x &= 10,000 \\ x - 4000 &= 10,000 - 4000 \\ &= 6000 \end{aligned}$$

Kevin should invest \$10,000 at 6% and \$6000 at 4%.

30. The percent increase is the increase in the number of dealers divided by the number of dealers in 1975.

Let $x =$ the percent increase.

$$x = \frac{269,712 - 161,927}{161,927}$$

$$x = \frac{107,785}{161,927}$$

$$x \approx .666 \text{ or } 66.6\%$$

There was an increase of about 66.6% in the number of firearm dealers.

31. Use the formula $d = rt$ or $r = \frac{d}{t}$. Here, d is about 400 mi and t is about 8 hr. Since $\frac{400}{8} = 50$, the best estimate is choice (a).

32. Use the formula $d = rt$.

(a) Here, $r = 53$ and $t = 10$.

$$d = 53(10) = 530$$

The distance is 530 mi.

(b) Here, $r = 164$ and $t = 2$.

$$d = 164(2) = 328$$

The distance is 328 mi.

33. Let $x =$ the time it takes for the trains to be 297 mi apart.

Use the formula $d = rt$.

	r	t	d
Passenger train	60	x	$60x$
Freight train	75	x	$75x$
			297

The total distance traveled is the sum of the distances traveled by each train.

$$60x + 75x = 297$$
$$135x = 297$$
$$x = 2.2$$

It will take the trains 2.2 hr before they are 297 mi apart.

34. Let $x =$ the speed of the slower car;
$x + 15 =$ the speed of the faster car.

	r	t	d
Slower car	x	2	$2x$
Faster car	$x+15$	2	$2(x+15)$
			230

The total distance traveled is the sum of the distances traveled by each car.

$$2x + 2(x+15) = 230$$
$$2x + 2x + 30 = 230$$
$$4x = 200$$
$$x = 50$$
$$x + 15 = 50 + 15 = 65$$

The slower car travels at 50 kph, while the faster car travels at 65 kph.

35. Let $x =$ the time traveled at the slower speed.

	r	t	d
Slower speed	45	x	$45x$
Faster speed	50	$4-x$	$50(4-x)$
			195

The total distance traveled was 195 mi.

$$45x + 50(4-x) = 195$$
$$45x + 200 - 50x = 195$$
$$-5x = -5$$
$$x = 1$$

The automobile traveled at 45 mph for 1 hr.

36. Let $x =$ the average speed for the first hour.

	r	t	d
First hour	x	1	$1x$ or x
Second hour	$x-7$	1	$1(x-7)$ or $x-7$
			85

The total distance traveled was 85 mi.

$$x + (x-7) = 85$$
$$2x - 7 = 85$$
$$2x = 92$$
$$x = 46$$

The Rodriguez family traveled 46 mph for the first hour.

37. Let $x =$ the number of reserved seats;
$1096 - x =$ the number of general admission seats.

	Number of seats	Cost per seat	Value
Reserved	x	15	$15x$
General admission	$1096-x$	12	$12(1096-x)$
			15,702

Multiply the number of seats by the cost of each seat. Add the results to get 15,702.

$$15x + 12(1096-x) = 15{,}702$$
$$15x + 13{,}152 - 12x = 15{,}702$$
$$3x = 2550$$
$$x = 850$$
$$1096 - x = 1096 - 850 = 246$$

850 reserved seats and 246 general admission seats were sold.

38. Let $x =$ the number of student tickets;

$311 - x =$ the number of nonstudent tickets.

Cost of each ticket	Number sold	Value
\$.25	x	$.25x$
\$.75	$311 - x$	$.75(311 - x)$
		108.75

Multiply the number of tickets by the cost per ticket, and add the results to get 108.75.

$$.25x + .75(311 - x) = 108.75$$

Multiply by 100.

$$\begin{aligned} 25x + 75(311 - x) &= 10,875 \\ 25x + 23,325 - 75x &= 10,875 \\ -50x &= -12,450 \\ x &= 249 \\ 311 - x = 311 - 249 &= 62 \end{aligned}$$

There were 249 student tickets sold and 62 nonstudent tickets sold.

39. The sum of the three angles is 180°.

$$\begin{aligned} (9x - 4) + (3x + 7) + (4x + 1) &= 180 \\ 16x + 4 &= 180 \\ 16x &= 176 \\ x &= 11 \end{aligned}$$

The measures of the angles are

$$\begin{aligned} (9x - 4)^\circ &= (9 \cdot 11 - 4)^\circ = 95^\circ, \\ (3x + 7)^\circ &= (3 \cdot 11 + 7)^\circ = 40^\circ, \\ (4x + 1)^\circ &= (4 \cdot 11 + 1)^\circ = 45^\circ. \end{aligned}$$

40. Based on the graph, 24.3% of the cars sold in 1993 were leased. This is about 25%. 25% of the 4 million total cars is 1 million. The best estimate is choice (c).

41. Since the angles are vertical angles, they have the same measure.

$$\begin{aligned} 10x - 15 &= 6x + 33 \\ 4x &= 48 \\ x &= 12 \end{aligned}$$

The angle measures are both

$$(10x - 15)^\circ = (10 \cdot 12 - 15)^\circ = 105^\circ.$$

Check the other angle measure.

$$(6x + 33)^\circ = (6 \cdot 12 + 33)^\circ = 105^\circ$$

42. Let $x =$ the smallest consecutive integer;

$x + 1 =$ the middle consecutive integer;

$x + 2 =$ the largest consecutive integer.

The sum of the smallest and largest integers is 47 more than the middle integer, so the equation is

$$\begin{aligned} x + (x + 2) &= 47 + (x + 1) \\ 2x + 2 &= 48 + x \\ x &= 46. \\ x + 1 &= 47 \\ x + 2 &= 48 \end{aligned}$$

The integers are 46, 47, and 48.

For Exercises 43-50, see the answer graphs in the back of the textbook.

43. $-\frac{2}{3}k < 6$

$-2k < 18$ *Multiply by* 3.

Divide by -2; reverse the inequality sign.

$$k > -9$$

Solution set: $(-9, \infty)$

44. $-5x - 4 \geq 11$

$-5x \geq 15$ *Add* 4.

Divide by -5; reverse the inequality sign.

$$x \leq -3$$

Solution set: $(-\infty, -3]$

45. $\dfrac{6a + 3}{-4} < -3$

Multiply by -4; reverse the inequality sign.

$$\begin{aligned} 6a + 3 &> 12 \\ 6a &> 9 \\ a &> \frac{9}{6} \text{ or } \frac{3}{2} \end{aligned}$$

Solution set: $(\frac{3}{2}, \infty)$

46. $\dfrac{9y + 5}{-3} > 3$

Multiply by -3; reverse the inequality sign.

$$\begin{aligned} 9y + 5 &< -9 \\ 9y &< -14 \\ y &< \frac{-14}{9} \end{aligned}$$

Solution set: $(-\infty, -\frac{14}{9})$

47. $5-(6-4k) \geq 2k-7$

$$\begin{aligned} 5-6+4k &\geq 2k-7 \\ 4k-1 &\geq 2k-7 \\ 2k &\geq -6 \\ k &\geq -3 \end{aligned}$$

Solution set: $[-3, \infty)$

48. $-6 \leq 2k \leq 24$

$-3 \leq k \leq 12$

Solution set: $[-3, 12]$

49. $8 \leq 3y-1 < 14$

$9 \leq 3y < 15$

$3 \leq y < 5$

Solution set: $[3, 5)$

50. $-4 < 3-2k < 9$

$-7 < -2k < 6$

Divide all parts by -2; reverse the inequality signs.

$$\frac{7}{2} > k > -3$$

or $$-3 < k < \frac{7}{2}$$

Solution set: $\left(-3, \frac{7}{2}\right)$

51. Let $x =$ the student's grade on the fifth test.

The average of the five test grades must be at least 70. The inequality is

$$\begin{aligned} \frac{75+79+64+71+x}{5} &\geq 70 \\ 75+79+64+71+x &\geq 350 \\ 289+x &\geq 350 \\ x &\geq 61. \end{aligned}$$

The student will pass algebra if he or she receives at least 61% on the fifth test.

52. The result, $-8 < -13$, is a false statement. There are no real numbers that make this inequality true. The solution set is $\emptyset$.

In Exercises 53 and 54, $A = \{1, 3, 5, 7, 9\}$ and $B = \{3, 6, 9, 12\}$.

53. The set $A \cap B$ is made up of all numbers that are in both set A and set B. Therefore,

$$A \cap B = \{3, 9\}.$$

54. The set $A \cup B$ is made up of all numbers that are in either set A or set B, or both. Therefore,

$$A \cup B = \{1, 3, 5, 6, 7, 9, 12\}.$$

For Exercises 55-60, see the answer graphs in the back of the textbook.

55. $x > 6$ and $x < 9$

The graph of the solution set will be all numbers greater than 6 and less than 9. The overlap is the numbers between 6 and 9, not including the endpoints.

Solution set: $(6, 9)$

56. $x+4 > 12$ and $x-2 < 12$

$x > 8$ and $x < 14$

The graph of the solution set will be all numbers between 8 and 14, not including the endpoints.

Solution set: $(8, 14)$

57. $x > 5$ or $x \leq -3$

The graph of the solution set will be all numbers that are either greater than 5 or less than or equal to -3.

Solution set: $(-\infty, -3] \cup (5, \infty)$

58. $x \geq -2$ or $x < 2$

The graph of the solution set will be all numbers that are either greater than or equal to -2 or less than 2. All real numbers satisfy these criteria.

Solution set: $(-\infty, \infty)$

59. $x-4 > 6$ and $x+3 \leq 10$

$x > 10$ and $x \leq 7$

The graph of the solution set will be all numbers that are both greater than 10 and less than or equal to 7. There are no real numbers satisfying these criteria.

Solution set: $\emptyset$

60. $-5x+1 \geq 11$ or $3x+5 \geq 26$

$-5x \geq 10$ $\qquad$ $3x \geq 21$

$x \leq -2$ or $x \geq 7$

The graph of the solution set will be all numbers that are either less than or equal to -2 or greater than or equal to 7.

Solution set: $(-\infty, -2] \cup [7, \infty)$

61. From the graph on the left, the number of U.S. AIDS cases exceeded 30,000 in the years 1987, 1988, 1989, 1990, 1991, and 1992.
From the graph on the right, the number of new AIDS cases among children under age thirteen exceeded 400 in the years 1988, 1989, 1990, 1991, and 1992.
And means *intersection*. Both of these events occurred during the years 1988, 1989, 1990, 1991, and 1992.

62. From the graph on the left, the number of U.S. AIDS cases was greater than 40,000 in 1988, 1989, 1990, 1991, and 1992.
From the graph on the right, the number of new AIDS cases among children under age thirteen was less than 200 in the years 1981, 1982, 1983, 1984, and 1985.
Or means *union*. Either one of these events occurred in the years 1981, 1982, 1983, 1984, 1985, 1988, 1989, 1990, 1991, and 1992.

63. $|x| = 7$

$$x = 7 \text{ or } x = -7$$

Solution set: $\{-7, 7\}$

64. $|y + 2| = 9$

$$y + 2 = 9 \quad \text{or} \quad y + 2 = -9$$
$$y = 7 \quad \text{or} \quad y = -11$$

Solution set: $\{-11, 7\}$

65. $|3k - 7| = 8$

$$3k - 7 = 8 \quad \text{or} \quad 3k - 7 = -8$$
$$3k = 15 \qquad 3k = -1$$
$$k = 5 \quad \text{or} \quad k = -\frac{1}{3}$$

Solution set: $\{-\frac{1}{3}, 5\}$

66. $|z - 4| = -12$

Since the absolute value of an expression can never be negative, there is no solution to this equation.

Solution set: $\emptyset$

67. $|2k - 7| + 4 = 11$
$|2k - 7| = 7$

$$2k - 7 = 7 \quad \text{or} \quad 2k - 7 = -7$$
$$2k = 14 \qquad 2k = 0$$
$$k = 7 \quad \text{or} \quad k = 0$$

Solution set: $\{0, 7\}$

68. $|4a + 2| - 7 = -3$
$|4a + 2| = 4$

$$4a + 2 = 4 \quad \text{or} \quad 4a + 2 = -4$$
$$4a = 2 \qquad 4a = -6$$
$$a = \frac{2}{4} \qquad a = -\frac{6}{4}$$
$$a = \frac{1}{2} \quad \text{or} \quad a = -\frac{3}{2}$$

Solution set: $\{-\frac{3}{2}, \frac{1}{2}\}$

69. $|3p + 1| = |p + 2|$

$$3p + 1 = p + 2$$
$$2p = 1$$
$$p = \frac{1}{2}$$

or

$$3p + 1 = -(p + 2)$$
$$3p + 1 = -p - 2$$
$$4p = -3$$
$$p = -\frac{3}{4}$$

Solution set: $\{-\frac{3}{4}, \frac{1}{2}\}$

70. $|2m - 1| = |2m + 3|$

$$2m - 1 = 2m + 3$$
$$0 = 4 \text{ } \textit{False}$$

or

$$2m - 1 = -(2m + 3)$$
$$2m - 1 = -2m - 3$$
$$4m = -2$$
$$m = -\frac{2}{4} \text{ or } -\frac{1}{2}$$

Solution set: $\{-\frac{1}{2}\}$

For Exercises 71-76, see the answer graphs in the back of the textbook.

71. $|p| < 14$

$-14 < p < 14$

Solution set: $(-14, 14)$

72. $|-y + 6| \leq 7$

$$-7 \leq -y + 6 \leq 7$$
$$-13 \leq -y \leq 1$$

Multiply all parts by -1; reverse the inequality signs.

$$13 \geq y \geq -1$$
$$\text{or } -1 \leq y \leq 13$$

Solution set: $[-1, 13]$

73. $|2p+5| \leq 1$

$$-1 \leq 2p+5 \leq 1$$
$$-6 \leq 2p \leq -4$$
$$-3 \leq p \leq -2$$

Solution set: $[-3, -2]$

74. $|x+1| \geq -3$

The absolute value of a number is never negative. Therefore, the inequality is true for all real numbers.

Solution set: $(-\infty, \infty)$

75. $|5r-1| > 9$

$$5r-1 > 9 \quad \text{or} \quad 5r-1 < -9$$
$$5r > 10 \qquad 5r < -8$$
$$r > 2 \quad \text{or} \quad r < -\frac{8}{5}$$

Solution set: $\left(-\infty, -\frac{8}{5}\right) \cup (2, \infty)$

76. $|3k+6| \geq 0$

The absolute value of a number is always greater than or equal to zero. Therefore, the inequality is true for all real numbers.

Solution set: $(-\infty, \infty)$

77. $(7-2k)+3(5-3k) \geq k+8$

$$7-2k+15-9k \geq k+8$$
$$-11k+22 \geq k+8$$
$$-12k \geq -14$$
$$k \leq \frac{-14}{-12}$$
$$k \leq \frac{7}{6}$$

Solution set: $\left(-\infty, \frac{7}{6}\right]$

78. $x < 5$ and $x \geq -4$

The solution set will be all numbers that are both less than 5 and greater than or equal to -4. The overlap is all numbers between -4 and 5, including -4 but not including 5.

Solution set: $[-4, 5)$

79. $-5(6p+4) - 2p = -32p + 14$

$$-30p - 20 - 2p = -32p + 14$$
$$-32p - 20 = -32p + 14$$
$$0 = 34 \quad \textit{False}$$

The equation is a contradiction.

Solution set: $\emptyset$

80. Let $x =$ the length of the shortest side;
$2x =$ the length of the middle side;
$3x - 2 =$ the length of the longest side.

The perimeter is 34 in. Add the lengths of the three sides.

$$x + 2x + (3x-2) = 34$$
$$6x - 2 = 34$$
$$6x = 36$$
$$x = 6$$
$$2x = 2(6) = 12$$
$$3x - 2 = 3(6) - 2 = 16$$

The sides have lengths 6 in, 12 in, and 16 in.

81. $-5r \geq -10$

$$r \leq 2$$

Solution set: $(-\infty, 2]$

82. $|7x-2| > 9$

$$7x-2 > 9 \quad \text{or} \quad 7x-2 < -9$$
$$7x > 11 \qquad 7x < -7$$
$$x > \frac{11}{7} \quad \text{or} \quad x < -1$$

Solution set: $(-\infty, -1) \cup \left(\frac{11}{7}, \infty\right)$

83. $|2x-10| = 20$

$$2x-10 = 20 \quad \text{or} \quad 2x-10 = -20$$
$$2x = 30 \qquad 2x = -10$$
$$x = 15 \quad \text{or} \quad x = -5$$

Solution set: $\{-5, 15\}$

84. $|m+3| \leq 13$

$$-13 \leq m+3 \leq 13$$
$$-16 \leq m \leq 10$$

Solution set: $[-16, 10]$

85. Let $x =$ the length of each side of the original square;
$x + 4 =$ the length of each side of the increased square.

The original perimeter is $4x$. The increased perimeter is $4(x+4)$. The increased perimeter is 8 in less than twice the original perimeter, so

$$\begin{aligned} 4(x+4) &= 2(4x) - 8 \\ 4x + 16 &= 8x - 8 \\ 24 &= 4x \\ 6 &= x. \end{aligned}$$

The length of a side of the original square is 6 in.

86. Let $x =$ the number of votes received by one candidate;
$x + 151 =$ the number of votes received by the other candidate.

The total votes cast was 1215.

$$\begin{aligned} x + (x + 151) &= 1215 \\ 2x + 151 &= 1215 \\ 2x &= 1064 \\ x &= 532 \\ x + 151 &= 532 + 151 = 683 \end{aligned}$$

One candidate received 532 votes, while the other candidate received 683 votes.

Chapter 2 Test

1. $$\begin{aligned} 3(2y-2) - 4(y+6) &= 3y + 8 + y \\ 6y - 6 - 4y - 24 &= 3y + 8 + y \\ 2y - 30 &= 4y + 8 \\ -2y &= 38 \\ y &= -19 \end{aligned}$$

Solution set: $\{-19\}$

2. $.08x + .06(x+9) = 1.24$

Multiply by 100.

$$\begin{aligned} 8x + 6(x+9) &= 124 \\ 8x + 6x + 54 &= 124 \\ 14x + 54 &= 124 \\ 14x &= 70 \\ x &= 5 \end{aligned}$$

Solution set: $\{5\}$

3. $\dfrac{x+6}{10} + \dfrac{x-4}{15} = \dfrac{x+2}{6}$

Multiply by the LCD, 30.

$$\begin{aligned} 3(x+6) + 2(x-4) &= 5(x+2) \\ 3x + 18 + 2x - 8 &= 5x + 10 \\ 5x + 10 &= 5x + 10 \end{aligned}$$

This equation is true for every value of x. The equation is an identity.

Solution set: $(-\infty, \infty)$

4. Solve $P = 2L + 2W$ for L.
Subtract $2W$; then divide by 2.

$$\begin{aligned} P - 2W &= 2L \\ \frac{P-2W}{2} &= \frac{2L}{2} \\ \frac{P-2W}{2} &= L \\ \text{or} \quad \frac{P}{2} - W &= L \end{aligned}$$

5. Solve $d = rt$ for t.

$$t = \frac{d}{r}$$

Substitute $d = 450$ and $r = 140.9$.

$$t = \frac{450}{140.9} \approx 3.2$$

Petty's time was about 3.2 hr.

6. Solve $I = prt$ for r.

$$r = \frac{I}{pt}$$

Substitute $I = 862.50$, $p = 23,000$, and $t = 1$.

$$r = \frac{862.50}{23,000(1)} = .0375 = 3.75\%$$

The rate of interest is 3.75%.

7. Let $x =$ the population of the county.
63.1% of x, or $.631x$, is the number of residents living in poverty.

$$\begin{aligned} .631x &= 6118 \\ x &= \frac{6118}{.631} \\ x &\approx 9696 \end{aligned}$$

There are about 9696 residents in the county.

8. Let $x =$ the amount invested at 3%;
$28,000 - x =$ the amount invested at 5%.

% as a decimal	Amount invested	Interest in 1 year
.03	x	$.03x$
.05	$28,000 - x$	$.05(28,000 - x)$
		1240

The last column gives the equation.

$$.03x + .05(28,000 - x) = 1240$$

Multiply by 100.

$$\begin{aligned} 3x + 5(28,000 - x) &= 124,000 \\ 3x + 140,000 - 5x &= 124,000 \\ -2x &= -16,000 \\ x &= 8000 \\ 28,000 - x &= 28,000 - 8000 \\ &= 20,000 \end{aligned}$$

Charles should invest \$8000 at 3% and \$20,000 at 5%.

9. Let $x =$ the speed of the slower car;
$x + 15 =$ the speed of the faster car.

Use the formula $d = rt$.

	r	t	d
Slower car	x	3	$3x$
Faster car	$x + 15$	3	$3(x + 15)$
			315

The total distance traveled is the sum of the distances traveled by each car.

$$\begin{aligned} 3x + 3(x + 15) &= 315 \\ 3x + 3x + 45 &= 315 \\ 6x &= 270 \\ x &= 45 \\ x + 15 &= 45 + 15 = 60 \end{aligned}$$

The slower car travels at 45 mph, while the faster car travels at 60 mph.

10. The sum of the measures must equal 180°.

$$\begin{aligned} (2x + 20) + x + x &= 180 \\ 4x + 20 &= 180 \\ 4x &= 160 \\ x &= 40 \end{aligned}$$

The two equal angles each measure 40°. The other angle measures

$$(2x + 20)° = (2 \cdot 40 + 20)° = 100°.$$

11. When multiplying or dividing both sides of an inequality by a negative number, we must reverse the direction of the inequality symbol.

For Exercises 12-14, see the answer graphs in the back of the textbook.

12. $$\begin{aligned} 4 - 6(x + 3) &\le -2 - 3(x + 6) + 3x \\ 4 - 6x - 18 &\le -2 - 3x - 18 + 3x \\ -6x - 14 &\le -20 \\ -6x &\le -6 \end{aligned}$$

Divide by -6; reverse the inequality sign.

$$x \ge 1$$

Solution set: $[1, \infty)$

13. $$-\frac{4}{7} > -16$$

$$-4x > -112$$

Divide by -4; reverse the inequality sign.

$$x < 28$$

Solution set: $(-\infty, 28)$

14. $$\begin{aligned} -6 \le \frac{4}{3}x - 2 &\le 2 \\ -18 \le 4x - 6 &\le 6 \\ -12 \le 4x &\le 12 \\ -3 \le x &\le 3 \end{aligned}$$

Solution set: $[-3, 3]$

15. Let $x =$ the grade the student must make on the third test.

To find the average of the three tests, add them and divide by 3. This average must be at least 90.

$$\begin{aligned} \frac{84 + 92 + x}{3} &\ge 90 \\ \frac{176 + x}{3} &\ge 90 \end{aligned}$$

Multiply by 3.

$$\begin{aligned} 176 + x &\ge 270 \\ x &\ge 94 \end{aligned}$$

The student must score at least 94 on the third test.

For Exercises 16-18, see the answer graphs in the back of the textbook.

16. $3k \geq 6$ and $k - 4 < 5$
$k \geq 2$ and $k < 9$

Solution set: $[2, 9)$

17. $|4x + 3| \leq 7$

$$-7 \leq 4x + 3 \leq 7$$
$$-10 \leq 4x \leq 4$$
$$-\frac{10}{4} \leq x \leq 1$$
$$-\frac{5}{2} \leq x \leq 1$$

Solution set: $\left[-\frac{5}{2}, 1\right]$

18. $|5 - 6x| > 12$

$$5 - 6x > 12 \quad \text{or} \quad 5 - 6x < -12$$
$$-6x > 7 \qquad\qquad -6x < -17$$
$$x < -\frac{7}{6} \quad \text{or} \quad x > \frac{17}{6}$$

Solution set: $\left(-\infty, -\frac{7}{6}\right) \cup \left(\frac{17}{6}, \infty\right)$

19. $|3k - 2| + 1 = 8$
$|3k - 2| = 7$

$$3k - 2 = 7 \quad \text{or} \quad 3k - 2 = -7$$
$$3k = 9 \qquad\qquad 3k = -5$$
$$k = 3 \quad \text{or} \quad k = -\frac{5}{3}$$

Solution set: $\left\{-\frac{5}{3}, 3\right\}$

20. $|3 - 5x| = |2x + 8|$

$$3 - 5x = 2x + 8$$
$$-7x = 5$$
$$x = -\frac{5}{7}$$

or

$$3 - 5x = -(2x + 8)$$
$$3 - 5x = -2x - 8$$
$$-3x = -11$$
$$x = \frac{11}{3}$$

Solution set: $\left\{-\frac{5}{7}, \frac{11}{3}\right\}$

Cumulative Review Exercises (Chapters 1-2)

Exercises 1-6 refer to set A.

$$A = \left\{-8, -\frac{2}{3}, -\sqrt{6}, 0, \frac{4}{5}, 9, \sqrt{36}\right\}$$

1. The elements 9 and $\sqrt{36}$ (or 6) are natural numbers.

2. The elements 0, 9, and $\sqrt{36}$ (or 6) are whole numbers.

3. The elements $-8, 0, 9$, and $\sqrt{36}$ (or 6) are integers.

4. The elements $-8, -\frac{2}{3}, 0, \frac{4}{5}, 9$, and $\sqrt{36}$ (or 6) are rational numbers.

5. The element $-\sqrt{6}$ is an irrational number.

6. All the elements in set A are real numbers.

7. $$-\frac{4}{3} - \left(-\frac{2}{7}\right) = -\frac{4}{3} + \frac{2}{7}$$
$$= -\frac{28}{21} + \frac{6}{21}$$
$$= -\frac{22}{21}$$

8. $$|-4| - |2| + |-6| = 4 - 2 + 6$$
$$= 2 + 6$$
$$= 8$$

9. $(-2)^4 + (-2)^3 = 16 + (-8) = 8$

10. $\sqrt{25} - \sqrt[3]{125} = 5 - 5 = 0$

11. $(-3)^5 = (-3)(-3)(-3)(-3)(-3) = -243$

12. $\left(\frac{6}{7}\right)^3 = \frac{6}{7} \cdot \frac{6}{7} \cdot \frac{6}{7} = \frac{216}{343}$

13. $(x^2 + 1)^0 = 1$

Any expression (except 0) raised to the power 0 is equal to 1.

14. $-4^6 = -4 \cdot 4 \cdot 4 \cdot 4 \cdot 4 \cdot 4 = -4096$

15. $-\sqrt{36} = -6$

$\sqrt{-36}$ is not a real number.

16. $\frac{4 - 4}{4 + 4} = \frac{0}{8} = 0$

$\frac{4 + 4}{4 - 4} = \frac{8}{0}$, which is undefined.

For Exercises 17-20, let $a = 2, b = -3$, and $c = 4$.

17. $$\begin{aligned} -3a + 2b - c &= -3(2) + 2(-3) - 4 \\ &= -6 - 6 - 4 \\ &= -16 \end{aligned}$$

18. $$\begin{aligned} -2b^2 - 4c &= -2(-3)^2 - 4(4) \\ &= -2(9) - 4(4) \\ &= -18 - 16 \\ &= -34 \end{aligned}$$

19. $$\begin{aligned} -8(a^2 + b^3) &= -8(2^2 + (-3)^3) \\ &= -8[4 + (-27)] \\ &= -8(-23) \\ &= 184 \end{aligned}$$

20. $$\begin{aligned} \frac{3a^3 - b}{4 + 3c} &= \frac{3(2)^3 - (-3)}{4 + 3(4)} \\ &= \frac{3(8) - (-3)}{4 + 3(4)} \\ &= \frac{24 + 3}{4 + 12} \\ &= \frac{27}{16} \end{aligned}$$

21. $$\begin{aligned} &-7r + 5 - 13r + 12 \\ &= -7r - 13r + 5 + 12 \\ &= (-7 - 13)r + (5 + 12) \\ &= -20r + 17 \end{aligned}$$

22. $$\begin{aligned} &-(3k + 8) - 2(4k - 7) + 3(8k + 12) \\ &= -3k - 8 - 8k + 14 + 24k + 36 \\ &= -3k - 8k + 24k - 8 + 14 + 36 \\ &= 13k + 42 \end{aligned}$$

23. $(a + b) + 4 = 4 + (a + b)$

The order of the terms $(a+b)$ and 4 have been reversed. This is an illustration of the commutative property.

24. $4x + 12x = (4 + 12)x$

The common variable, x, has been factored from each term. This is an illustration of the distributive property.

25. $-9 + 9 = 0$

The sum of a number and its opposite is equal to 0. This is an illustration of the inverse property.

26. The product of a number and its reciprocal must equal 1. Given the number $-\frac{2}{3}$, its reciprocal is $-\frac{3}{2}$, since

$$\left(-\frac{2}{3}\right)\left(-\frac{3}{2}\right) = \frac{6}{6} = 1.$$

27. $$\begin{aligned} -4x + 7(2x + 3) &= 7x + 36 \\ -4x + 14x + 21 &= 7x + 36 \\ 10x + 21 &= 7x + 36 \\ 3x &= 15 \\ x &= 5 \end{aligned}$$

Solution set: $\{5\}$

28. $-\frac{3}{5}x + \frac{2}{3}x = 2$

Multiply by the LCD, 15.

$$\begin{aligned} 3(-3x) + 5(2x) &= 30 \\ -9x + 10x &= 30 \\ x &= 30 \end{aligned}$$

Solution set: $\{30\}$

29. $.06x + .03(100 + x) = 4.35$

Multiply by 100.

$$\begin{aligned} 6x + 3(100 + x) &= 435 \\ 6x + 300 + 3x &= 435 \\ 9x + 300 &= 435 \\ 9x &= 135 \\ x &= 15 \end{aligned}$$

Solution set: $\{15\}$

30. Solve $P = a + b + c$ for b.

Subtract $a + c$.

$$\begin{aligned} P - (a - c) &= a + b + c - (a + c) \\ P - a - c &= b \text{ or} \\ b &= P - a - c \end{aligned}$$

For Exercises 31-34, see the answer graphs in the back of the textbook.

31. $$\begin{aligned} 3 - 2(x + 7) &\leq -x + 3 \\ 3 - 2x - 14 &\leq -x + 3 \\ -2x - 11 &\leq -x + 3 \\ -x &\leq 14 \end{aligned}$$

Multiply by -1; reverse the inequality sign.

$$x \geq -14$$

Solution set: $[-14, \infty)$

32. $-4 < 5 - 3x \leq 0$
$-9 < -3x \leq -5$

Divide all parts by -3; reverse the inequality signs.

$$3 > x \geq \frac{5}{3} \quad \text{or}$$
$$\frac{5}{3} \leq x < 3$$

Solution set: $[\frac{5}{3}, 3)$

33. $2x + 1 > 5$ or $2 - x > 2$
$2x > 4$ $\quad -x > 0$
$x > 2$ or $x < 0$

Solution set: $(-\infty, 0) \cup (2, \infty)$

34. $|-7k + 3| \geq 4$

$-7k + 3 \geq 4$ or $-7k + 3 \leq -4$
$-7k \geq 1$ $\quad -7k \leq -7$
$k \leq -\frac{1}{7}$ or $k \geq 1$

Solution set: $(-\infty, -\frac{1}{7}] \cup [1, \infty)$

35. The median earnings for men are less than \$900 includes managerial and professional specialty, waiters, and bus drivers.
The median earnings for women are greater than \$500 includes managerial and professional specialty and mathematical and computer scientists.
Find the intersection. The only occupation in both groups is managerial and professional specialty.

36. The median earnings for men are greater than \$900 includes mathematical and computer scientists.
The median earnings for women are greater than \$600 includes mathematical and computer scientists.
The sets are the same, so the union would be the occupation mathematical and computer scientists.

37. Let $x =$ the amount of pure alcohol that should be added.

Strength	Liters of solution	Liters of pure alcohol
100%	x	$1.00x$
10%	7	$.10(7)$
30%	$x + 7$	$.30(x + 7)$

The last column gives the equation.

$$1.00x + .10(7) = .30(x + 7)$$

Multiply by 10.

$$10x + 1(7) = 3(x + 7)$$
$$10x + 7 = 3x + 21$$
$$7x = 14$$
$$x = 2$$

2 L of pure alcohol should be added to the solution.

38. Let $x =$ the number of nickels;
$x - 4 =$ the number of quarters.

The number of cents is

$$29 - x - (x - 4) = 33 - 2x.$$

Denomination	Number of coins	Value
\$.01	$33 - 2x$	$.01(33 - 2x)$
\$.05	x	$.05x$
\$.25	$x - 4$	$.25(x - 4)$
		2.69

Multiply the number of coins by the denominations, and add the results to get 2.69.

$$.01(33 - 2x) + .05x + .25(x - 4) = 2.69$$

Multiply by 100.

$$1(33 - 2x) + 5x + 25(x - 4) = 269$$
$$33 - 2x + 5x + 25x - 100 = 269$$
$$28x - 67 = 269$$
$$28x = 336$$
$$x = 12$$
$$x - 4 = 12 - 4 = 8$$
$$33 - 2x = 33 - 2(12) = 9$$

There are 9 cents, 12 nickels, and 8 quarters.

Chapter 3

LINEAR EQUATIONS AND INEQUALITIES IN TWO VARIABLES; FUNCTIONS

3.1 The Rectangular Coordinate System

3.1 Margin Exercises

1. **(a)** To plot $(-4, 2)$, go four units from zero to the left along the x-axis, and then go two units up parallel to the y-axis.

   (b) To plot $(3, -2)$, go three units from zero to the right along the x-axis, and then go two units down parallel to the y-axis.

   (c) To plot $(-5, -6)$, go five units from zero to the left along the x-axis, and then go six units down parallel to the y-axis.

   (d) To plot $(4, 6)$, go four units from zero to the right along the x-axis, and then go six units up parallel to the y-axis.

   (e) To plot $(-3, 0)$, go three units to the left along the x-axis. Do not move up or down parallel to the y-axis since the y-coordinate is 0.

   (f) To plot $(0, -5)$, do not move along the x-axis at all since the x-coordinate is 0. Move five units down along the y-axis.

   See the answer graph for the margin exercises in the textbook.

2. **(a)** $3x - 4y = 12$

   To complete the ordered pairs, substitute the given values for x or y in the equation.

   For $(0, \quad)$, let $x = 0$.

$$\begin{aligned} 3x - 4y &= 12 \\ 3(0) - 4y &= 12 \\ 0 - 4y &= 12 \\ 4y &= 12 \\ y &= -3 \end{aligned}$$

   The ordered pair is $(0, -3)$.

   For $(\quad, 0)$, let $y = 0$.

$$\begin{aligned} 3x - 4y &= 12 \\ 3x - 4(0) &= 12 \\ 3x - 0 &= 12 \\ 3x &= 12 \\ x &= 4 \end{aligned}$$

   The ordered pair is $(4, 0)$.

   For $(\quad, -2)$, let $y = -2$.

$$\begin{aligned} 3x - 4y &= 12 \\ 3x - 4(-2) &= 12 \\ 3x + 8 &= 12 \\ 3x &= 4 \\ x &= \frac{4}{3} \end{aligned}$$

   The ordered pair is $(\frac{4}{3}, -2)$.

   For $(-4, \quad)$, let $x = -4$.

$$\begin{aligned} 3x - 4y &= 12 \\ 3(-4) - 4y &= 12 \\ -12 - 4y &= 12 \\ -4y &= 24 \\ y &= -6 \end{aligned}$$

   The ordered pair is $(-4, -6)$.

   (b) To find one possible answer, let $x = -6$.

$$\begin{aligned} 3x - 4y &= 12 \\ 3(-6) - 4y &= 12 \\ -18 - 4y &= 12 \\ -4y &= 30 \\ y &= \frac{30}{4} \\ y &= -\frac{15}{2} \end{aligned}$$

   The ordered pair is $(-6, -\frac{15}{2})$.

For Exercises 3-5, see the answer graphs for the margin exercises in the textbook.

3. Plot the intercepts, $(4,0)$ and $(0,-3)$, from Exercise 2 and draw the line through them.

4. $2x - y = 4$

   To find the x-intercept, let $y = 0$.

$$\begin{aligned} 2x - y &= 4 \\ 2x - 0 &= 4 \\ 2x &= 4 \\ x &= 2 \end{aligned}$$

   The x-intercept is $(2,0)$.

   To find the y-intercept, let $x = 0$.

$$\begin{aligned} 2x - y &= 4 \\ 2(0) - y &= 4 \\ -y &= 4 \\ y &= -4 \end{aligned}$$

   The y-intercept is $(0,-4)$.

   Plot the intercepts, and draw the line through them.

5. **(a)** $y + 4 = 0$

   In standard form, the equation is $0x + y = -4$. Every value of x gives $y = -4$, so the y-intercept is $(0,-4)$. There is no x-intercept. The graph is the horizontal line through $(0,-4)$.

   (b) $x = 2$

   In standard form, the equation is $x + 0y = 2$. Every value of y gives $x = 2$, so the x-intercept is $(2,0)$. There is no y-intercept. The graph is the vertical line through $(2,0)$.

3.1 Section Exercises

3. Another name for the rectangular coordinate system is the Cartesian system, named after Rene Descartes.

7. The x-intercept is the point where a line crosses the x-axis. To find the x-intercept of a line, we let y equal 0 and solve for x.

11. **(a)** Point $(1,6)$ is located in quadrant I, since the x- and y-coordinates are both positive.

    (b) Point $(-4,-2)$ is located in quadrant III, since the x- and y-coordinates are both negative.

    (c) Point $(-3,6)$ is located in quadrant II, since the x-coordinate is negative and the y-coordinate is positive.

    (d) Point $(7,-5)$ is located in quadrant IV, since the x-coordinate is positive and the y-coordinate is negative.

    (e) Point $(-3,0)$ is located on the x-axis, so it does not belong to any quadrant.

For Exercises 15-27 and 35-43, see the answer graphs in the back of the textbook.

15. To plot $(2,3)$, go two units from zero to the right along the x-axis, and then go three units up parallel to the y-axis.

19. To plot $(0,5)$, do not move along the x-axis at all since the x-coordinate is 0. Move five units up along the y-axis.

23. To plot $(-2,0)$, go two units to the left along the x-axis. Do not move up or down parallel to the y-axis since the y-coordinate is 0.

27. $x + 2y = 5$

    To complete the ordered pairs, substitute the given values for x or y in the equation.

    For $(0,\ \ )$, let $x = 0$.

$$\begin{aligned} x + 2y &= 5 \\ 0 + 2y &= 5 \\ 2y &= 5 \\ y &= \frac{5}{2} \end{aligned}$$

    The ordered pair is $(0, \frac{5}{2})$.

    For $(\ \ ,0)$, let $y = 0$.

$$\begin{aligned} x + 2y &= 5 \\ x + 2(0) &= 5 \\ x + 0 &= 5 \\ x &= 5 \end{aligned}$$

    The ordered pair is $(5,0)$.

    For $(2,\ \ )$, let $x = 2$.

$$\begin{aligned} x + 2y &= 5 \\ 2 + 2y &= 5 \\ 2y &= 3 \\ y &= \frac{3}{2} \end{aligned}$$

    The ordered pair is $(2, \frac{3}{2})$.

For (,2), let $y = 2$.

$$\begin{aligned} x + 2y &= 5 \\ x + 2(2) &= 5 \\ x + 4 &= 5 \\ x &= 1 \end{aligned}$$

The ordered pair is $(1, 2)$.

Plot the four ordered pairs, and draw the line through them.

31. The y-axis is a vertical line, so its equation must have the form $x = k$, where k is a constant. Every point of the y-axis has x-coordinate 0. The equation is $x = 0$.

35. $x - 3y = 6$

To find the x-intercept, let $y = 0$.

$$\begin{aligned} x - 3y &= 6 \\ x - 3(0) &= 6 \\ x - 0 &= 6 \\ x &= 6 \end{aligned}$$

The x-intercept is $(6, 0)$.

To find the y-intercept, let $x = 0$.

$$\begin{aligned} x - 3y &= 6 \\ 0 - 3y &= 6 \\ -3y &= 6 \\ y &= -2 \end{aligned}$$

The y-intercept is $(0, -2)$.

Plot the intercepts, and draw the line through them.

39. $y = 5$

There is no x-intercept, since the equation $y = 5$ represents a horizontal line. The y-intercept is $(0, 5)$.

43. $x + 5y = 0$

To find the x-intercept, let $y = 0$.

$$\begin{aligned} x + 5y &= 0 \\ x + 5(0) &= 0 \\ x &= 0 \end{aligned}$$

The x-intercept is $(0, 0)$.

To find the y-intercept, let $x = 0$.

$$\begin{aligned} x + 5y &= 0 \\ 0 + 5y &= 0 \\ y &= 0 \end{aligned}$$

The y-intercept is $(0, 0)$.

Since both intercepts are the same ordered pair, $(0, 0)$, another point is needed to graph the line. Choose any number for x, say $x = 5$, and solve the equation for y.

$$\begin{aligned} x + 5y &= 0 \\ 5 + 5y &= 0 \\ 5y &= -5 \\ y &= -1 \end{aligned}$$

This gives the ordered pair $(5, -1)$. Plot $(5, -1)$ and the intercept point $(0, 0)$, and draw the line through them.

47. The graph goes through the point $(2, 6)$ which satisfies only equation (c). The correct equation is (c).

3.2 The Slope of a Line

3.2 Margin Exercises

1. **(a)** Let $(-2, 7) = (x_1,\ y_1)$ and $(4, -3) = (x_2,\ y_2)$. Then,

$$m = \frac{y_2 - y_1}{x_2 - x_1} = \frac{-3 - 7}{4 - (-2)} = \frac{-10}{6} = -\frac{5}{3}.$$

The slope is $-\frac{5}{3}$.

(b) Let $(1, 2) = (x_1,\ y_1)$ and $(8, 5) = (x_2,\ y_2)$. Then,

$$m = \frac{y_2 - y_1}{x_2 - x_1} = \frac{5 - 2}{8 - 1} = \frac{3}{7}.$$

The slope is $\frac{3}{7}$.

(c) Let $(8, -2) = (x_1,\ y_1)$ and $(3, -2) = (x_2,\ y_2)$. Then,

$$m = \frac{y_2 - y_1}{x_2 - x_1} = \frac{-2 - (-2)}{3 - 8} = \frac{0}{-5} = 0.$$

The slope is 0.

2. **(a)** To find the slope of

$$2x + y = 6,$$

find the intercepts. Replace y with 0 to find that the x-intercept is $(3, 0)$. Replace x with 0 to find that the y-intercept is $(0, 6)$. The slope is then

$$m = \frac{6-0}{0-3} = \frac{6}{-3} = -2.$$

The slope is -2.

(b) To find the slope of

$$3x - 4y = 12,$$

find the intercepts. The x-intercept is $(4, 0)$, and the y-intercept is $(0, -3)$. The slope is then

$$m = \frac{-3-0}{0-4} = \frac{-3}{-4} = \frac{3}{4}.$$

The slope is $\frac{3}{4}$.

(c) To find the slope of

$$x = -6,$$

select two different points on the line, such as $(-6, 0)$ and $(-6, 3)$, and use the definition of slope.

$$m = \frac{3-0}{-6-(-6)} = \frac{3}{0}$$

Since division by zero is undefined, the slope is undefined.

(d) To find the slope of

$$y + 5 = 0$$

select two different points on the line, such as $(0, -5)$ and $(2, -5)$, and use the definition of slope.

$$m = \frac{-5-(-5)}{2-0} = \frac{0}{2} = 0$$

The slope is 0.

For Exercise 3, see the answer graphs for the margin exercises in the textbook.

3. **(a)** Through $(1, -3)$; $m = -\frac{3}{4}$

Locate the point $(1, -3)$ on the graph. Use the definition of slope to find a second point on the line, writing $-\frac{3}{4}$ as $\frac{-3}{4}$.

$$m = \frac{\text{change in } y}{\text{change in } x} = \frac{-3}{4}$$

From $(1, -3)$, move 3 units down and then 4 units to the right to $(5, -6)$. Draw the line through the two points.

(b) Through $(-1, -4)$; $m = 2$

Locate $(-1, -4)$ on the graph. Use the definition of slope to find a second point on the line, writing 2 as $\frac{2}{1}$.

$$m = \frac{\text{change in } y}{\text{change in } x} = \frac{2}{1}$$

From $(-1, -4)$, move 2 units up and then 1 unit to the right to $(0, -2)$. Draw the line through the two points.

4. **(a)** Find the slope of each line. The line through $(-1, 2)$ and $(3, 5)$ has slope

$$m = \frac{5-2}{3-(-1)} = \frac{3}{4}.$$

The line through $(4, 7)$ and $(8, 10)$ has slope

$$m = \frac{10-7}{8-4} = \frac{3}{4}.$$

The slopes are the same, so the lines are parallel.

(b) Find the slope of each line. The line through $(5, -9)$ and $(3, 7)$ has slope

$$m = \frac{7-(-9)}{3-5} = \frac{16}{-2} = -8.$$

The line through $(0, 2)$ and $(8, 3)$ has slope

$$m = \frac{3-2}{8-0} = \frac{1}{8}.$$

The product of the slopes is

$$(-8)\left(\frac{1}{8}\right) = -1,$$

so the lines are perpendicular.

(c) $2x - y = 4$ has intercepts $(2, 0)$ and $(0, -4)$, so the slope is

$$m_1 = \frac{-4-0}{0-2} = \frac{-4}{-2} = 2.$$

$2x + y = 6$ has intercepts $(3, 0)$ and $(0, 6)$, so the slope is

$$m_2 = \frac{6-0}{0-3} = -2.$$

Since $m_1 \neq m_2$, the lines are not parallel. Since $m_1 m_2 = 2(-2) = -4$, the lines are not perpendicular either. Therefore the answer is "neither."

(d) $3x+5y=6$ has intercepts $(2,0)$ and $(0,\frac{6}{5})$, so the slope is

$$m_1=\frac{\frac{6}{5}-0}{0-2}=\frac{6}{5}\left(-\frac{1}{2}\right)=-\frac{6}{10}=-\frac{3}{5}.$$

$5x-3y=2$ has intercepts $(\frac{2}{5},0)$ and $(0,-\frac{2}{3})$, so the slope is

$$m_2=\frac{-\frac{2}{3}-0}{0-\frac{2}{5}}=\left(-\frac{2}{3}\right)\left(-\frac{5}{2}\right)=\frac{5}{3}.$$

Since $m_1m_2=\left(-\frac{3}{5}\right)\left(\frac{5}{3}\right)=-1$, the lines are perpendicular.

3.2 Section Exercises

3. Count the number of units of rise from C to D. The rise is 7. Count the number of units of run from C to D. The run is 0.

$$\text{slope of } CD=\frac{\text{rise}}{\text{run}}=\frac{7}{0}$$

The slope of CD is undefined.

7. Let $(-2,-3)=(x_1,\ y_1)$ and $(-1,5)=(x_2,\ y_2)$. Then,

$$m=\frac{y_2-y_1}{x_2-x_1}=\frac{5-(-3)}{-1-(-2)}=\frac{8}{1}=8.$$

The slope is 8.

11. Let $(2,4)=(x_1,\ y_1)$ and $(-4,4)=(x_2,\ y_2)$. Then,

$$m=\frac{y_2-y_1}{x_2-x_1}=\frac{4-4}{-4-2}=\frac{0}{-6}=0.$$

The slope is 0.

15. The graph of the line $y=2$ is a horizontal line with y-intercept $(0,2)$. The slope of a horizontal line is always 0. To check this, select two points on the line, say $(0,2)$ and $(2,2)$, and use the definition of slope.

$$m=\frac{2-2}{2-0}=\frac{0}{2}=0$$

See the answer graph in the back of the textbook.

19. Line A is a horizontal line and has slope 0.

For Exercises 23-35, see the answer graphs in the back of the textbook.

23. To find the slope of

$$-x+y=4,$$

first find the intercepts. Replace y with 0 to find that the x-intercept is $(-4,0)$; replace x with 0 to find that the y-intercept is $(0,4)$. The slope is then

$$m=\frac{4-0}{0-(-4)}=\frac{4}{4}=1.$$

To sketch the graph, plot the intercepts and draw the line through them.

27. To find the slope of

$$5x-2y=10,$$

first find the intercepts. Replace y with 0 to find that the x-intercept is $(2,0)$; replace x with 0 to find that the y-intercept is $(0,-5)$. The slope is then

$$m=\frac{-5-0}{0-2}=\frac{-5}{-2}=\frac{5}{2}.$$

To sketch the graph, plot the intercepts and draw the line through them.

31. $y-3=0$
$y=3$

The graph of $y=3$ is the horizontal line with y-intercept $(0,3)$. The slope of a horizontal line is 0.

35. To graph the line through $(0,-2)$ with slope $m=-\frac{2}{3}$, locate the point $(0,-2)$ on the graph. To find a second point on the line, use the definition of slope, writing $-\frac{2}{3}$ as $\frac{-2}{3}$.

$$m=\frac{\text{change in } y}{\text{change in } x}=\frac{-2}{3}$$

From $(0,-2)$, move 2 units down and then 3 units to the right to $(3,-4)$. Draw a line through $(3,-4)$ and $(0,-2)$. (Note that the slope could also be written as $\frac{2}{-3}$. In this case, move 2 units up and 3 units to the left to get another point on the same line.)

39. To decide whether

$$2x + 5y = -7 \text{ and } 5x - 2y = 1$$

are parallel, perpendicular, or neither, find the slope of each line by first finding two points on each line. The intercepts of $2x + 5y = -7$ are $(0, -\frac{7}{5})$ and $(-\frac{7}{2}, 0)$, so the slope is

$$m_1 = \frac{0 - (-\frac{7}{5})}{-\frac{7}{2} - 0} = \frac{\frac{7}{5}}{-\frac{7}{2}} = -\frac{2}{5}.$$

The intercepts of $5x - 2y = 1$ are $(0, -\frac{1}{2})$ and $(\frac{1}{5}, 0)$, so the slope is

$$m_2 = \frac{0 - (-\frac{1}{2})}{\frac{1}{5} - 0} = \frac{\frac{1}{2}}{\frac{1}{5}} = \frac{5}{2}.$$

Since

$$m_1 m_2 = -\frac{2}{5}\left(\frac{5}{2}\right) = -1,$$

the lines are perpendicular.

43. To decide whether

$$2x + y = 6 \text{ and } x - y = 4$$

are parallel, perpendicular, or neither, find the slope of each line by first finding two points on each line. The intercepts of $2x + y = 6$ are $(0, 6)$ and $(3, 0)$, so

$$m_1 = \frac{0 - 6}{3 - 0} = \frac{-6}{3} = -2.$$

The intercepts of $x - y = 4$ are $(0, -4)$ and $(4, 0)$, so

$$m_2 = \frac{0 - (-4)}{4 - 0} = \frac{4}{4} = 1.$$

The slopes of m_1 and m_2 are not the same; they are not negative reciprocals of each other either. Therefore, the lines are neither parallel nor perpendicular.

47. **(a)** Let $(x_1, y_1) = (1980, 7.85)$ and $(x_2, y_2) = (1992, 18.85)$. Then

$$m = \frac{18.85 - 7.85}{1992 - 1980} = \frac{11}{12} \approx .92.$$

The average rate of change is about \$.92/yr.

(b) A positive rate of change means an *increase* in price.

51. Let $(x_2, y_1) = (3, 1)$ and $(x_2, y_2) = (6, 2)$. Then

$$m = \frac{2 - 1}{6 - 3} = \frac{1}{3}.$$

The slope of AB is $\frac{1}{3}$.

52. Let $(x_1, y_1) = (6, 2)$ and $(x_2, y_2) = (9, 3)$. Then

$$m = \frac{3 - 2}{9 - 6} = \frac{1}{3}.$$

The slope of BC is $\frac{1}{3}$.

53. Let $(x_1, y_1) = (3, 1)$ and $(x_2, y_2) = (9, 3)$. Then

$$m = \frac{3 - 1}{9 - 3} = \frac{2}{6} = \frac{1}{3}.$$

The slope of AC is $\frac{1}{3}$.

54. The slope of AB = slope of BC
= slope of AC
$= \frac{1}{3}$.

55. Consider the points $A(1, -2)$, $B(3, -1)$, and $C(5, 0)$. Find the slopes of segments AB, BC, and AC. The slope of AB is

$$m = \frac{-1 - (-2)}{3 - 1} = \frac{1}{2}.$$

The slope of BC is

$$m = \frac{0 - (-1)}{5 - 3} = \frac{1}{2}.$$

The slope of AC is

$$m = \frac{0 - (-2)}{5 - 1} = \frac{2}{4} = \frac{1}{2}.$$

Since the slope of AB = slope of BC = slope of $AC = \frac{1}{2}$, then A, B, and C are collinear.

56. Consider the points $A(0, 6)$, $B(4, -5)$, and $C(-2, 12)$. The slope of AB is

$$m = \frac{-5 - 6}{4 - 0} = \frac{-11}{4} = -\frac{11}{4}.$$

The slope of BC is

$$m = \frac{12 - (-5)}{-2 - 4} = \frac{17}{-6} = -\frac{17}{6}.$$

The slope of AC is

$$m = \frac{12 - 6}{-2 - 0} = \frac{6}{-2} = -3.$$

Since the slopes are not equal, the points A, B, and C are not collinear.

3.3 Linear Equations in Two Variables

3.3 Margin Exercises

1. **(a)** Through $(-2,7)$, $m=3$

Use the point-slope form with $(x_1,\ y_1)=(-2,7)$ and $m=3$.

$$\begin{aligned} y-y_1&=m(x-x_1)\\ y-7&=3[x-(-2)]\\ y-7&=3(x+2)\\ y-7&=3x+6\\ y&=3x+13 \end{aligned}$$

In standard form $Ax+By=C$,

$$3x-y=-13.$$

(b) Through $(1,3)$, $m=-\frac{5}{4}$

$$\begin{aligned} y-y_1&=m(x-x_1)\\ y-3&=-\frac{5}{4}(x-1) \end{aligned}$$

Multiply by 4 to clear the fraction. Then write the equation in standard form.

$$\begin{aligned} 4y-12&=-5(x-1)\\ 4y-12&=-5x+5\\ 5x+4y&=17 \end{aligned}$$

2. **(a)** Through $(8,-2)$, $m=0$

A horizontal line has slope $m=0$. A horizontal line through the point (x,k) has equation $y=k$. Here $k=-2$, so the equation is $y=-2$.

(b) The vertical line through $(3,5)$

A vertical line through the point (k,y) has equation $x=k$. Here $k=3$, so the equation is $x=3$.

3. First find the slope of the line. Then use the slope and one of the points in the point-slope form.

(a) Through $(-1,2)$ and $(5,7)$

$$m=\frac{7-2}{5-(-1)}=\frac{5}{6}$$

Let $(x_1,\ y_1)=(5,7)$.

$$\begin{aligned} y-y_1&=m(x-x_1)\\ y-7&=\frac{5}{6}(x-5)\\ 6y-42&=5x-25\\ 5x-6y&=-17 \end{aligned}$$

(b) Through $(-2,6)$ and $(1,4)$

$$m=\frac{4-6}{1-(-2)}=\frac{-2}{3}=-\frac{2}{3}$$

Let $(x_1,\ y_1)=(1,4)$.

$$\begin{aligned} y-y_1&=m(x-x_1)\\ y-4&=-\frac{2}{3}(x-1)\\ 3y-12&=-2x+2\\ 2x+3y&=14 \end{aligned}$$

4. **(a)** Slope 2; y-intercept $(0,-3)$

Here $m=2$ and $b=-3$. Substitute these values in the slope-intercept form.

$$\begin{aligned} y&=mx+b\\ y&=2x+(-3)\\ y&=2x-3 \end{aligned}$$

In standard form, the equation is $2x-y=3$.

(b) Slope $-\frac{2}{3}$; y-intercept $(0,0)$

Here $m=-\frac{2}{3}$ and $b=0$.

$$\begin{aligned} y&=mx+b\\ y&=-\frac{2}{3}x+0\\ y&=-\frac{2}{3}x\\ 3y&=-2x \end{aligned}$$

In standard form, $2x+3y=0$.

(c) Slope 0; y-intercept $(0,3)$

Here $m=0$ and $b=3$.

$$\begin{aligned} y&=mx+b\\ y&=0x+3 \end{aligned}$$

In standard form, $y=3$.

5. **(a)** $x+y=2$

Solve for y to put the equation in slope-intercept form.

$$y=-x+2$$

So, $m=-1$ and $b=2$. The slope is -1, and the y-intercept is $(0,2)$.

(b) $2x - 5y = 1$

Solve for y.

$$-5y = -2x + 1$$
$$y = \frac{2}{5}x - \frac{1}{5}$$

So, $m = \frac{2}{5}$ and $b = -\frac{1}{5}$. The slope is $\frac{2}{5}$, and the y-intercept is $(0, -\frac{1}{5})$.

6. Through $(5, 7)$; parallel to $2x - 5y = 15$
Find the slope of

$$2x - 5y = 15.$$
$$-5y = -2x + 15$$
$$y = \frac{2}{5}x - 3$$

The slope is $\frac{2}{5}$, so a line parallel to it also has slope $\frac{2}{5}$. Use $m = \frac{2}{5}$ and $(x_1, y_1) = (5, 7)$ in the point-slope form.

$$y - y_1 = m(x - x_1)$$
$$y - 7 = \frac{2}{5}(x - 5)$$
$$5y - 35 = 2x - 10$$
$$-2x + 5y = 25$$
$$2x - 5y = -25$$

7. **(a)** Through $(1, 6)$; perpendicular to $x + y = 9$

Find the slope of

$$x + y = 9$$
$$y = -x + 9$$

The slope is -1. The negative reciprocal of -1 is 1, so the slope of the line through $(1, 6)$ is 1. Use the point-slope form.

$$y - y_1 = m(x - x_1)$$
$$y - 6 = 1(x - 1)$$
$$y - 6 = x - 1$$
$$x - y = -5$$

(b) Through $(-8, 3)$; perpendicular to $2x - 3y = 10$

Find the slope of

$$2x - 3y = 10.$$
$$-3y = -2x + 10$$
$$y = \frac{2}{3}x - \frac{10}{3}$$

The slope is $\frac{2}{3}$. The negative reciprocal of $\frac{2}{3}$ is $-\frac{3}{2}$, so the slope of the line through $(-8, 3)$ is $-\frac{3}{2}$.

$$y - y_1 = m(x - x_1)$$
$$y - 3 = -\frac{3}{2}[(x - (-8)]$$
$$y - 3 = -\frac{3}{2}(x + 8)$$
$$2y - 6 = -3x - 24$$
$$3x + 2y = -18$$

8. Use the equation $y = 1.20x$ from Example 7.

When $x = 5.5$, $y = 1.20(5.5) = 6.6$.
Ordered pair: $(5.5, 6.6)$

This ordered pair is interpreted as follows: When 5.5 gal have been pumped, the price is \$6.60.

9. Since the price you will pay is \$10/day plus a flat rate of \$15, the equation for x days is

$$y = 10x + 15.$$

3.3 Section Exercises

3. The line $y = -2x - 3$ has slope -2. A negative slope indicates that the line goes down from left to right. The line also has y-intercept $(0, -3)$. The answer is graph C.

7. The line $y = 3$ is a horizontal line with y-intercept $(0, 3)$. The answer is graph B.

11. Through $(5, 8)$; $m = -2$

Use the point-slope form with $(x_1, y_1) = (5, 8)$ and $m = -2$. Then write the equation in standard form $Ax + By = C$.

$$y - y_1 = m(x - x_1)$$
$$y - 8 = -2(x - 5)$$
$$y - 8 = -2x + 10$$
$$2x + y = 18$$

15. Through $(-4, 12)$; horizontal

A horizontal line through the point (x, k) has equation $y = k$. Here $k = 12$, so the equation is $y = 12$.

19. Through $(.5, .2)$; vertical

A vertical line through the point (k, y) has equation $x = k$. Here $k = .5$, so the equation is $x = .5$.

23. $(6,1)$ and $(-2,5)$

Find the slope.

$$m = \frac{5-1}{-2-6} = \frac{4}{-8} = -\frac{1}{2}$$

Use the point-slope form with $(x_1, y_1) = (6,1)$ and $m = -\frac{1}{2}$.

$$y - y_1 = m(x - x_1)$$
$$y - 1 = -\frac{1}{2}(x - 6)$$

Multiply by 2 to clear the fraction. Then write the equation in standard form.

$$2y - 2 = -x + 6$$
$$x + 2y = 8$$

27. $(2,5)$ and $(1,5)$

Find the slope.

$$m = \frac{5-5}{1-2} = \frac{0}{-1} = 0$$

A line with slope 0 is horizontal. A horizontal line through the point (x,k) has equation $y = k$, so the equation is $y = 5$.

31. Solve for y to put the equation in slope-intercept form.

$$5x + 2y = 20$$
$$2y = -5x + 20$$
$$y = -\frac{5}{2}x + 10$$

The slope is $-\frac{5}{2}$; the y-intercept is $(0,10)$.

35. $m = 5; b = 15$

Substitute these values in the slope-intercept form.

$$y = mx + b$$
$$y = 5x + 15$$

39. Slope $\frac{2}{5}$; y-intercept $(0,5)$

Here, $m = \frac{2}{5}$ and $b = 5$. Substitute these values in the slope-intercept form.

$$y = mx + b$$
$$y = \frac{2}{5}x + 5$$

43. Through $(-2,-2)$; parallel to $-x + 2y = 10$

Find the slope of

$$-x + 2y = 10.$$
$$2y = x + 10$$
$$y = \frac{1}{2}x + 5$$

The slope is $\frac{1}{2}$, so a line parallel to it also has slope $\frac{1}{2}$. Use $m = \frac{1}{2}$ and $(x_1, y_1) = (-2,-2)$ in the point-slope form.

$$y - y_1 = m(x - x_1)$$
$$y - (-2) = \frac{1}{2}[x - (-2)]$$
$$y + 2 = \frac{1}{2}(x + 2)$$

Multiply by 2 to clear the fraction. Then write the equation in standard form.

$$2y + 4 = x + 2$$
$$-x + 2y = -2$$
$$x - 2y = 2$$

47. Through $(-2,7)$; perpendicular to $x = 9$

The equation $x = 9$ is for a vertical line with undefined slope. A horizontal line, which has slope 0, would be perpendicular to a vertical line. Use $m = 0$ and $(x_1, y_1) = (-2,7)$ in the point-slope form.

$$y - y_1 = m(x - x_1)$$
$$y - 7 = 0[x - (-2)]$$
$$y - 7 = 0$$
$$y = 7$$

51. The equation is $y = 1.30x$.

When $x = 0$, $y = 1.30(0) = 0$.
Ordered pair: $(0,0)$

When $x = 5$, $y = 1.30(5) = 6.50$.
Ordered pair: $(5, 6.50)$

When $x = 10$, $y = 1.30(10) = 13.00$.
Ordered pair: $(10, 13.00)$

55. Since the rental car costs \$.10 per mile plus an additional \$25.00, the equation for x miles is

$$y = .10x + 25.00.$$

When $x = 0$, $y = .10(0) + 25.00 = 25.00$.
Ordered pair: $(0, 25.00)$

When $x = 5$, $y = .10(5) + 25.00 = 25.50$.
Ordered pair: $(5, 25.50)$

When $x = 10$, $y = .10(10) + 25.00 = 26.00$.
Ordered pair: $(10, 26.00)$

59. The equation from Exercise 55 is

$$y = .10x + 25.00.$$

Since y represents the rental car cost, substitute 42.30 for y in the equation and solve for x, the number of miles.

$$\begin{aligned} 42.30 &= .10x + 25.00 \\ 17.30 &= .10x \\ 173 &= x \end{aligned}$$

The car was driven 173 mi.

63. Line A goes through the point $(0, 3)$ which satisfies the equation for y_1. So, its equation must be $y_1 = -2x + 3$. Line B goes through $(0, -4)$, which satisfies the equation for y_2. So, its equation is $y_2 = 3x - 4$.

67. When $\text{C} = 0°$, $\text{F} = 32°$, and when $\text{C} = 100°$, $\text{F} = 212°$.

68. The two points would be $(0, 32)$ and $(100, 212)$.

69. $m = \dfrac{212 - 32}{100 - 0} = \dfrac{180}{100} = \dfrac{9}{5}$

70. Let $m = \frac{9}{5}$ and $(x_1, y_1) = (0, 32)$. If C replaces x and F replaces y in the point-slope form, then

$$\begin{aligned} y - y_1 &= m(x - x_1) \\ \text{F} - 32 &= \frac{9}{5}(\text{C} - 0) \\ \text{F} - 32 &= \frac{9}{5}\text{C} \\ \text{F} &= \frac{9}{5}\text{C} + 32. \end{aligned}$$

71.
$$\begin{aligned} \text{F} &= \frac{9}{5}\text{C} + 32 \\ \text{F} - 32 &= \frac{9}{5}\text{C} \\ \frac{5}{9}(\text{F} - 32) &= \text{C} \end{aligned}$$

72. A temperature of 50°C corresponds to a temperature of 122°F.

3.4 Linear Inequalities in Two Variables

3.4 Margin Exercises

For Exercises 1-5, see the answer graphs for the margin exercises in the textbook.

1. **(a)** $x + y \leq 4$

Graph the line, $x + y = 4$, which has intercepts $(4, 0)$ and $(0, 4)$, as a solid line since the inequality involves "$\leq$."

Test $(0, 0)$.

$$\begin{aligned} x + y &\leq 4 \\ 0 + 0 &\leq 4 \\ 0 &\leq 4 \quad \textit{True} \end{aligned}$$

Since the result is true, shade the region that contains $(0, 0)$.

(b) $3x + y \geq 6$

Graph the line $3x + y = 6$ as a solid line through $(2, 0)$ and $(0, 6)$.

Test $(0, 0)$.

$$\begin{aligned} 3x + y &\geq 6 \\ 3(0) + 0 &\geq 0 \\ 0 &> 6 \quad \textit{False} \end{aligned}$$

Since the result is false, shade the region that does not contain $(0, 0)$.

2. **(a)** $x - y > 2$

Graph $x - y = 2$, through $(2, 0)$ and $(0, -2)$, as a dashed line since the inequality involves "$>$." Test $(0, 0)$.

$$\begin{aligned} x - y &> 2 \\ 0 - 0 &> 2 \\ 0 &> 2 \quad \textit{False} \end{aligned}$$

Shade the region that does not contain $(0, 0)$.

(b) $3x + 4y < 12$

Graph $3x + 4y = 12$ as a dashed line through $(4, 0)$ and $(0, 3)$. Test $(0, 0)$.

$$\begin{aligned} 3x + 4y &< 12 \\ 3(0) + 4(0) &< 12 \\ 0 &< 12 \quad \textit{True} \end{aligned}$$

Shade the region that contains $(0, 0)$.

3. $x - y \leq 4$ and $x \geq -2$

Graph $x - y = 4$, which has intercepts $(4, 0)$ and $(0, -4)$, as a solid line since the inequality involves "$\leq$." Test $(0, 0)$, which yields $0 \leq 4$, a true statement. Shade the region that includes $(0, 0)$.

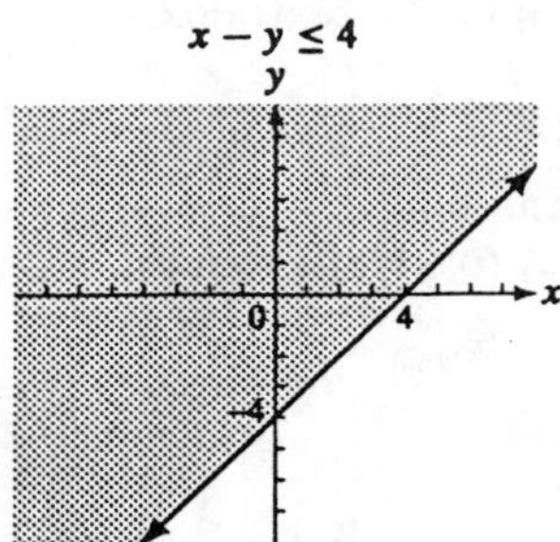

Graph $x = -2$ as a solid vertical line through $(-2, 0)$. Shade the region to the right of $x = -2$.

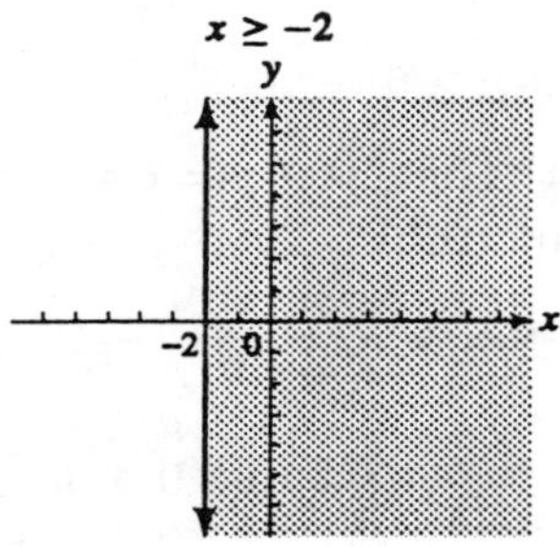

The graph of the intersection is the region common to both graphs.

4. $7x - 3y < 21$ or $x > 2$

Graph $7x - 3y = 21$ as a dashed line through intercepts $(3, 0)$ and $(0, -7)$. Test $(0, 0)$, which yields $0 < 21$, a true statement. Shade the region that includes $(0, 0)$.

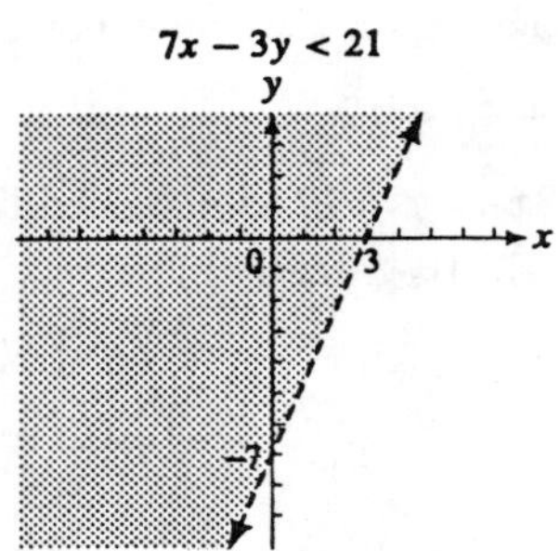

Graph $x = 2$ as a dashed vertical line through $(2, 0)$. Shade the region to the right of $x = 2$.

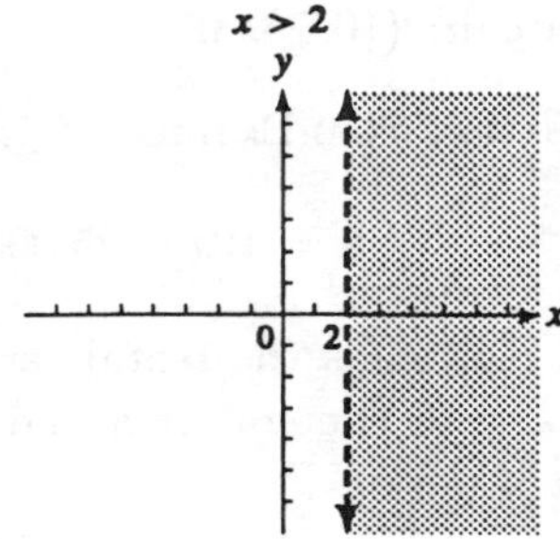

The graph of the union is the region that includes all the points in both graphs.

5. **(a)** $|x| \geq 4$

$|x| = 4$ can be rewritten as

$$x = -4 \quad \text{or} \quad x = 4.$$

The graph consists of the two solid vertical lines $x = -4$ and $x = 4$.

Test points in the three regions bounded by these lines, such as $(-5, 0)$, $(0, 0)$, and $(5, 0)$.

$$|-5| \geq 4 \qquad\qquad |0| \geq 4$$
$$5 \geq 4 \quad \textit{True} \qquad 0 \geq 4 \quad \textit{False}$$

$$|5| \geq 4$$
$$5 \geq 4 \quad \textit{True}$$

The graph includes the regions that contain $(-5, 0)$ and $(5, 0)$.

(b) $|x - 2| < 1$

$|x - 2| < 1$ can be rewritten as

$$-1 < x - 2 < 1$$
$$1 < x < 3.$$

Graph the boundary lines $x = 1$ and $x = 3$ as dashed vertical lines. The graph is the region between these dashed lines.

3.4 Section Exercises

3. The boundary of the graph of $y > -x + 2$ will be a *dashed* line, and the shading will be *above* the line.

For Exercises 7-31, see the answer graphs in the back of the textbook.

7. $4x - y < 4$

Graph the line $4x - y = 4$, which has intercepts $(1, 0)$ and $(0, -4)$, as a dashed line since the inequality involves "<." To decide which side of the line belongs to the graph, test $(0, 0)$. Substitute 0 for x and 0 for y in the original inequality.

$$4x - y < 4$$
$$4(0) - 0 < 4$$
$$0 < 4 \quad \textit{True}$$

Since the result is true, shade the region that includes $(0, 0)$.

11. $x + y > 0$

Graph the line $x+y = 0$, which includes the points $(0, 0)$ and $(2, -2)$, as a dashed line since the inequality involves ">." Since $(0, 0)$ is on the line, test another point, say $(3, 3)$.

$$x + y > 0$$
$$3 + 3 > 0$$
$$6 > 0 \quad \textit{True}$$

Since the result is true, shade the region that includes $(3, 3)$.

15. $y < x$

Graph the line $y = x$, which includes the points $(0, 0)$ and $(-3, -3)$, as a dashed line since the inequality involves "<." Use $(4, 2)$ as a test point.

$$y < x$$
$$2 < 4 \quad \textit{True}$$

Since the result is true, shade the region that includes $(4, 2)$.

19. $x + y \leq 1$ and $x \geq 1$

Graph $x + y = 1$, which has intercepts $(1, 0)$ and $(0, 1)$, as a solid line since the inequality involves "$\leq$." Test $(0, 0)$, which yields $0 \leq 1$, a true statement. Shade the region that includes $(0, 0)$.

Graph $x = 1$ as a solid vertical line since the inequality involves "$\geq$." Shade the region to the right of $x = 1$.

The required graph of the intersection is the region common to both graphs.

23. $x + y > -5$ and $y < -2$

Graph $x + y = -5$, which has intercepts $(-5, 0)$ and $(0, -5)$, as a dashed line. Test $(0, 0)$, which yields $0 > -5$, a true statement. Shade the region that includes $(0, 0)$.

Graph $y = -2$ as a dashed horizontal line. Shade the region below $y = -2$.

The required graph of the intersection is the region common to both graphs.

27. $x - 2 > y$ or $x < 1$

Graph $x - 2 = y$, which has intercepts $(2, 0)$ and $(0, -2)$, as a dashed line. Test $(0, 0)$, which yields $-2 > 0$, a false statement. Shade the region that does not include $(0, 0)$.

Graph $x = 1$ as a dashed vertical line. Shade the region to the left of $x = 1$.

The required graph of the union includes all the shaded regions, that is, all the points that satisfy either inequality.

31. $|x| \geq 3$

Rewrite $|x| \geq 3$ as $x \geq 3$ or $x \leq -3$. The graph consists of the region to the left of the solid vertical line $x = -3$ and to the right of the solid vertical line $x = 3$.

35. $y \leq 3x - 6$

The boundary line, $y = 3x - 6$, has slope 3 and y-intercept -6. This would be graph B or graph C. Since the inequality sign is $\leq$, we want the region on or below the boundary line. The answer is graph C.

3.5 An Introduction to Functions

3.5 Margin Exercises

1. Answers will vary in parts (a)-(c).

(a) From the table, two other ordered pairs are $(1927, 2)$ and $(1975, 4)$.

(b) Use the equation

$$y = .25x + 40.$$

To find two other ordered pairs, substitute two values for x and find the corresponding values for y.

When $x = 44$, $y = .25(44) + 40 = 51$.
Ordered pair: $(44, 51)$

When $x = 48$, $y = .25(48) + 40 = 52$.
Ordered pair: $(48, 52)$

(c) From the graph, two other ordered pairs are $(1991, 3.665)$ and $(1992, 4.065)$.

2. **(a)** $\{(1,2),(2,4),(3,3),(4,2)\}$

The relation is a function because for each x-value of the ordered pairs there is exactly one y-value.

(b) $\{(0,3),(-1,2),(-1,3)\}$

The relation is not a function because there is an x-value, -1, that has two y-values, 2 and 3, associated with it.

(c) The relation is not a function because there is a value, 10, in the set on the left, that has two values, A and B, in the set on the right.

(d) The relation is a function because each value on the left has exactly one value on the right.

3. **(a)** $\{(1,2),(2,4),(3,3),(4,2)\}$

The domain is the set of x-values, $\{1,2,3,4\}$. The range is the set of y-values, $\{2,3,4\}$.

(b) The domain is the set of values on the left, $\{-3,4,6,7\}$. The range is the set of values on the right, {A,B,C}.

4. **(a)** The x-values of the points on the graph include all real numbers beginning with -2. The domain is $[-2, \infty)$.

The arrowheads indicate that the graph extends infinitely upward and downward. The range is the set of all real numbers, $(-\infty, \infty)$.

(b) The domain is the set of all real numbers $(-\infty, \infty)$. The greatest y-value is 0, so the range is $(-\infty, 0]$.

5. **(a)** $y = 6x+12$ is a function because each value of x corresponds to exactly one value of y. Its domain is the set of all real numbers, $(-\infty, \infty)$.

(b) $y \leq 4x$ is not a function because if $x = 0$, then $y \leq 0$. Thus, the x-value 0 corresponds to many y-values. Its domain is the set of all real numbers, $(-\infty, \infty)$.

(c) $y = -\sqrt{3x-2}$ is a function because each value of x corresponds to exactly one value of y. Since the quantity under the radical sign must be nonnegative, the domain is the set of real numbers that satisfies the condition

$$3x - 2 \geq 0$$
$$x \geq \frac{2}{3}.$$

Therefore, the domain is $[\frac{2}{3}, \infty)$.

(d) $y^2 = 25x$ is not a function. If $x = 1$, for example, $y^2 = 25$ and $y = 5$ or $y = -5$. Since y^2 must be nonnegative, the domain is the set of nonnegative real numbers, $[0, \infty)$.

6. **(a)** Any vertical line would intersect the graph at most once, so the graph represents a function.

(b) The graph does not represent a function, since a vertical line, such as the y-axis, intersects the graph in more than one point.

(c) No vertical line would intersect the graph in more than one point, so the graph represents a function.

7. **(a)**
$$f(x) = 6x - 2$$
$$f(-3) = 6(-3) - 2 = -20$$
$$f(p) = 6p - 2$$

(b)
$$f(x) = \frac{-3x+5}{2}$$
$$f(-3) = \frac{-3(-3)+5}{2} = \frac{14}{2} = 7$$
$$f(p) = \frac{-3p+5}{2}$$

(c)
$$f(x) = \frac{1}{6}x^2 - 1$$
$$f(-3) = \frac{1}{6}(-3)^2 - 1 = \frac{3}{2} - 1 = \frac{1}{2}$$
$$f(p) = \frac{1}{6}p^2 - 1$$

8. $f(x) = 3x + 1$

To graph this linear function, replace $f(x)$ with y. Then graph the linear equation $y = 3x + 1$, which is a line with slope 3 and y-intercept $(0, 1)$. See the answer graph for the margin exercises in the textbook.

The domain and range are both $(-\infty, \infty)$.

3.5 Section Exercises

3. From the figure, the ordered pairs with first entries 1990, 1992, and 1993 would be $(1990, 285.7)$, $(1992, 318.8)$, and $(1993, 339.9)$.

7. $\{(5,1), (3,2), (4,9), (7,3)\}$

The relation is a function, because for each x-value, there is only one y-value. The domain, the set of x-values, is $\{5, 3, 4, 7\}$; the range, the set of y-values, is $\{1, 2, 9, 3\}$.

11. The relation is not a function since the value, 2, in the set on the left has two different values, 15 and 19, in the set on the right. The domain is $\{1, 2, 3, 5\}$; the range is $\{10, 15, 19, -27\}$.

15. Since a vertical line intersects the graph of the relation in more than one point, the relation is not a function. The domain, the x-values of the points on the graph, is $[-4, 4]$. The range, the y-values of the points on the graph, is $[-3, 3]$.

19. $x = y^6$

The ordered pairs $(64, 2)$ and $(64, -2)$ both satisfy the equation. Since one value of x, 64, corresponds to two values of y, 2 and -2, the equation does not define a function. Because x is equal to the sixth power of y, the values of x must always be nonnegative. The domain is $[0, \infty)$.

23. $y = \sqrt{x}$

For any value of x, there is exactly one corresponding value for y, so this equation defines a function. Since the number under the radical sign must be nonnegative, x must always be nonnegative. The domain is $[0, \infty)$.

27. $y = 2x - 6$

For any value of x, there is exactly one value of y, so this equation defines a function. It is also in the form $f(x) = mx + b$, so it is a linear function. The domain is the set of all real numbers $(-\infty, \infty)$.

31. $y = \dfrac{2}{x - 9}$

There is exactly one value of y for each x-value, so the equation represents a function. The domain includes all real numbers except those that make the denominator zero.

$$\begin{aligned} x - 9 &= 0 \\ x &= 9 \end{aligned}$$

The domain includes all real numbers except 9. The domain is $(-\infty, 9) \cup (9, \infty)$.

In Exercises 35-43,

$$f(x) = -3x + 4 \text{ and } g(x) = -x^2 + 4x + 1.$$

35. $$\begin{aligned} g(-2) &= -(-2)^2 + 4(-2) + 1 \\ &= -4 - 8 + 1 \\ &= -11 \end{aligned}$$

39. $f(-x) = -3(-x) + 4 = 3x + 4$

43. $f(g(1))$

First find $g(1)$.

$$\begin{aligned} g(1) &= -1^2 + 4(1) + 1 \\ &= -1 + 4 + 1 \\ &= 4 \end{aligned}$$

Since $g(1) = 4$, $f(g(1)) = f(4)$.
Find f(4).

$$f(4) = -3(4) + 4 = -12 + 4 = -8$$

So, $f(g(1)) = -8$.

47. The equation $2x + y = 4$ has a straight *line* as its graph.

To find y in $(3, y)$, let $x = 3$ in the equation.

$$\begin{aligned} 2x + y &= 4 \\ 2(3) + y &= 4 \\ 6 + y &= 4 \\ y &= -2 \end{aligned}$$

One point that lies on the graph is $(3, -2)$.
If we solve the equation for y and use function notation, we have a *linear* function.

$$y = -2x + 4$$

Replace y with $f(x)$ to get

$$f(x) = -2x + 4.$$

For this function,

$$f(3) = -2(3) + 4 = -2,$$

meaning that the point (*3*, *-2*) lies on the graph of the function.

51. $h(x) = \frac{1}{2}x + 2$

The graph will be a line. The intercepts are $(0, 2)$ and $(-4, 0)$. See the answer graph in the back of the textbook.

The domain is $(-\infty, \infty)$, and the range is $(-\infty, \infty)$.

55. $f(x) = -123x + 29,685$

(a) For 1985, $x = 1$ $(1985 - 1984 = 1)$.

$$\begin{aligned} f(1) &= -123(1) + 29,685 \\ &= -123 + 29,685 \\ &= 29,562 \end{aligned}$$

There were 29,562 post offices in 1985.

(b) For 1987, $x = 3$ $(1987 - 1984 = 3)$.

$$\begin{aligned} f(3) &= -123(3) + 29,685 \\ &= -369 + 29,685 \\ &= 29,316 \end{aligned}$$

There were 29,316 post offices in 1987.

(c) For 1990, $x = 6$ $(1990 - 1984 = 6)$.

$$\begin{aligned} f(6) &= -123(6) + 29,685 \\ &= -738 + 29,685 \\ &= 28,947 \end{aligned}$$

There were 28,947 post offices in 1990.

59. The graph shows $x = 3$ and $y = 7$. In function notation, this is $f(3) = 7$.

3.6 An Application of Functions; Variation

3.6 Margin Exercises

1. Let $d =$ the number of days she worked and
$E =$ her earnings.

$$E = kd,$$

where k represents Vicki's daily wage.

Let $d = 17$ and $E = 1334.50$.

$$\begin{aligned} 1334.50 &= 17k \\ k &= 78.50 \end{aligned}$$

Her daily wage is \$78.50. Thus,

$$E = 78.50d.$$

2. From Example 2, the variables y and z are related by the equation

$$y = \frac{1}{2}z.$$

(a) Substitute 80 for z.

$$\begin{aligned} y &= \frac{1}{2}z \\ y &= \frac{1}{2}(80) \\ y &= 40 \end{aligned}$$

(b) Substitute 6 for z.

$$\begin{aligned} y &= \frac{1}{2}z \\ y &= \frac{1}{2}(6) \\ y &= 3 \end{aligned}$$

3. Let $c =$ the cost and
$h =$ the number of kilowatt hours.

Use $c = kh$ with $c = 52$ and $h = 800$ to find k.

$$\begin{aligned} c &= kh \\ 52 &= k(800) \\ \frac{52}{800} &= k \\ \frac{13}{200} &= k \end{aligned}$$

Therefore,

$$c = \frac{13}{200}h.$$

(a) Let $h = 1000$. Find c.

$$c = \frac{13}{200}(1000) = 65$$

1000 kilowatt-hours could cost \$65.

(b) Let $h = 650$. Find c.

$$c = \frac{13}{200}(650) = \frac{169}{4} \text{ or } 42\frac{1}{4}$$

650 kilowatt-hours would cost \$42.25.

4. **(a)** Let A = the area of a circle and r = its radius.

A varies directly as r^2, so

$$A = kr^2$$

for some constant k. Since $A = 28.278$ when $r = 3$, substitute these values in the equation and solve for k.

$$\begin{aligned} A &= kr^2 \\ 28.278 &= k(3)^2 \\ 28.278 &= 9k \\ 3.142 &= k \end{aligned}$$

So, $A = 3.142r^2$.

(b) Let $r = 4.1$. Find A.

$$\begin{aligned} A &= 3.142r^2 \\ A &= 3.142(4.1)^2 \\ A &= 52.817 \end{aligned}$$

(to the nearest thousandth)

The area is 52.817 in^2.

5. **(a)** Let V = volume, and P = pressure.

V varies inversely as P, so

$$V = \frac{k}{P}$$

for some constant k. Since $V = 10$ when $P = 6$, find k.

$$\begin{aligned} V &= \frac{k}{P} \\ 10 &= \frac{k}{6} \\ 60 &= k \end{aligned}$$

So, $V = \dfrac{60}{P}$.

(b) Let $P = 12$. Find V.

$$V = \frac{60}{P} = \frac{60}{12} = 5$$

The volume is 5 cm^3.

6. Let L = the maximum load, d = the diameter of the cross section, and h = the height.

L varies directly as d^4 and inversely as h^2, so

$$L = \frac{kd^4}{h^2}$$

for some constant k. Since $L = 8$ when $h = 9$ and $d = 1$, find k.

$$\begin{aligned} L &= \frac{kd^4}{h^2} \\ 8 &= \frac{k(1)^4}{9^2} \\ 648 &= k \end{aligned}$$

So, $L = \dfrac{648d^4}{h^2}$.

Let $h = 12$ and $d = \frac{2}{3}$.

$$\begin{aligned} L &= \frac{648\left(\frac{2}{3}\right)^4}{12^2} \\ L &= \frac{648\left(\frac{16}{81}\right)}{144} \\ L &= \frac{128}{144} = \frac{8}{9} \end{aligned}$$

The column can support $\frac{8}{9}$ metric ton.

3.6 Section Exercises

3. "z varies inversely as w" means

$$z = \frac{k}{w}$$

for some constant k. Since $z = 10$ when $w = .5$, substitute these values in the equation and solve for k.

$$\begin{aligned} z &= \frac{k}{w} \\ 10 &= \frac{k}{.5} \\ 5 &= k \end{aligned}$$

So, $z = \dfrac{5}{w}$.

To find z when $w = 8$, substitute 8 for w in the equation.

$$z = \frac{5}{w} = \frac{5}{8} \text{ or } .625$$

7. For $k > 0$, if y varies directly as x, when x increases, y *increases,* and when x decreases, y *decreases.*

11. Let y = the weight of an object on earth. and x = the weight of the object on the moon.

y varies directly as x, so

$$y = kx$$

for some constant k. Since $y = 200$ when $x = 32$, substitute these values in the equation and solve for k.

$$\begin{aligned} y &= kx \\ 200 &= k(32) \\ 6.25 &= k \end{aligned}$$

So, $y = 6.25x$.

To find x when $y = 50$, substitute 50 for y in the equation

$$\begin{aligned} y &= 6.25x \\ 50 &= 6.25x \\ 8 &= x. \end{aligned}$$

The dog would weight 8 lb on the moon.

15. Let $I =$ the illumination produced by a light source
and $d =$ the distance from the source.

I varies inversely as d^2, so

$$I = \frac{k}{d^2}$$

for some constant k. Since $I = 48$ when $d = 4$, substitute these values in the equation and solve for k.

$$\begin{aligned} I &= \frac{k}{d^2} \\ 48 &= \frac{k}{4^2} \\ 48 &= \frac{k}{16} \\ 768 &= k \end{aligned}$$

So, $I = \dfrac{768}{d^2}$.

When $d = 16$,

$$I = \frac{768}{d^2} = \frac{768}{16^2} = \frac{768}{256} = 3.$$

The illumination produced by the light source is 3 footcandles.

19. Let $A =$ the amount of water emptied by a pipe
and $d =$ the diameter of the pipe.

A varies directly as d^2, so

$$A = kd^2$$

for some constant k. Since $A = 200$ when $d = 6$, substitute these values in the equation and solve for k.

$$\begin{aligned} A &= kd^2 \\ 200 &= k(6)^2 \\ 200 &= 36k \\ 5.556 &\approx k \text{ (to the nearest thousandth)} \end{aligned}$$

So, $A = 5.556d^2$.

When $d = 12$,

$$\begin{aligned} A &= 5.556d^2 \\ A &= 5.556(12)^2 \\ A &= 5.556(144) \\ A &\approx 800 \\ &\text{(to the nearest gallon).} \end{aligned}$$

A 12-inch pipe would empty 800 gal of water.

23. $f(x) = kx$

From the graph, let $f(x) = 2$ and $x = 8$. Then

$$\begin{aligned} 2 &= k(8) \\ \frac{1}{4} &= k. \end{aligned}$$

So, $f(x) = \frac{1}{4}x$.

Therefore,

$$f(36) = \frac{1}{4}(36) = 9.$$

25. The ordered pairs are $(0,0)$ and $(1,1.25)$.

26. Let $(x_1,\ y_1) = (0,0)$ and $(x_2,\ y_2) = (1,1.25)$. Then

$$m = \frac{1.25 - 0}{1 - 0} = 1.25.$$

The slope is 1.25.

27. Since $m = 1.25$ and $b = 0$, the equation is

$$\begin{aligned} y &= 1.25x + 0 \\ \text{or } y &= 1.25x. \end{aligned}$$

28. If $f(x) = ax + b$, then $a = 1.25$ and $b = 0$.

29. The value of a, 1.25, is the price in dollars per gallon of gasoline. It is also the slope of the line.

30. Since $f(x) = 1.25x$, it fits the form for a direct variation, that is, $y = kx$. The value of a, 1.25, is the constant of variation ($k = a$).

Chapter 3 Review Exercises

For Exercises 1-4, see the answer graphs in the back of the textbook.

1. $3x + 2y = 6$

To complete the ordered pairs, substitute the given values for x or y in the equation.

For (0,), let $x = 0$.

$$\begin{aligned} 3x + 2y &= 6 \\ 3(0) + 2y &= 6 \\ 2y &= 6 \\ y &= 3 \end{aligned}$$

The ordered pair is $(0, 3)$.

For (,0), let $y = 0$.

$$\begin{aligned} 3x + 2y &= 6 \\ 3x + 2(0) &= 6 \\ 3x &= 6 \\ x &= 2 \end{aligned}$$

The ordered pair is $(2, 0)$.

For (2,), let $x = 2$.

$$\begin{aligned} 3x + 2y &= 6 \\ 3(2) + 2y &= 6 \\ 6 + 2y &= 6 \\ 2y &= 0 \\ y &= 0 \end{aligned}$$

The ordered pair is $(2, 0)$.

For (,−2), let $y = -2$.

$$\begin{aligned} 3x + 2y &= 6 \\ 3x + 2(-2) &= 6 \\ 3x - 4 &= 6 \\ 3x &= 10 \\ x &= \frac{10}{3} \end{aligned}$$

The ordered pair is $(\frac{10}{3}, -2)$.

Plot the ordered pairs, and draw the line through them.

2. $x - y = 6$

To complete the ordered pairs, substitute the given values for x or y in the equation.

For (2,), let $x = 2$.

$$\begin{aligned} x - y &= 6 \\ 2 - y &= 6 \\ -y &= 4 \\ y &= -4 \end{aligned}$$

The ordered pair is $(2, -4)$.

For (,−3), let $y = -3$.

$$\begin{aligned} x - y &= 6 \\ x - (-3) &= 6 \\ x + 3 &= 6 \\ x &= 3 \end{aligned}$$

The ordered pair is $(3, -3)$.

For (1,), let $x = 1$.

$$\begin{aligned} x - y &= 6 \\ 1 - y &= 6 \\ -y &= 5 \\ y &= -5 \end{aligned}$$

The ordered pair is $(1, -5)$.

For (,−2), let $y = -2$.

$$\begin{aligned} x - y &= 6 \\ x - (-2) &= 6 \\ x + 2 &= 6 \\ x &= 4 \end{aligned}$$

The ordered pair is $(4, -2)$.

Plot the four ordered pairs, and draw the line through them.

3. $4x + 3y = 12$

To find the x-intercept, let $y = 0$.

$$\begin{aligned} 4x + 3y &= 12 \\ 4x + 3(0) &= 12 \\ 4x &= 12 \\ x &= 3 \end{aligned}$$

The x-intercept is $(3, 0)$.

To find the y-intercept, let $x = 0$.

$$\begin{aligned} 4x + 3y &= 12 \\ 4(0) + 3y &= 12 \\ 3y &= 12 \\ y &= 4 \end{aligned}$$

The y-intercept is $(0, 4)$.

Plot the intercepts, and draw the line through them.

4. $5x + 7y = 15$

To find the x-intercept, let $y = 0$.

$$\begin{aligned} 5x + 7y &= 15 \\ 5x + 7(0) &= 15 \\ 5x &= 15 \\ x &= 3 \end{aligned}$$

The x-intercept is $(3, 0)$.

To find the y-intercept, let $x = 0$.

$$\begin{aligned} 5x + 7y &= 15 \\ 5(0) + 7y &= 15 \\ 7y &= 15 \\ y &= \frac{15}{7} \end{aligned}$$

The y-intercept is $(0, \frac{15}{7})$.

Plot the intercepts, and draw the line through them.

5. Let $(-1, 2) = (x_1,\ y_1)$ and $(4, -6) = (x_2,\ y_2)$. Then,

$$m = \frac{y_2 - y_1}{x_2 - x_1} = \frac{-6 - 2}{4 - (-1)} = \frac{-8}{5} = -\frac{8}{5}.$$

The slope is $-\frac{8}{5}$.

6. The equation $y = 2x + 3$ is in slope-intercept form $y = mx + b$, so the slope, m, is 2.

7. To find the slope of

$$-3x + 4y = 5,$$

write the equation in slope-intercept form by solving for y.

$$\begin{aligned} -3x + 4y &= 5 \\ 4y &= 3x + 5 \\ y &= \frac{3}{4}x + \frac{5}{4} \end{aligned}$$

The slope is $\frac{3}{4}$.

8. The graph of $y = 4$ is the horizontal line with y-intercept $(0, 4)$. The slope of a horizontal line is 0.

9. The line will have the same slope as $3y = -2x + 5$ since the two lines are parallel.

$$\begin{aligned} 3y &= -2x + 5 \\ y &= -\frac{2}{3}x + \frac{5}{3} \end{aligned}$$

The slope of the line is $-\frac{2}{3}$.

10. The slope of the line will be the negative reciprocal of the slope of $3x - y = 6$ since the two lines are perpendicular.

$$\begin{aligned} 3x - y &= 6 \\ -y &= -3x + 6 \\ y &= 3x - 6 \end{aligned}$$

The slope of $3x - y = 6$ is 3. The negative reciprocal of 3 is $-\frac{1}{3}$, since

$$-\frac{1}{3}(3) = -1.$$

The slope of the line is $-\frac{1}{3}$.

11. A line with a positive slope rises from left to right.

12. A line with a negative slope falls from left to right.

13. A horizontal line has 0 slope.

14. A vertical line has undefined slope.

15. Slope $\frac{3}{5}$; y-intercept $(0, -8)$

Here, $m = \frac{3}{5}$ and $b = -8$. Substitute these values in the slope-intercept form.

$$\begin{aligned} y &= mx + b \\ y &= \frac{3}{5}x - 8 \end{aligned}$$

16. Slope $-\frac{1}{3}$; y-intercept $(0, 5)$

Here, $m = -\frac{1}{3}$ and $b = 5$. Substitute these values in the slope-intercept form.

$$\begin{aligned} y &= mx + b \\ y &= -\frac{1}{3}x + 5 \end{aligned}$$

17. Slope 0; y-intercept $(0, 12)$

A horizontal line $y = k$ has slope 0. Here $k = 12$, so the line has equation $y = 12$.

18. Undefined slope; through $(2, 7)$

A vertical line $x = k$ has undefined slope. Here $k = 2$, so the line has equation $x = 2$.

19. Horizontal; through $(-1, 4)$

A horizontal line has equation $y = k$. Here $k = 4$, so the line has equation $y = 4$.

20. Vertical; through (.3, .6)

A vertical line has equation $x = k$. Here $k = .3$, so the line has equation $x = .3$.

21. Through $(2, -5)$ and $(1, 4)$

Find the slope.

$$m = \frac{4-(-5)}{1-2} = \frac{9}{-1} = -9$$

Use the point-slope form with $(x_1,\ y_1) = (1, 4)$ and $m = -9$ to find the equation of the line.

$$\begin{aligned} y - y_1 &= m(x - x_1) \\ y - 4 &= -9(x - 1) \end{aligned}$$

Write in standard form.

$$\begin{aligned} y - 4 &= -9x + 9 \\ y &= -9x + 13 \\ 9x + y &= 13 \end{aligned}$$

22. Through $(-3, -1)$ and $(2, 6)$

Find the slope.

$$m = \frac{6-(-1)}{2-(-3)} = \frac{7}{5}$$

Use the point-slope form with $(x_1,\ y_1) = (2, 6)$ and $m = \frac{7}{5}$ to find the equation of the line.

$$\begin{aligned} y - y_1 &= m(x - x_1) \\ y - 6 &= \frac{7}{5}(x - 2) \end{aligned}$$

Multiply by 5 to clear the fraction. Then write the equation in standard form.

$$\begin{aligned} 5y - 30 &= 7(x - 2) \\ 5y - 30 &= 7x - 14 \\ -7x + 5y &= 16 \\ 7x - 5y &= -16 \end{aligned}$$

23. Parallel to $4x - y = 3$ and through $(6, -2)$

Find the slope of

$$\begin{aligned} 4x - y &= 3. \\ -y &= -4x + 3 \\ y &= 4x - 3 \end{aligned}$$

The slope is 4, so a line parallel to it also has slope 4. Let $m = 4$ and $(x_1,\ y_1) = (6, -2)$ in the point-slope form.

$$\begin{aligned} y - y_1 &= m(x - x_1) \\ y - (-2) &= 4(x - 6) \\ y + 2 &= 4x - 24 \\ -4x + y &= -26 \\ 4x - y &= 26 \end{aligned}$$

24. Perpendicular to $2x - 5y = 7$ and through $(0, 1)$

Find the slope of

$$\begin{aligned} 2x - 5y &= 7. \\ -5y &= -2x + 7 \\ y &= \frac{2}{5}x - \frac{7}{5} \end{aligned}$$

The slope is $\frac{2}{5}$, so a line perpendicular to it has a slope that is the negative reciprocal of $\frac{2}{5}$, that is, $-\frac{5}{2}$. Let $m = -\frac{5}{2}$ and $(x_1,\ y_1) = (0, 1)$ in the point-slope form.

$$\begin{aligned} y - y_1 &= m(x - x_1) \\ y - 1 &= -\frac{5}{2}(x - 0) \end{aligned}$$

Multiply by 2 to clear the fraction. Then write the equation in standard form.

$$\begin{aligned} 2y - 2 &= -5x \\ 5x + 2y &= 2 \end{aligned}$$

25. **(a)** If $x = 0$ corresponds to 1980, then $x = 19$ corresponds to 1999 $(1999 - 1980 = 19)$. Substitute 19 for x in the equation and solve for y.

$$\begin{aligned} y &= 382.75x + 1742 \\ y &= 382.75(19) + 1742 \\ y &= 7272.25 + 1742 \\ y &= 9014.25 \end{aligned}$$

In 1999, the national average family health care cost would be \$9014.25.

(b) To find the year, substitute 3273 for y in the equation and solve for x.

$$\begin{aligned} y &= 382.75x + 1742 \\ 3273 &= 382.75x + 1742 \\ 1531 &= 382.75x \\ 4 &= x \end{aligned}$$

Since $x = 0$ corresponds to 1980, $x = 4$ corresponds to 1984. In 1984, the cost was \$3273.

(c) Since

$$y = 382.75x + 1742,$$

three points on the line are $(0, 1742)$, $(19, 9014.25)$ from part (a), and $(4, 3273)$ from part (b). Plot these points, and draw the line through them. See the answer graph in the back of the textbook.

26. **(a)** $y = 2.503x + 198.729$ where $x = 0$ corresponds to 1980, $x = 12$ corresponds to 1992, and y is in miles per hour.
In 1987, $x = 7$.

$$\begin{aligned} y &= 2.503(7) + 198.729 \\ &= 216.25 \end{aligned}$$

Unser's speed, from the linear equation model, would have been 216.25 mph.

(b) Unser's actual speed was 215.390 mph, a discrepancy of $216.25 - 215.390 = .860$ mph. There is a discrepancy because the equation is only an *approximate* linear model.

For Exercises 27-32, see the answer graphs in the back of the textbook.

27. $3x - 2y \leq 12$

Graph the line through $3x - 2y = 12$, which has intercepts $(4, 0)$ and $(0, -6)$, as a solid line since the inequality involves "$\leq$." To decide which side of the line belongs to the graph, test $(0, 0)$.

$$\begin{aligned} 3x - 2y &\leq 12 \\ 3(0) - 2(0) &\leq 12 \\ 0 &\leq 12 \quad \textit{True} \end{aligned}$$

Since the result is true, shade the region that includes $(0, 0)$.

28. $5x - y > 6$

Graph the line through $5x - y = 6$, which has intercepts $(\frac{6}{5}, 0)$ and $(0, -6)$, as a dashed line since the inequality involves "$>$." Use $(0, 0)$ as a test point.

$$\begin{aligned} 5x - y &> 6 \\ 5(0) - 0 &> 6 \\ 0 &> 6 \quad \textit{False} \end{aligned}$$

Since the result is false, shade the region that does not include $(0, 0)$.

29. $x \geq 2$

Graph $x = 2$ as a solid vertical line through $(2, 0)$. Shade the region to the right of $x = 2$.

30. $2x + y \leq 1$ and $x \geq 2y$

Graph $2x + y = 1$, which has intercepts $(\frac{1}{2}, 0)$ and $(0, 1)$, as a solid line. Test $(0, 0)$, which yields $0 \leq 1$, a true statement. Shade the region that includes $(0, 0)$.

Graph $x = 2y$, which includes points $(0, 0)$ and $(2, 1)$, as a solid line. Because $(0, 0)$ is on the line, test another point, say $(3, 3)$. A false statement, $3 \geq 6$, results. Shade the region that does not include $(3, 3)$.

The required graph of the intersection is the region common to both graphs.

31. $x - 2y < 4$ or $x + y < 3$

Graph $x - 2y = 4$, which has intercepts $(4, 0)$ and $(0, -2)$, as a dashed line. Test $(0, 0)$, which yields $0 < 4$, a true statement. Shade the region that includes $(0, 0)$.

Graph $x + y = 3$, which has intercepts $(3, 0)$ and $(0, 3)$, as a dashed line. Test $(0, 0)$, which yields $0 < 3$, a true statement. Shade the region that includes $(0, 0)$.

The required graph of the union includes all the shaded regions, that is, all the points that satisfy either inequality.

32. $|x - 1| < 4$
Rewrite $|x - 1| < 4$ as

$$\begin{aligned} -4 < x - 1 < 4 \\ -3 < x < 5. \end{aligned}$$

The graph consists of the two dashed boundary lines $x = -3$ and $x = 5$ and the region between them.

33. $\{(-4, 2), (-4, -2), (1, 5), (1, -5)\}$

The domain, the set of x-values, is $\{-4, 1\}$.
The range, the set of y-values, is $\{2, -2, 5, -5\}$.
Since each x-value has more than one y-value, the relation is not a function.

34. The domain is $\{-14, 91, 17, 75, -23\}$.
The range is $\{9, 12, 18, 70, 56, 5\}$.

The x-value, 17, has two y-values, 12 and 18. The x-value, 75, also has two y-values, 70 and 56. Therefore, the relation is not a function.

35. The domain, the x-values of the points on the graph, is $[-4, 4]$. The range, the y-values of the points on the graph, is $[0, 2]$. Since a vertical line intersects the graph of the relation in at most one point, the relation is a function.

In Exercises 36-39,

$$f(x) = -2x^2 + 3x - 6.$$

36. $f(0) = -2(0)^2 + 3(0) - 6 = -6$

37. $f(3) = -2(3)^2 + 3(3) - 6$
$= -18 + 9 - 6 = -15$

38. $f[f(0)]$

First find $f(0)$. (See Exercise 36.)

$$f(0) = -2(0)^2 + 3(0) - 6 = -6$$

Since $f(0) = -6$, $f[f(0)] = f(-6)$.

Find $f(-6)$.

$$\begin{aligned} f(-6) &= -2(-6)^2 + 3(-6) - 6 \\ &= -72 - 18 - 6 = -96 \end{aligned}$$

So, $f[f(0)] = -96$.

39. $f(2p) = -2(2p)^2 + 3(2p) - 6$
$= -8p^2 + 6p - 6$

40. $y = 3x - 3$

For any value of x, there is exactly one value of y, so the equation defines a function. Since the equation is in the form $y = mx + b$, it also defines a linear function. The domain is the set of all real numbers $(-\infty, \infty)$.

41. $y < x + 2$

For any value of x, there are many values of y. For example, $(1, 0)$ and $(1, 1)$ are both solutions of the inequality that have the same x-value but different y-values. The equation does not define a function. The domain is the set of all real numbers $(-\infty, \infty)$.

42. $y = |x - 4|$

For any value of x, there is exactly one value of y, so the equation defines a function. The domain is the set of all real numbers $(-\infty, \infty)$.

43. $y = \sqrt{4x + 7}$

Given any value of x, y is found by multiplying x by 4, adding 7, and taking the square root of the result. This process produces exactly one value of y for each x-value, so the equation defines a function. Since the quantity under the radical sign must be nonnegative,

$$\begin{aligned} 4x + 7 &\geq 0 \\ 4x &\geq -7 \\ x &\geq -\frac{7}{4}. \end{aligned}$$

The domain is $[-\frac{7}{4}, \infty)$.

44. $x = y^2$

The ordered pairs $(4, 2)$ and $(4, -2)$ both satisfy the equation. Since one value of x, 4, corresponds to two values of y, 2 and -2, the equation does not define a function. Because x is equal to the square of y, the values of x must always be nonnegative. The domain is $[0, \infty)$.

45. $y = \dfrac{7}{x - 36}$

There is exactly one value of y for each x-value, so the equation defines a function. The domain includes all real numbers except those that make the denominator zero.

$$\begin{aligned} x - 36 &= 0 \\ x &= 36 \end{aligned}$$

The domain is $(-\infty, 36) \cup (36, \infty)$.

46. $f(x) = -\dfrac{3}{2}x + \dfrac{7}{2}$

Replace $f(x)$ with y and graph the equation $y = -\frac{3}{2}x + \frac{7}{2}$, which has slope $-\frac{3}{2}$ and y-intercept $(0, \frac{7}{2})$. Use this information to graph the line. See the answer graph in the back of the textbook.

47. "m varies directly as p^2 and inversely as q" means

$$m = \frac{kp^2}{q}$$

for some constant k. Since $m = 32$ when $p = 8$ and $q = 10$, substitute these values in the equation and solve for k.

$$m = \frac{kp^2}{q}$$
$$32 = \frac{k(8)^2}{10}$$
$$32 = \frac{64k}{10}$$
$$320 = 64k$$
$$5 = k$$

So, $m = \dfrac{5p^2}{q}$.

To find q when $p = 12$ and $m = 48$, substitute these values in the equation and solve for q.

$$m = \frac{5p^2}{q}$$
$$48 = \frac{5(12)^2}{q}$$
$$48 = \frac{720}{q}$$
$$q = 15$$

48. "x varies jointly as y and z and inversely as $\sqrt{w}$" means

$$x = \frac{kyz}{\sqrt{w}}$$

for some constant k. Since $x = 12$ when $y = 3, z = 8$, and $w = 36$, substitute these values in the equation and solve for k.

$$x = \frac{kyz}{\sqrt{w}}$$
$$12 = \frac{k(3)(8)}{\sqrt{36}}$$
$$12 = \frac{24k}{6}$$
$$12 = 4k$$
$$3 = k$$

So, $x = \dfrac{3yz}{\sqrt{w}}$.

To find y when $x = 12$, $z = 4$, and $w = 25$, substitute these values in the equation and solve for y.

$$x = \frac{3yz}{\sqrt{w}}$$
$$12 = \frac{3y(4)}{\sqrt{25}}$$
$$12 = \frac{12y}{5}$$
$$60 = 12y$$
$$5 = y$$

49. Let $R =$ the resistance in ohms and $d =$ the temperature in degrees Kelvin.

R varies directly as d, so

$$R = kd$$

for some constant k. Since $R = .646$ when $d = 190$, substitute these values in the equation and solve for k.

$$R = kd$$
$$.646 = k(190)$$
$$.0034 = k$$

So, $R = .0034d$.
When $d = 250$,

$$R = .0034d = .0034(250) = .850.$$

The resistance is .850 ohm.

50. Let $v =$ the viewing distance and $e =$ the amount of enlargement.

v varies directly as e, so

$$v = ke$$

for some constant k. Since $v = 250$ when $e = 5$, substitute these values in the equation and solve for k.

$$v = ke$$
$$250 = k(5)$$
$$50 = k$$

So, $v = 50e$.
When $e = 8.6$,

$$v = 50e = 50(8.6) = 430.$$

It should be viewed from 430 mm.

51. Let $v =$ the velocity of the meteorite, and $d =$ the distance from the center of the earth.

v is inversely proportional to $\sqrt{d}$, so

$$v = \frac{k}{\sqrt{d}}$$

for some constant k. Since $v = 5$ when $d = 8100$, substitute these values in the equation and solve for k.

$$\begin{aligned} v &= \frac{k}{\sqrt{d}} \\ 5 &= \frac{k}{\sqrt{8100}} \\ 5 &= \frac{k}{90} \\ 450 &= k \end{aligned}$$

So, $v = \dfrac{450}{\sqrt{d}}$.

When $d = 6400$,

$$v = \frac{450}{\sqrt{d}} = \frac{450}{\sqrt{6400}} = 5.625.$$

The velocity is 5.625 km per sec.

52. Let $P =$ the period of the pendulum,
$L =$ the length, and
$g =$ the acceleration due to gravity.

P varies directly as $\sqrt{L}$ and inversely as $\sqrt{g}$, so

$$P = \frac{k\sqrt{L}}{\sqrt{g}}$$

for some constant k. Since $P = 1.06\pi$ when $L = 9$ and $g = 32$, substitute these values in the equation and solve for k.

$$\begin{aligned} 1.06\pi &= \frac{k\sqrt{9}}{\sqrt{32}} \\ 1.06\pi &= \frac{3k}{\sqrt{32}} \\ 2.00\pi &\approx k \end{aligned}$$

So, $P = \dfrac{2.00\pi\sqrt{L}}{\sqrt{g}}$.

Let $L = 4$ and $g = 32$.

$$P = \frac{2.00\pi\sqrt{4}}{\sqrt{32}} = \frac{2\pi \cdot 2}{4\sqrt{2}} = \frac{\pi}{\sqrt{2}} = \frac{\sqrt{2}}{2}\pi \approx .71\pi$$

The period is about $.71\pi$ sec.

Chapter 3 Test

1. Let $(x_1,\ y_1) = (6, 4)$ and $(x_2,\ y_2) = (-4, -1)$. Then

$$m = \frac{y_2 - y_1}{x_2 - x_1} = \frac{-1-4}{-4-6} = \frac{-5}{-10} = \frac{1}{2}.$$

The slope is $\frac{1}{2}$.

2. To find the slope and y-intercept of

$$3x - 2y = 13,$$

write the equation in slope-intercept form by solving for y.

$$\begin{aligned} 3x - 2y &= 13 \\ -2y &= -3x + 13 \\ y &= \frac{3}{2}x - \frac{13}{2} \end{aligned}$$

The slope is $\frac{3}{2}$, and the y-intercept is $(0, -\frac{13}{2})$.

To find the x-intercept, substitute 0 for y in the original equation and solve for x.

$$\begin{aligned} 3x - 2y &= 13 \\ 3x - 2(0) &= 13 \\ 3x &= 13 \\ x &= \frac{13}{3} \end{aligned}$$

The x-intercept is $(\frac{13}{3}, 0)$.

3. The graph of $y = 5$ is the horizontal line with slope 0 and y-intercept $(0, 5)$. There is no x-intercept.

4. The graph of a line with undefined slope is the graph of a vertical line.

5. Through $(-3, 14)$; horizontal

A horizontal line has equation $y = k$. Here $k = 14$, so the line has equation $y = 14$.

6. Through $(4, -1)$; $m = -5$

Let $m = -5$ and $(x_1, y_1) = (4, -1)$ in the point-slope form.

$$\begin{aligned} y - y_1 &= m(x - x_1) \\ y - (-1) &= -5(x - 4) \\ y + 1 &= -5x + 20 \\ 5x + y &= 19 \end{aligned}$$

7. Through $(-7, 2)$

(a) Parallel to $3x + 5y = 6$

To find the slope of $3x+5y = 6$, write the equation in slope-intercept form by solving for y.

$$\begin{aligned} 3x + 5y &= 6 \\ 5y &= -3x + 6 \\ y &= -\frac{3}{5}x + \frac{6}{5} \end{aligned}$$

The slope is $-\frac{3}{5}$, so a line parallel to it also has slope $-\frac{3}{5}$. Let $m = -\frac{3}{5}$ and $(x_1, y_1) = (-7, 2)$ in the point-slope form.

$$y - y_1 = m(x - x_1)$$
$$y - 2 = -\frac{3}{5}[x - (-7)]$$

Multiply by 5 to clear the fraction. Then write the equation in standard form.

$$5y - 10 = -3(x + 7)$$
$$5y - 10 = -3x - 21$$
$$3x + 5y = -11$$

(b) Perpendicular to $y = 2x$

Since $y = 2x$ is in slope-intercept form ($b = 0$), the slope, m, of $y = 2x$ is 2. A line perpendicular to it has a slope that is the negative reciprocal of 2, that is, $-\frac{1}{2}$. Let $m = -\frac{1}{2}$ and $(x_1, y_1) = (-7, 2)$ in the point-slope form.

$$y - y_1 = m(x - x_1)$$
$$y - 2 = -\frac{1}{2}[x - (-7)]$$
$$y - 2 = -\frac{1}{2}(x + 7)$$

Multiply by 2 to clear the fraction. Then write the equation in standard form.

$$2y - 4 = -x - 7$$
$$x + 2y = -3$$

For Exercises 8-11, see the answer graphs in the back of the textbook.

8. $4x - 3y = -12$

   The intercepts are $(-3, 0)$ and $(0, 4)$. To sketch the graph, plot the intercepts and draw the line through them.

9. $y - 2 = 0$
   $y = 2$

   The graph of $y = 2$ is the horizontal line with y-intercept $(0, 2)$.

10. $f(x) = -2x$

    Replace $f(x)$ with y, and graph the equation $y = -2x$. The intercept is $(0, 0)$. Another ordered pair that satisfies the equation is $(1, -2)$. To sketch the graph, plot the two points and draw the line through them.

11. $3x - 2y > 6$

    Graph the line $3x - 2y = 6$, which has intercepts $(2, 0)$ and $(0, -3)$, as a dashed line since the inequality involves "$>$." Test $(0, 0)$, which yields $0 > 6$, a false statement. Shade the region that does not include $(0, 0)$.

12. Choice D is the only graph that passes the vertical line test.

13. $f(x) = \sqrt{x - 3}$

    (a) $f(7) = \sqrt{7 - 3} = \sqrt{4} = 2$

    (b) In this function, $x - 3$ must be nonnegative, so

    $$x - 3 \geq 0$$
    $$x \geq 3.$$

    The domain is $[3, \infty)$.

14. For 1989, $x = 4$ ($1989 - 1985 = 4$).

    $$f(x) = 825x + 8689$$
    $$f(x) = 825(4) + 8689$$
    $$f(x) = 11{,}989$$

    There were 11,989 cases.

15. Let d = distance
    and t = time.

    d varies directly as t^2, so

    $$d = kt^2$$

    for some constant k. Since $d = 576$ when $t = 6$, substitute these values in the equation and solve for k.

    $$d = kt^2$$
    $$576 = k(6)^2$$
    $$576 = 36k$$
    $$16 = k$$

    So, $d = 16t^2$.

    When $t = 4$,

    $$d = 16t^2 = 16(4)^2 = 16(16) = 256.$$

    It fell a distance of 256 ft.

16. Let f = the force of the wind,
    A = the area of the surface,
    and v = the velocity of the wind.

    f varies jointly as A and v^2, so

    $$f = kAv^2$$

for some constant k. Since $f = 50$ when $A = \frac{1}{2}$ and $v = 40$, substitute these values in the equation to find k.

$$\begin{aligned} f &= kAv^2 \\ 50 &= k\left(\frac{1}{2}\right)(40)^2 \\ 50 &= 800k \\ .0625 &= k \end{aligned}$$

So, $f = .0625Av^2$.

When $A = 2$ and $v = 80$,

$$f = .0625Av^2 = .0625(2)(80)^2 = 800.$$

The force of the wind is 800 lb.

Cumulative Review Exercises (Chapters 1-3)

1. $5 \cdot 6 \geq |32 - 20|$
   $30 \geq 12$ *True*

   The statement is true.

2. $5 - |-4| \leq 9$
   $5 - 4 \leq 9$
   $1 \leq 9$ *True*

   The statement is true.

3. $-4(4 - 8) \geq |-20|$
   $-4(-4) \geq 20$
   $16 \geq 20$ *False*

   This statement is false.

4. $$\begin{aligned} &-|-2| - 4 + |-3| + 7 \\ &= -2 - 4 + 3 + 7 \\ &= -6 + 3 + 7 \\ &= -3 + 7 \\ &= 4 \end{aligned}$$

5. $(-.8)^2 = (-.8)(-.8) = .64$

6. $\sqrt[3]{-64} = -4$, since $(-4)^3 = -64$.

7. $-\frac{2}{3}\left(-\frac{12}{5}\right)$

   The product of two numbers with the same signs is positive.

   $$-\frac{2}{3}\left(-\frac{12}{5}\right) = \frac{2 \cdot 12}{3 \cdot 5} = \frac{24}{15} = \frac{8}{5}$$

8. $$\begin{aligned} -2(m - 3) &= -2(m) - 2(-3) \\ &= -2m + 6 \end{aligned}$$

9. $$\begin{aligned} -(-4m + 3) &= -(-4m) - 3 \\ &= 4m - 3 \end{aligned}$$

10. $$\begin{aligned} &3x^2 - 4x + 4 + 9x - x^2 \\ &= 3x^2 - x^2 - 4x + 9x + 4 \\ &= 2x^2 + 5x + 4 \end{aligned}$$

11. $\{x \mid x > 2\}$

    This is the set of numbers greater than 2. In interval notation, the set is $(2, \infty)$. The parenthesis at 2 indicates that 2 is not included in the interval.

12. $\{x \mid x \leq 1\}$

    This is the set of numbers less than or equal to 1. In interval notation, the set is $(-\infty, 1]$. The bracket at 1 indicates that 1 is included in the interval.

13. $\{x \mid -3 < x \leq 5\}$

    This is the set of numbers between -3 and 5, not including -3 (use a parenthesis), but including 5 (use a bracket). In interval notation, the set is $(-3, 5]$.

14. $\sqrt{\frac{-2+4}{-5}} = \sqrt{\frac{2}{-5}} = \sqrt{-\frac{2}{5}}$

    This is not a real number since the number under the radical sign is negative.

For Exercises 15-19, let $p = -4$, $q = -2$, and $r = 5$.

15. $$\begin{aligned} -3(2q - 3p) &= -3[2(-2) - 3(-4)] \\ &= -3(-4 + 12) \\ &= -3(8) \\ &= -24 \end{aligned}$$

16. $$\begin{aligned} 8r^2 + q^2 &= 8(5)^2 + (-2)^2 \\ &= 8(25) + 4 \\ &= 200 + 4 \\ &= 204 \end{aligned}$$

17. $$\begin{aligned} |p|^3 - |q^3| &= |-4|^3 - |(-2)^3| \\ &= 4^3 - |-8| \\ &= 64 - 8 \\ &= 56 \end{aligned}$$

18. $\dfrac{\sqrt{r}}{-p+2q} = \dfrac{\sqrt{5}}{-(-4)+2(-2)}$

$= \dfrac{\sqrt{5}}{4-4}$

$= \dfrac{\sqrt{5}}{0}$

This expression is undefined since the denominator is zero.

19. $\dfrac{5p+6r^2}{p^2+q-1} = \dfrac{5(-4)+6(5)^2}{(-4)^2+(-2)-1}$

$= \dfrac{-20+6(25)}{16-2-1}$

$= \dfrac{-20+150}{13}$

$= \dfrac{130}{13}$

$= 10$

20. $2z-5+3z = 4-(z+2)$

$5z-5 = 4-z-2$

$5z-5 = 2-z$

$6z = 7$

$z = \dfrac{7}{6}$

Solution set: $\{\frac{7}{6}\}$

21. $\dfrac{3a-1}{5} + \dfrac{a+2}{2} = -\dfrac{3}{10}$

Multiply both sides by 10.

$$10\left(\frac{3a-1}{5}+\frac{a+2}{2}\right) = 10\left(-\frac{3}{10}\right)$$

$2(3a-1)+5(a+2) = -3$

$6a-2+5a+10 = -3$

$11a+8 = -3$

$11a = -11$

$a = -1$

Solution set: $\{-1\}$

22. $-\dfrac{4}{3}d \geq -5$

Multiply by -3; reverse the direction of the inequality symbol.

$$-3\left(-\frac{4}{3}d\right) \leq -3(-5)$$

$4d \leq 15$

$d \leq \dfrac{15}{4}$

Solution set: $(-\infty, \frac{15}{4}]$

23. $3-2(m+3) < 4m$

$3-2m-6 < 4m$

$-2m-3 < 4m$

$-6m < 3$

Multiply by $-\frac{1}{6}$; reverse the direction of the inequality symbol.

$-\dfrac{1}{6}(-6m) > -\dfrac{1}{6}(3)$

$m > -\dfrac{3}{6}$

$m > -\dfrac{1}{2}$

Solution set: $(-\frac{1}{2}, \infty)$

24. $2k+4 < 10$ and $3k-1 > 5$

$2k < 6$ $\qquad 3k > 6$

$k < 3$ and $k > 2$

The solution set of these inequalities is the set of numbers that are both less than 3 and greater than 2. This is the set of all numbers between 2 and 3.

Solution set: $(2,3)$

25. $2k+4 > 10$ or $3k-1 < 5$

$2k > 6$ $\qquad 3k < 6$

$k > 3$ or $k < 2$

The solution set is the set of numbers that are either greater than 3 or less than 2.

Solution set: $(-\infty, 2) \cup (3, \infty)$

26. $|5x+3| = 13$

$5x+3 = 13$ or $5x+3 = -13$

$5x = 10$ $\qquad 5x = -16$

$x = 2$ or $x = -\dfrac{16}{5}$

Solution set: $\{-\frac{16}{5}, 2\}$

27. $|x+2| < 9$

$-9 < x+2 < 9$

Subtract 2 from each part of the inequality.

$-11 < x < 7$

Solution set: $(-11, 7)$

28. $|2y-5| \geq 9$

$2y-5 \geq 9$ or $2y-5 \leq -9$

$2y \geq 14$ $\qquad 2y \leq -4$

$y \geq 7$ or $y \leq -2$

Solution set: $(-\infty, -2] \cup [7, \infty)$

29. Let $x =$ the time it takes for the planes to be 2100 mi apart.

Make a table. Use the formula $d = rt$.

	r	t	d
Eastbound plane	550	x	$550x$
Westbound plane	500	x	$500x$

The total distance is 2100 mi.
Solve the equation.

$$550x + 500x = 2100$$
$$1050x = 2100$$
$$x = 2$$

It will take 2 hr for the planes to be 2100 mi apart.

30. Let x = the number of white pills;
$2x$ = the number of yellow pills.

Make a table.

	Strength	Number of pills	Amount of medication
White	3	x	$3x$
Yellow	3	$2x$	$3(2x)$
			30

The total amount of medication must be at least 30 units. Solve the inequality.

$$3x + 3(2x) \geq 30$$
$$3x + 6x \geq 30$$
$$9x \geq 30$$
$$x \geq \frac{30}{9}$$
$$x \geq \frac{10}{3} \text{ or } 3\frac{1}{3}$$

Since Ms. Bell must take a whole number of pills, she must take at least 4 white pills.

31. $3x - 4y = 12$

To complete the ordered pairs, substitute the given values for x or y in the equation.

For $(0, \quad)$, let $x = 0$.

$$3x - 4y = 12$$
$$3(0) - 4y = 12$$
$$0 - 4y = 12$$
$$-4y = 12$$
$$y = -3$$

The ordered pair is $(0, -3)$.

For $(\quad, 0)$, let $y = 0$.

$$3x - 4y = 12$$
$$3x - 4(0) = 12$$
$$3x - 0 = 12$$
$$3x = 12$$
$$x = 4$$

The ordered pair is $(4, 0)$.

For $(2, \quad)$, let $x = 2$.

$$3x - 4y = 12$$
$$3(2) - 4y = 12$$
$$6 - 4y = 12$$
$$-4y = 6$$
$$y = -\frac{6}{4}$$
$$y = -\frac{3}{2}$$

The ordered pair is $\left(2, -\frac{3}{2}\right)$.

32. $-4x + 2y = 8$

To find the x-intercept of the equation, let $y = 0$.

$$-4x + 2y = 8$$
$$-4x + 2(0) = 8$$
$$-4x = 8$$
$$x = -2$$

The x-intercept is $(-2, 0)$.

To find the y-intercept of the equation, let $x = 0$.

$$-4x + 2y = 8$$
$$-4(0) + 2y = 8$$
$$2y = 8$$
$$y = 4$$

The y-intercept is $(0, 4)$.

Plot the intercepts, and draw the line through them. See the answer graph in the back of the textbook.

33. Through $(-5, 8)$ and $(-1, 2)$

Let $(x_1, y_1) = (-5, 8)$ and $(x_2, y_2) = (-1, 2)$.

$$m = \frac{y_2 - y_1}{x_2 - x_1} = \frac{2 - 8}{-1 - (-5)}$$
$$= \frac{-6}{4} = -\frac{3}{2}$$

The slope is $-\frac{3}{2}$.

34. Perpendicular to $4x - 3y = 12$

To find the slope of the line perpendicular to the given line, solve for y to write the equation in slope-intercept form.

$$4x - 3y = 12$$
$$-3y = -4x + 12$$
$$y = \frac{4}{3}x - 4$$

The slope is $\frac{4}{3}$. Perpendicular lines have slopes that are negative reciprocals of each other. The negative reciprocal of $\frac{4}{3}$ is $-\frac{3}{4}$, since

$$\frac{4}{3}\left(-\frac{3}{4}\right) = -1.$$

The slope of a line perpendicular to the given line is $-\frac{3}{4}$.

35. Slope $-\frac{3}{4}$; y-intercept $(0, -1)$

To write an equation of this line, let $m = -\frac{3}{4}$ and $b = -1$ in the slope-intercept form.

$$y = mx + b$$
$$y = -\frac{3}{4}x - 1$$

Multiply by 4 to clear the fraction. Then write the equation in standard form $Ax + By = C$.

$$4y = 4\left(-\tfrac{3}{4}x - 1\right)$$
$$4y = -3x - 4$$
$$3x + 4y = -4$$

36. Horizontal; through $(2, -2)$

A horizontal line through the point (x, k) has equation $y = k$. Here $k = -2$, so the equation of the line is $y = -2$.

37. Through $(4, -3)$ and $(1, 1)$

To write an equation of the line through these points, first find the slope of the line.

$$m = \frac{1 - (-3)}{1 - 4} = -\frac{4}{3}.$$

Now substitute $(x_1, y_1) = (4, -3)$ and the slope $m = -\frac{4}{3}$ in the point-slope form.

$$y - y_1 = m(x - x_1)$$
$$y - (-3) = -\frac{4}{3}(x - 4)$$
$$y + 3 = -\frac{4}{3}(x - 4)$$

Multiply by 3 to clear the fraction. Then write the equation in standard form.

$$3(y + 3) = -4(x - 4)$$
$$3y + 9 = -4x + 16$$
$$4x + 3y = 7$$

38. $f(x) = -4x + 10$

(a) The domain includes the set of all real numbers $(-\infty, \infty)$.

(b) $f(-3) = -4(-3) + 10$
$= 12 + 10$
$= 22$

39. From the graph, use the ordered pairs (1988, 575,000) and (1991, 892,000).

$$m = \frac{892{,}000 - 575{,}000}{1991 - 1988} = \frac{317{,}000}{3} = 105{,}666\tfrac{2}{3}$$

40. The segment from 1988 to 1991 rises more sharply and has a greater slope.

Chapter 4

SYSTEMS OF LINEAR EQUATIONS

4.1 Linear Systems of Equations in Two Variables

4.1 Margin Exercises

1. **(a)** To determine whether $(-4, 2)$ is a solution of the system

$$2x + y = -6$$
$$x + 3y = 2,$$

replace x with -4 and y with 2 in each equation.

$$2x + y = -6$$
$$2(-4) + 2 = -6 \; ?$$
$$-8 + 2 = -6 \; ?$$
$$-6 = -6 \quad \textit{True}$$

$$x + 3y = 2$$
$$-4 + 3(2) = 2 \; ?$$
$$-4 + 6 = 2 \; ?$$
$$2 = 2 \quad \textit{True}$$

Since $(-4, 2)$ makes both equations true, $(-4, 2)$ is a solution of the system.

(b) To determine whether $(-1, 5)$ is a solution of the system

$$9x - y = -4$$
$$4x + 3y = 11,$$

replace x with -1 and y with 5 in each equation.

$$9x - y = -4$$
$$9(-1) - 5 = -4 \; ?$$
$$-9 - 5 = -4 \; ?$$
$$-14 = -4 \quad \textit{False}$$

Since $(-1, 5)$ does not satisfy the first equation, it is not necessary to check if it satisfies the second equation. $(-1, 5)$ is not a solution of the system.

For Exercise 2, see the answer graphs for the margin exercises in the textbook.

2. **(a)** When the equations of the system

$$x - y = 3$$
$$2x - y = 4$$

are graphed, the point of intersection is $(1, -2)$. To check, substitute 1 for x and -2 for y in each equation of the system. Since $(1, -2)$ makes both equations true, the solution set of the system is $\{(1, -2)\}$.

(b) When the equations of the system

$$2x + y = -5$$
$$-x + 3y = 6$$

are graphed, the point of intersection is $(-3, 1)$. To check, substitute -3 for x and 1 for y in each equation of the system. Since $(-3, 1)$ makes both equations true, the solution set is $\{(-3, 1)\}$.

In Exercises 3 and 4, check each solution in the equations of the original system.

3. **(a)** $3x - y = -7 \quad (1)$
$2x + y = -3 \quad (2)$

Eliminate y by adding equations (1) and (2).

$$\begin{aligned} 3x - y &= -7 \quad (1) \\ 2x + y &= -3 \quad (2) \\ \hline 5x \quad &= -10 \\ x &= -2 \end{aligned}$$

To find y, replace x with -2 in either equation (1) or (2).

$$3x - y = -7 \quad (1)$$
$$3(-2) - y = -7$$
$$-6 - y = -7$$
$$-y = -1$$
$$y = 1$$

Solution set: $\{(-2, 1)\}$

(b) $2x - 3y = 12 \quad (1)$
$-2x + y = -4 \quad (2)$

To eliminate x, add equations (1) and (2).

$$\begin{array}{rl} 2x - 3y = 12 & (1) \\ -2x + y = -4 & (2) \\ \hline -2y = 8 & \\ y = -4 & \end{array}$$

To find x, replace y with -4 in equation (1) or (2).

$$\begin{aligned} 2x - 3y &= 12 \quad (1) \\ 2x - 3(-4) &= 12 \\ 2x + 12 &= 12 \\ 2x &= 0 \\ x &= 0 \end{aligned}$$

Solution set: $\{(0, -4)\}$

4. **(a)** $x + 3y = 8 \quad (1)$
$2x - 5y = -17 \quad (2)$

To eliminate x, multiply equation (1) by -2 and add the result to equation (2).

$$\begin{array}{rl} -2x - 6y = -16 & \\ 2x - 5y = -17 & (2) \\ \hline -11y = -33 & \\ y = 3 & \end{array}$$

To find x, substitute 3 for y in equation (1) or (2).

$$\begin{aligned} x + 3y &= 8 \quad (1) \\ x + 3(3) &= 8 \\ x + 9 &= 8 \\ x &= -1 \end{aligned}$$

Solution set: $\{(-1, 3)\}$

(b) $4x - y = 14 \quad (1)$
$3x + 4y = 20 \quad (2)$

To eliminate y, multiply equation (1) by 4 and add the result to equation (2).

$$\begin{array}{rl} 16x - 4y = 56 & \\ 3x + 4y = 20 & (2) \\ \hline 19x \qquad = 76 & \\ x = 4 & \end{array}$$

To find y, substitute 4 for x in equation (1) or (2).

$$\begin{aligned} 3x + 4y &= 20 \quad (2) \\ 3(4) + 4y &= 20 \\ 12 + 4y &= 20 \\ 4y &= 8 \\ y &= 2 \end{aligned}$$

Solution set: $\{(4, 2)\}$

(c) $2x + 3y = 19 \quad (1)$
$3x - 7y = -6 \quad (2)$

To eliminate x, multiply equation (1) by 3 and equation (2) by -2. Then add the results.

$$\begin{array}{r} 6x + 9y = 57 \\ -6x + 14y = 12 \\ \hline 23y = 69 \\ y = 3 \end{array}$$

To find x, substitute 3 for y in equation (1) or (2).

$$\begin{aligned} 2x + 3y &= 19 \quad (1) \\ 2x + 3(3) &= 19 \\ 2x + 9 &= 19 \\ 2x &= 10 \\ x &= 5 \end{aligned}$$

Solution set: $\{(5, 3)\}$

For Exercises 5 and 6, see the answer graphs for the margin exercises in the textbook.

5. $2x + y = 6 \quad (1)$
$-8x - 4y = -24 \quad (2)$

Multiply equation (1) by 4, and add the result to equation (2).

$$\begin{array}{rl} 8x + 4y = 24 & \\ -8x - 4y = -24 & (2) \\ \hline 0 = 0 & \textit{True} \end{array}$$

Multiplying equation (1) by -4 gives equation (2). Therefore, equations (1) and (2) are equivalent and have the same line as graph. The solution is the set of all points on the line.

Solution set: $\{(x, y) | 2x + y = 6\}$

6. $4x - 3y = 8$ (1)
$8x - 6y = 14$ (2)

Multiply equation (1) by -2 and add the result to equation (2).

$$\begin{array}{rl} -8x + 6y = -16 & \\ 8x - 6y = 14 & (2) \\ \hline 0 = -2 & \textit{False} \end{array}$$

The system is inconsistent. The graphs of the original equations are parallel lines. No ordered pairs satisfy both equations.

Solution set: $\emptyset$

7. $2x - y = 3$
$6x - 3y = 9$

An equation of the form $y = mx + b$ is in slope-intercept form. Solve for y in each equation.

$$\begin{aligned} 2x - y &= 3 \\ -y &= -2x + 3 \\ y &= 2x - 3 \end{aligned}$$

$$\begin{aligned} 6x - 3y &= 9 \\ -3y &= -6x + 9 \\ y &= 2x - 3 \end{aligned}$$

Replace y with $f(x)$ for function notation. Both equations can be written as $f(x) = 2x - 3$.

8. $x + 3y = 4$
$-2x - 6y = 3$

Write each equation in slope-intercept form by solving for y.

$$\begin{aligned} x + 3y &= 4 \\ 3y &= -x + 4 \\ y &= -\frac{1}{3}x + \frac{4}{3} \end{aligned}$$

$$\begin{aligned} -2x - 6y &= 3 \\ -6y &= 2x + 3 \\ y &= -\frac{1}{3}x - \frac{1}{2} \end{aligned}$$

In each case, replace y with $f(x)$ for function notation. The equations are

$$f(x) = -\frac{1}{3}x + \frac{4}{3} \text{ and } f(x) = -\frac{1}{3}x - \frac{1}{2}.$$

In Exercises 9-11, check each solution in the equations of the original system.

9. $3x - y = 10$ (1)
$2x + 5y = 1$ (2)

To use the substitution method, first solve one of the equations for x or y. Since the coefficient of y in equation (1) is -1, it is easiest to solve for y in this equation.

$$\begin{aligned} 3x - y &= 10 \quad (1) \\ -y &= -3x + 10 \\ y &= 3x - 10 \end{aligned}$$

Substitute $3x - 10$ for y in equation (2) and solve for x.

$$\begin{aligned} 2x + 5y &= 1 \quad (2) \\ 2x + 5(3x - 10) &= 1 \\ 2x + 15x - 50 &= 1 \\ 17x &= 51 \\ x &= 3 \end{aligned}$$

Since $y = 3x - 10$ and $x = 3$,

$$y = 3(3) - 10 = 9 - 10 = -1.$$

Solution set: $\{(3, -1)\}$

10. $\frac{x}{5} + \frac{2y}{3} = -\frac{8}{5}$ (1)
$3x - y = 9$ (2)

To eliminate the fractions in equation (1), multiply by 15. The new system is

$$\begin{aligned} 3x + 10y &= -24 \quad (3) \\ 3x - y &= 9. \quad (2) \end{aligned}$$

Solve equation (2) for y.

$$\begin{aligned} 3x - y &= 9 \quad (2) \\ -y &= -3x + 9 \\ y &= 3x - 9 \end{aligned}$$

Substitute $3x - 9$ for y in equation (3) and solve for x.

$$\begin{aligned} 3x + 10y &= -24 \quad (3) \\ 3x + 10(3x - 9) &= -24 \\ 3x + 30x - 90 &= -24 \\ 33x - 90 &= -24 \\ 33x &= 66 \\ x &= 2 \end{aligned}$$

Since $y = 3x - 9$ and $x = 2$,

$$y = 3(2) - 9 = 6 - 9 = -3.$$

Solution set: $\{(2, -3)\}$

11. **(a)** $7x - 2y = -2$ (*1*)
$y = 3x$ (*2*)

Since equation (2) is given in terms of y, substitute $3x$ for y in equation (1).

$$\begin{aligned} 7x - 2y &= -2 \quad (1) \\ 7x - 2(3x) &= -2 \\ 7x - 6x &= -2 \\ x &= -2 \end{aligned}$$

Since $y = 3x$ and $x = -2$,

$$y = 3(-2) = -6.$$

Solution set: $\{(-2, -6)\}$

(b) $2x - 3y = 1$ (*1*)
$3x = 2y + 9$ (*2*)

Solve equation (2) for x by dividing both sides by 3.

$$\begin{aligned} 3x &= 2y + 9 \quad (2) \\ x &= \frac{2}{3}y + 3 \end{aligned}$$

Substitute $\frac{2}{3}y + 3$ for x in equation (1) and solve for y.

$$\begin{aligned} 2x - 3y &= 1 \quad (1) \\ 2\left(\frac{2}{3}y + 3\right) - 3y &= 1 \\ \frac{4}{3}y + 6 - 3y &= 1 \end{aligned}$$

Multiply by 3 to clear the fraction.

$$\begin{aligned} 4y + 18 - 9y &= 3 \\ 18 - 5y &= 3 \\ -5y &= -15 \\ y &= 3 \end{aligned}$$

Since $x = \frac{2}{3}y + 3$ and $y = 3$,

$$x = \frac{2}{3}(3) + 3 = 2 + 3 = 5.$$

Solution set: $\{(5, 3)\}$

4.1 Section Exercises

3. If the solution process leads to a false statement such as $0 = 5$ when solving a system, the solution set is $\emptyset$.

7. To determine whether $(5, 1)$ is a solution of the system

$$\begin{aligned} x + y &= 6 \\ x - y &= 4, \end{aligned}$$

replace x with 5 and y with 1 in each equation.

$$\begin{aligned} x + y &= 6 \\ 5 + 1 &= 6\,? \\ 6 &= 6 \quad \textit{True} \end{aligned}$$

$$\begin{aligned} x - y &= 4 \\ 5 - 1 &= 4\,? \\ 4 &= 4 \quad \textit{True} \end{aligned}$$

Since $(5, 1)$ makes both equations true, $(5, 1)$ is a solution of the system.

11. When the equations of the system

$$\begin{aligned} x + y &= 4 \\ 2x - y &= 2 \end{aligned}$$

are graphed, the point of intersection is $(2, 2)$. To check, substitute 2 for x and 2 for y in each equation of the system. Since $(2, 2)$ makes both equations true, the solution set of the system is $\{(2, 2)\}$. See the answer graph in the back of the textbook.

In Exercises 15-35, check each solution in the equations of the original system.

15. $3x + 4y = -6$ (*1*)
$5x + 3y = 1$ (*2*)

To eliminate x, multiply equation (1) by 5 and equation (2) by -3. Then add the results.

$$\begin{aligned} 15x + 20y &= -30 \\ -15x - 9y &= -3 \\ \hline 11y &= -33 \\ y &= -3 \end{aligned}$$

To find x, substitute -3 for y in equation (2).

$$\begin{aligned} 5x + 3y &= 1 \quad (2) \\ 5x + 3(-3) &= 1 \\ 5x - 9 &= 1 \\ 5x &= 10 \\ x &= 2 \end{aligned}$$

Solution set: $\{(2, -3)\}$

19. $\begin{aligned} 7x + 2y &= 6 \quad (1) \\ -14x - 4y &= -12 \quad (2) \end{aligned}$

To eliminate y, multiply equation (1) by 2 and add the result to equation (2).

$$\begin{aligned} 14x + 4y &= 12 \\ -14x - 4y &= -12 \quad (2) \\ \hline 0 &= 0 \quad \textit{True} \end{aligned}$$

Multiplying equation (1) by -2 gives equation (2). The equations are dependent, and the solution is the set of all points on the line.

Solution set: $\{(x, y) | 7x + 2y = 6\}$

23. $\begin{aligned} 5x - 5y &= 3 \quad (1) \\ x - y &= 12 \quad (2) \end{aligned}$

To eliminate x, multiply equation (2) by -5 and add the result to equation (1).

$$\begin{aligned} -5x + 5y &= -60 \\ 5x - 5y &= 3 \quad (1) \\ \hline 0 &= -57 \quad \textit{False} \end{aligned}$$

The system is inconsistent. Since graphing the equations would yield parallel lines, there is no solution for the system.

Solution set: $\emptyset$

27. $\begin{aligned} 2x &= -3y + 1 \\ 6x &= -9y + 3 \end{aligned}$

Write each equation in slope-intercept form by solving for y.

$$\begin{aligned} 2x &= -3y + 1 \\ 3y &= -2x + 1 \\ y &= -\frac{2}{3}x + \frac{1}{3} \end{aligned}$$

$$\begin{aligned} 6x &= -9y + 3 \\ 9y &= -6x + 3 \\ y &= -\frac{2}{3}x + \frac{1}{3} \end{aligned}$$

The graphs of the equations are the same line, so the system has infinitely many solutions.

31. $\begin{aligned} 3x - 4y &= -22 \quad (1) \\ -3x + y &= 0 \quad (2) \end{aligned}$

Solve equation (2) for y.

$$y = 3x$$

Substitute $3x$ for y in equation (1) and solve for x.

$$\begin{aligned} 3x - 4y &= -22 \quad (1) \\ 3x - 4(3x) &= -22 \\ 3x - 12x &= -22 \\ -9x &= -22 \\ x &= \frac{22}{9} \end{aligned}$$

Since $y = 3x$ and $x = \frac{22}{9}$,

$$y = 3\left(\frac{22}{9}\right) = \frac{22}{3}.$$

Solution set: $\{(\frac{22}{9}, \frac{22}{3})\}$

35. $\begin{aligned} x &= 3y + 5 \quad (1) \\ x &= \frac{3}{2}y \quad (2) \end{aligned}$

Both equations are given in terms of x. Choose equation (2), and substitute $\frac{3}{2}y$ for x in equation (1).

$$\begin{aligned} x &= 3y + 5 \quad (1) \\ \frac{3}{2}y &= 3y + 5 \end{aligned}$$

Multiply by 2 to clear the fraction.

$$\begin{aligned} 3y &= 6y + 10 \\ -3y &= 10 \\ y &= -\frac{10}{3} \end{aligned}$$

Since $x = \frac{3}{2}y$ and $y = -\frac{10}{3}$,

$$x = \frac{3}{2}\left(-\frac{10}{3}\right) = -5.$$

Solution set: $\{(-5, -\frac{10}{3})\}$

39. $$\begin{aligned} 3x + y &= 6 \quad (1) \\ -2x + 3y &= 7 \quad (2) \end{aligned}$$

Multiply equation (1) by -3 and add the result to equation (2).

$$\begin{array}{rl} -9x - 3y = -18 & \\ -2x + 3y = 7 & (2) \\ \hline -11x = -11 & \\ x = 1 & \end{array}$$

To find y, substitute 1 for x in equation (1).

$$\begin{aligned} 3x + y &= 6 \quad (1) \\ 3(1) + y &= 6 \\ y &= 3 \end{aligned}$$

Solution set: $\{(1, 3)\}$

40. $3x + y = 6$
$$y = -3x + 6$$

Replace y with $f(x)$.

$$f(x) = -3x + 6$$

Since f is in the form $f(x) = mx + b$, it is a linear function.

41. $-2x + 3y = 7$
$$3y = 2x + 7$$
$$y = \frac{2}{3}x + \frac{7}{3}$$

Replace y with $g(x)$.

$$g(x) = \frac{2}{3}x + \frac{7}{3}$$

Since g is in the form $g(x) = mx + b$, it is a linear function.

42. Because the graphs of f and g are straight lines that are neither parallel nor coincide, they intersect in exactly *one* point. The coordinates of the point are $(1, 3)$. Using function notation, this is given by $f(1) = 3$ and $g(1) = 3$.

43. **(a)** Since the fixed rate mortgage starts out more expensive and decreases slowly, its monthly payment would be more for years 0-10.

(b) In year 10, when the two lines on the graph intersect, the payment would be the same. The payment would be slightly less than \$700, say \$690.

47. From the graph, it is obvious that the lines will intersect in quadrant I. That fits only choice (b).

4.2 Linear Systems of Equations in Three Variables

4.2 Margin Exercises

Check each solution in the equations of the original system.

1. $$\begin{aligned} 13x + 31y &= -8 \quad (4) \\ -6x - 19y &= -1 \quad (5) \end{aligned}$$

To eliminate x, multiply equation (4) by 6 and equation (5) by 13. Then add the results.

$$\begin{array}{r} 78x + 186y = -48 \\ -78x - 247y = -13 \\ \hline -61y = -61 \\ y = 1 \end{array}$$

To find x, substitute 1 for y in equation (5).

$$\begin{aligned} -6x - 19y &= -1 \quad (5) \\ -6x - 19(1) &= -1 \\ -6x - 19 &= -1 \\ -6x &= 18 \\ x &= -3 \end{aligned}$$

Solution set: $\{(-3, 1)\}$

2. $$\begin{aligned} x + y + z &= 2 \quad (1) \\ x - y + 2z &= 2 \quad (2) \\ -x + 2y - z &= 1 \quad (3) \end{aligned}$$

Eliminate y by adding equations (1) and (2).

$$\begin{array}{rl} x + y + z = 2 & (1) \\ x - y + 2z = 2 & (2) \\ \hline 2x + 3z = 4 & (4) \end{array}$$

To eliminate y again, multiply equation (2) by 2 and add the result to equation (3).

$$\begin{array}{rl} 2x - 2y + 4z = 4 & \\ -x + 2y - z = 1 & (3) \\ \hline x + 3z = 5 & (5) \end{array}$$

Use equations (4) and (5) to eliminate z. Multiply equation (5) by -1 and add the result to equation (4).

$$\begin{array}{rl} -x - 3z = -5 & \\ 2x + 3z = 4 & (4) \\ \hline x = -1 & \end{array}$$

Substitute -1 for x in equation (5) to find z.

$$\begin{aligned} x + 3z &= 5 \quad (5) \\ -1 + 3z &= 5 \\ 3z &= 6 \\ z &= 2 \end{aligned}$$

Substitute -1 for x and 2 for z in equation (1) to find y.

$$\begin{aligned} x + y + z &= 2 \quad (1) \\ -1 + y + 2 &= 2 \\ y + 1 &= 2 \\ y &= 1 \end{aligned}$$

Solution set: $\{(-1, 1, 2)\}$

3. $$\begin{aligned} x - y \quad\quad &= 6 \quad (1) \\ 2y + 5z &= 1 \quad (2) \\ 3x \quad\quad - 4z &= 8 \quad (3) \end{aligned}$$

Since equation (3) is missing y, eliminate y again from equations (1) and (2). Multiply equation (1) by 2 and add the result to equation (2).

$$\begin{array}{rcl} 2x - 2y \quad\quad &=& 12 \\ 2y + 5z &=& 1 \quad (2) \\ \hline 2x \quad\quad + 5z &=& 13 \quad (4) \end{array}$$

Use equation (4) together with equation (3) to eliminate x. Multiply equation (4) by 3 and equation (3) by -2. Then add the results.

$$\begin{array}{rcl} 6x + 15z &=& 39 \\ -6x + 8z &=& -16 \\ \hline 23z &=& 23 \\ z &=& 1 \end{array}$$

Substitute 1 for z in equation (2) to find y.

$$\begin{aligned} 2y + 5z &= 1 \quad (2) \\ 2y + 5(1) &= 1 \\ 2y + 5 &= 1 \\ 2y &= -4 \\ y &= -2 \end{aligned}$$

Substitute -2 for y in equation (1) to find x.

$$\begin{aligned} x - y &= 6 \quad (1) \\ x - (-2) &= 6 \\ x + 2 &= 6 \\ x &= 4 \end{aligned}$$

Solution set: $\{(4, -2, 1)\}$

4. **(a)** $$\begin{aligned} 3x - 5y + 2z &= 1 \quad (1) \\ 5x + 8y - z &= 4 \quad (2) \\ -6x + 10y - 4z &= 5 \quad (3) \end{aligned}$$

Multiply equation (1) by 2 and add the result to equation (3).

$$\begin{array}{rcl} 6x - 10y + 4z &=& 2 \\ -6x + 10y - 4z &=& 5 \quad (3) \\ \hline 0 &=& 7 \quad \textit{False} \end{array}$$

Since a false statement results, the system is inconsistent.

Solution set: $\emptyset$

(b) $$\begin{aligned} 7x - 9y + 2z &= 0 \quad (1) \\ y + z &= 0 \quad (2) \\ 8x \quad\quad - z &= 0 \quad (3) \end{aligned}$$

To eliminate z, add equations (2) and (3).

$$\begin{array}{rcl} y + z &=& 0 \quad (2) \\ 8x \quad\quad - z &=& 0 \quad (3) \\ \hline 8x + y \quad\quad &=& 0 \quad (4) \end{array}$$

To eliminate z again, multiply equation (3) by 2 and add the result to equation (1).

$$\begin{array}{rcl} 16x \quad\quad - 2z &=& 0 \\ 7x - 9y + 2z &=& 0 \quad (1) \\ \hline 23x - 9y \quad\quad &=& 0 \quad (5) \end{array}$$

Use equations (4) and (5) to eliminate y. Multiply equation (4) by 9 and add the result to equation (5).

$$\begin{array}{rcl} 72x + 9y &=& 0 \\ 23x - 9y &=& 0 \quad (5) \\ \hline 95x \quad\quad &=& 0 \\ x &=& 0 \end{array}$$

Substitute 0 for x in equation (3) to find z.

$$\begin{aligned} 8x - z &= 0 \quad (3) \\ 8(0) - z &= 0 \\ -z &= 0 \\ z &= 0 \end{aligned}$$

Substitute 0 for z in equation (2) to find y.

$$\begin{aligned} y + z &= 0 \quad (2) \\ y + 0 &= 0 \\ y &= 0 \end{aligned}$$

Solution set: $\{(0, 0, 0)\}$

5.
$$\begin{aligned} x - y + z &= 4 \quad (1) \\ -3x + 3y - 3z &= -12 \quad (2) \\ 2x - 2y + 2z &= 8 \quad (3) \end{aligned}$$

Since equation (2) is -3 times equation (1) and equation (3) is 2 times equation (1), the three equations are dependent. All three have the same graph.

Solution set: $\{(x,y,z)|x - y + z = 4\}$

4.2 Section Exercises

Check each solution in the equations of the original system.

3.
$$\begin{aligned} 3x + 2y + z &= 8 \quad (1) \\ 2x - 3y + 2z &= -16 \quad (2) \\ x + 4y - z &= 20 \quad (3) \end{aligned}$$

Eliminate z by adding equations (1) and (3).

$$\begin{aligned} 3x + 2y + z &= 8 \quad (1) \\ x + 4y - z &= 20 \quad (3) \\ \hline 4x + 6y &= 28 \quad (4) \end{aligned}$$

To get another equation without z, multiply equation (3) by 2 and add the result to equation (2).

$$\begin{aligned} 2x + 8y - 2z &= 40 \\ 2x - 3y + 2z &= -16 \quad (2) \\ \hline 4x + 5y &= 24 \quad (5) \end{aligned}$$

Use equations (4) and (5) to eliminate x. Multiply equation (4) by -1 and add the result to equation (5).

$$\begin{aligned} -4x - 6y &= -28 \\ 4x + 5y &= 24 \quad (5) \\ \hline -y &= -4 \\ y &= 4 \end{aligned}$$

Substitute 4 for y in equation (5) to find x.

$$\begin{aligned} 4x + 5y &= 24 \quad (5) \\ 4x + 5(4) &= 24 \\ 4x + 20 &= 24 \\ 4x &= 4 \\ x &= 1 \end{aligned}$$

Substitute 1 for x and 4 for y in equation (3) to find z.

$$\begin{aligned} x + 4y - z &= 20 \quad (3) \\ 1 + 4(4) - z &= 20 \\ 1 + 16 - z &= 20 \\ 17 - z &= 20 \\ -z &= 3 \\ z &= -3 \end{aligned}$$

Solution set: $\{(1, 4, -3)\}$

7.
$$\begin{aligned} x + y - z &= -2 \quad (1) \\ 2x - y + z &= -5 \quad (2) \\ -x + 2y - 3z &= -4 \quad (3) \end{aligned}$$

Eliminate y and z by adding equations (1) and (2).

$$\begin{aligned} x + y - z &= -2 \quad (1) \\ 2x - y + z &= -5 \quad (2) \\ \hline 3x &= -7 \\ x &= -\frac{7}{3} \end{aligned}$$

To get another equation without y, multiply equation (2) by 2 and add the result to equation (3).

$$\begin{aligned} 4x - 2y + 2z &= -10 \\ -x + 2y - 3z &= -4 \quad (3) \\ \hline 3x - z &= -14 \quad (4) \end{aligned}$$

Substitute $-\frac{7}{3}$ for x in equation (4) to find z.

$$\begin{aligned} 3x - z &= -14 \quad (4) \\ 3\left(-\frac{7}{3}\right) - z &= -14 \\ -7 - z &= -14 \\ -z &= -7 \\ z &= 7 \end{aligned}$$

Substitute $-\frac{7}{3}$ for x and 7 for z in equation (1) to find y.

$$\begin{aligned} x + y - z &= -2 \quad (1) \\ -\frac{7}{3} + y - 7 &= -2 \end{aligned}$$

Multiply by 3 to clear the fraction.

$$\begin{aligned} -7 + 3y - 21 &= -6 \\ 3y - 28 &= -6 \\ 3y &= 22 \\ y &= \frac{22}{3} \end{aligned}$$

Solution set: $\{(-\frac{7}{3}, \frac{22}{3}, 7)\}$

11.
$$\begin{aligned} 4x + 2y - 3z &= 6 \quad (1) \\ x - 4y + z &= -4 \quad (2) \\ -x + 2z &= 2 \quad (3) \end{aligned}$$

Equation (3) is missing y. Eliminate y again by multiplying equation (1) by 2 and adding the result to equation (2).

$$\begin{array}{rl} 8x + 4y - 6z = 12 & \\ x - 4y + z = -4 & (2) \\ \hline 9x \quad - 5z = 8 & (4) \end{array}$$

Use equations (3) and (4) to eliminate x. Multiply equation (3) by 9 and add the result to equation (4).

$$\begin{array}{rl} -9x + 18z = 18 & \\ 9x - 5z = 8 & (4) \\ \hline 13z = 26 & \\ z = 2 & \end{array}$$

Substitute 2 for z in equation (3) to find x.

$$\begin{aligned} -x + 2z &= 2 \quad (3) \\ -x + 2(2) &= 2 \\ -x + 4 &= 2 \\ -x &= -2 \\ x &= 2 \end{aligned}$$

Substitute 2 for x and 2 for z in equation (2) to find y.

$$\begin{aligned} x - 4y + z &= -4 \quad (2) \\ 2 - 4y + 2 &= -4 \\ -4y + 4 &= -4 \\ -4y &= -8 \\ y &= 2 \end{aligned}$$

Solution set: $\{(2,2,2)\}$

15. Answers will vary.

(a) Two perpendicular walls and the ceiling in a normal room intersect in a single point.

(b) The floors of three different levels of an office building do not intersect.

(c) Three pages of this book intersect in infinitely many points (since they intersect in the spine).

19. $$\begin{array}{rl} -5x + 5y - 20z = -40 & (1) \\ x - y + 4z = 8 & (2) \\ 3x - 3y + 12z = 24 & (3) \end{array}$$

Dividing equation (1) by -5 gives equation (2). Dividing equation (3) by 3 also gives equation (2). The resulting equations are the same, so the three equations are dependent.

Solution set: $\{(x,y,z) | x - y + 4z = 8\}$

23. $$\begin{array}{rl} x + y - 2z = 0 & (1) \\ 3x - y + z = 0 & (2) \\ 4x + 2y - z = 0 & (3) \end{array}$$

Eliminate z by adding equations (2) and (3).

$$\begin{array}{rl} 3x - y + z = 0 & (2) \\ 4x + 2y - z = 0 & (3) \\ \hline 7x + y \quad = 0 & (4) \end{array}$$

To get another equation without z, multiply equation (2) by 2 and add the result to equation (1).

$$\begin{array}{rl} 6x - 2y + 2z = 0 & \\ x + y - 2z = 0 & (1) \\ \hline 7x - y \quad = 0 & (5) \end{array}$$

Add equations (4) and (5) to find x.

$$\begin{array}{rl} 7x + y = 0 & (4) \\ 7x - y = 0 & (5) \\ \hline 14x \quad = 0 & \\ x = 0 & \end{array}$$

Substitute 0 for x in equation (4) to find y.

$$\begin{aligned} 7x + y &= 0 \quad (4) \\ 7(0) + y &= 0 \\ 0 + y &= 0 \\ y &= 0 \end{aligned}$$

Substitute 0 for x and 0 for y in equation (1) to find z.

$$\begin{aligned} x + y - 2z &= 0 \quad (1) \\ 0 + 0 - 2z &= 0 \\ -2z &= 0 \\ z &= 0 \end{aligned}$$

Solution set: $\{(0,0,0)\}$

In Exercises 25-30,

$$f(x) = ax^2 + bx + c \ (a \neq 0).$$

25. $$\begin{aligned} f(1) &= a(1)^2 + b(1) + c \\ &= a + b + c \end{aligned}$$

Since $f(1) = 128$, the first equation is

$$a + b + c = 128.$$

26. $$\begin{aligned} f(1.5) &= a(1.5)^2 + b(1.5) + c \\ &= 2.25a + 1.5b + c \end{aligned}$$

Since $f(1.5) = 140$, the second equation is

$$2.25a + 1.5b + c = 140.$$

27. $f(3) = a(3)^2 + b(3) + c$
$= 9a + 3b + c$

Since $f(3) = 80$, the third equation is

$$9a + 3b + c = 80.$$

28. Using the three equations from Exercises 25-27, the system is

$$\begin{aligned} a + b + c &= 128 \quad (1) \\ 2.25a + 1.5b + c &= 140 \quad (2) \\ 9a + 3b + c &= 80. \quad (3) \end{aligned}$$

Multiply equation (2) by -1 and add the result to equation (1).

$$\begin{aligned} a + b + c &= 128 \quad (1) \\ -2.25a - 1.5b - c &= -140 \\ \hline -1.25a - .5b &= -12 \quad (4) \end{aligned}$$

Multiply equation (3) by -1 and add the result to equation (1).

$$\begin{aligned} a + b + c &= 128 \quad (1) \\ -9a - 3b - c &= -80 \\ \hline -8a - 2b &= 48 \quad (5) \end{aligned}$$

Use equations (4) and (5) to eliminate b. Multiply equation (4) by -4 and add the result to equation (5).

$$\begin{aligned} 5a + 2b &= 48 \\ -8a - 2b &= 48 \quad (5) \\ \hline -3a &= 96 \\ a &= -32 \end{aligned}$$

To find b, substitute $a = -32$ into equation (5).

$$\begin{aligned} -8a - 2b &= 48 \quad (5) \\ -8(-32) - 2b &= 48 \\ 256 - 2b &= 48 \\ 2b &= 208 \\ b &= 104 \end{aligned}$$

To find c, substitute $a = -32$ and $b = 104$ into equation (1).

$$\begin{aligned} a + b + c &= 128 \quad (1) \\ -32 + 104 + c &= 128 \\ 72 + c &= 128 \\ c &= 56 \end{aligned}$$

Solution set: $\{(-32, 104, 56)\}$

29. If $(a, b, c) = (-32, 104, 56)$, then

$$f(x) = -32x^2 + 104x + 56.$$

30. $f(x) = -32x^2 + 104x + 56$
$f(0) = -32(0)^2 + 104(0) + 56$
$= 56$

The initial height was 56 ft.

31.
$$\begin{aligned} x + y + z - w &= 5 \quad (1) \\ 2x + y - z + w &= 3 \quad (2) \\ x - 2y + 3z + w &= 18 \quad (3) \\ x + y - z - 2w &= -8 \quad (4) \end{aligned}$$

Eliminate w. Add equations (1) and (2).

$$\begin{aligned} x + y + z - w &= 5 \quad (1) \\ 2x + y - z + w &= 3 \quad (2) \\ \hline 3x + 2y &= 8 \quad (5) \end{aligned}$$

Eliminate w again. Add equations (1) and (3).

$$\begin{aligned} x + y + z - w &= 5 \quad (1) \\ x - 2y + 3z + w &= 18 \quad (3) \\ \hline 2x - y + 4z &= 23 \quad (6) \end{aligned}$$

Eliminate w again. Multiply equation (2) by 2. Add the result to equation (4).

$$\begin{aligned} 4x + 2y - 2z + 2w &= 6 \\ x + y - z - 2w &= -8 \quad (4) \\ \hline 5x + 3y - 3z &= -2 \quad (7) \end{aligned}$$

Eliminate z. Multiply equation (6) by 3 and equation (7) by 4. Then add the results.

$$\begin{aligned} 6x - 3y + 12z &= 69 \\ 20x + 12y - 12z &= -8 \\ \hline 26x + 9y &= 61 \quad (8) \end{aligned}$$

Eliminate y. Multiply equation (5) by 9 and equation (8) by -2. Then add the results.

$$\begin{aligned} 27x + 18y &= 72 \\ -52x - 18y &= -122 \\ \hline -25x &= -50 \\ x &= 2 \end{aligned}$$

To find y substitute $x = 2$ into equation (5).

$$\begin{aligned} 3x + 2y &= 8 \quad (5) \\ 3(2) + 2y &= 8 \\ 2y &= 2 \\ y &= 1 \end{aligned}$$

To find z, substitute $x = 2$ and $y = 1$ into equation (6).

$$\begin{aligned} 2x - y + 4z &= 23 \quad (6) \\ 2(2) - 1 + 4z &= 23 \\ 4z &= 20 \\ z &= 5 \end{aligned}$$

To find w, substitute $x = 2, y = 1$, and $z = 5$ into equation (1).

$$\begin{aligned} x + y + z - w &= 5 \quad (1) \\ 2 + 1 + 5 - w &= 5 \\ -w &= -3 \\ w &= 3 \end{aligned}$$

Solution set: $\{(2, 1, 5, 3)\}$

4.3 Applications of Linear Systems

4.3 Margin Exercises

1. Solve $x = 6 + y$ (1)
$48 = 2x + 2y.$ (2)

Substitute $6 + y$ for x in equation (2).

$$\begin{aligned} 48 &= 2x + 2y \\ 48 &= 2(6 + y) + 2y \\ 48 &= 12 + 2y + 2y \\ 36 &= 4y \\ 9 &= y \end{aligned}$$

Since $x = 6 + y$ and $y = 9$,

$$x = 6 + 9 = 15$$

The length of the house is 15 m, and the width is 9 m.

2. Let $x =$ the width of the rectangle and $y =$ the length.

The perimeter of a rectangle is $P = 2L + 2W$. Here $P = 76$, $L = y$, and $W = x$, so

$$76 = 2y + 2x$$
$$\text{or} \quad 76 = 2x + 2y.$$

Double the width, $2x$, is 13 in more than the length, y, so

$$2x = y + 13.$$

The system of equation is

$$\begin{aligned} 2x + 2y &= 76 \\ 2x &= y + 13. \end{aligned}$$

In Exercises 3-7, check each solution in the equations of the original system.

3. **(a)** Let $p =$ the price per pound for peaches and $a =$ the price per pound for apricots.

Solve the system.

$$\begin{aligned} 4p + 2a &= 5.00 \quad (1) \\ 7p + 3a &= 8.25 \quad (2) \end{aligned}$$

To eliminate a, multiply equation (1) by -3 and equation (2) by 2. Then add the results.

$$\begin{aligned} -12p - 6a &= -15.00 \\ 14p + 6a &= 16.50 \\ \hline 2p &= 1.50 \\ p &= .75 \end{aligned}$$

To find a, substitute .75 for p in equation (1).

$$\begin{aligned} 4p + 2a &= 5.00 \quad (1) \\ 4(.75) + 2a &= 5.00 \\ 3.00 + 2a &= 5.00 \\ 2a &= 2.00 \\ a &= 1.00 \end{aligned}$$

Peaches cost \$.75 per lb, and apricots cost \$1.00 per lb.

(b) Let $x =$ the number of \$10 bills and $y =$ the number of \$20 bills.

Make a table.

Bill	Number	Value
Tens	x	$10x$
Twenties	y	$20y$
Total	98	1260

Solve the system.

$$\begin{aligned} x + y &= 98 \quad (1) \\ 10x + 20y &= 1260 \quad (2) \end{aligned}$$

To eliminate x, multiply equation (1) by -10 and add the result to equation (2).

$$\begin{aligned} -10x - 10y &= -980 \\ 10x + 20y &= 1260 \quad (2) \\ \hline 10y &= 280 \\ y &= 28 \end{aligned}$$

Since $y = 28$,

$$x + y = 98 \quad (1)$$
$$x + 28 = 98$$
$$x = 70.$$

There are 70 \$10-bills and 28 \$20-bills.

4. **(a)** Let $x =$ the \$4 per lb coffee and $y =$ the \$8 per lb coffee.

Value of Coffee	Pounds	Price
\$4	x	$4x$
\$8	y	$8y$
\$5.60	50	$5.6(50) = 280$

Solve the system.

$$x + y = 50 \quad (1)$$
$$4x + 8y = 280 \quad (2)$$

To eliminate x, multiply equation (1) by -4 and add the result to equation (2).

$$-4x - 4y = -200$$
$$4x + 8y = 280 \quad (2)$$
$$4y = 80$$
$$y = 20$$

Since $y = 20$,

$$x + y = 50 \quad (1)$$
$$x + 20 = 50$$
$$x = 30.$$

To mix the coffee, 30 lb of \$4 coffee and 20 lb of \$8 coffee should be used.

(b) Let $x =$ the amount of 40% alcohol solution and $y =$ the amount of 80% alcohol solution.

Kind of solution	Liters of solution	Amount of alcohol
$40\% = .40$	x	$.4x$
$80\% = .80$	y	$.8y$
$50\% = .50$	200	$.50(200) = 100$

Solve the system.

$$x + y = 200 \quad (1)$$
$$.4x + .8y = 100 \quad (2)$$

To eliminate x, multiply equation (1) by -4 and equation (2) by 10. Then add the results.

$$-4x - 4y = -800$$
$$4x + 8y = 1000$$
$$4y = 200$$
$$y = 50$$

Since $y = 50$,

$$x + y = 200 \quad (1)$$
$$x + 50 = 200$$
$$x = 150.$$

The mixture should contain 150 L of 40% and 50 L of 80% solution.

5. Let $x =$ the train's speed and $y =$ the truck's speed.

Make a table.

	d	r	t
Train	600	x	$\frac{600}{x}$
Truck	520	y	$\frac{520}{y}$

Since both times are the same,

$$\frac{600}{x} = \frac{520}{y}$$
$$600y = 520x$$
$$-520x + 600y = 0. \quad (1)$$

Since the train's average speed, x, is 8 mph faster than the truck's speed, y,

$$x = y + 8. \quad (2)$$

Substitute $y + 8$ for x in equation (1) to find y.

$$-520x + 600y = 0 \quad (1)$$
$$-520(y + 8) + 600y = 0$$
$$-520y - 4160 + 600y = 0$$
$$80y = 4160$$
$$y = 52$$

Since $x = y + 8$ and $y = 52$,

$$x = 52 + 8 = 60.$$

The train's speed is 60 mph, and the truck's speed is 52 mph.

6. Let $x =$ the number of bottles of cheap, at \$8
$y =$ the number of bottles of better, at \$15,
and $z =$ the number of bottles of best, at \$32.

There are 10 more bottles of cheap than better, so

$$x = y + 10. \quad (1)$$

There are 3 fewer bottles of best than better, so

$$z = y - 3. \quad (2)$$

The total value is \$589, so

$$8x + 15y + 32z = 589. \quad (3)$$

Substitute $y + 10$ for x and $y - 3$ for z in equation (3) to find y.

$$\begin{aligned} 8(y + 10) + 15y + 32(y - 3) &= 589 \\ 8y + 80 + 15y + 32y - 96 &= 589 \\ 55y - 16 &= 589 \\ 55y &= 605 \\ y &= 11 \end{aligned}$$

Since $y = 11$,

$$\begin{aligned} x &= y + 10 = 11 + 10 = 21 \\ \text{and} \quad z &= y - 3 = 11 - 3 = 8. \end{aligned}$$

There are 21 bottles of cheap perfume, 11 of better, and 8 of best.

7. Let $x =$ the number of tons of newsprint,
$y =$ the number of tons of bond,
and $z =$ the number of tons of copy machine paper.

The amount of recycled paper used to make all three items is 4200, so

$$3x + 2y + 2z = 4200. \quad (1)$$

The amount of wood pulp used to make all three items is 5800, so

$$x + 4y + 3z = 5800. \quad (2)$$

The amount of rags used to make all three items is 3900, so

$$3y + 2z = 3900. \quad (3)$$

Solve the system of equations (1), (2), and (3). To eliminate z, multiply equation (3) by -1 and add the result to equation (1).

$$\begin{array}{rcrl} -3y - 2z &=& -3900 & \\ 3x + 2y + 2z &=& 4200 & (1) \\ \hline 3x - y \quad &=& 300 & (4) \end{array}$$

To eliminate z again, multiply equation (2) by -2 and equation (3) by 3 and add the results.

$$\begin{array}{rcrl} -2x - 8y - 6z &=& -11,600 & \\ 9y + 6z &=& 11,700 & \\ \hline -2x + y \quad &=& 100 & (5) \end{array}$$

Add equations (4) and (5) to eliminate y.

$$\begin{array}{rcll} 3x - y &=& 300 & (4) \\ -2x + y &=& 100 & (5) \\ \hline x &=& 400 & \end{array}$$

Since $x = 400$,

$$\begin{aligned} 3x - y &= 300 \quad (4) \\ 3(400) - y &= 300 \\ 1200 - y &= 300 \\ 900 &= y. \end{aligned}$$

Since $y = 900$,

$$\begin{aligned} 3y + 2z &= 3900 \quad (3) \\ 3(900) + 2z &= 3900 \\ 2700 + 2z &= 3900 \\ 2z &= 1200 \\ z &= 600 \end{aligned}$$

The paper mill can make 400 tons of newsprint, 900 tons of bond, and 600 tons of copy machine paper.

4.3 Section Exercises

3. From the figure in the text, the angles marked y and $3x + 10$ are supplementary, so

$$3x + 10 + y = 180. \quad (1)$$

Also, angles x and y are complementary, so

$$x + y = 90. \quad (2)$$

Solve equation (2) for y to get

$$y = 90 - x. \quad (3)$$

Substitute $90 - x$ for y in equation (1).

$$\begin{aligned} 3x + 10 + y &= 180 \quad (1) \\ 3x + 10 + (90 - x) &= 180 \\ 2x + 100 &= 180 \\ 2x &= 80 \\ x &= 40 \end{aligned}$$

Substitute $x = 40$ into equation (3) to get

$$\begin{aligned} y &= 90 - x \quad (3) \\ y &= 90 - 40 \\ y &= 50. \end{aligned}$$

The angles measure 40° and 50°.

7. Let $x =$ the price of a CGA monitor and $y =$ the price of a VGA monitor.

For the first purchase, the total cost of 4 CGA monitors and 6 VGA monitors is \$4600, so

$$4x + 6y = 4600. \quad (1)$$

For the other purchase, the total cost of 6 CGA monitors and 4 VGA monitors is \$4400, so

$$6x + 4y = 4400. \quad (2)$$

To solve the system, multiply equation (1) by -2 and equation (2) by 3. Then add the results.

$$\begin{aligned} -8x - 12y &= -9200 \\ 18x + 12y &= 13,200 \\ \hline 10x \quad\quad &= 4000 \\ x &= 400 \end{aligned}$$

Since $x = 400$,

$$\begin{aligned} 4x + 6y &= 4600 \quad (1) \\ 4(400) + 6y &= 4600 \\ 1600 + 6y &= 4600 \\ 6y &= 3000 \\ y &= 500. \end{aligned}$$

A CGA monitor costs \$400, and a VGA monitor costs \$500.

11. Let $x =$ the cost of a kilogram of dark clay and $y =$ the cost of a kilogram of light clay.

Solve the system.

$$\begin{aligned} 2x + 3y &= 22 \quad (1) \\ x + 2y &= 13 \quad (2) \end{aligned}$$

Solve equation (2) for x.

$$\begin{aligned} x + 2y &= 13 \quad (2) \\ x &= 13 - 2y \end{aligned}$$

Substitute $13 - 2y$ for x in equation (1) and solve for y.

$$\begin{aligned} 2x + 3y &= 22 \quad (1) \\ 2(13 - 2y) + 3y &= 22 \\ 26 - 4y + 3y &= 22 \\ 26 - y &= 22 \\ -y &= -4 \\ y &= 4 \end{aligned}$$

Since $x = 13 - 2y$ and $y = 4$,

$$x = 13 - 2(4) = 13 - 8 = 5.$$

The dark clay costs \$5 per kg; the light clay costs \$4 per kg.

15. The cost is the price per pound, \$.89, times the number of pounds, x, or $\$.89x$.

19. Let $x =$ the amount of pure acid and $y =$ the amount of 10% acid.

Make a table.

Kind	Amount	Pure acid
100%	x	$1.00x = x$
10%	y	$.10y$
20%	54	$.20(54) = 10.8$

Solve the system.

$$\begin{aligned} x + \quad y &= 54 \quad (1) \\ x + .10y &= 10.8 \quad (2) \end{aligned}$$

Multiply equation (2) by 10 to clear the decimals.

$$10x + y = 108 \quad (3)$$

To eliminate y, multiply equation (1) by -1 and add the result to equation (3).

$$\begin{aligned} -x - y &= -54 \\ 10x + y &= 108 \quad (3) \\ \hline 9x \quad\quad &= 54 \\ x &= 6 \end{aligned}$$

Since $x = 6$,

$$\begin{aligned} x + y &= 54 \quad (1) \\ 6 + y &= 54 \\ y &= 48. \end{aligned}$$

Use 6 L of pure acid and 48 L of 10% acid.

23. Let $x =$ the number of general admission tickets
and $y =$ the number of student tickets.

Ticket	Number	Value of tickets
General	x	$5.00x = 5x$
Student	y	$4.00y = 4y$
Total	184	812

Solve the system.

$$\begin{aligned} x + y &= 184 \quad (1) \\ 5x + 4y &= 812 \quad (2) \end{aligned}$$

To eliminate x, multiply equation (1) by -5 and add the result to equation (2).

$$\begin{aligned} -5x - 5y &= -920 \\ 5x + 4y &= 812 \quad (2) \\ \hline -y &= -108 \\ y &= 108 \end{aligned}$$

Since $y = 108$,

$$\begin{aligned} x + y &= 184 \quad (1) \\ x + 108 &= 184 \\ x &= 76. \end{aligned}$$

76 general admission tickets and 108 student tickets were sold.

27. Let $x =$ the amount invested at 2%
and $y =$ the amount invested at 4%.

Make a table.

Rate	Amount	Interest
2%	x	$.02x$
4%	y	$.04y$
Total	$3000	$100

Solve the system.

$$\begin{aligned} x + y &= 3000 \quad (1) \\ .02x + .04y &= 100 \quad (2) \end{aligned}$$

Multiply equation (2) by 100 to clear the decimals.

$$2x + 4y = 10,000 \quad (3)$$

To eliminate x, multiply equation (1) by -2 and add the result to equation (3).

$$\begin{aligned} -2x - 2y &= -6000 \\ 2x + 4y &= 10,000 \quad (3) \\ \hline 2y &= 4000 \\ y &= 2000 \end{aligned}$$

Since $y = 2000$,

$$\begin{aligned} x + y &= 3000 \quad (1) \\ x + 2000 &= 3000 \\ x &= 1000. \end{aligned}$$

$1000 is invested at 2%, and $2000 is invested at 4%.

31. Let $x =$ the speed of the freight train
and $y =$ the speed of the express train.

Make a table.

	r	t	d
Freight train	x	3	$3x$
Express train	y	3	$3y$

Since $d = rt$ and the trains are 390 km apart,

$$3x + 3y = 390. \quad (1)$$

The freight train travels 30 km/hr slower than the express train, so

$$x = y - 30. \quad (2)$$

Substitute $y - 30$ for x in equation (1) and solve for y.

$$\begin{aligned} 3x + 3y &= 390 \quad (1) \\ 3(y - 30) + 3y &= 390 \\ 3y - 90 + 3y &= 390 \\ 6y - 90 &= 390 \\ 6y &= 480 \\ y &= 80 \end{aligned}$$

Since $x = y - 30$ and $y = 80$,

$$x = 80 - 30 = 50.$$

The freight train travels at 50 km/hr, while the express train travels at 80 km/hr.

35. Let $x =$ the measure of one angle,
$y =$ the measure of another angle,
and $z =$ the measure of the last angle.

Two equations are given, so

$$z = x + 10$$
$$\text{or} \quad -x + z = 10 \qquad (1)$$
$$\text{and} \quad x + y = 100. \qquad (2)$$

Since the sum of the measures of the angles of a triangle is 180°, the third equation of the system is

$$x + y + z = 180. \quad (3)$$

Equation (1) is missing y. To eliminate y again, multiply equation (2) by -1 and add the result to equation (3).

$$\begin{array}{rl} -x - y &= -100 \\ x + y + z &= 180 \quad (3) \\ \hline z &= 80 \end{array}$$

Since $z = 80$,

$$\begin{aligned} -x + z &= 10 \quad (1) \\ -x + 80 &= 10 \\ -x &= -70 \\ x &= 70. \end{aligned}$$

Since $x = 70$,

$$\begin{aligned} x + y &= 100 \quad (2) \\ 70 + y &= 100 \\ y &= 30. \end{aligned}$$

The measures of the angles are 70°, 30°, and 80°.

39. Let $x =$ the length of the longest side,
$y =$ the length of the middle side,
and $z =$ the length of the shortest side.

Perimeter is the sum of the measures of the sides, so

$$x + y + z = 70. \quad (1)$$

The longest side is 4 cm less than the sum of the other sides, so

$$x = y + z - 4$$
$$\text{or} \quad x - y - z = -4. \quad (2)$$

Twice the shortest side is 9 cm less than the longest side, so

$$2z = x - 9$$
$$\text{or} \quad -x + 2z = -9. \quad (3)$$

Add equations (1) and (2) to eliminate y and z.

$$\begin{array}{rl} x + y + z &= 70 \quad (1) \\ x - y - z &= -4 \quad (2) \\ \hline 2x &= 66 \\ x &= 33 \end{array}$$

Since $x = 33$,

$$\begin{aligned} -x + 2z &= -9 \quad (3) \\ -33 + 2z &= -9 \\ 2z &= 24 \\ z &= 12. \end{aligned}$$

Since $x = 33$ and $z = 12$,

$$\begin{aligned} x + y + z &= 70 \quad (1) \\ 33 + y + 12 &= 70 \\ y + 45 &= 70 \\ y &= 25. \end{aligned}$$

The shortest side is 12 cm long, the middle side is 25 cm long, and the longest side is 33 cm long.

In Exercises 43-45,

$$x^2 + y^2 + ax + by + c = 0.$$

43. Let $x = 2$ and $y = 1$.

$$\begin{aligned} 2^2 + 1^2 + a(2) + b(1) + c &= 0 \\ 4 + 1 + 2a + b + c &= 0 \\ 2a + b + c &= -5 \end{aligned}$$

44. Let $x = -1$ and $y = 0$.

$$\begin{aligned} (-1)^2 + 0^2 + a(-1) + b(0) + c &= 0 \\ 1 - a + c &= 0 \\ -a + c &= -1 \\ \text{or} \qquad a - c &= 1 \end{aligned}$$

45. Let $x = 3$ and $y = 3$.

$$\begin{aligned} 3^2 + 3^2 + a(3) + b(3) + c &= 0 \\ 9 + 9 + 3a + 3b + c &= 0 \\ 3a + 3b + c &= -18 \end{aligned}$$

46. Use the equations from Exercises 43-45 to form a system of equations.

$$\begin{array}{rl} 2a + b + c &= -5 \quad (1) \\ a - c &= 1 \quad (2) \\ 3a + 3b + c &= -18 \quad (3) \end{array}$$

Add equations (1) and (2).

$$\begin{array}{rl} 2a+b+c=-5 & (1) \\ \underline{a \quad -c= \ \ 1} & (2) \\ 3a+b \quad =-4 & (4) \end{array}$$

Add equations (2) and (3).

$$\begin{array}{rl} a \quad -c= \ \ 1 & (2) \\ \underline{3a+3b+c=-18} & (3) \\ 4a+3b \quad =-17 & (5) \end{array}$$

Multiply equation (4) by -3 and add the result to equation (5).

$$\begin{array}{rl} -9a-3b= \ \ 12 & \\ \underline{4a+3b=-17} & (5) \\ -5a \quad = \ -5 & \\ a=1 & \end{array}$$

Substitute $a=1$ into equation (4).

$$\begin{array}{rl} 3a+b=-4 & (4) \\ 3(1)+b=-4 & \\ b=-7 & \end{array}$$

Substitute $a=1$ into equation (2).

$$\begin{array}{rl} a-c=1 & (2) \\ 1-c=1 & \\ c=0 & \end{array}$$

Since $a=1$, $b=-7$, and $c=0$, the equation of the circle is

$$x^2+y^2+ax+by+c=0$$
$$x^2+y^2+x-7y=0.$$

47. The relation is not a function because it does not pass the vertical line test, that is, a vertical line intersects the graph in more than one point.

4.4 Determinants

4.4 Margin Exercises

1. $\begin{vmatrix} a & b \\ c & d \end{vmatrix} = ad-bc$

(a) $\begin{vmatrix} -4 & 6 \\ 2 & 3 \end{vmatrix} = -4(3)-6(2)$
$= -12-12=-24$

(b) $\begin{vmatrix} 3 & -1 \\ 0 & 2 \end{vmatrix} = 3(2)-(-1)0$
$= 6-0=6$

(c) $\begin{vmatrix} -2 & 5 \\ 1 & 5 \end{vmatrix} = -2(5)-5(1)$
$= -10-5=-15$

2. (a) $\begin{vmatrix} 0 & -1 & 0 \\ 2 & 4 & 2 \\ 3 & 1 & 5 \end{vmatrix}$

Expand by minors about column 1.

$$= 0\begin{vmatrix} 4 & 2 \\ 1 & 5 \end{vmatrix} - 2\begin{vmatrix} -1 & 0 \\ 1 & 5 \end{vmatrix} + 3\begin{vmatrix} -1 & 0 \\ 4 & 2 \end{vmatrix}$$
$$= 0-2[-1(5)-0(1)]+3[-1(2)-0(4)]$$
$$= -2(-5)+3(-2)$$
$$= 10-6=4$$

(b) $\begin{vmatrix} 2 & 1 & 4 \\ -3 & 0 & 2 \\ -2 & 1 & 5 \end{vmatrix}$

Expand by minors about column 1.

$$= 2\begin{vmatrix} 0 & 2 \\ 1 & 5 \end{vmatrix} - (-3)\begin{vmatrix} 1 & 4 \\ 1 & 5 \end{vmatrix} + (-2)\begin{vmatrix} 1 & 4 \\ 0 & 2 \end{vmatrix}$$
$$= 2[0(5)-2(1)]+3[1(5)-4(1)]$$
$$-2[1(2)-4(0)]$$
$$= 2(-2)+3(1)-2(2)$$
$$= -4+3-4=-5$$

3. (a) $\begin{vmatrix} 2 & 1 & 3 \\ -1 & 0 & 4 \\ 2 & 4 & 3 \end{vmatrix}$

Expand about column 2.

$$= -1\begin{vmatrix} -1 & 4 \\ 2 & 3 \end{vmatrix} + 0\begin{vmatrix} 2 & 3 \\ 2 & 3 \end{vmatrix} - 4\begin{vmatrix} 2 & 3 \\ -1 & 4 \end{vmatrix}$$
$$= -1[-1(3)-4(2)]+0-4[2(4)-3(-1)]$$
$$= -1(-11)-4(11)$$
$$= 11-44=-33$$

(b) $\begin{vmatrix} 5 & -1 & 2 \\ 0 & 4 & 3 \\ -1 & 2 & 0 \end{vmatrix}$

Expand about column 2.

$$= -(-1)\begin{vmatrix} 0 & 3 \\ -1 & 0 \end{vmatrix} + 4\begin{vmatrix} 5 & 2 \\ -1 & 0 \end{vmatrix} - 2\begin{vmatrix} 5 & 2 \\ 0 & 3 \end{vmatrix}$$
$$= 1[0(0)-3(-1)]+4[5(0)-2(-1)]$$
$$-2[5(3)-2(0)]$$
$$= 1(3)+4(2)-2(15)=3+8-30=-19$$

4. $\begin{vmatrix} 1 & 0 & 2 & 0 \\ 3 & 0 & 0 & 4 \\ 0 & -1 & 1 & 0 \\ 2 & 0 & -1 & 0 \end{vmatrix}$

Expand about row 1.

$$= 1\begin{vmatrix} 0 & 0 & 4 \\ -1 & 1 & 0 \\ 0 & -1 & 0 \end{vmatrix} - 0 + 2\begin{vmatrix} 3 & 0 & 4 \\ 0 & -1 & 0 \\ 2 & 0 & 0 \end{vmatrix} - 0$$

Expand the first determinant about row 1 and the second determinant about row 3.

$$= 0 - 0 + 4\begin{vmatrix} -1 & 1 \\ 0 & -1 \end{vmatrix} + 2\left[2\begin{vmatrix} 0 & 4 \\ -1 & 0 \end{vmatrix} - 0 + 0\right]$$
$$= 4(1) + 2[2(4)]$$
$$= 4 + 16 = 20$$

4.4 Section Exercises

3. The determinant $\begin{vmatrix} a & b \\ c & d \end{vmatrix}$ is equal to $ad - bc$.

This is the definition of the value of a 2×2 determinant.

True

7. $\begin{vmatrix} -2 & 5 \\ -1 & 4 \end{vmatrix} = -2(4) - 5(-1)$
$= -8 + 5 = -3$

11. $\begin{vmatrix} 0 & 4 \\ 0 & 4 \end{vmatrix} = 0(4) - 4(0)$
$= 0 - 0 = 0$

15. $\begin{vmatrix} 1 & 0 & -2 \\ 0 & 2 & 3 \\ 1 & 0 & 5 \end{vmatrix}$

Expand about column 1.

$$= 1\begin{vmatrix} 2 & 3 \\ 0 & 5 \end{vmatrix} - 0\begin{vmatrix} 0 & -2 \\ 0 & 5 \end{vmatrix} + 1\begin{vmatrix} 0 & -2 \\ 2 & 3 \end{vmatrix}$$
$$= 1[2(5) - 3(0)] - 0 + 1[0(3) - (-2)(2)]$$
$$= 1(10) + 1(4)$$
$$= 10 + 4 = 14$$

19. $\begin{vmatrix} 4 & 4 & 2 \\ 1 & -1 & -2 \\ 1 & 0 & 2 \end{vmatrix}$

Expand about column 2.

$$= -4\begin{vmatrix} 1 & -2 \\ 1 & 2 \end{vmatrix} + (-1)\begin{vmatrix} 4 & 2 \\ 1 & 2 \end{vmatrix} - 0\begin{vmatrix} 4 & 2 \\ 1 & -2 \end{vmatrix}$$
$$= -4[1(2) - (-2)(1)] - 1[4(2) - 2(1)] - 0$$
$$= -4(4) - 1(6)$$
$$= -16 - 6 = -22$$

23. $\begin{vmatrix} -6 & 3 & 5 \\ -3 & 2 & 2 \\ 0 & 0 & 0 \end{vmatrix}$

All the elements of row 3 are 0. If we expand about row 3, each minor will be multiplied by 0. Therefore, the determinant equals 0.

27. $\begin{vmatrix} 1 & 3 & -2 \\ -1 & 4 & 5 \\ 2 & 6 & -4 \end{vmatrix}$

Expand about column 1.

$$= 1\begin{vmatrix} 4 & 5 \\ 6 & -4 \end{vmatrix} - (-1)\begin{vmatrix} 3 & -2 \\ 6 & -4 \end{vmatrix} + 2\begin{vmatrix} 3 & -2 \\ 4 & 5 \end{vmatrix}$$
$$= 1[4(-4) - 5(6)] + 1[3(-4) - (-2)(6)]$$
$$+ 2[3(5) - (-2)(4)]$$
$$= -46 + 0 + 2(23)$$
$$= -46 + 46$$
$$= 0$$

31. $\begin{vmatrix} 4 & 2 & 2 & 2 \\ 1 & -1 & 0 & 0 \\ 2 & 1 & 1 & 1 \\ 0 & 0 & -3 & -2 \end{vmatrix}$

Expand about row 2 leaving out 0 times a 3 × 3 determinant.

$$= -1\begin{vmatrix} 2 & 2 & 2 \\ 1 & 1 & 1 \\ 0 & -3 & -2 \end{vmatrix} + (-1)\begin{vmatrix} 4 & 2 & 2 \\ 2 & 1 & 1 \\ 0 & -3 & -2 \end{vmatrix}$$

Expand about row 3.

$$= -1\left[0\begin{vmatrix} 2 & 2 \\ 1 & 1 \end{vmatrix} - (-3)\begin{vmatrix} 2 & 2 \\ 1 & 1 \end{vmatrix} + (-2)\begin{vmatrix} 2 & 2 \\ 1 & 1 \end{vmatrix}\right]$$
$$-1\left[0\begin{vmatrix} 2 & 2 \\ 1 & 1 \end{vmatrix} - (-3)\begin{vmatrix} 4 & 2 \\ 2 & 1 \end{vmatrix} + (-2)\begin{vmatrix} 4 & 2 \\ 2 & 1 \end{vmatrix}\right]$$
$$= -1[0 + 3(0) - 2(0)] - [0 + 3(0) - 2(0)]$$
$$= 0$$

33. The slope through the points (x_1, y_1) and (x_2, y_2) is

$$m = \frac{y_2 - y_1}{x_2 - x_1}.$$

34. $y - y_1 = m(x - x_1)$

Let $m = \dfrac{y_2 - y_1}{x_2 - x_1}$ from Exercise 33.

$$y - y_1 = \frac{y_2 - y_1}{x_2 - x_1}(x - x_1)$$

35. $y - y_1 = \dfrac{y_2 - y_1}{x_2 - x_1}(x - x_1)$

Multiply both sides by $x_2 - x_1$.

$(x_2 - x_1)(y - y_1) = (y_2 - y_1)(x - x_1)$

Subtract $(y_2 - y_1)(x - x_1)$ from both sides.

$(x_2 - x_1)(y - y_1) - (y_2 - y_1)(x - x_1) = 0$

Multiply and collect terms.

$$\begin{aligned} x_2y - x_1y - x_2y_1 + x_1y_1 \\ - (xy_2 - x_1y_2 - xy_1 + x_1y_1) &= 0 \\ x_2y - x_1y - x_2y_1 + x_1y_1 \\ - xy_2 + x_1y_2 + xy_1 - x_1y_1 &= 0 \\ x_2y - x_1y - x_2y_1 - xy_2 + x_1y_2 + xy_1 &= 0 \end{aligned}$$

36. $\begin{vmatrix} x & y & 1 \\ x_1 & y_1 & 1 \\ x_2 & y_2 & 1 \end{vmatrix} = 0$

Expand about row 1.

$$x\begin{vmatrix} y_1 & 1 \\ y_2 & 1 \end{vmatrix} - y\begin{vmatrix} x_1 & 1 \\ x_2 & 1 \end{vmatrix} + 1\begin{vmatrix} x_1 & y_1 \\ x_2 & y_2 \end{vmatrix}$$

$$\begin{aligned} &= x(y_1 - y_2) - y(x_1 - x_2) + 1(x_1y_2 - x_2y_1) \\ &= xy_1 - xy_2 - x_1y + x_2y + x_1y_2 - x_2y_1 \end{aligned}$$

Thus,

$xy_1 - xy_2 - x_1y + x_2y + x_1y_2 - x_2y_1 = 0$

or

$x_2y - x_1y - x_2y_1 - xy_2 + x_1y_2 + xy_1 = 0.$

This is the same equation as that found in Exercise 35.

39. $\begin{vmatrix} 2 & -1 & 5 \\ -4 & 1 & 2 \\ -3 & 0 & 6 \end{vmatrix}$

Expand about row 3.

$$\begin{aligned} &= -3\begin{vmatrix} -1 & 5 \\ 1 & 2 \end{vmatrix} - 0 + 6\begin{vmatrix} 2 & -1 \\ -4 & 1 \end{vmatrix} \\ &= -3[-1(2) - 5(1)] + 6[2(1) - (-1)(-4)] \\ &= -3(-7) + 6(-2) \\ &= 21 - 12 \\ &= 9 \end{aligned}$$

4.5 Solution of Linear Systems of Equations By Determinants—Cramer's Rule

4.5 Margin Exercises

Check each solution in the equations of the given system.

1. **(a)** $x + y = 5$
$x - y = 1$

$D = \begin{vmatrix} 1 & 1 \\ 1 & -1 \end{vmatrix} = 1(-1) - 1(1) = -2$

$D_x = \begin{vmatrix} 5 & 1 \\ 1 & -1 \end{vmatrix} = 5(-1) - 1(1) = -6$

$D_y = \begin{vmatrix} 1 & 5 \\ 1 & 1 \end{vmatrix} = 1(1) - 5(1) = -4$

$x = \dfrac{D_x}{D} = \dfrac{-6}{-2} = 3$

$y = \dfrac{D_y}{D} = \dfrac{-4}{-2} = 2$

Solution set: $\{(3, 2)\}$

(b) $2x - 3y = -26$
$3x + 4y = 12$

$D = \begin{vmatrix} 2 & -3 \\ 3 & 4 \end{vmatrix} = 2(4) - (-3)(3) = 17$

$D_x = \begin{vmatrix} -26 & -3 \\ 12 & 4 \end{vmatrix} = -26(4) - (-3)(12) = -68$

$D_y = \begin{vmatrix} 2 & -26 \\ 3 & 12 \end{vmatrix} = 2(12) - (-26)(3) = 102$

$x = \dfrac{D_x}{D} = \dfrac{-68}{17} = -4$

$y = \dfrac{D_y}{D} = \dfrac{102}{17} = 6$

Solution set: $\{(-4, 6)\}$

(c) $4x - 5y = -8$
$3x + 7y = -6$

$D = \begin{vmatrix} 4 & -5 \\ 3 & 7 \end{vmatrix} = 4(7) - (-5)(3) = 43$

$D_x = \begin{vmatrix} -8 & -5 \\ -6 & 7 \end{vmatrix} = -8(7) - (-5)(-6) = -86$

$$D_y = \begin{vmatrix} 4 & -8 \\ 3 & -6 \end{vmatrix} = 4(-6) - (-8)(3) = 0$$

$$x = \frac{D_x}{D} = \frac{-86}{43} = -2$$

$$y = \frac{D_y}{D} = \frac{0}{43} = 0$$

Solution set: $\{(-2, 0)\}$

2. $D_y = \begin{vmatrix} 1 & -2 & -1 \\ 2 & -5 & 1 \\ 1 & 4 & 3 \end{vmatrix}$

Expand about column 1.

$$= 1\begin{vmatrix} -5 & 1 \\ 4 & 3 \end{vmatrix} - 2\begin{vmatrix} -2 & -1 \\ 4 & 3 \end{vmatrix} + 1\begin{vmatrix} -2 & -1 \\ -5 & 1 \end{vmatrix}$$

$$= 1(-19) - 2(-2) + 1(-7)$$
$$= -19 + 4 - 7 = -22$$

$$D_z = \begin{vmatrix} 1 & 1 & -2 \\ 2 & -1 & -5 \\ 1 & -2 & 4 \end{vmatrix}$$

Expand about column 1.

$$= 1\begin{vmatrix} -1 & -5 \\ -2 & 4 \end{vmatrix} - 2\begin{vmatrix} 1 & -2 \\ -2 & 4 \end{vmatrix} + 1\begin{vmatrix} 1 & -2 \\ -1 & -5 \end{vmatrix}$$

$$= 1(-14) - 2(0) + 1(-7)$$
$$= -14 - 7 = -21$$

3. **(a)**
$$\begin{aligned} x + y + z &= 2 \\ 2x - z &= -3 \\ y + 2z &= 4 \end{aligned}$$

$$D = \begin{vmatrix} 1 & 1 & 1 \\ 2 & 0 & -1 \\ 0 & 1 & 2 \end{vmatrix}$$

Expand about column 2.

$$= -1\begin{vmatrix} 2 & -1 \\ 0 & 2 \end{vmatrix} + 0\begin{vmatrix} 1 & 1 \\ 0 & 2 \end{vmatrix} - 1\begin{vmatrix} 1 & 1 \\ 2 & -1 \end{vmatrix}$$

$$= -1(4) + 0 - 1(-3)$$
$$= -4 + 3 = -1$$

$$D_x = \begin{vmatrix} 2 & 1 & 1 \\ -3 & 0 & -1 \\ 4 & 1 & 2 \end{vmatrix}$$

Expand about column 2.

$$= -1\begin{vmatrix} -3 & -1 \\ 4 & 2 \end{vmatrix} + 0\begin{vmatrix} 2 & 1 \\ 4 & 2 \end{vmatrix} - 1\begin{vmatrix} 2 & 1 \\ -3 & -1 \end{vmatrix}$$

$$= -1(-2) + 0 - 1(1)$$
$$= 2 - 1 = 1$$

$$D_y = \begin{vmatrix} 1 & 2 & 1 \\ 2 & -3 & -1 \\ 0 & 4 & 2 \end{vmatrix}$$

Expand about column 1.

$$= 1\begin{vmatrix} -3 & -1 \\ 4 & 2 \end{vmatrix} - 2\begin{vmatrix} 2 & 1 \\ 4 & 2 \end{vmatrix} + 0\begin{vmatrix} 2 & 1 \\ -3 & -1 \end{vmatrix}$$

$$= 1(-2) - 2(0) + 0 = -2$$

$$D_z = \begin{vmatrix} 1 & 1 & 2 \\ 2 & 0 & -3 \\ 0 & 1 & 4 \end{vmatrix}$$

Expand about column 2.

$$= -1\begin{vmatrix} 2 & -3 \\ 0 & 4 \end{vmatrix} + 0\begin{vmatrix} 1 & 2 \\ 0 & 4 \end{vmatrix} - 1\begin{vmatrix} 1 & 2 \\ 2 & -3 \end{vmatrix}$$

$$= -1(8) + 0 - 1(-7)$$
$$= -8 + 7 = -1$$

$$x = \frac{D_x}{D} = \frac{1}{-1} = -1$$

$$y = \frac{D_y}{D} = \frac{-2}{-1} = 2$$

$$z = \frac{D_z}{D} = \frac{-1}{-1} = 1$$

Solution set: $\{(-1, 2, 1)\}$

(b)
$$\begin{aligned} 3x - 2y + 4z &= 5 \\ 4x + y + z &= 14 \\ x - y - z &= 1 \end{aligned}$$

$$D = \begin{vmatrix} 3 & -2 & 4 \\ 4 & 1 & 1 \\ 1 & -1 & -1 \end{vmatrix}$$

Expand about row 3.

$$= 1\begin{vmatrix} -2 & 4 \\ 1 & 1 \end{vmatrix} - (-1)\begin{vmatrix} 3 & 4 \\ 4 & 1 \end{vmatrix} + (-1)\begin{vmatrix} 3 & -2 \\ 4 & 1 \end{vmatrix}$$

$$= -6 + (-13) - 11 = -30$$

$$D_x = \begin{vmatrix} 5 & -2 & 4 \\ 14 & 1 & 1 \\ 1 & -1 & -1 \end{vmatrix}$$

Expand about row 3.

$$= 1\begin{vmatrix} -2 & 4 \\ 1 & 1 \end{vmatrix} - (-1)\begin{vmatrix} 5 & 4 \\ 14 & 1 \end{vmatrix} + (-1)\begin{vmatrix} 5 & -2 \\ 14 & 1 \end{vmatrix}$$

$$= -6 + (-51) - 33 = -90$$

$$D_y = \begin{vmatrix} 3 & 5 & 4 \\ 4 & 14 & 1 \\ 1 & 1 & -1 \end{vmatrix}$$

Expand about row 3.

$$= 1\begin{vmatrix} 5 & 4 \\ 14 & 1 \end{vmatrix} - 1\begin{vmatrix} 3 & 4 \\ 4 & 1 \end{vmatrix} + (-1)\begin{vmatrix} 3 & 5 \\ 4 & 14 \end{vmatrix}$$
$$= -51 - (-13) - 22 = -60$$

$$D_z = \begin{vmatrix} 3 & -2 & 5 \\ 4 & 1 & 14 \\ 1 & -1 & 1 \end{vmatrix}$$

Expand about row 3.

$$= 1\begin{vmatrix} -2 & 5 \\ 1 & 14 \end{vmatrix} - (-1)\begin{vmatrix} 3 & 5 \\ 4 & 14 \end{vmatrix} + 1\begin{vmatrix} 3 & -2 \\ 4 & 1 \end{vmatrix}$$
$$= -33 + 22 + 11 = 0$$

$$x = \frac{D_x}{D} = \frac{-90}{-30} = 3$$
$$y = \frac{D_y}{D} = \frac{-60}{-30} = 2$$
$$z = \frac{D_z}{D} = \frac{0}{-30} = 0$$

Solution set: $\{(3, 2, 0)\}$

4. $x - y + z = 6$
$3x + 2y + z = 4$
$2x - 2y + 2z = 14$

$$D = \begin{vmatrix} 1 & -1 & 1 \\ 3 & 2 & 1 \\ 2 & -2 & 2 \end{vmatrix}$$

Expand about row 3.

$$= 2\begin{vmatrix} -1 & 1 \\ 2 & 1 \end{vmatrix} - (-2)\begin{vmatrix} 1 & 1 \\ 3 & 1 \end{vmatrix} + 2\begin{vmatrix} 1 & -1 \\ 3 & 2 \end{vmatrix}$$
$$= 2(-3) + 2(-2) + 2(5)$$
$$= -6 - 4 + 10 = 0$$

Cramer's rule does not apply since $D = 0$.

4.5 Section Exercises

Check each solution in the equations of the given system.

3. $3x + 5y = -5$
$-2x + 3y = 16$

$$D = \begin{vmatrix} 3 & 5 \\ -2 & 3 \end{vmatrix} = 3(3) - 5(-2) = 19$$

$$D_x = \begin{vmatrix} -5 & 5 \\ 16 & 3 \end{vmatrix} = -5(3) - 5(16) = -95$$

$$D_y = \begin{vmatrix} 3 & -5 \\ -2 & 16 \end{vmatrix} = 3(16) - (-5)(-2) = 38$$

$$x = \frac{D_x}{D} = \frac{-95}{19} = -5$$
$$y = \frac{D_y}{D} = \frac{38}{19} = 2$$

Solution set: $\{(-5, 2)\}$

7. $8x + 3y = 1$
$6x - 5y = 2$

$$D = \begin{vmatrix} 8 & 3 \\ 6 & -5 \end{vmatrix} = 8(-5) - 3(6) = -58$$

$$D_x = \begin{vmatrix} 1 & 3 \\ 2 & -5 \end{vmatrix} = 1(-5) - 3(2) = -11$$

$$D_y = \begin{vmatrix} 8 & 1 \\ 6 & 2 \end{vmatrix} = 8(2) - 1(6) = 10$$

$$x = \frac{D_x}{D} = \frac{-11}{-58} = \frac{11}{58}$$
$$y = \frac{D_y}{D} = \frac{10}{-58} = -\frac{5}{29}$$

Solution set: $\{(\frac{11}{58}, -\frac{5}{29})\}$

11. $2x + 2y + z = 10$
$4x - y + z = 20$
$-x + y - 2z = -5$

Expand about row 1 to find D, D_x, D_y, and D_z.

$$D = \begin{vmatrix} 2 & 2 & 1 \\ 4 & -1 & 1 \\ -1 & 1 & -2 \end{vmatrix}$$

$$= 2\begin{vmatrix} -1 & 1 \\ 1 & -2 \end{vmatrix} - 2\begin{vmatrix} 4 & 1 \\ -1 & -2 \end{vmatrix} + 1\begin{vmatrix} 4 & -1 \\ -1 & 1 \end{vmatrix}$$
$$= 2(1) - 2(-7) + 1(3)$$
$$= 2 + 14 + 3 = 19$$

$$D_x = \begin{vmatrix} 10 & 2 & 1 \\ 20 & -1 & 1 \\ -5 & 1 & -2 \end{vmatrix}$$

$$= 10\begin{vmatrix} -1 & 1 \\ 1 & -2 \end{vmatrix} - 2\begin{vmatrix} 20 & 1 \\ -5 & -2 \end{vmatrix} + 1\begin{vmatrix} 20 & -1 \\ -5 & 1 \end{vmatrix}$$
$$= 10(1) - 2(-35) + 1(15)$$
$$= 10 + 70 + 15 = 95$$

$$D_y = \begin{vmatrix} 2 & 10 & 1 \\ 4 & 20 & 1 \\ -1 & -5 & -2 \end{vmatrix}$$

$$= 2\begin{vmatrix} 20 & 1 \\ -5 & -2 \end{vmatrix} - 10\begin{vmatrix} 4 & 1 \\ -1 & -2 \end{vmatrix} + 1\begin{vmatrix} 4 & 20 \\ -1 & -5 \end{vmatrix}$$

$$= 2(-35) - 10(-7) + 1(0)$$
$$= -70 + 70 = 0$$

$$D_z = \begin{vmatrix} 2 & 2 & 10 \\ 4 & -1 & 20 \\ -1 & 1 & -5 \end{vmatrix}$$

$$= 2\begin{vmatrix} -1 & 20 \\ 1 & -5 \end{vmatrix} - 2\begin{vmatrix} 4 & 20 \\ -1 & -5 \end{vmatrix} + 10\begin{vmatrix} 4 & -1 \\ -1 & 1 \end{vmatrix}$$

$$= 2(-15) - 2(0) + 10(3)$$
$$= -30 + 30 = 0$$

$$x = \frac{D_x}{D} = \frac{95}{19} = 5$$

$$y = \frac{D_y}{D} = \frac{0}{19} = 0$$

$$z = \frac{D_z}{D} = \frac{0}{19} = 0$$

Solution set: $\{(5, 0, 0)\}$

15.
$$\begin{aligned} 3x \qquad + 5z &= 0 \\ 2x + 3y \qquad &= 1 \\ -y + 2z &= -11 \end{aligned}$$

Expand about row 1 to find D, D_x, D_y, and D_z.

$$D = \begin{vmatrix} 3 & 0 & 5 \\ 2 & 3 & 0 \\ 0 & -1 & 2 \end{vmatrix}$$

$$= 3\begin{vmatrix} 3 & 0 \\ -1 & 2 \end{vmatrix} - 0 + 5\begin{vmatrix} 2 & 3 \\ 0 & -1 \end{vmatrix}$$

$$= 3(6) + 5(-2)$$
$$= 18 - 10 = 8$$

$$D_x = \begin{vmatrix} 0 & 0 & 5 \\ 1 & 3 & 0 \\ -11 & -1 & 2 \end{vmatrix}$$

$$= 0 - 0 + 5\begin{vmatrix} 1 & 3 \\ -11 & -1 \end{vmatrix}$$

$$= 5(32) = 160$$

$$D_y = \begin{vmatrix} 3 & 0 & 5 \\ 2 & 1 & 0 \\ 0 & -11 & 2 \end{vmatrix}$$

$$= 3\begin{vmatrix} 1 & 0 \\ -11 & 2 \end{vmatrix} - 0 + 5\begin{vmatrix} 2 & 1 \\ 0 & -11 \end{vmatrix}$$

$$= 3(2) + 5(-22)$$
$$= 6 - 110 = -104$$

$$D_z = \begin{vmatrix} 3 & 0 & 0 \\ 2 & 3 & 1 \\ 0 & -1 & -11 \end{vmatrix}$$

$$= 3\begin{vmatrix} 3 & 1 \\ -1 & -11 \end{vmatrix} - 0 + 0$$

$$= 3(-32) = -96$$

$$x = \frac{D_x}{D} = \frac{160}{8} = 20$$

$$y = \frac{D_y}{D} = \frac{-104}{8} = -13$$

$$z = \frac{D_z}{D} = \frac{-96}{8} = -12$$

Solution set: $\{(20, -13, -12)\}$

19.
$$\begin{aligned} 3x + 2y \qquad - w &= 0 \quad (1) \\ 2x \qquad + z + 2w &= 5 \quad (2) \\ x + 2y - z \qquad &= -2 \quad (3) \\ 2x - y + z + w &= 2 \quad (4) \end{aligned}$$

Use equation (1) to eliminate w from the system. Then use 3×3 determinants and Cramer's rule.

$$3x + 2y - w = 0 \qquad (1)$$
$$w = 3x + 2y$$

Substitute $3x + 2y$ for w in equation (2).

$$2x + z + 2w = 5 \qquad (2)$$
$$2x + z + 2(3x + 2y) = 5$$
$$2x + z + 6x + 4y = 5$$
$$8x + 4y + z = 5 \qquad (5)$$

Substitute $3x + 2y$ for w in equation (4).

$$2x - y + z + w = 2 \qquad (4)$$
$$2x - y + z + (3x + 2y) = 2$$
$$5x + y + z = 2 \qquad (6)$$

Solve the system.

$$\begin{aligned} x + 2y - z &= -2 \quad (3) \\ 8x + 4y + z &= 5 \quad (5) \\ 5x + y + z &= 2 \quad (6) \end{aligned}$$

Expand about row 1 in each case.

$$D = \begin{vmatrix} 1 & 2 & -1 \\ 8 & 4 & 1 \\ 5 & 1 & 1 \end{vmatrix}$$

$$= 1\begin{vmatrix} 4 & 1 \\ 1 & 1 \end{vmatrix} - 2\begin{vmatrix} 8 & 1 \\ 5 & 1 \end{vmatrix} + (-1)\begin{vmatrix} 8 & 4 \\ 5 & 1 \end{vmatrix}$$

$$= 1(3) - 2(3) - 1(-12)$$
$$= 3 - 6 + 12 = 9$$

$$D_x = \begin{vmatrix} -2 & 2 & -1 \\ 5 & 4 & 1 \\ 2 & 1 & 1 \end{vmatrix}$$

$$= -2\begin{vmatrix} 4 & 1 \\ 1 & 1 \end{vmatrix} - 2\begin{vmatrix} 5 & 1 \\ 2 & 1 \end{vmatrix} + (-1)\begin{vmatrix} 5 & 4 \\ 2 & 1 \end{vmatrix}$$

$$= -2(3) - 2(3) - 1(-3)$$
$$= -6 - 6 + 3 = -9$$

$$D_y = \begin{vmatrix} 1 & -2 & -1 \\ 8 & 5 & 1 \\ 5 & 2 & 1 \end{vmatrix}$$

$$= 1\begin{vmatrix} 5 & 1 \\ 2 & 1 \end{vmatrix} - (-2)\begin{vmatrix} 8 & 1 \\ 5 & 1 \end{vmatrix} + (-1)\begin{vmatrix} 8 & 5 \\ 5 & 2 \end{vmatrix}$$

$$= 1(3) + 2(3) - 1(-9)$$
$$= 3 + 6 + 9 = 18$$

$$x = \frac{D_x}{D} = \frac{-9}{9} = -1$$
$$y = \frac{D_y}{D} = \frac{18}{9} = 2$$

To find z, substitute -1 for x and 2 for y in equation (3).

$$\begin{aligned} x + 2y - z &= -2 \quad (3) \\ -1 + 2(2) - z &= -2 \\ -1 + 4 - z &= -2 \\ 3 - z &= -2 \\ -z &= -5 \\ z &= 5 \end{aligned}$$

To find w, substitute -1 for x and 2 for y in equation (1).

$$\begin{aligned} 3x + 2y - w &= 0 \quad (1) \\ 3(-1) + 2(2) - w &= 0 \\ -3 + 4 - w &= 0 \\ 1 - w &= 0 \\ -w &= -1 \\ w &= 1 \end{aligned}$$

Solution set: $\{(-1, 2, 5, 1)\}$

22. See the answer graph in the back of the textbook.

23. Use $(0,0)$ as $(x_1, y_1), (-3,-4)$ as (x_2, y_2), and $(2,-2)$ as (x_3, y_3). Then

$$\frac{1}{2}\begin{vmatrix} x_1 & y_1 & 1 \\ x_2 & y_2 & 1 \\ x_3 & y_3 & 1 \end{vmatrix} = \frac{1}{2}\begin{vmatrix} 0 & 0 & 1 \\ -3 & -4 & 1 \\ 2 & -2 & 1 \end{vmatrix}.$$

24. $\frac{1}{2}\begin{vmatrix} 0 & 0 & 1 \\ -3 & -4 & 1 \\ 2 & -2 & 1 \end{vmatrix}$

Expand about row 1.

$$= \frac{1}{2}\left(0 - 0 + 1\begin{vmatrix} -3 & -4 \\ 2 & -2 \end{vmatrix}\right)$$
$$= \frac{1}{2}[1(6+8)]$$
$$= \frac{1}{2}(14)$$
$$= 7$$

The area is 7.

Chapter 4 Review Exercises

When solving systems, check each solution in the equations of the original system. For Exercises 1 and 2, see the answer graphs in the back of the textbook.

1. $\begin{aligned} x + 3y &= 8 \quad (1) \\ 2x - \ \ y &= 2 \quad (2) \end{aligned}$

To eliminate y, multiply equation (2) by 3 and add the result to equation (1).

$$\begin{aligned} 6x - 3y &= \ \ 6 \\ x + 3y &= \ \ 8 \quad (1) \\ \hline 7x \quad\quad &= 14 \\ x &= 2 \end{aligned}$$

Substitute 2 for x in equation (1) to find y.

$$\begin{aligned} x + 3y &= 8 \quad (1) \\ 2 + 3y &= 8 \\ 3y &= 6 \\ y &= 2 \end{aligned}$$

Solution set: $\{(2, 2)\}$

2. $x - 4y = -4$ (1)
$3x + y = 1$ (2)

To eliminate y, multiply equation (2) by 4 and add the result to equation (1).

$$\begin{aligned} 12x + 4y &= 4 \\ x - 4y &= -4 \quad (1) \\ \hline 13x \quad &= 0 \\ x &= 0 \end{aligned}$$

Substitute 0 for x in equation (1) to find y.

$$\begin{aligned} x - 4y &= -4 \quad (1) \\ 0 - 4y &= -4 \\ y &= 1 \end{aligned}$$

Solution set: $\{(0,1)\}$

3. $6x + 5y = 4$ (1)
$-4x + 2y = 8$ (2)

To eliminate x, multiply equation (1) by 2 and equation (2) by 3. Then add the results.

$$\begin{aligned} 12x + 10y &= 8 \\ -12x + 6y &= 24 \\ \hline 16y &= 32 \\ y &= 2 \end{aligned}$$

Since $y = 2$,

$$\begin{aligned} -4x + 2y &= 8 \quad (2) \\ -4x + 2(2) &= 8 \\ -4x + 4 &= 8 \\ -4x &= 4 \\ x &= -1. \end{aligned}$$

Solution set: $\{(-1,2)\}$

4. $\frac{x}{6} + \frac{y}{6} = -\frac{1}{2}$ (1)
$x - y = -9$ (2)

Multiply equation (1) by 6 to clear the fractions. Add the result to equation (2) to eliminate y.

$$\begin{aligned} x + y &= -3 \\ x - y &= -9 \quad (2) \\ \hline 2x \quad &= -12 \\ x &= -6 \end{aligned}$$

Since $x = -6$,

$$\begin{aligned} x - y &= -9 \quad (2) \\ -6 - y &= -9 \\ -y &= -3 \\ y &= 3. \end{aligned}$$

Solution set: $\{(-6,3)\}$

5. $4x + 5y = 9$ (1)
$3x + 7y = -1$ (2)

To eliminate x, multiply equation (1) by 3 and equation (2) by -4. Then add the results.

$$\begin{aligned} 12x + 15y &= 27 \\ -12x - 28y &= 4 \\ \hline -13y &= 31 \\ y &= -\frac{31}{13} \end{aligned}$$

Since $y = -\frac{31}{13}$,

$$\begin{aligned} 3x + 7y &= -1 \quad (2) \\ 3x + 7\left(-\frac{31}{13}\right) &= -1 \end{aligned}$$

Multiply by 13 to clear the fraction.

$$\begin{aligned} 39x + 7(-31) &= -13 \\ 39x - 217 &= -13 \\ 39x &= 204 \\ x &= \frac{204}{39} \text{ or } \frac{68}{13} \end{aligned}$$

Solution set: $\{(\frac{68}{13}, -\frac{31}{13})\}$

6. $9x - y = -4$ (1)
$y = x + 4$
or $-x + y = 4$ (2)

To eliminate y, add equations (1) and (2).

$$\begin{aligned} 9x - y &= -4 \quad (1) \\ -x + y &= 4 \quad (2) \\ \hline 8x \quad &= 0 \\ x &= 0 \end{aligned}$$

Since $x = 0$,

$$\begin{aligned} -x + y &= 4 \quad (2) \\ 0 + y &= 4 \\ y &= 4. \end{aligned}$$

Solution set: $\{(0,4)\}$

7. $-3x + y = 6$ (1)
$y = 6 + 3x$ (2)

Since equation (2) can be rewritten as

$$-3x + y = 6,$$

the two equations are the same. The equations are dependent.

Solution set: $\{(x,y) \mid -3x + y = 6\}$

8. $5x - 4y = 2 \quad (1)$
$-10x + 8y = 7 \quad (2)$

Multiply equation (1) by 2 and add the result to equation (2).

$$\begin{aligned} 10x - 8y &= 4 \\ -10x + 8y &= 7 \quad (2) \\ \hline 0 &= 11 \quad \textit{False} \end{aligned}$$

Since a false statement results, the system is inconsistent.

Solution set: $\emptyset$

9. $3x + y = -4 \quad (1)$
$x = \frac{2}{3}y \quad (2)$

Substitute $\frac{2}{3}y$ for x in equation (1) and solve for y.

$$\begin{aligned} 3x + y &= -4 \quad (1) \\ 3\left(\frac{2}{3}y\right) + y &= 4 \\ 2y + y &= -4 \\ 3y &= -4 \\ y &= -\frac{4}{3} \end{aligned}$$

Since $x = \frac{2}{3}y$ and $y = -\frac{4}{3}$,

$$x = \frac{2}{3}\left(-\frac{4}{3}\right) = -\frac{8}{9}.$$

Solution set: $\{(-\frac{8}{9}, -\frac{4}{3})\}$

10. $-5x + 2y = -2 \quad (1)$
$x + 6y = 26 \quad (2)$

Solve equation (2) for x.

$$\begin{aligned} x + 6y &= 26 \quad (2) \\ x &= 26 - 6y \end{aligned}$$

Substitute $26 - 6y$ for x in equation (1).

$$\begin{aligned} -5x + 2y &= -2 \quad (1) \\ -5(26 - 6y) + 2y &= -2 \\ -130 + 30y + 2y &= -2 \\ -130 + 32y &= -2 \\ 32y &= 128 \\ y &= 4 \end{aligned}$$

Since $x = 26 - 6y$ and $y = 4$,

$$x = 26 - 6(4) = 26 - 24 = 2.$$

Solution set: $\{(2, 4)\}$

For Exercises 11-13, answers may vary. See the answer graphs in the back of the textbook.

11. If the system has a single solution, the graphs of the two linear equations of the system will intersect in one point.

12. If the system has no solution, the graphs of the two linear equations do not intersect but are two parallel lines.

13. If the system has infinitely many solutions, the graphs of the two linear equations are the same line.

14. The equations are given in slope-intercept form, $y = mx + b$. Since the equations have the same slope, 3, but different y-intercepts, $(0, 2)$ and $(0, -4)$, the lines when graphed are parallel. There is no ordered pair that would satisfy both equations so the solution set is $\emptyset$.

15. $2x + 3y - z = -16 \quad (1)$
$x + 2y + 2z = -3 \quad (2)$
$-3x + y + z = -5 \quad (3)$

To eliminate z, add equations (1) and (3).

$$\begin{aligned} 2x + 3y - z &= -16 \quad (1) \\ -3x + y + z &= -5 \quad (3) \\ \hline -x + 4y &= -21 \quad (4) \end{aligned}$$

To eliminate z again, multiply equation (1) by 2 and add the result to equation (2).

$$\begin{aligned} 4x + 6y - 2z &= -32 \\ x + 2y + 2z &= -3 \quad (2) \\ \hline 5x + 8y &= -35 \quad (5) \end{aligned}$$

Use equations (4) and (5) to eliminate x. Multiply equation (4) by 5 and add the result to equation (5).

$$\begin{aligned} -5x + 20y &= -105 \\ 5x + 8y &= -35 \quad (5) \\ \hline 28y &= -140 \\ y &= -5 \end{aligned}$$

Substitute -5 for y in equation (4) to find x.

$$\begin{aligned} -x + 4y &= -21 \quad (4) \\ -x + 4(-5) &= -21 \\ -x - 20 &= -21 \\ -x &= -1 \\ x &= 1 \end{aligned}$$

Substitute 1 for x and -5 for y in equation (2) to find z.

$$\begin{aligned} x+2y+2z&=-3 \quad (2)\\ 1+2(-5)+2z&=-3\\ 1-10+2z&=-3\\ -9+2z&=-3\\ 2z&=6\\ z&=3 \end{aligned}$$

Solution set: $\{(1,-5,3)\}$

16. $$\begin{aligned} 3x-y-z&=-8 \quad (1)\\ 4x+2y+3z&=15 \quad (2)\\ -6x+2y+2z&=10 \quad (3) \end{aligned}$$

To eliminate y, multiply equation (1) by 2 and add the result to equation (3).

$$\begin{aligned} 6x-2y-2z&=-16\\ -6x+2y+2z&=10 \quad (3)\\ \hline 0&=-6 \quad \textit{False} \end{aligned}$$

Since a false statement results, equations (1) and (3) have no common solution. The system is inconsistent.

Solution set: $\emptyset$

17. $$\begin{aligned} 4x-y&=2 \quad (1)\\ 3y+z&=9 \quad (2)\\ x+2z&=7 \quad (3) \end{aligned}$$

To eliminate y, multiply equation (1) by 3 and add the result to equation (2).

$$\begin{aligned} 12x-3y&=6\\ 3y+z&=9 \quad (2)\\ \hline 12x+z&=15 \quad (4) \end{aligned}$$

To eliminate z, multiply equation (4) by -2 and add the result to equation (3).

$$\begin{aligned} -24x-2z&=-30\\ x+2z&=7 \quad (3)\\ \hline -23x&=-23\\ x&=1 \end{aligned}$$

Substitute 1 for x in equation (3) to find z.

$$\begin{aligned} x+2z&=7 \quad (3)\\ 1+2z&=7\\ 2z&=6\\ z&=3 \end{aligned}$$

Substitute 1 for x in equation (1) to find y.

$$\begin{aligned} 4x-y&=2 \quad (1)\\ 4(1)-y&=2\\ 4-y&=2\\ -y&=-2\\ y&=2 \end{aligned}$$

Solution set: $\{(1,2,3)\}$

18. $$\begin{aligned} x+5y-3z&=0 \quad (1)\\ 2x+6y+z&=0 \quad (2)\\ 3x-y+4z&=0 \quad (3) \end{aligned}$$

To eliminate x, multiply equation (1) by -2 and add the result to equation (2).

$$\begin{aligned} -2x-10y+6z&=0\\ 2x+6y+z&=0 \quad (2)\\ \hline -4y+7z&=0 \quad (4) \end{aligned}$$

To eliminate x again, multiply equation (1) by -3 and add the result to equation (3).

$$\begin{aligned} -3x-15y+9z&=0\\ 3x-y+4z&=0 \quad (3)\\ \hline -16y+13z&=0 \quad (5) \end{aligned}$$

Use equations (4) and (5) to eliminate y. Multiply equation (4) by -4 and add the result to equation (5).

$$\begin{aligned} 16y-28z&=0\\ -16y+13z&=0 \quad (5)\\ \hline -15z&=0\\ z&=0 \end{aligned}$$

Substitute 0 for z in equation (5) to find y.

$$\begin{aligned} -16y+13z&=0 \quad (5)\\ -16y+13(0)&=0\\ -16y+0&=0\\ -16y&=0\\ y&=0 \end{aligned}$$

Substitute 0 for y and 0 for z in equation (1) to find x.

$$\begin{aligned} x+5y-3z&=0 \quad (1)\\ x+5(0)-3(0)&=0\\ x+0-0&=0\\ x&=0 \end{aligned}$$

Solution set: $\{(0,0,0)\}$

19. Let $x =$ the length of the table top and $y =$ the width of the table top.

The length, x, is 2 ft longer than the width, y, so

$$x = y + 2. \quad (1)$$

The perimeter is $P = 2L + 2W$. Here $P = 20$, $L = x$, $W = y$, so

$$20 = 2x + 2y. \quad (2)$$

Solve the system.

$$x = y + 2 \quad (1)$$
$$20 = 2x + 2y \quad (2)$$

Since equation (1) is given in terms of x, substitute $y + 2$ for x in equation (2) and solve for y.

$$\begin{aligned} 20 &= 2x + 2y \quad (2) \\ 20 &= 2(y + 2) + 2y \\ 20 &= 2y + 4 + 2y \\ 20 &= 4y + 4 \\ 16 &= 4y \\ 4 &= y \end{aligned}$$

Since $x = y + 2$ and $y = 4$,

$$x = 4 + 2 = 6.$$

The length of the table top is 6 ft, and the width is 4 ft.

20. Let $x =$ the number of days at the weekday rate and $y =$ the number of days at the weekend rate.

Since the business trip lasted 8 days,

$$x + y = 8. \quad (1)$$

The total bill for x days at \$32 and y days at \$19 was \$217, so

$$32x + 19y = 217. \quad (2)$$

To solve the system, solve equation (1) for x.

$$x + y = 8 \quad (1)$$
$$x = 8 - y$$

Substitute $8 - y$ for x in equation (2) and solve for y.

$$\begin{aligned} 32x + 19y &= 217 \quad (2) \\ 32(8 - y) + 19y &= 217 \\ 256 - 32y + 19y &= 217 \\ 256 - 13y &= 217 \\ -13y &= -39 \\ y &= 3 \end{aligned}$$

Since $x = 8 - y$ and $y = 3$,

$$x = 8 - 3 = 5.$$

He rented the car for 5 weekdays and 3 weekend days.

21. Let $x =$ the speed of the plane and $y =$ the speed of the wind.

Make a table.

	r	t	d
With wind	$x + y$	1.75	$1.75(x + y)$
Against wind	$x - y$	2	$2(x - y)$

The distance each way is 560 miles. From the table,

$$1.75(x + y) = 560.$$

Divide by 1.75.

$$x + y = 320 \quad (1)$$

From the table,

$$2(x - y) = 560.$$
$$x - y = 280 \quad (2)$$

Solve the system by adding equations (1) and (2) to eliminate y.

$$\begin{aligned} x + y &= 320 \quad (1) \\ x - y &= 280 \quad (2) \\ \hline 2x \quad &= 600 \\ x &= 300 \end{aligned}$$

Substitute 300 for x in equation (1) to find y.

$$\begin{aligned} x + y &= 320 \quad (1) \\ 300 + y &= 320 \\ y &= 20 \end{aligned}$$

The speed of the plane was 300 mph, and the speed of the wind was 20 mph.

22. Let $x =$ the amount of \$2-a-pound candy and $y =$ the amount of \$1-a-pound candy.

Make a table.

Ingredient	Amount	Value of the ingredients
\$2-a-pound candy	x	$2x$
\$1-a-pound candy	y	$1y = y$
Mixed	100	$1.30(100) = 130$

Solve the system.

$$x + y = 100 \quad (1)$$
$$2x + y = 130 \quad (2)$$

Solve equation (1) for y.

$$x + y = 100 \quad (1)$$
$$y = 100 - x$$

Substitute $100 - x$ for y in equation (2).

$$2x + y = 130 \quad (2)$$
$$2x + (100 - x) = 130$$
$$x + 100 = 130$$
$$x = 30$$

Since $y = 100 - x$ and $x = 30$,

$$y = 100 - 30 = 70.$$

She should use 30 lb of \$2-a-pound candy and 70 lb of \$1-a-pound candy.

23. Let $x =$ the measure of the largest angle,
$y =$ the measure of the middle-sized angle,
and $z =$ the measure of the smallest angle.

Since the sum of the measures of the angles of a triangle is 180°,

$$x + y + z = 180. \quad (1)$$

Since one angle measures 10° less than the sum of the other two,

$$x = y + z - 10$$
$$\text{or} \quad x - y - z = -10. \quad (2)$$

Since the measure of the middle-sized angle is the average of the other two,

$$y = \frac{x + z}{2}$$
$$2y = x + z$$
$$-x + 2y - z = 0. \quad (3)$$

Solve the system.

$$x + y + z = 180 \quad (1)$$
$$x - y - z = -10 \quad (2)$$
$$-x + 2y - z = 0 \quad (3)$$

Add equations (1) and (2) to find x.

$$\begin{array}{rl} x + y + z = 180 & (1) \\ x - y - z = -10 & (2) \\ \hline 2x \quad\quad = 170 & \\ x = 85 & \end{array}$$

Add equations (1) and (3) to find y.

$$\begin{array}{rl} x + y + z = 180 & (1) \\ -x + 2y - z = 0 & (3) \\ \hline 3y \quad\quad = 180 & \\ y = 60 & \end{array}$$

Substitute 85 for x and 60 for y in equation (1) to find z.

$$x + y + z = 180 \quad (1)$$
$$85 + 60 + z = 180$$
$$145 + z = 180$$
$$z = 35$$

The three angles are $85°, 60°$, and $35°$.

24. Let $x =$ the value of sales at 10%,
$y =$ the value of sales at 6%,
and $z =$ the value of sales at 5%.

Since his total sales were \$280,000,

$$x + y + z = 280,000. \quad (1)$$

Since his commissions on the sales totaled \$17,000,

$$.10x + .06y + .05z = 17,000.$$

Multiply by 100 to clear the decimals, so

$$10x + 6y + 5z = 1,700,000. \quad (2)$$

Since the 5% sale amounted to the sum of the other two sales,

$$z = x + y. \quad (3)$$

Solve the system.

$$x + y + z = 280,000 \quad (1)$$
$$10x + 6y + 5z = 1,700,000 \quad (2)$$
$$z = x + y \quad (3)$$

Since equation (3) is given in terms of x and y, substitute $x + y$ for z in equations (1) and (2).

$$x + y + z = 280,000 \quad (1)$$

$$x+y+(x+y)=280,000$$
$$2x+2y=280,000 \qquad (4)$$

$$10x+6y+5z=1,700,000 \qquad (2)$$
$$10x+6y+5(x+y)=1,700,000$$
$$10x+6y+5x+5y=1,700,000$$
$$15x+11y=1,700,000 \qquad (5)$$

To eliminate x, multiply equation (4) by -15 and equation (5) by 2. Then add the results.

$$\begin{aligned} -30x-30y &= -4,200,000 \\ 30x+22y &= 3,400,000 \\ \hline -8y &= -800,000 \\ y &= 100,000 \end{aligned}$$

Substitute 100,000 for y in equation (4) to find x.

$$\begin{aligned} 2x+2y &= 280,000 \qquad (4) \\ 2x+2(100,000) &= 280,000 \\ 2x+200,000 &= 280,000 \\ 2x &= 80,000 \\ x &= 40,000 \end{aligned}$$

Substitute 40,000 for x and 100,000 for y in equation (3) to find z.

$$\begin{aligned} z &= x+y \qquad (3) \\ z &= 40,000+100,000 \\ z &= 140,000 \end{aligned}$$

He sold \$40,000 at 10%, \$100,000 at 6%, and \$140,000 at 5%.

25. Let $x=$ the amount of jelly beans,
$y=$ the amount of chocolate eggs,
and $z=$ the amount of marshmallow chicks.

The manager plans to make 15 lb of the mixture, so

$$x+y+z=15. \quad (1)$$

She uses twice as many pounds of jelly beans as eggs and chicks, so

$$x=2(y+z),$$
$$\text{or} \quad x-2y-2z=0, \qquad (2)$$

and five times as many pounds of jelly beans as eggs, so

$$x=5y. \quad (3)$$

Solve the system of equations (1), (2), and (3). Multiply equation (1) by 2 and add the result to equation (2).

$$\begin{aligned} 2x+2y+2z &= 30 \\ x-2y-2z &= 0 \quad (2) \\ \hline 3x \qquad\quad &= 30 \\ x &= 10 \end{aligned}$$

Since $x=10$,

$$\begin{aligned} x &= 5y \quad (3) \\ 10 &= 5y \\ 2 &= y. \end{aligned}$$

Since $x=10$ and $y=2$,

$$\begin{aligned} x+y+z &= 15 \quad (1) \\ 10+2+z &= 15 \\ 12+z &= 15 \\ z &= 3. \end{aligned}$$

She should use 10 lb of jelly beans, 2 lb of chocolate eggs, and 3 lb of marshmallow chicks.

26. Let $x=$ the number of \$20 fish,
$y=$ the number of \$40 fish,
and $z=$ the number of \$65 fish.

Since the number of \$40 fish is one less than twice the number of \$20 fish,

$$y=2x-1. \qquad (1)$$

There are 29 fish in all, so

$$x+y+z=29. \qquad (2)$$

The total value of the fish is \$1150, so

$$20x+40y+65z=1150. \quad (3)$$

Eliminate z using equations (2) and (3). Multiply equation (2) by -65 and add the result to equation (3).

$$\begin{aligned} -65x-65y-65z &= -1885 \\ 20x+40y+65z &= 1150 \quad (3) \\ \hline -45x-25y \qquad &= -735 \\ 45x+25y \qquad &= 735 \quad (4) \end{aligned}$$

Substitute $2x-1$ for y from equation (1) into equation (4).

$$\begin{aligned} 45x+25y &= 735 \quad (4) \\ 45x+25(2x-1) &= 735 \\ 45x+50x-25 &= 735 \\ 95x &= 760 \\ x &= 8 \end{aligned}$$

From equation (1),

$$y = 2x - 1 = 2(8) - 1 = 15.$$

From equation (2),

$$\begin{aligned} x + y + z &= 29 \quad (2) \\ 8 + 15 + z &= 29 \\ 23 + z &= 29 \\ z &= 6. \end{aligned}$$

There are 8 \$20-fish, 15 \$40-fish, and 6 \$65-fish in the collection.

27. $\begin{vmatrix} 2 & -9 \\ 8 & 4 \end{vmatrix} = 2(4) - (-9)(8)$
$= 8 + 72 = 80$

28. $\begin{vmatrix} 7 & 0 \\ 5 & -3 \end{vmatrix} = 7(-3) - 0(5)$
$= -21$

29. $\begin{vmatrix} 2 & 10 & 4 \\ 0 & 1 & 3 \\ 0 & 6 & -1 \end{vmatrix}$

Expand about column 1.

$$\begin{aligned} &= 2\begin{vmatrix} 1 & 3 \\ 6 & -1 \end{vmatrix} - 0 + 0 \\ &= 2[1(-1) - 3(6)] \\ &= 2(-19) = -38 \end{aligned}$$

30. $\begin{vmatrix} 0 & 0 & 0 \\ 0 & 2 & 5 \\ -1 & 3 & 6 \end{vmatrix}$

All the elements of row 1 are 0. If we expand about row 1, each minor will be multiplied by 0. Therefore, the determinant equals 0.

31. $\begin{vmatrix} 0 & 0 & 2 \\ 2 & 1 & 0 \\ -1 & 0 & 0 \end{vmatrix}$

Expand about row 1.

$$\begin{aligned} &= 0 - 0 + 2\begin{vmatrix} 2 & 1 \\ -1 & 0 \end{vmatrix} \\ &= 2[2(0) - 1(-1)] \\ &= 2(1) = 2 \end{aligned}$$

32. $\begin{vmatrix} 1 & 3 & -2 \\ 2 & 6 & -4 \\ 5 & 0 & 1 \end{vmatrix}$

Expand about row 3.

$$\begin{aligned} &= 5\begin{vmatrix} 3 & -2 \\ 6 & -4 \end{vmatrix} - 0 + 1\begin{vmatrix} 1 & 3 \\ 2 & 6 \end{vmatrix} \\ &= 5[3(-4) - (-2)6] + [1(6) - 3(2)] \\ &= 5(0) + 0 \\ &= 0 \end{aligned}$$

33. If $D = 0$, Cramer's rule does not apply.

34. $3x - 4y = 5$
$2x + y = 8$

$$D = \begin{vmatrix} 3 & -4 \\ 2 & 1 \end{vmatrix} = 3(1) - (-4)(2) = 11$$

$$D_x = \begin{vmatrix} 5 & -4 \\ 8 & 1 \end{vmatrix} = 5(1) - (-4)(8) = 37$$

$$D_y = \begin{vmatrix} 3 & 5 \\ 2 & 8 \end{vmatrix} = 3(8) - 5(2) = 14$$

$$x = \frac{D_x}{D} = \frac{37}{11}$$

$$y = \frac{D_y}{D} = \frac{14}{11}$$

Solution set: $\{(\frac{37}{11}, \frac{14}{11})\}$

35. $-4x + 3y = -12$
$2x + 6y = 15$

$$D = \begin{vmatrix} -4 & 3 \\ 2 & 6 \end{vmatrix} = -4(6) - 3(2) = -30$$

$$D_x = \begin{vmatrix} -12 & 3 \\ 15 & 6 \end{vmatrix} = -12(6) - 3(15) = -117$$

$$D_y = \begin{vmatrix} -4 & -12 \\ 2 & 15 \end{vmatrix} = -4(15) - (-12)(2) = -36$$

$$x = \frac{D_x}{D} = \frac{-117}{-30} = \frac{39}{10}$$

$$y = \frac{D_y}{D} = \frac{-36}{-30} = \frac{6}{5}$$

Solution set: $\{(\frac{39}{10}, \frac{6}{5})\}$

36. $4x + y + z = 11$
$x - y - z = 4$
$y + 2z = 0$

Expand about row 3 to find D, D_x, D_y, and D_z.

$$D = \begin{vmatrix} 4 & 1 & 1 \\ 1 & -1 & -1 \\ 0 & 1 & 2 \end{vmatrix}$$

$$= 0 - 1\begin{vmatrix} 4 & 1 \\ 1 & -1 \end{vmatrix} + 2\begin{vmatrix} 4 & 1 \\ 1 & -1 \end{vmatrix}$$

$$= -1(-5) + 2(-5)$$
$$= 5 - 10 = -5$$

$$D_x = \begin{vmatrix} 11 & 1 & 1 \\ 4 & -1 & -1 \\ 0 & 1 & 2 \end{vmatrix}$$

$$= 0 - 1\begin{vmatrix} 11 & 1 \\ 4 & -1 \end{vmatrix} + 2\begin{vmatrix} 11 & 1 \\ 4 & -1 \end{vmatrix}$$

$$= -1(-15) + 2(-15)$$
$$= 15 - 30 = -15$$

$$D_y = \begin{vmatrix} 4 & 11 & 1 \\ 1 & 4 & -1 \\ 0 & 0 & 2 \end{vmatrix} = 0 - 0 + 2\begin{vmatrix} 4 & 11 \\ 1 & 4 \end{vmatrix}$$
$$= 2(5) = 10$$

$$D_z = \begin{vmatrix} 4 & 1 & 11 \\ 1 & -1 & 4 \\ 0 & 1 & 0 \end{vmatrix} = 0 - 1\begin{vmatrix} 4 & 11 \\ 1 & 4 \end{vmatrix} + 0$$
$$= -1(5) = -5$$

$$x = \frac{D_x}{D} = \frac{-15}{-5} = 3$$

$$y = \frac{D_y}{D} = \frac{10}{-5} = -2$$

$$z = \frac{D_z}{D} = \frac{-5}{-5} = 1$$

Solution set: $\{(3, -2, 1)\}$

37. $-x + 3y - 4z = 2$
$2x + 4y + z = 3$
$3x - z = 9$

Expand about row 3 in each case.

$$D = \begin{vmatrix} -1 & 3 & -4 \\ 2 & 4 & 1 \\ 3 & 0 & -1 \end{vmatrix}$$

$$= 3\begin{vmatrix} 3 & -4 \\ 4 & 1 \end{vmatrix} - 0 + (-1)\begin{vmatrix} -1 & 3 \\ 2 & 4 \end{vmatrix}$$

$$= 3(19) - 1(-10)$$
$$= 57 + 10 = 67$$

$$D_x = \begin{vmatrix} 2 & 3 & -4 \\ 3 & 4 & 1 \\ 9 & 0 & -1 \end{vmatrix}$$

$$= 9\begin{vmatrix} 3 & -4 \\ 4 & 1 \end{vmatrix} - 0 + (-1)\begin{vmatrix} 2 & 3 \\ 3 & 4 \end{vmatrix}$$

$$= 9(19) - 1(-1)$$
$$= 171 + 1 = 172$$

$$D_y = \begin{vmatrix} -1 & 2 & -4 \\ 2 & 3 & 1 \\ 3 & 9 & -1 \end{vmatrix}$$

$$= 3\begin{vmatrix} 2 & -4 \\ 3 & 1 \end{vmatrix} - 9\begin{vmatrix} -1 & -4 \\ 2 & 1 \end{vmatrix} + (-1)\begin{vmatrix} -1 & 2 \\ 2 & 3 \end{vmatrix}$$

$$= 3(14) - 9(7) - 1(-7)$$
$$= 42 - 63 + 7 = -14$$

$$D_z = \begin{vmatrix} -1 & 3 & 2 \\ 2 & 4 & 3 \\ 3 & 0 & 9 \end{vmatrix}$$

$$= 3\begin{vmatrix} 3 & 2 \\ 4 & 3 \end{vmatrix} - 0 + 9\begin{vmatrix} -1 & 3 \\ 2 & 4 \end{vmatrix}$$

$$= 3(1) + 9(-10)$$
$$= 3 - 90 = -87$$

$$x = \frac{D_x}{D} = \frac{172}{67}$$

$$y = \frac{D_y}{D} = \frac{-14}{67} \text{ or } -\frac{14}{67}$$

$$z = \frac{D_z}{D} = \frac{-87}{67} \text{ or } -\frac{87}{67}$$

Solution set: $\{(\frac{172}{67}, -\frac{14}{67}, -\frac{87}{67})\}$

38. $\frac{2}{3}x + \frac{1}{6}y = \frac{19}{2}$ (1)

$\frac{1}{3}x - \frac{2}{9}y = 2$ (2)

Multiply equation (1) by 6 and equation (2) by 9 to clear the fractions.

$$4x + y = 57 \quad (3)$$
$$3x - 2y = 18 \quad (4)$$

To eliminate y, multiply equation (3) by 2 and add the result to equation (4).

$$8x + 2y = 114$$
$$3x - 2y = 18 \quad (4)$$
$$11x = 132$$
$$x = 12$$

Substitute 12 for x in equation (3) to find y.

$$4x + y = 57 \quad (3)$$
$$4(12) + y = 57$$
$$48 + y = 57$$
$$y = 9$$

Solution set: $\{(12, 9)\}$

39.
$$\begin{aligned} 2x + 5y - z &= 12 \quad (1) \\ -x + y - 4z &= -10 \quad (2) \\ -8x - 20y + 4z &= 31 \quad (3) \end{aligned}$$

Multiply equation (1) by 4 and add the result to equation (3).

$$\begin{aligned} 8x + 20y - 4z &= 48 \\ -8x - 20y + 4z &= 31 \quad (3) \\ \hline 0 &= 79 \quad \textit{False} \end{aligned}$$

Since a false statement results, the system is inconsistent.

Solution set: $\emptyset$

40.
$$\begin{aligned} x &= 7y + 10 \quad (1) \\ 2x + 3y &= 3 \quad (2) \end{aligned}$$

Since equation (1) is given in terms of x, substitute $7y + 10$ for x in equation (2) and solve for y.

$$\begin{aligned} 2x + 3y &= 3 \quad (2) \\ 2(7y + 10) + 3y &= 3 \\ 14y + 20 + 3y &= 3 \\ 17y + 20 &= 3 \\ 17y &= -17 \\ y &= -1 \end{aligned}$$

Since $x = 7y + 10$ and $y = -1$,

$$x = 7(-1) + 10 = -7 + 10 = 3.$$

Solution set: $\{(3, -1)\}$

41.
$$\begin{aligned} x + 4y &= 17 \quad (1) \\ -3x + 2y &= -9 \quad (2) \end{aligned}$$

To eliminate x, multiply equation (1) by 3 and add the result to equation (2).

$$\begin{aligned} 3x + 12y &= 51 \\ -3x + 2y &= -9 \quad (2) \\ \hline 14y &= 42 \\ y &= 3 \end{aligned}$$

Substitute 3 for y in equation (1) to find x.

$$\begin{aligned} x + 4y &= 17 \quad (1) \\ x + 4(3) &= 17 \\ x + 12 &= 17 \\ x &= 5 \end{aligned}$$

Solution set: $\{(5, 3)\}$

42.
$$\begin{aligned} -7x + 3y &= 12 \quad (1) \\ 5x + 2y &= 8 \quad (2) \end{aligned}$$

To eliminate y, multiply equation (1) by 2 and equation (2) by -3. Then add the results.

$$\begin{aligned} -14x + 6y &= 24 \\ -15x - 6y &= -24 \\ \hline -19x \quad &= 0 \\ x &= 0 \end{aligned}$$

Substitute 0 for x in equation (1) to find y.

$$\begin{aligned} -7x + 3y &= 12 \quad (1) \\ -7(0) + 3y &= 12 \\ 0 + 3y &= 12 \\ y &= 4 \end{aligned}$$

Solution set: $\{(0, 4)\}$

43.
$$\begin{aligned} 2x - 5y &= 8 \quad (1) \\ 3x + 4y &= 10 \quad (2) \end{aligned}$$

To eliminate y, multiply equation (1) by 4 and equation (2) by 5 and add the results.

$$\begin{aligned} 8x - 20y &= 32 \\ 15x + 20y &= 50 \\ \hline 23x \quad &= 82 \\ x &= \frac{82}{23} \end{aligned}$$

Substitute $\frac{82}{23}$ for x in equation (1) to find y.

$$\begin{aligned} 2x - 5y &= 8 \quad (1) \\ 2\left(\frac{82}{23}\right) - 5y &= 8 \\ \frac{164}{23} - 5y &= 8 \end{aligned}$$

Multiply by 23 to clear the fraction.

$$\begin{aligned} 164 - 115y &= 184 \\ -115y &= 20 \\ y &= -\frac{20}{115} \text{ or } -\frac{4}{23} \end{aligned}$$

Solution set: $\{(\frac{82}{23}, -\frac{4}{23})\}$

44. Let $x =$ the number of liters of 5% solution and $y =$ the number of liters of 10% solution.

Make a table.

Kind	Amount	Pure acid
5%	x	$.05x$
20%	10	$.20(10) = 2$
10%	y	$.10y$

Solve the system.

$$\begin{aligned} x + 10 &= y \quad (1) \\ .05x + \ 2 &= .10y \quad (2) \end{aligned}$$

Multiply equation (2) by 100 to clear the decimals.

$$5x + 200 = 10y \quad (3)$$

Substitute $x + 10$ for y in equation (3) to find x.

$$\begin{aligned} 5x + 200 &= 10y \quad (3) \\ 5x + 200 &= 10(x + 10) \\ 5x + 200 &= 10x + 100 \\ 100 &= 5x \\ 20 &= x \end{aligned}$$

He should use 20 L of 5% solution.

45. Let $x =$ the measure of the smallest angle, $y =$ the measure of the largest angle, and $z =$ the measure of the third angle.

From the problem,

$$\begin{aligned} x + y + z &= 180 \quad (1) \\ y &= 2x \quad (2) \\ z &= y - 10. \quad (3) \end{aligned}$$

From equation (3), substitute $y - 10$ for z in equation (1).

$$\begin{aligned} x + y + z &= 180 \quad (1) \\ x + y + (y - 10) &= 180 \\ x + 2y &= 190 \quad (3) \end{aligned}$$

From equation (2), substitute $2x$ for y in equation (3) to find x.

$$\begin{aligned} x + 2y &= 190 \quad (3) \\ x + 2(2x) &= 190 \\ 5x &= 190 \\ x &= 38 \end{aligned}$$

If $x = 38$, then

$$y = 2x = 2(38) = 76$$

and

$$z = y - 10 = 76 - 10 = 66.$$

The angles measure $38°, 66°$, and $76°$.

Chapter 4 Test

When solving systems, check each solution in the equations of the original system.

1. When each equation of the system

$$\begin{aligned} x + y &= 7 \\ x - y &= 5 \end{aligned}$$

is graphed, the point of intersection is $(6, 1)$. To check, substitute 6 for x and 1 for y in each of the equations. Since $(6, 1)$ makes both equations true, the solution set of the system is $\{(6, 1)\}$. See the answer graph in the back of the textbook.

2. $$\begin{aligned} 3x + y &= 12 \quad (1) \\ 2x - y &= \ 3 \quad (2) \end{aligned}$$

To eliminate y, add equations (1) and (2).

$$\begin{aligned} 3x + y &= 12 \quad (1) \\ 2x - y &= \ 3 \quad (2) \\ \hline 5x \quad &= 15 \\ x &= 3 \end{aligned}$$

Substitute 3 for x in equation (1) to find y.

$$\begin{aligned} 3x + y &= 12 \quad (1) \\ 3(3) + y &= 12 \\ 9 + y &= 132 \\ y &= \end{aligned}$$

Solution set: $\{(3, 3)\}$

3. $$\begin{aligned} -5x + 2y &= -4 \quad (1) \\ 6x + 3y &= -6 \quad (2) \end{aligned}$$

To eliminate x, multiply equation (1) by 6 and equation (2) by 5. Then add the results.

$$\begin{aligned} -30x + 12y &= -24 \\ 30x + 15y &= -30 \\ \hline 27y &= -54 \\ y &= -2 \end{aligned}$$

Substitute -2 for y in equation (1) to find x.

$$\begin{aligned} -5x + 2y &= -4 \quad (1) \\ -5x + 2(-2) &= -4 \\ -5x - 4 &= -4 \\ -5x &= 0 \\ x &= 0 \end{aligned}$$

Solution set: $\{(0, -2)\}$

4. $$\begin{aligned} 3x + 4y &= 8 && (1) \\ 8y &= 7 - 6x \\ \text{or} \quad 6x + 8y &= 7 && (2) \end{aligned}$$

Multiply equation (1) by -2 and add the result to equation (2).

$$\begin{aligned} -6x - 8y &= -16 \\ 6x + 8y &= 7 \quad (2) \\ \hline 0 &= -9 \quad \textit{False} \end{aligned}$$

Since a false statement results, the system is inconsistent.

Solution set: $\emptyset$

5. $$\begin{aligned} 3x + 5y + 3z &= 2 && (1) \\ 6x + 5y + z &= 0 && (2) \\ 3x + 10y - 2z &= 6 && (3) \end{aligned}$$

To eliminate x, multiply equation (1) by -1 and add the result to equation (3).

$$\begin{aligned} -3x - 5y - 3z &= -2 \\ 3x + 10y - 2z &= 6 \quad (3) \\ \hline 5y - 5z &= 4 \end{aligned}$$

To eliminate x again, multiply equation (1) by -2 and add the result to equation (2).

$$\begin{aligned} -6x - 10y - 6z &= -4 \\ 6x + 5y + z &= 0 \quad (4) \\ \hline -5y - 5z &= -4 \quad (5) \end{aligned}$$

To eliminate y, add equations (4) and (5).

$$\begin{aligned} 5y - 5z &= 4 \quad (4) \\ -5y - 5z &= -4 \quad (5) \\ \hline -10z &= 0 \\ z &= 0 \end{aligned}$$

Substitute 0 for z in equation (4) to find y.

$$\begin{aligned} 5y - 5z &= 4 \quad (4) \\ 5y - 5(0) &= 4 \\ 5y - 0 &= 4 \\ 5y &= 4 \\ y &= \frac{4}{5} \end{aligned}$$

Substitute $\frac{4}{5}$ for y and 0 for z in equation (1) to find x.

$$\begin{aligned} 3x + 5y + 3z &= 2 \quad (1) \\ 3x + 5\left(\frac{4}{5}\right) + 3(0) &= 2 \\ 3x + 4 &= 2 \\ 3x &= -2 \\ x &= -\frac{2}{3} \end{aligned}$$

Solution set: $\{(-\frac{2}{3}, \frac{4}{5}, 0)\}$

6. $$\begin{aligned} 2x - 3y &= 24 && (1) \\ y &= -\frac{2}{3}x && (2) \end{aligned}$$

Since equation (2) is given in terms of y, substitute $-\frac{2}{3}x$ for y in equation (1) and solve for x.

$$\begin{aligned} 2x - 3y &= 24 \quad (1) \\ 2x - 3\left(-\frac{2}{3}x\right) &= 24 \\ 2x + 2x &= 24 \\ 4x &= 24 \\ x &= 6 \end{aligned}$$

Since $y = -\frac{2}{3}x$ and $x = 6$,

$$y = -\frac{2}{3}(6) = -4.$$

Solution set: $\{(6, -4)\}$

7. $$\begin{aligned} 12x - 5y &= 8 && (1) \\ 3x &= \frac{5}{4}y + 2 \\ \text{or} \quad x &= \frac{5}{12}y + \frac{2}{3} && (2) \end{aligned}$$

Substitute $\frac{5}{12}y + \frac{2}{3}$ for x in equation (1) and solve for y.

$$\begin{aligned} 12x - 5y &= 8 \quad (1) \\ 12\left(\frac{5}{12}y + \frac{2}{3}\right) - 5y &= 8 \\ 5y + 8 - 5y &= 8 \\ 8 &= 8 \quad \textit{True} \end{aligned}$$

Equations (1) and (2) are dependent.

Solution set: $\{(x, y) | 12x - 5y = 8\}$

8. Let $x =$ the number of votes for one candidate
and $y =$ the number of votes for the other candidate.

Solve the system of equations.

$$x = 45 + y \quad (1)$$
$$x + y = 405 \quad (2)$$

Substitute $45 + y$ for x in equation (2) to find y.

$$x + y = 405 \quad (2)$$
$$(45 + y) + y = 405$$
$$2y = 360$$
$$y = 180$$

If $y = 180$, then

$$x = 45 + y = 45 + 180 = 225.$$

One candidate received 180 votes, and the other received 225 votes.

9. Let $x =$ the number of liters of 20% solution
and $y =$ the number of liters of 50% solution.

Make a table.

Kind of solution	Liters of solution	Liters of pure alcohol
20%	x	$.20x$
50%	y	$.50y$
40%	12	$.40(12) = 4.8$

Since 12 L of the mixture are needed,

$$x + y = 12. \quad (1)$$

Since the amount of pure alcohol in the 20% solution plus the amount of pure alcohol in the 50% solution must equal the amount of alcohol in the mixture,

$$.2x + .5y = 4.8.$$

Multiply by 10 to clear the decimals.

$$2x + 5y = 48 \quad (2)$$

Solve the system.

$$x + y = 12 \quad (1)$$
$$2x + 5y = 48 \quad (2)$$

Multiply equation (1) by -2 and add the result to equation (2).

$$\begin{aligned} -2x - 2y &= -24 \\ 2x + 5y &= 48 \quad (2) \\ \hline 3y &= 24 \\ y &= 8 \end{aligned}$$

Substitute 8 for y in equation (1) to find x.

$$x + y = 12 \quad (1)$$
$$x + 8 = 12$$
$$x = 4$$

4 L of 20% solution and 8 L of 50% solution will be needed.

10. Let $x =$ the price of an AC adaptor
and $y =$ the price of a rechargeable flashlight.

Since 7 AC adaptors and 2 rechargeable flashlights cost \$86,

$$7x + 2y = 86. \quad (1)$$

Since 3 AC adaptors and 4 rechargeable flashlights cost \$84,

$$3x + 4y = 84. \quad (2)$$

Solve the system.

$$7x + 2y = 86 \quad (1)$$
$$3x + 4y = 84 \quad (2)$$

To eliminate y, multiply equation (1) by -2 and add the result to equation (2).

$$\begin{aligned} -14x - 4y &= -172 \\ 3x + 4y &= 84 \quad (2) \\ \hline -11x \quad &= -88 \\ x &= 8 \end{aligned}$$

Substitute 8 for x in equation (1) to find y.

$$7x + 2y = 86 \quad (1)$$
$$7(8) + 2y = 86$$
$$56 + 2y = 86$$
$$2y = 30$$
$$y = 15$$

An AC adapter costs \$8, and a rechargeable flashlight costs \$15.

11. Let $x =$ the amount of Orange Pekoe,
$y =$ the amount of Irish Breakfast,
$z =$ the amount of Earl Grey.

The owner wants 100 oz of tea, so

$$x + y + z = 100. \quad (1)$$

The tea sells for

$$.80x + .85y + .95z = .83(100).$$

Multiply by 100 to clear the decimals.

$$80x + 85y + 95z = 8300 \quad (2)$$

The mixture must use twice as much Orange Pekoe as Irish Breakfast, so

$$x = 2y. \quad (3)$$

To eliminate z, multiply equation (1) by -95 and add the result to equation (2).

$$\begin{array}{rcl} -95x - 95y - 95z &=& -9500 \\ \underline{80x + 85y + 95z} &=& \underline{8300} \quad (2) \\ -15x - 10y \qquad &=& -1200 \end{array}$$

Divide by -5.

$$3x + 2y = 240 \quad (4)$$

Substitute $2y$ for x in equation (4) to find y.

$$\begin{aligned} 3x + 2y &= 240 \quad (4) \\ 3(2y) + 2y &= 240 \\ 8y &= 240 \\ y &= 30 \end{aligned}$$

If $x = 2y$ and $y = 30$, then

$$x = 2(30) = 60.$$

Substitute 60 for x and 30 for y in equation (1) to find z.

$$\begin{aligned} x + y + z &= 100 \quad (1) \\ 60 + 30 + z &= 100 \\ z &= 10 \end{aligned}$$

He should use 60 oz of Orange Pekoe, 30 oz of Irish Breakfast, and 10 oz of Earl Grey.

12. $\begin{vmatrix} 6 & -3 \\ 5 & -2 \end{vmatrix} = 6(-2) - (-3)(5)$
$= -12 + 15 = 3$

13. $\begin{vmatrix} 4 & 1 & 0 \\ -2 & 7 & 3 \\ 0 & 5 & 2 \end{vmatrix}$

Expand about row 1.

$$= 4\begin{vmatrix} 7 & 3 \\ 5 & 2 \end{vmatrix} - 1\begin{vmatrix} -2 & 3 \\ 0 & 2 \end{vmatrix} + 0$$
$$= 4(-1) - (-4)$$
$$= -4 + 4 = 0$$

14. $3x - \ y = -8$
$2x + 6y = \ 3$

$$D = \begin{vmatrix} 3 & -1 \\ 2 & 6 \end{vmatrix} = 18 - (-2) = 20$$

$$D_x = \begin{vmatrix} -8 & -1 \\ 3 & 6 \end{vmatrix} = -48 - (-3) = -45$$

$$D_y = \begin{vmatrix} 3 & -8 \\ 2 & 3 \end{vmatrix} = 9 - (-16) = 25$$

$$x = \frac{D_x}{D} = -\frac{45}{20} = -\frac{9}{4}$$

$$y = \frac{D_y}{D} = \frac{25}{20} = \frac{5}{4}$$

Solution set: $\{(-\frac{9}{4}, \frac{5}{4})\}$

15. $\begin{aligned} x + y + z &= 4 \\ -2x \qquad + z &= 5 \\ 3y + z &= 9 \end{aligned}$

Expand about column 1 in each case.

$$D = \begin{vmatrix} 1 & 1 & 1 \\ -2 & 0 & 1 \\ 0 & 3 & 1 \end{vmatrix}$$

$$= 1\begin{vmatrix} 0 & 1 \\ 3 & 1 \end{vmatrix} - (-2)\begin{vmatrix} 1 & 1 \\ 3 & 1 \end{vmatrix} + 0$$
$$= 1(-3) + 2(-2)$$
$$= -3 - 4 = -7$$

$$D_x = \begin{vmatrix} 4 & 1 & 1 \\ 5 & 0 & 1 \\ 9 & 3 & 1 \end{vmatrix}$$

$$= 4\begin{vmatrix} 0 & 1 \\ 3 & 1 \end{vmatrix} - 5\begin{vmatrix} 1 & 1 \\ 3 & 1 \end{vmatrix} + 9\begin{vmatrix} 1 & 1 \\ 0 & 1 \end{vmatrix}$$
$$= 4(-3) - 5(-2) + 9(1)$$
$$= -12 + 10 + 9 = 7$$

$$D_y = \begin{vmatrix} 1 & 4 & 1 \\ -2 & 5 & 1 \\ 0 & 9 & 1 \end{vmatrix}$$

$$= 1\begin{vmatrix} 5 & 1 \\ 9 & 1 \end{vmatrix} - (-2)\begin{vmatrix} 4 & 1 \\ 9 & 1 \end{vmatrix} + 0$$

$$= 1(-4) + 2(-5)$$
$$= -4 - 10 = -14$$

$$D_z = \begin{vmatrix} 1 & 1 & 4 \\ -2 & 0 & 5 \\ 0 & 3 & 9 \end{vmatrix}$$

$$= 1\begin{vmatrix} 0 & 5 \\ 3 & 9 \end{vmatrix} - (-2)\begin{vmatrix} 1 & 4 \\ 3 & 9 \end{vmatrix} + 0$$

$$= 1(-15) + 2(-3)$$
$$= -15 - 6 = -21$$

$$x = \frac{D_x}{D} = \frac{7}{-7} = -1$$

$$y = \frac{D_y}{D} = \frac{-14}{-7} = 2$$

$$z = \frac{D_z}{D} = \frac{-21}{-7} = 3$$

Solution set: $\{(-1, 2, 3)\}$

Cumulative Review Exercises (Chapters 1-4)

1. $(-3)^4 = (-3)(-3)(-3)(-3) = 81$

2. $-3^4 = -(3)(3)(3)(3) = -81$

3. $-(-3)^4 = -(-3)(-3)(-3)(-3) = -81$

4. $\sqrt{.49} = .7$, since $(.7)^2 = .49$.

5. $-\sqrt{.49} = -.7$, since $(.7)^2 = .49$ and the negative sign in front of the radical must be applied.

6. $\sqrt{-.49}$ is not a real number because of the negative sign under the radical. No real number squared is negative.

7. $\sqrt[3]{64} = 4$, since $4^3 = 64$.

8. $\sqrt[3]{-64} = -4$, since $(-4)^3 = -64$.

For Exercises 9-11, let $x = -4, y = 3$, and $z = 6$.

9. $$|2x| + 3y - z^3 = |2(-4)| + 3(3) - 6^3$$
$$= |-8| + 9 - 216$$
$$= 8 + 9 - 216$$
$$= -199$$

10. $$-5(x^3 - y^3) = -5[(-4)^3 - 3^3]$$
$$= -5(-64 - 27)$$
$$= -5(-91)$$
$$= 455$$

11. $$\frac{2x^2 - x + z}{3y - z} = \frac{2(-4)^2 - (-4) + 6}{3(3) - 6}$$
$$= \frac{2(16) + 4 + 6}{9 - 6}$$
$$= \frac{32 + 4 + 6}{9 - 6}$$
$$= \frac{42}{3}$$
$$= 14$$

12. $$7(2x + 3) - 4(2x + 1) = 2(x + 1)$$
$$14x + 21 - 8x - 4 = 2x + 2$$
$$6x + 17 = 2x + 2$$
$$4x = -15$$
$$x = -\frac{15}{4}$$

Solution set: $\{-\frac{15}{4}\}$

13. $|6x - 8| = 4$

$6x - 8 = 4$ or $6x - 8 = -4$

$6x = 12$ $\quad$ $6x = 4$

$x = 2$ or $x = \frac{4}{6} = \frac{2}{3}$

Solution set: $\{\frac{2}{3}, 2\}$

14. $ax + by = cx + d$

To solve for x, get all terms with x alone on one side of the equals sign.

$$ax - cx = d - by$$
$$x(a - c) = d - by$$
$$x = \frac{d - by}{a - c}$$

or

$$x = \frac{(-1)(by - d)}{(-1)(c - a)}$$
$$x = \frac{by - d}{c - a}$$

15. $.04x + .06(x - 1) = 1.04$

Multiply by 100 to clear the decimals.

$$\begin{aligned} 4x + 6(x-1) &= 104 \\ 4x + 6x - 6 &= 104 \\ 10x - 6 &= 104 \\ 10x &= 110 \\ x &= 11 \end{aligned}$$

Solution set: $\{11\}$

16. $\frac{2}{3}y + \frac{5}{12}y \le 20$

Multiply both sides by 12.

$$\begin{aligned} 12\left(\frac{2}{3}y + \frac{5}{12}y\right) &\le 12(20) \\ 8y + 5y &\le 240 \\ 13y &\le 240 \\ y &\le \frac{240}{13} \end{aligned}$$

Solution set: $(-\infty, \frac{240}{13}]$

17. $|3x + 2| \le 4$

$$\begin{aligned} -4 \le 3x + 2 \le 4 \\ -6 \le 3x \le 2 \\ -2 \le x \le \frac{2}{3} \end{aligned}$$

Solution set: $[-2, \frac{2}{3}]$

18. $|12t + 7| \ge 0$

The solution set is $(-\infty, \infty)$ since the absolute value of any number is greater than or equal to 0.

19. Let $x =$ the speed of the fast car and
$x - 30 =$ the speed of the slow car.

Make a table.

	r	t	d
Fast car	x	3.5	$3.5x$
Slow car	$x - 30$	3.5	$3.5(x - 30)$

The car travels a total of 420 mi.

$$\begin{aligned} 3.5x + 3.5(x - 30) &= 420 \\ 3.5x + 3.5x - 105 &= 420 \\ 7x &= 525 \\ x &= 75 \\ x - 30 = 75 - 30 &= 45 \end{aligned}$$

The fast car is traveling at 75 mph, and the slow car is traveling at 45 mph.

20. Let $h =$ the height of the triangle.

Here, $A = 42$ and $b = 14$, so substitute these values in the formula for the area of a triangle and solve for h.

$$\begin{aligned} A &= \frac{1}{2}bh \\ 42 &= \frac{1}{2}(14)h \\ 42 &= 7h \\ 6 &= h \end{aligned}$$

The height is 6 m.

21. Let $x =$ the number of nickels,
$x + 1 =$ the number of dimes, and
$x + 6 =$ the number of pennies.

The total value is \$4.80, so

$$.05x + .10(x + 1) + .01(x + 6) = 4.80.$$

Multiply by 100 to clear the decimals.

$$\begin{aligned} 5x + 10(x + 1) + 1(x + 6) &= 480 \\ 5x + 10x + 10 + x + 6 &= 480 \\ 16x + 16 &= 480 \\ 16x &= 464 \\ x &= 29 \end{aligned}$$

Then $x + 1 = 29 + 1 = 30$

and $x + 6 = 29 + 6 = 35$.

There are 35 pennies, 29 nickels, and 30 dimes.

22. Let $x =$ the measure of the equal angles and
$2x - 4 =$ the measure of the third angle.

The sum of the measures of the angles in a triangle is 180, so

$$\begin{aligned} x + x + 2x - 4 &= 180 \\ 4x - 4 &= 180 \\ 4x &= 184 \\ x &= 46. \end{aligned}$$

So,

$$2x - 4 = 2(46) - 4 = 92 - 4 = 88.$$

The measures of the angles are $46°, 46°$, and $88°$.

In Exercises 23-28, point A has coordinates $(-2, 6)$ and point B has coordinates $(4, -2)$.

23. A horizontal line through the point (x, k) has equation $y = k$. Since point A has coordinates $(-2, 6)$, $k = 6$. The equation of the horizontal line through A is $y = 6$.

24. A vertical line through the point (k, y) has equation $x = k$. Since point B has coordinates $(4, -2)$, $k = 4$. The equation of the vertical line through B is $x = 4$.

25. Let $(x_1, y_1) = (-2, 6)$ and $(x_2, y_2) = (4, -2)$. Then,

$$m = \frac{y_2 - y_1}{x_2 - x_1} = \frac{-2-6}{4-(-2)} = \frac{-8}{6} = -\frac{4}{3}.$$

The slope is $-\frac{4}{3}$.

26. Perpendicular lines have slopes that are negative reciprocals of each other, so their product is -1. Since the slope of line AB is $-\frac{4}{3}$ (from Exercise 25), the negative reciprocal of $-\frac{4}{3}$ is $\frac{3}{4}$, and

$$-\frac{4}{3}\left(\frac{3}{4}\right) = -1.$$

The slope of the line perpendicular to line AB is $\frac{3}{4}$.

27. To find the standard form of the equation of line AB, let $m = -\frac{4}{3}$ (from Exercise 25) and $(x_1, y_1) = (4, -2)$ in the point-slope form.

$$\begin{aligned} y - y_1 &= m(x - x_1) \\ y - (-2) &= -\frac{4}{3}(x - 4) \\ y + 2 &= -\frac{4}{3}x + \frac{16}{3} \end{aligned}$$

Multiply by 3 to clear the fractions. Then write the equation in standard form $Ax + By = C$.

$$\begin{aligned} 3(y+2) &= 3\left(-\frac{4}{3}x + \frac{16}{3}\right) \\ 3y + 6 &= -4x + 16 \\ 4x + 3y &= 10 \end{aligned}$$

28. Write the equation of the line in slope-intercept form.

$$\begin{aligned} 4x + 3y &= 10 \\ 3y &= -4x + 10 \\ y &= -\frac{4}{3}x + \frac{10}{3} \end{aligned}$$

Replace y with $f(x)$.

$$f(x) = -\frac{4}{3}x + \frac{10}{3}$$

29. Let $(x_1, y_1) = (-3, 0)$ and $(x_2, y_2) = (0, 5)$.

$$m = \frac{y_2 - y_1}{x_2 - x_1} = \frac{5-0}{0-(-3)} = \frac{5}{3}.$$

Now let $m = \frac{5}{3}$ and $b = 5$ (from y-intercept $(0, 5)$) in the slope-intercept form.

$$\begin{aligned} y &= mx + b \\ y &= \frac{5}{3}x + 5 \end{aligned}$$

Multiply by 3 to clear the fraction. Then write the equation in standard form $Ax + By = C$.

$$\begin{aligned} 3y &= 3\left(\frac{5}{3}x + 5\right) \\ 3y &= 5x + 15 \\ -5x + 3y &= 15 \\ 5x - 3y &= -15 \end{aligned}$$

For Exercises 30 and 31, see the answer graphs in the back of the textbook.

30. First locate the point $(-1, -3)$ on a graph. Then use the definition of slope to find a second point on the line.

$$m = \frac{\text{change in } y}{\text{change in } x} = \frac{2}{3}$$

From $(-1, -3)$, move 2 units up and 3 units to the right. The line through $(-1, -3)$ and the new point $(2, -1)$ is the graph.

31. To graph

$$-3x - 2y \le 6,$$

graph the line $-3x - 2y = 6$ as a solid line through $(-2, 0)$ and $(0, -3)$ since the inequality involves "$\le$." To determine the region that belongs to the graph, test $(0, 0)$.

$$\begin{aligned} -3x - 2y &\le 6 \\ -3(0) - 2(0) &\le 6 \\ 0 &\le 6 \quad \textit{True} \end{aligned}$$

Since the result is true, shade the region that includes $(0, 0)$.

32.
$$\begin{aligned} -2x + 3y &= -15 \quad (1) \\ 4x - y &= 15 \quad (2) \end{aligned}$$

To eliminate x, multiply equation (1) by 2 and add the result to equation (2).

$$\begin{aligned} -4x + 6y &= -30 \\ 4x - y &= 15 \quad (2) \\ \hline 5y &= -15 \\ y &= -3 \end{aligned}$$

Substitute -3 for y in equation (2) to find x.

$$\begin{aligned} 4x - y &= 15 \quad (2) \\ 4x - (-3) &= 15 \\ 4x + 3 &= 15 \\ 4x &= 12 \\ x &= 3 \end{aligned}$$

Solution set: $\{(3, -3)\}$

33. $$\begin{aligned} x + y + z &= 10 \quad (1) \\ x - y - z &= 0 \quad (2) \\ -x + y - z &= -4 \quad (3) \end{aligned}$$

Add equations (1) and (2) to eliminate y and z. The result is

$$\begin{aligned} 2x &= 10 \\ x &= 5. \end{aligned}$$

Add equations (2) and (3) to eliminate x and y. The result is

$$\begin{aligned} -2z &= -4 \\ z &= 2. \end{aligned}$$

Substitute 5 for x and 2 for z in equation (1) to find y.

$$\begin{aligned} x + y + z &= 10 \quad (1) \\ 5 + y + 2 &= 10 \\ y + 7 &= 10 \\ y &= 3 \end{aligned}$$

Solution set: $\{(5, 3, 2)\}$

34. $$\begin{vmatrix} 1 & 2 & 3 \\ 0 & 5 & 1 \\ -1 & 0 & 4 \end{vmatrix}$$

Expand about row 2.

$$\begin{aligned} &= 0 + 5\begin{vmatrix} 1 & 3 \\ -1 & 4 \end{vmatrix} - 1\begin{vmatrix} 1 & 2 \\ -1 & 0 \end{vmatrix} \\ &= 5(7) - 1(2) \\ &= 35 - 2 \\ &= 33 \end{aligned}$$

35. Let $x =$ the amount of oranges and $y =$ the amount of apples.

Since she bought 6 lb of fruit,

$$x + y = 6. \quad (1)$$

The total cost of x lb of oranges at \$.90 per lb and y lb of apples at \$.70 per lb is \$5.20, so

$$.90x + .70y = 5.20.$$

Multiply by 10 to clear the decimals.

$$9x + 7y = 52 \quad (2)$$

The system of equations is

$$\begin{aligned} x + y &= 6 \quad (1) \\ 9x + 7y &= 52. \end{aligned}$$

To solve the system, solve equation (1) for x.

$$\begin{aligned} x + y &= 6 \quad (1) \\ x &= 6 - y \end{aligned}$$

Substitute $6 - y$ for x in equation (2) and solve for y.

$$\begin{aligned} 9x + 7y &= 52 \quad (2) \\ 9(6 - y) + 7y &= 52 \\ 54 - 9y + 7y &= 52 \\ -2y &= -2 \\ y &= 1 \end{aligned}$$

Since $x = 6 - y$ and $y = 1$,

$$x = 6 - 1 = 5.$$

She bought 5 lb of oranges and 1 lb of apples.

36. Let $x =$ the amount invested in a video rental firm,
$y =$ the amount invested in a tax-free bond,
and $z =$ the amount invested in a money market fund.

Since she inherited \$80,000,

$$x + y + z = 80{,}000. \quad (1)$$

Her annual return is \$5200 on her investments of 7%, 6%, and 4% respectively, so

$$.07x + .06y + .04z = 5200.$$

Multiply by 100 to clear the decimals.

$$7x + 6y + 4z = 520{,}000 \quad (2)$$

The amounts invested at 6% and 4% are the same, so

$$\begin{aligned} y &= z \\ \text{or} \quad y - z &= 0. \quad (3) \end{aligned}$$

Solve the system using Cramer's rule.

$$\begin{aligned} x + y + z &= 80{,}000 \quad (1) \\ 7x + 6y + 4z &= 520{,}000 \quad (2) \\ y - z &= 0 \quad (3) \end{aligned}$$

Expand about row 3 in each case.

$$D = \begin{vmatrix} 1 & 1 & 1 \\ 7 & 6 & 4 \\ 0 & 1 & -1 \end{vmatrix}$$

$$= 0 - 1\begin{vmatrix} 1 & 1 \\ 7 & 4 \end{vmatrix} + (-1)\begin{vmatrix} 1 & 1 \\ 7 & 6 \end{vmatrix}$$

$$= -1(-3) - 1(-1)$$
$$= 3 + 1$$
$$= 4$$

Just find D_y.

$$D_y = \begin{vmatrix} 1 & 80,000 & 1 \\ 7 & 520,000 & 4 \\ 0 & 0 & -1 \end{vmatrix}$$

$$= 0 - 0 + (-1)\begin{vmatrix} 1 & 80,000 \\ 7 & 520,000 \end{vmatrix}$$

$$= -1(-40,000)$$
$$= 40,000$$

$$y = \frac{D_y}{D} = \frac{40,000}{4} = 10,000$$

Equation (3) written in terms of y is

$$y = z,$$

so $y = 10{,}000$ and $z = 10{,}000$. Substitute 10,000 for y and 10,000 for z in equation (1) to find x.

$$\begin{aligned} x + y + z &= 80,000 \quad (1) \\ x + 10,000 + 10,000 &= 80,000 \\ x + 20,000 &= 80,000 \\ x &= 60,000 \end{aligned}$$

She invested \$60,000 in the video rental firm and \$10,000 each in the bond and money market fund.

37. The lines intersect at $(8, 3000)$, so cost equals revenue at $x = 8$ (which is 800 parts). The revenue is \$3000.

38. $$\begin{aligned} \text{Profit} &= \text{Revenue} - \text{Cost} \\ &\approx 4100 - 3600 \\ &= 500 \end{aligned}$$

The profit is about \$500.

Chapter 5

POLYNOMIALS

5.1 Integer Exponents

5.1 Margin Exercises

1. **(a)** $-2p^8$

   The base is p since the exponent refers only to the factor closest to it.

   (b) $(-2p)^8$

   The base is $-2p$ because of the parentheses.

   (c) $3y^7$

   The base is y since the exponent refers only to the factor closest to it.

   (d) $-m^{10}$

   The base is m since there are no parentheses.

2. Use the product rule for exponents.

   (a) $m^8 \cdot m^6 = m^{8+6} = m^{14}$

   (b) $r^7 \cdot r = r^7 \cdot r^1 = r^{7+1} = r^8$

   (c) $k^4 k^3 k^6 = k^{4+3+6} = k^{13}$

   (d) $m^5 \cdot p^4$ cannot be simplified further because the bases m and p are not the same.

   (e) Use the commutative and associative properties to group constants together and variables together.

$$\begin{aligned}(-4a^3)(6a^2) &= (-4)(6)(a^3a^2)\\ &= -24a^{3+2}\\ &= -24a^5\end{aligned}$$

   (f) $$\begin{aligned}(-5p^4)(-9p^5) &= (-5)(-9)(p^4p^5)\\ &= 45p^{4+5}\\ &= 45p^9\end{aligned}$$

3. **(a)** $46^0 = 1$

   Any real number (except 0) raised to the power zero is equal to 1.

   (b) $(-29)^0 = 1$

   (c) -29^0

   First, find 29^0; then, take the opposite of the answer.

$$-29^0 = -(29^0) = -1$$

   (d) $8^0 + 15^0 = 1 + 1 = 2$

   (e) $(-15p^5)^0 = 1$

   Since $p \neq 0, -15p^5$ will not equal zero. Therefore, $-15p^5$ raised to the power zero will equal 1.

4. **(a)** $8^{-1} = \dfrac{1}{8^1} = \dfrac{1}{8}$

   (b) $6^{-3} = \dfrac{1}{6^3} = \dfrac{1}{216}$

   (c) $-3^{-2} = -(3^{-2}) = -\left(\dfrac{1}{3^2}\right) = -\dfrac{1}{9}$

   (d) $y^{-3} = \dfrac{1}{y^3} \quad (y \neq 0)$

   (e) $(2m^4)^{-3} = \dfrac{1}{(2m^4)^3} \quad (m \neq 0)$

5. **(a)** $3^{-1} + 5^{-1} = \dfrac{1}{3} + \dfrac{1}{5} = \dfrac{5}{15} + \dfrac{3}{15} = \dfrac{8}{15}$

   (b) $4^{-1} - 2^{-1} = \dfrac{1}{4} - \dfrac{1}{2} = \dfrac{1}{4} - \dfrac{2}{4} = -\dfrac{1}{4}$

   (c) $\dfrac{4^{-1}}{2^{-2}} = \dfrac{\frac{1}{4}}{\frac{1}{2^2}} = \dfrac{1}{4} \cdot \dfrac{2^2}{1} = \dfrac{1}{4} \cdot \dfrac{4}{1} = \dfrac{4}{4} = 1$

   (d) $\dfrac{3^{-3}}{9^{-1}} = \dfrac{\frac{1}{3^3}}{\frac{1}{9}} = \dfrac{1}{3^3} \cdot \dfrac{9}{1} = \dfrac{1}{27} \cdot \dfrac{9}{1} = \dfrac{9}{27} = \dfrac{1}{3}$

6. Use the quotient rule for exponents.

   (a) $\dfrac{6^9}{6^4} = 6^{9-4} = 6^5$

   (b) $\dfrac{9^{12}}{9^7} = 9^{12-7} = 9^5$

   (c) $\dfrac{15^7}{15^{10}} = 15^{7-10} = 15^{-3} = \dfrac{1}{15^3}$

   (d) $\dfrac{m^3}{m^8} = m^{3-8} = m^{-5} = \dfrac{1}{m^5} \quad (m \neq 0)$

   (e) $\dfrac{x^3}{y^5}$ $(y \neq 0)$ cannot be simplified because the bases x and y are not the same.

7. (a) $\frac{2^{-4}}{2^2} = 2^{-4-2} = 2^{-6} = \frac{1}{2^6}$

(b) $\frac{8^{-2}}{8^{-6}} = 8^{-2-(-6)} = 8^{-2+6} = 8^4$

(c) $\frac{9^{-5}}{9^{-2}} = 9^{-5-(-2)} = 9^{-5+2} = 9^{-3} = \frac{1}{9^3}$

(d) $\frac{7^{-1}}{7} = \frac{7^{-1}}{7^1} = 7^{-1-1} = 7^{-2} = \frac{1}{7^2}$

(e) $\frac{k^4}{k^{-5}} = k^{4-(-5)} = k^{4+5} = k^9 \quad (k \neq 0)$

5.1 Section Exercises

3. In -5^4, the exponent is 4. The base is 5 since the exponent refers only to the factor closest to it.

7. In $-3p^{-1}$, the exponent is -1. The base is p.

11. -7^0

First, find 7^0; then take the opposite of the answer.

$$-7^0 = -(7^0) = -1$$

15. $-4^0 - m^0 = -1 - 1 = -2$

19. $(-7)^{-2} = \frac{1}{(-7)^2}$ or $\frac{1}{7^2}$

Since $(-7)^2$ results in a positive number, it can be written as 7^2.

23. $\frac{1}{5^2} = 5^{-2}$

27. $\frac{1}{(-4)^3}$

The base is -4. Since $(-4)^{-3}$ results in a negative number, it can be written as -4^{-3}.

$$\frac{1}{(-4)^3} = (-4)^{-3} = -4^{-3}$$

31. $\left(\frac{2}{3}\right)^2 = \frac{2}{3} \cdot \frac{2}{3} = \frac{4}{9}$

35. $-4^{-3} = -\frac{1}{4^3} = -\frac{1}{64}$

39. $\frac{1}{3^{-2}} = \frac{1}{\frac{1}{3^2}} = 3^2 = 9$

43. $\left(\frac{2}{3}\right)^{-3} = \frac{1}{\left(\frac{2}{3}\right)^3} = \left(\frac{3}{2}\right)^3 = \frac{27}{8}$

47. $2^6 \cdot 2^{10} = 2^{6+10} = 2^{16}$

51. $\frac{3^{-5}}{3^{-2}} = 3^{-5-(-2)} = 3^{-5+2} = 3^{-3} = \frac{1}{3^3}$

55. $t^5 t^{-12} = t^{5+(-12)} = t^{5-12} = t^{-7} = \frac{1}{t^7}$

59. $a^{-3}a^2a^{-4} = a^{-3+2-4} = a^{-5} = \frac{1}{a^5}$

63. $$\begin{aligned}\frac{r^3r^{-4}}{r^{-2}r^{-5}} = \frac{r^{3-4}}{r^{-2-5}} &= \frac{r^{-1}}{r^{-7}} \\ &= r^{-1-(-7)} \\ &= r^{-1+7} \\ &= r^6\end{aligned}$$

67. Let $x = 2$ and $y = 3$.

$(x+y)^{-1} = (2+3)^{-1} = 5^{-1} = \frac{1}{5}$

$x^{-1} + y^{-1} = 2^{-1} + 3^{-1} = \frac{1}{2} + \frac{1}{3} = \frac{3}{6} + \frac{2}{6} = \frac{5}{6}$

$\frac{1}{5} \neq \frac{5}{6}$

Therefore, $(x+y)^{-1} \neq x^{-1} + y^{-1}$.

5.2 Further Properties of Exponents

5.2 Margin Exercises

1. (a) Use the first power rule.

$$(m^5)^4 = m^{5(4)} = m^{20}$$

(b) $(x^3)^9 = x^{3(9)} = x^{27}$

(c) Use the third power rule.

$$\left(\frac{3}{8}\right)^7 = \frac{3^7}{8^7}$$

(d) Use the second power rule.

$$(2r)^{10} = 2^{10}r^{10}$$

(e) Use both the first and second power rules.

$$(-3y^5)^2 = (-3)^2(y^5)^2 = 9y^{5(2)} = 9y^{10}$$

(f) Use all three power rules.

$$\left(\frac{5p^2}{r^3}\right)^4 = \frac{(5p^2)^4}{(r^3)^4} = \frac{5^4p^{2(4)}}{r^{3(4)}} = \frac{5^4p^8}{r^{12}} \quad (r \neq 0)$$

2. (a) $4^{-3} = \left(\frac{1}{4}\right)^3 = \frac{1}{64}$

(b) $\left(\frac{2}{3}\right)^{-4} = \left(\frac{3}{2}\right)^4 = \frac{81}{16}$

(c) $\left(\frac{1}{6}\right)^{-3} = 6^3 = 216$

3. (a) $5^2 \cdot 5^{-4} = 5^{2+(-4)} = 5^{-2} = \frac{1}{5^2}$ or $\frac{1}{25}$

(b) $p^{-5} \cdot p^{-3} \cdot p^4 = p^{-5+(-3)+4} = p^{-4} = \frac{1}{p^4}$ $(p \neq 0)$

(c) $(4^2)^{-5} = 4^{2(-5)} = 4^{-10} = \frac{1}{4^{10}}$

(d) $(m^{-3})^{-4} = m^{-3(-4)} = m^{12}$ $(m \neq 0)$

(e) $(4a)^5 = 4^5 a^5$

4. (a) $\frac{a^{-3}b^5}{a^4 b^{-2}} = \frac{a^{-3}}{a^4} \cdot \frac{b^5}{b^{-2}}$

$= a^{-3-4} b^{5-(-2)}$

$= a^{-7} b^7$

$= \frac{b^7}{a^7}$

(b) $(3^2 k^{-4})^{-1} = \left(\frac{3^2}{k^4}\right)^{-1} = \frac{k^4}{3^2}$ or $\frac{k^4}{9}$

5. (a) $400,000 = 4_\wedge 00,000.$ ←Decimal point
Place a caret after the first nonzero digit, 4. Count 5 places from the caret to the decimal point (understood to be after the last 0). Use a positive exponent on 10 since $400,000 > 4$.

$$400,000 = 4 \times 10^5$$

(b) $29,800,000 = 2_\wedge 9,800,000.$
Count 7 places.

Use a positive exponent on 10 since $29,800,000 > 2.98$.

$$29,800,000 = 2.98 \times 10^7$$

(c) $-6083 = -6_\wedge 083.$
Count 3 places.

Use a positive exponent on 10 since $6083 > 6.083$.

$$-6083 = -6.083 \times 10^3$$

(d) $.00172 = .001_\wedge 72$
Count 3 places.

Use a negative exponent on 10 since $.00172 < 1.72$.

$$.00172 = 1.72 \times 10^{-3}$$

(e) $.0000000503 = .00000005_\wedge 03$
Count 8 places.

Use a negative exponent on 10 since $.0000000503 < 5.03$.

$$.0000000503 = 5.03 \times 10^{-8}$$

6. (a) $3.7 \times 10^8 = 3_\wedge 70000000. = 370,000,000$

Since the exponent is positive, move the decimal point 8 places to the right. Attach extra zeros as needed.

(b) $2.51 \times 10^3 = 2_\wedge 510.$
$= 2510$

Move the decimal point 3 places to the right. Attach an extra zero.

(c) $4.6 \times 10^{-5} = .00004_\wedge 6$
$= .000046$

Since the exponent is negative, move the decimal point 5 places to the left. Attach extra zeros.

(d) $9.372 \times 10^{-6} = .000009_\wedge 372$
$= .000009372$

Move the decimal point 6 places to the left.

(e) $8.5 \times 10^{-1} = .8_\wedge 5 = .85$

Move the decimal point 1 place to the left.

7. (a) $\frac{8 \times 10^3}{2 \times 10^2} = \frac{8}{2} \times \frac{10^3}{10^2} = 4 \times 10^1 = 40$

(b) $\frac{9 \times 10^3}{3 \times 10^{-2}} = \frac{9}{3} \times \frac{10^3}{10^{-2}} = 3 \times 10^5 = 300,000$

(c) $\frac{.06}{.003} = \frac{6 \times 10^{-2}}{3 \times 10^{-3}}$

$= \frac{6}{3} \times \frac{10^{-2}}{10^{-3}}$

$= 2 \times 10^1$

$= 20$

(d) $\frac{200,000 \times .0003}{.06}$

$= \frac{2 \times 10^5 \times 3 \times 10^{-4}}{6 \times 10^{-2}}$

$= \frac{2 \times 3}{6} \times \frac{10^5 \times 10^{-4}}{10^{-2}}$

$= 1 \times 10^3$

$= 1000$

8. A light year is about 5.880×10^{12} mi.

$10.6(5.880 \times 10^{12})$
$= (1.06 \times 10^1)(5.880 \times 10^{12})$
$= 1.06 \times 5.880 \times 10^1 \times 10^{12}$
$= 6.2328 \times 10^{13}$

Pollux is 6.2328×10^{13} mi from Earth.

5.2 Section Exercises

3. $4a^{-1} = \frac{4}{a^1} = \frac{4}{a}$

True

7. $(ab)^2 = a^2b^2$ by a power rule.

Since $a^2b^2 \neq ab^2$, the expression $(ab)^2 = ab^2$ has been simplified incorrectly.

11. $\left(\frac{4}{a}\right)^5 = \frac{4^5}{a^5}$ by a power rule.

Since $\frac{4^5}{a^5} \neq \frac{4^5}{a}$, the expression $\left(\frac{4}{a}\right)^5 = \frac{4^5}{a}$ has been simplified incorrectly.

15. $\left(\frac{3}{4}\right)^{-2} = \left(\frac{4}{3}\right)^2$

19. $(2^{-3} \cdot 5^{-1})^3 = 2^{-3(3)} \cdot 5^{-1(3)} = 2^{-9} \cdot 5^{-3} = \frac{1}{2^9 \cdot 5^3}$

23. $(k^2)^{-3}k^4 = k^{2(-3)}k^4 = k^{-6}k^4 = k^{-2} = \frac{1}{k^2}$

27. $$\begin{aligned}(5a^{-1})^4(a^2)^{-3} &= 5^4a^{-1(4)}a^{2(-3)}\\ &= 5^4a^{-4}a^{-6}\\ &= 5^4a^{-10}\\ &= \frac{5^4}{a^{10}}\end{aligned}$$

31. $\frac{(p^{-2})^3}{5p^4} = \frac{p^{-2(3)}}{5p^4} = \frac{p^{-6}}{5p^4} = \frac{1}{5p^4p^6} = \frac{1}{5p^{10}}$

35. $\frac{(-y^{-4})^2}{6(y^{-5})^{-1}} = \frac{y^{-4(2)}}{6y^{-5(-1)}} = \frac{y^{-8}}{6y^5} = \frac{1}{6y^5y^8} = \frac{1}{6y^{13}}$

39. To rewrite a fraction raised to a negative power, write the fraction as its reciprocal raised to the negative of the negative power. For example,

$$\left(\frac{2}{3}\right)^{-5} = \left(\frac{3}{2}\right)^5.$$

43. $.830 = .8_\wedge30$

Place a caret after the first nonzero digit, 8. Count 1 place from the caret to the decimal point. Use a negative exponent on 10 since $.830 < 8.3$.

$$.830 = 8.3 \times 10^{-1}$$

47. $-38,500 = -3_\wedge8,500.$

Place a caret after the first nonzero digit, 3. Count 4 places from the caret to the decimal point. Use a positive exponent on 10 since $38,500 > 3.85$.

$$-38,500 = -3.85 \times 10^4$$

51. $2.54 \times 10^{-3} = .002_\wedge54 = .00254$

Since the exponent is negative, move the decimal point 3 places to the left. Attach extra zeros.

55. $1.2 \times 10^{-5} = .00001_\wedge2 = .000012$

Since the exponent is negative, move the decimal point 5 places to the left. Attach extra zeros.

59. $$\begin{aligned}\frac{.05 \times 1600}{.0004} &= \frac{5 \times 10^{-2} \times 1.6 \times 10^3}{4 \times 10^{-4}}\\ &= \frac{5 \times 1.6 \times 10^{-2} \times 10^3}{4 \times 10^{-4}}\\ &= \frac{8}{4} \times 10^{-2+3-(-4)}\\ &= 2 \times 10^5\\ &= 200,000\end{aligned}$$

63. To solve the problem, divide the number of different ways to choose the numbers by the number of people ($1000 = 10^3$).

$$\begin{aligned}\frac{3.83838 \times 10^6}{10^3} &= 3.83838 \times 10^3\\ &= 3838.38\end{aligned}$$

Therefore, each person will buy about 3838.38 tickets at \$1 per ticket, or \$3838.38.

5.3 Addition and Subtraction of Polynomials

5.3 Margin Exercises

1. **(a)** In $-9m^5$, the coefficient is -9.

(b) In $12y^2x$, the coefficient is 12.

(c) In x, the coefficient is 1 since $x = 1 \cdot x$.

(d) In $-y$, the coefficient is -1 since $-y = -1 \cdot y$.

2. **(a)** $-4 + 9y + y^3$ is written in descending powers as $y^3 + 9y - 4$.

(b) $-3z^4 + 2z^3 + z^5 - 6z$ is written in descending powers as $z^5 - 3z^4 + 2z^3 - 6z$.

(c) $-12m^{10} + 8m^9 + 10m^{12}$ is written in descending powers as $10m^{12} - 12m^{10} + 8m^9$.

3. **(a)** $12m^4 - 6m^2$ is a binomial since it has two terms.

(b) $-6y^3 + 2y^2 - 8y$ is a trinomial since it has three terms.

(c) $3a^5$ is a monomial since it has one term.

(d) $-2k^{10}+2k^9-8k^5+2k$ has four terms so it is none of the choices.

4. The degree of a polynomial is the highest degree of any of the terms.

(a) $9y^4+8y^3-6$ has degree 4 since the largest exponent is 4.

(b) $-12m^7+11m^3+m^9$ has degree 9.

(c) $-2k$ has degree 1 since $-2k=-2k^1$.

(d) 10 has degree 0 since $10=10x^0$.

(d) $3mn^2+2m^3n$ has degree 4.

Since the degree of $3mn^2$ or $3m^1n^2$ is $1+2$ or 3 and the degree of $2m^3n$ or $2m^3n^1$ is $3+1$ or 4, the degree of the polynomial is the highest degree, 4.

5. **(a)**
$$\begin{aligned}&11x+12x-7x-3x\\&=(11+12-7-3)x\\&=13x\end{aligned}$$

(b)
$$\begin{aligned}&11p^5+4p^5-6p^3+8p^3\\&=(11+4)p^5+(-6+8)p^3\\&=15p^5+2p^3\end{aligned}$$

(c)
$$\begin{aligned}&2y^2z^4+3y^4+5y^4-9y^4z^2\\&=2y^2z^4+(3+5)y^4-9y^4z^2\\&=2y^2z^4+8y^4-9y^4z^2\end{aligned}$$

(d) $2x^2y^3+5x^3y^2$ cannot be simplified because $2x^2y^3$ and $5x^3y^2$ are unlike terms.

6. **(a)** Add in columns.

$$\begin{array}{r}-4x^3+2x^2\\ 8x^3-6x^2\\ \hline 4x^3-4x^2\end{array}$$

(b)
$$\begin{aligned}&(-5p^3+6p^2)+(8p^3-12p^2)\\&=-5p^3+8p^3+6p^2-12p^2\\&=3p^3-6p^2\end{aligned}$$

(c) Add in columns.

$$\begin{array}{r}-6r^5+2r^3-\ \ r^2\\ 8r^5-2r^3+5r^2\\ \hline 2r^5+0r^3+4r^2\end{array}$$

The answer can be written as $2r^5+4r^2$.

(d)
$$\begin{aligned}&(12y^2-7y+9)+(-4y^2-11y+5)\\&=12y^2-4y^2-7y-11y+9+5\\&=8y^2-18y+14\end{aligned}$$

7. **(a)** $(2a^2-a)-(5a^2+8a)$

Remove parentheses. The terms in the second polynomial change sign due to the negative sign in front of the parentheses.

$$\begin{aligned}&=2a^2-a-5a^2-8a\\&=2a^2-5a^2-a-8a\\&=-3a^2-9a\end{aligned}$$

(b)
$$\begin{aligned}&(6y^3-9y^2+8)-(2y^3+y^2+5)\\&=6y^3-9y^2+8-2y^3-y^2-5\\&=6y^3-2y^3-9y^2-y^2+8-5\\&=4y^3-10y^2+3\end{aligned}$$

8. **(a)** Subtract.

$$\begin{array}{r}m^4-2m^3\\ \hline 5m^4+8m^3\end{array}$$

Change all the signs in the second polynomial and add.

$$\begin{array}{r}m^4-\ \ 2m^3\\ -5m^4-\ \ 8m^3\\ \hline -4m^4-10m^3\end{array}$$

(b) Subtract.

$$\begin{array}{r}6y^3-2y^2+\ \ 5y\\ \hline -2y^3+8y^2-11y\end{array}$$

Change signs in the second polynomial and add.

$$\begin{array}{r}6y^3-\ \ 2y^2+\ \ 5y\\ 2y^3-\ \ 8y^2+11y\\ \hline 8y^3-10y^2+16y\end{array}$$

9. $P(x)=-x^2+5x-11$

(a)
$$\begin{aligned}P(1)&=-1^2+5(1)-11\\&=-1+5-11\\&=-7\end{aligned}$$

(b)
$$\begin{aligned}P(-4)&=-(-4)^2+5(-4)-11\\&=-16-20-11\\&=-47\end{aligned}$$

(c)
$$\begin{aligned}P(5)&=-5^2+5(5)-11\\&=-25+25-11\\&=-11\end{aligned}$$

(d)
$$\begin{aligned}P(0)&=-0^2+5(0)-11\\&=0+0-11\\&=-11\end{aligned}$$

5.3 Section Exercises

3. Only choice (a) is a trinomial (it has 3 terms) with descending powers and having degree 6.

7. In $-15p^2$, the coefficient is -15, and the degree is 2.

11. In $-mn^5$, the coefficient is -1 since $-mn^5 = -1 \cdot mn^5$. The degree is 6 since the sum of the exponents on m and n is $1+5$ or 6.

15. In $2x^3+x-3x^2$, the degrees of the terms are 3, 1, and 2 from left to right. Therefore, this polynomial is in neither ascending nor descending powers.

19. In $-m^3+5m^2+3m+10$, the degrees of the terms are 3, 2, 1, and 0 from left to right. Therefore, this polynomial is written in descending powers.

23. $7m-21$ is a binomial since there are two terms. The degree is 1.

27. $-6p^4q-3p^3q^2+2pq^3-q^4$ has four terms so it is none of the choices. The degree of the polynomial is 5 since the degrees of the terms are $4+1 = 5, 3+2=5, 1+3=4$, and 4.

31. $-m^3+2m^3+6m^3=(-1+2+6)m^3=7m^3$

35. m^4-3m^2+m is already simplified.

39. $$\begin{aligned} &2k+3k^2+5k^2-7 \\ &= 3k^2+5k^2+2k-7 \\ &= (3+5)k^2+2k-7 \\ &= 8k^2+2k-7 \end{aligned}$$

43. Add in columns.

$$\begin{array}{r} -12p^2+\ 4p-1 \\ 3p^2+\ 7p-8 \\ \hline -9p^2+11p-9 \end{array}$$

47. Subtract.

$$\begin{array}{r} 6m^2-11m+5 \\ \underline{-8m^2+\ 2m-1} \end{array}$$

Change all the signs in the second polynomial and add.

$$\begin{array}{r} 6m^2-11m+5 \\ 8m^2-\ 2m+1 \\ \hline 14m^2-13m+6 \end{array}$$

51. Add.

$$\begin{array}{r} 6y^3-9y^2\qquad+8 \\ 4y^3+2y^2+5y\qquad \\ \hline 10y^3-7y^2+5y+8 \end{array}$$

55. $(3r+8)-(2r-5)$

Change all signs in the second polynomial and add.

$$\begin{aligned} &= (3r+8)+(-2r+5) \\ &= 3r+8-2r+5 \\ &= 3r-2r+8+5 \\ &= r+13 \end{aligned}$$

59. $$\begin{aligned} &(2a^2+3a-1)-(4a^2+5a+6) \\ &= (2a^2+3a-1)+(-4a^2-5a-6) \\ &= 2a^2-4a^2+3a-5a-1-6 \\ &= -2a^2-2a-7 \end{aligned}$$

63. $$\begin{aligned} &(9y^4-5y^2+10)+(2-5y^2-3y^4) \\ &= 9y^4-3y^4-5y^2-5y^2+10+2 \\ &= 6y^4-10y^2+12 \end{aligned}$$

67. $P(x)=6x-4$

(a) $P(-1)=6(-1)-4=-6-4=-10$

(b) $P(2)=6(2)-4=12-4=8$

71. $P(x)=5x^4-3x^2+6$

(a) $$\begin{aligned} P(-1) &= 5(-1)^4-3(-1)^2+6 \\ &= 5(1)-3(1)+6 \\ &= 5-3+6 \\ &= 8 \end{aligned}$$

(b) $$\begin{aligned} P(2) &= 5(2)^4-3(2)^2+6 \\ &= 5(16)-3(4)+6 \\ &= 80-12+6 \\ &= 74 \end{aligned}$$

In Exercises 75 and 79,

$$P(x)=1.06x^3-6.00x^2+7.86x+31.6.$$

75. Since $x=0$ corresponds to 1990, substitute 0 for x in the function $P(x)$.

$$\begin{aligned} P(0) &= 1.06(0)^3-6.00(0)^2+7.86(0)+31.6 \\ &= 31.6 \end{aligned}$$

In 1990, the percent was 31.6.

79. Since $x = 0$ corresponds to 1990, $x = 8$ corresponds to 1998 ($1998 - 1990 = 8$). Find $P(8)$ by substituting 8 for x in the function.

$$\begin{aligned} P(8) &= 1.06(8)^3 - 6.00(8)^2 + 7.86(8) + 31.6 \\ &= 1.06(512) - 6.00(64) + 62.88 + 31.6 \\ &= 542.72 - 384 + 62.88 + 31.6 \\ &= 253.2 \end{aligned}$$

In 1998, the percent would be 253.2. This answer is not reasonable since it is more than 100%. The mathematical model only applies to the years 1990 through 1994.

5.4 Multiplication of Polynomials

5.4 Margin Exercises

1. **(a)** $$\begin{aligned} -6m^5(2m^4) &= (-6)(2)m^5 \cdot m^4 \\ &= -12m^{5+4} \\ &= -12m^9 \end{aligned}$$

   (b) $$\begin{aligned} &(8k^3y)(9ky^3) \\ &= (8)(9)k^3 \cdot k^1 \cdot y^1 \cdot y^3 \\ &= 72k^{3+1}y^{1+3} \\ &= 72k^4y^4 \end{aligned}$$

2. **(a)** $$\begin{aligned} -2r(9r - 5) &= -2r(9r) - 2r(-5) \\ &= -18r^2 + 10r \end{aligned}$$

   (b) $$\begin{aligned} &3p^2(5p^3 + 2p^2 - 7) \\ &= 3p^2(5p^3) + 3p^2(2p^2) + 3p^2(-7) \\ &= 15p^5 + 6p^4 - 21p^2 \end{aligned}$$

3. **(a)**
$$\begin{array}{rrrl} & 2m & -\ 5 & \\ & 3m & +\ 4 & \\ \hline & 8m & -\ 20 & \leftarrow 4(2m - 5) \\ 6m^2 & -\ 15m & & \leftarrow 3m(2m - 5) \\ \hline 6m^2 & -\ 7m & -\ 20 & \text{Add.} \end{array}$$

   (b) $(4a - 5)(3a + 6)$

   Rewrite the polynomials vertically.

$$\begin{array}{rrrl} & 4a & -\ 5 & \\ & 3a & +\ 6 & \\ \hline & 24a & -\ 30 & \leftarrow 6(4a - 5) \\ 12a^2 & -\ 15a & & \leftarrow 3a(4a - 5) \\ \hline 12a^2 & +\ 9a & -\ 30 & \text{Add.} \end{array}$$

   (c) $(2k - 5m)(3k + 2m)$

   Rewrite vertically.

$$\begin{array}{rrrl} & 2k & -\ 5m & \\ & 3k & +\ 2m & \\ \hline & 4km & -\ 10m^2 & \leftarrow 2m(2k - 5m) \\ 6k^2 & -\ 15km & & \leftarrow 3k(2k - 5m) \\ \hline 6k^2 & -\ 11km & -\ 10m^2 & \text{Add.} \end{array}$$

4. **(a)** $(r^4 - 2r^3 + 6)(3r - 1)$

$$\begin{array}{rrrrr} & r^4 & -\ 2r^3 & & +\ 6 \\ & & & & 3r - 1 \\ \hline & -r^4 & +\ 2r^3 & & -\ 6 \\ 3r^5 & -\ 6r^4 & & +\ 18r & \\ \hline 3r^5 & -\ 7r^4 & +\ 2r^3 & +\ 18r & -\ 6 \end{array}$$

   (b)
$$\begin{array}{rrrrr} & 5a^3 & -\ 6a^2 & +\ 2a & -\ 3 \\ & & & 2a & -\ 5 \\ \hline & -25a^3 & +\ 30a^2 & -\ 10a & +\ 15 \\ 10a^4 & -\ 12a^3 & +\ 4a^2 & -\ 6a & \\ \hline 10a^4 & -\ 37a^3 & +\ 34a^2 & -\ 16a & +\ 15 \end{array}$$

5. **(a)** $$\begin{aligned} &(3z - 2)(z + 1) \\ &= \underset{\text{F}}{3z \cdot z} + \underset{\text{O}}{3z \cdot 1} - \underset{\text{I}}{2 \cdot z} - \underset{\text{L}}{2 \cdot 1} \\ &= 3z^2 + 3z - 2z - 2 \\ &= 3z^2 + z - 2 \end{aligned}$$

   (b) $$\begin{aligned} &(5r + 3)(2r - 5) \\ &= \underset{\text{F}}{10r^2} - \underset{\text{O}}{25r} + \underset{\text{I}}{6r} - \underset{\text{L}}{15} \\ &= 10r^2 - 19r - 15 \end{aligned}$$

6. **(a)** $$\begin{aligned} &(4p + 5q)(3p - 2q) \\ &= 12p^2 - 8pq + 15pq - 10q^2 \\ &= 12p^2 + 7pq - 10q^2 \end{aligned}$$

   (b) $$\begin{aligned} &(4y - z)(2y + 3z) \\ &= 8y^2 + 12yz - 2yz - 3z^2 \\ &= 8y^2 + 10yz - 3z^2 \end{aligned}$$

   (c) $$\begin{aligned} &(8r + 1)(8r - 1) \\ &= 64r^2 - 8r + 8r - 1 \\ &= 64r^2 - 1 \end{aligned}$$

   (d) $$\begin{aligned} &(3p + 5)(3p + 5) \\ &= 9p^2 + 15p + 15p + 25 \\ &= 9p^2 + 30p + 25 \end{aligned}$$

7. Use $(x + y)(x - y) = x^2 - y^2$.

   (a) $$\begin{aligned} (m + 5)(m - 5) &= m^2 - 5^2 \\ &= m^2 - 25 \end{aligned}$$

   (b) $$\begin{aligned} (x - 4y)(x + 4y) &= x^2 - (4y)^2 \\ &= x^2 - 16y^2 \end{aligned}$$

(c) $(7m-2n)(7m+2n) = (7m)^2 - (2n)^2$
$= 49m^2 - 4n^2$

(d) $(6a+b)(6a-b) = (6a)^2 - b^2$
$= 36a^2 - b^2$

8. Use $(x+y)^2 = x^2 + 2xy + y^2$
or $(x-y)^2 = x^2 - 2xy + y^2$.

(a) $(a+2)^2 = a^2 + 2(a)(2) + 2^2$
$= a^2 + 4a + 4$

(b) $(2m-5)^2$
$= (2m)^2 - 2(2m)(5) + 5^2$
$= 4m^2 - 20m + 25$

(c) $(y+6z)^2$
$= y^2 + 2(y)(6z) + (6z)^2$
$= y^2 + 12yz + 36z^2$

(d) $(3k-2n)^2$
$= (3k)^2 - 2(3k)(2n) + (2n)^2$
$= 9k^2 - 12kn + 4n^2$

9. (a) $[(m-2n)-3] \cdot [(m-2n)+3]$

This is the product of a sum and a difference.

$= (m-2n)^2 - 3^2$

The first term is the square of a binomial.

$= m^2 - 2(m)(2n) + (2n)^2 - 3^2$
$= m^2 - 4mn + 4n^2 - 9$

(b) $[(k-5h)+2]^2$

This is the square of a binomial, with $(k-5h)$ as the first term.

$= (k-5h)^2 + 2(k-5h)(2) + 2^2$
$= k^2 - 2(k)(5h) + (5h)^2 + 4(k-5h) + 4$
$= k^2 - 10kh + 25h^2 + 4k - 20h + 4$

(c) $(p+2q)^3$
$= (p+2q)^2(p+2q)$

Square $p+2q$.

$= (p^2 + 4pq + 4q^2)(p+2q)$
$= p^3 + 4p^2q + 4pq^2 + 2p^2q + 8pq^2 + 8q^3$
$= p^3 + 6p^2q + 12pq^2 + 8q^3$

5.4 Section Exercises

3. The product of any two polynomials is found by multiplying each *term* of the first by each *term* of the second, and then combining *like terms.*

7. $-8m^3(3m^2) = -8(3)m^3 \cdot m^2 = -24m^5$

11. $-q^3(2+3q) = -q^3(2) - q^3(3q) = -2q^3 - 3q^4$

15. $(2m+3)(3m^2 - 4m - 1)$
$= 2m(3m^2) + 2m(-4m) + 2m(-1)$
$+ 3(3m^2) + 3(-4m) + 3(-1)$
$= 6m^3 - 8m^2 - 2m + 9m^2 - 12m - 3$
$= 6m^3 - 8m^2 + 9m^2 - 2m - 12m - 3$
$= 6m^3 + m^2 - 14m - 3$

19.
$$\begin{array}{r} 2y + 3 \\ \underline{3y - 4} \\ -8y - 12 \\ \underline{6y^2 + 9y \phantom{{}-12}} \\ 6y^2 + y - 12 \end{array}$$

23.
$$\begin{array}{r} 5m \quad -3n \\ \underline{5m \quad +3n} \\ 15mn - 9n^2 \\ \underline{25m^2 - 15mn \phantom{{}-9n^2}} \\ 25m^2 \qquad\quad -9n^2 \end{array}$$

or $25m^2 - 9n^2$

27.
$$\begin{array}{r} 2p^2 + 3p + 6 \\ \underline{3p^2 - 4p - 1} \\ -2p^2 - 3p - 6 \\ -8p^3 - 12p^2 - 24p \phantom{{}-6} \\ \underline{6p^4 + 9p^3 + 18p^2 \phantom{{}-24p-6}} \\ 6p^4 + p^3 + 4p^2 - 27p - 6 \end{array}$$

31. $(4k+3)(3k-2)$
$= \underset{F}{12k^2} - \underset{O}{8k} + \underset{I}{9k} - \underset{L}{6}$
$= 12k^2 + k - 6$

35. $(6c-d)(2c+3d)$
$= 12c^2 + 18cd - 2cd - 3d^2$
$= 12c^2 + 16cd - 3d^2$

39. $\left(3r + \frac{1}{4}y\right)(r - 2y)$

$= 3r^2 - 6ry + \frac{1}{4}ry - \frac{1}{2}y^2$

$= 3r^2 - \frac{23}{4}ry - \frac{1}{2}y^2$

43. $(2p-3)(2p+3) = (2p)^2 - 3^2$
$= 4p^2 - 9$

47. $(3a+2c)(3a-2c) = (3a)^2 - (2c)^2$
$= 9a^2 - 4c^2$

51. $(4m+7n^2)(4m-7n^2) = (4m)^2 - (7n^2)^2$
$= 16m^2 - 49n^4$

55. $(y-5)^2 = y^2 - 2(y)(5) + 5^2$
$= y^2 - 10y + 25$

59. $(4n-3m)^2 = (4n)^2 - 2(4n)(3m) + (3m)^2$
$= 16n^2 - 24nm + 9m^2$

63. $(x+y)^2 = x^2+2xy+y^2$, a perfect square trinomial. x^2+y^2 is the sum of two squares, x^2 and y^2. It does not include the term $2xy$.

67. $[(2a+b)-3][(2a+b)+3]$
$= (2a+b)^2 - 3^2$
$= 4a^2 + 4ab + b^2 - 9$

71. $(5r-s)^3$
$= (5r-s)^2(5r-s)$
$= (25r^2 - 10rs + s^2)(5r-s)$
$= 125r^3 - 50r^2s + 5rs^2 - 25r^2s + 10rs^2 - s^3$
$= 125r^3 - 75r^2s + 15rs^2 - s^3$

75. Use

$$A = \frac{1}{2}bh$$

with $b = x+2y$ and $h = x-2y$.

$$\begin{aligned} A &= \frac{1}{2}(x+2y)(x-2y) \\ &= \frac{1}{2}[x^2 - (2y)^2] \\ &= \frac{1}{2}(x^2 - 4y^2) \\ &= \frac{1}{2}x^2 - 2y^2 \end{aligned}$$

79. Substitute 2 for x and 3 for y.

$$(x+y)^3 = (2+3)^3 = 5^3 = 125$$

Again, substitute 2 for x and 3 for y.

$$x^3 + y^3 = 2^3 + 3^3 = 8 + 27 = 35$$

Since $125 \neq 35, (x+y)^3 \neq x^3 + y^3$.

The correct product is

$(x+y)^3$
$= (x+y)^2(x+y)$
$= (x^2+2xy+y^2)(x+y)$
$= x^3 + 2x^2y + xy^2 + x^2y + 2xy^2 + y^3$
$= x^3 + 3x^2y + 3xy^2 + y^3$.

5.5 Division of Polynomials

5.5 Margin Exercises

1. (a) $\dfrac{12p+30}{6} = \dfrac{12p}{6} + \dfrac{30}{6} = 2p + 5$

(b) $\dfrac{9y^3 - 4y^2 + 8y}{2y^2} = \dfrac{9y^3}{2y^2} - \dfrac{4y^2}{2y^2} + \dfrac{8y}{2y^2}$
$= \dfrac{9y}{2} - 2 + \dfrac{4}{y}$

(c) $\dfrac{8a^2b^2 - 20ab^3}{4a^3b} = \dfrac{8a^2b^2}{4a^3b} - \dfrac{20ab^3}{4a^3b}$
$= \dfrac{2b}{a} - \dfrac{5b^2}{a^2}$

2. (a) $\dfrac{2r^2 + r - 21}{r-3}$

$$\begin{array}{r} 2r+7 \\ r-3\overline{)2r^2 + r - 21} \\ \underline{2r^2 - 6r} \\ 7r - 21 \\ \underline{7r - 21} \\ 0 \end{array}$$

Answer: $2r + 7$

(b) $\dfrac{2r^2 + 17k + 30}{2k+5}$

$$\begin{array}{r} k + 6 \\ 2k+5\overline{)2k^2 + 17k + 30} \\ \underline{2k^2 + 5k} \\ 12k + 30 \\ \underline{12k + 30} \\ 0 \end{array}$$

Answer: $k + 6$

(c) $\dfrac{10y^3 - 39y^2 + 41y - 10}{2y-5}$

$$\begin{array}{r} 5y^2 - 7y + 3 \\ 2y-5\overline{)10y^3 - 39y^2 + 41y - 10} \\ \underline{10y^3 - 25y^2} \\ -14y^2 + 41y \\ \underline{-14y^2 + 35y} \\ 6y - 10 \\ \underline{6y - 15} \\ 5 \end{array}$$

Remainder

Answer: $5y^2 - 7y + 3 + \dfrac{5}{2y-5}$

3. **(a)** $\dfrac{2r^2+2}{2r-4}$

Add a term with a 0 coefficient for the missing r-term.

$$\begin{array}{r}
r+2 \\
2r-4\overline{)2r^2+0r+2} \\
\underline{2r^2-4r} \\
4r+2 \\
\underline{4r-8} \\
10 \\
\textit{Remainder}
\end{array}$$

Answer: $r+2+\dfrac{10}{2r-4}$

(b) $\dfrac{3k^3+9k-12}{3k-3}$

Add $0k^2$ for the missing k^2-term.

$$\begin{array}{r}
k^2+k+4 \\
3k-3\overline{)3k^3+0k^2+9k-12} \\
\underline{3k^3-3k^2} \\
3k^2+9k \\
\underline{3k^2-3k} \\
12k-12 \\
\underline{12k-12} \\
0
\end{array}$$

Answer: k^2+k+4

(c) $\dfrac{5x^4-13x^3+11x^2-33x}{5x-3}$

Add a 0 for the missing constant term.

$$\begin{array}{r}
x^3-2x^2+x-6 \\
5x-3\overline{)5x^4-13x^3+11x^2-33x+0} \\
\underline{5x^4-3x^3} \\
-10x^3+11x^2 \\
\underline{-10x^3+6x^2} \\
5x^2-33x \\
\underline{5x^2-3x} \\
-30x+0 \\
\underline{-30x+18} \\
-18 \\
\textit{Remainder}
\end{array}$$

Answer: $x^3-2x^2+x-6+\dfrac{-18}{5x-3}$

4. **(a)** $\dfrac{4x^4-7x^2+x+5}{2x^2-x}$

The polynomial has a missing x^3-term.

$$\begin{array}{r}
2x^2+x-3 \\
2x^2-x\overline{)4x^4+0x^3-7x^2+x+5} \\
\underline{4x^4-2x^3} \\
2x^3-7x^2 \\
\underline{2x^3-x^2} \\
-6x^2+x \\
\underline{-6x^2+3x} \\
-2x+5 \\
\textit{Remainder}
\end{array}$$

Answer: $2x^2+x-3+\dfrac{-2x+5}{2x^2-x}$

(b) $\dfrac{3r^5-15r^4-2r^3+19r^2-7}{3r^2-2}$

Each polynomial has a missing r-term.

$$\begin{array}{r}
r^3-5r^2+3 \\
3r^2+0r-2\overline{)3r^5-15r^4-2r^3+19r^2+0r-7} \\
\underline{3r^5+0r^4-2r^3} \\
-15r^4+0r^3+19r^2 \\
\underline{-15r^4+0r^3+10r^2} \\
9r^2+0r-7 \\
\underline{9r^2+0r-6} \\
-1 \\
\textit{Remainder}
\end{array}$$

Answer: $r^3-5r^2+3+\dfrac{-1}{3r^2-2}$

5.5 Section Exercises

3. When dividing polynomials that are not monomials, the first step is to write them in *descending powers*.

7. $$\begin{aligned}\frac{15m^3+25m^2+30m}{5m^2} &= \frac{15m^3}{5m^2}+\frac{25m^2}{5m^2}+\frac{30m}{5m^2} \\ &= 3m+5+\frac{6}{m}\end{aligned}$$

11. $$\begin{array}{r}
r^2-7r+6 \\
3r-1\overline{)3r^3-22r^2+25r-6} \\
\underline{3r^3-r^2} \\
-21r^2+25r \\
\underline{-21r^2+7r} \\
18r-6 \\
\underline{18r-6} \\
0
\end{array}$$

Answer: r^2-7r+6

15. $\dfrac{3t^2+17t+10}{3t+2}$

$$\begin{array}{r} t+5 \\ 3t+2\overline{)3t^2+17t+10} \\ \underline{3t^2+2t} \\ 15t+10 \\ \underline{15t+10} \\ 0 \end{array}$$

Answer: $t+5$

19. $(4x^3+9x^2-10x+3)\div(4x+1)$

$$\begin{array}{r} x^2+2x-3 \\ 4x+1\overline{)4x^3+9x^2-10x+3} \\ \underline{4x^3+x^2} \\ 8x^2-10x \\ \underline{8x^2+2x} \\ -12x+3 \\ \underline{-12x-3} \\ 6 \\ \textit{Remainder} \end{array}$$

Answer: $x^2+2x-3+\dfrac{6}{4x+1}$

23. $\dfrac{4k^4+6k^3+3k-1}{2k^2+1}$

Each polynomial has a missing term.

$$\begin{array}{r} 2k^2+3k-1 \\ 2k^2+0k+1\overline{)4k^4+6k^3+0k^2+3k-1} \\ \underline{4k^4+0k^3+2k^2} \\ 6k^3-2k^2+3k \\ \underline{6k^3+0k^2+3k} \\ -2k^2+0k-1 \\ \underline{-2k^2-0k-1} \\ 0 \end{array}$$

Answer: $2k^2+3k-1$

27. $\left(2x^2-\dfrac{7}{3}x-1\right)\div(3x+1)$

$$\begin{array}{r} \frac{2}{3}x-1 \\ 3x+1\overline{)2x^2-\frac{7}{3}x-1} \\ \underline{2x^2+\frac{2}{3}x} \\ -3x-1 \\ \underline{-3x-1} \\ 0 \end{array}$$

Answer: $\dfrac{2}{3}x-1$

31. $P(x)=2x^3+15x^2+28x$; $Q(x)=x^2+4x$

$$\left(\frac{P}{Q}\right)(x)=\frac{P(x)}{Q(x)}=\frac{2x^3+15x^2+28x}{x^2+4x}$$

$$\begin{array}{r} 2x+7 \\ x^2+4x\overline{)2x^3+15x^2+28x} \\ \underline{2x^3+8x^2} \\ 7x^2+28x \\ \underline{7x^2+28x} \\ 0 \end{array}$$

$$\left(\frac{P}{Q}\right)(x)=2x+7$$

33. If $t=10$, then

$$\begin{aligned} 1654 &= 1000+600+50+4 \\ &= 10^3+6(10)^2+5(10)+4 \\ &= t^3+6t^2+5t+4. \end{aligned}$$

34. If $t=10$, then

$$14=10+4=t+4.$$

35.

$$\begin{array}{r} t^2+2t-3 \\ t+4\overline{)t^3+6t^2+5t+4} \\ \underline{t^3+4t^2} \\ 2t^2+5t \\ \underline{2t^2+8t} \\ -3t+4 \\ \underline{-3t-12} \\ 16 \end{array}$$

Answer: $t^2+2t-3+\dfrac{16}{t+4}$

36.

$$\begin{array}{r} 118 \\ 14\overline{)1654} \\ \underline{14} \\ 25 \\ \underline{14} \\ 114 \\ \underline{112} \\ 2 \end{array}$$

Answer: $118\frac{2}{14}$ or $118\frac{1}{7}$

37. If $t=10$, then

$$\begin{aligned} t^2+2t-3+\frac{16}{t+4} &= 10^2+2(10)-3+\frac{16}{10+4} \\ &= 100+20-3+\frac{16}{14} \\ &= 117+\frac{16}{14} \\ &= 118\frac{2}{14}\ or 118\frac{1}{7}. \end{aligned}$$

38. The quotients in Exercises 36 and 37 are the same.

5.6 Synthetic Division

5.6 Margin Exercises

1. **(a)** $\dfrac{3z^2 + 10z - 8}{z + 4}$

$$\begin{array}{r|rrr} -4 & 3 & 10 & -8 \\ & & -12 & 8 \\ \hline & 3 & -2 & 0 \end{array}$$

Remainder

Read the quotient from the bottom row.

Answer: $3z - 2$

(b) $(2x^2 + 3x - 5) \div (x + 1)$

$$\begin{array}{r|rrr} -1 & 2 & 3 & -5 \\ & & -2 & -1 \\ \hline & 2 & 1 & -6 \end{array}$$

Remainder

Answer: $2x + 1 + \dfrac{-6}{x + 1}$

2. **(a)** $\dfrac{3a^3 - 2a + 21}{a + 2}$

Insert a 0 for the missing a^2-term.

$$\begin{array}{r|rrrr} -2 & 3 & 0 & -2 & 21 \\ & & -6 & 12 & -20 \\ \hline & 3 & -6 & 10 & 1 \end{array}$$

Answer: $3a^2 - 6a + 10 + \dfrac{1}{a + 2}$

(b) $(-4x^4 + 3x^3 + 18x + 2) \div (x - 2)$

Insert a 0 for the missing x^2-term.

$$\begin{array}{r|rrrrr} 2 & -4 & 3 & 0 & 18 & 2 \\ & & -8 & -10 & -20 & -4 \\ \hline & -4 & -5 & -10 & -2 & -2 \end{array}$$

Answer: $-4x^3 - 5x^2 - 10x - 2 + \dfrac{-2}{x - 2}$

3. Let $P(x) = x^3 - 5x^2 + 7x - 3$.

To find $P(a)$, divide the polynomial by $x - a$ using synthetic division. $P(a)$ will be the remainder.

(a) Find $P(1)$.

Divide $P(x)$ by $x - 1$.

$$\begin{array}{r|rrrr} 1 & 1 & -5 & 7 & -3 \\ & & 1 & -4 & 3 \\ \hline & 1 & -4 & 3 & 0 \end{array}$$

Remainder

The remainder is 0, so by the remainder theorem, $P(1) = 0$.

(b) Find $P(-2)$.

Divide $P(x)$ by $x - (-2) = x + 2$.

$$\begin{array}{r|rrrr} -2 & 1 & -5 & 7 & -3 \\ & & -2 & 14 & -42 \\ \hline & 1 & -7 & 21 & -45 \end{array}$$

Remainder

By the remainder theorem, $P(-2) = -45$.

4. To check whether 2 is a solution to $P(x) = 0$, divide the polynomial by $x - 2$. If the remainder is 0, then 2 is a solution. Otherwise, it is not.

(a) $3r^3 - 11r^2 + 17r - 14 = 0$

$$\begin{array}{r|rrrr} 2 & 3 & -11 & 17 & -14 \\ & & 6 & -10 & 14 \\ \hline & 3 & -5 & 7 & 0 \end{array}$$

Remainder

Since the remainder is 0, 2 is a solution to the given equation.

(b) $4k^5 - 7k^4 - 11k^2 + 2k + 6 = 0$

Insert a 0 for the missing k^3-term.

$$\begin{array}{r|rrrrrr} 2 & 4 & -7 & 0 & -11 & 2 & 6 \\ & & 8 & 2 & 4 & -14 & -24 \\ \hline & 4 & 1 & 2 & -7 & -12 & -18 \end{array}$$

Remainder

Since the remainder is not 0, 2 is not a solution to the given equation.

5.6 Section Exercises

3. Since the variables are not present, a missing term will not be noticed in synthetic division, so the quotient will be wrong if placeholders are not inserted.

7. $\dfrac{4m^2 + 19m - 5}{m + 5}$

$$\begin{array}{r|rrr} -5 & 4 & 19 & -5 \\ & & -20 & 5 \\ \hline & 4 & -1 & 0 \end{array}$$

Answer: $4m - 1$

11. $(p^2 - 3p + 5) \div (p + 1)$

$$\begin{array}{r|rrr} -1 & 1 & -3 & 5 \\ & & -1 & 4 \\ \hline & 1 & -4 & 9 \end{array}$$

Remainder

Answer: $p - 4 + \dfrac{9}{p + 1}$

15. $(x^5 - 2x^3 + 3x^2 - 4x - 2) \div (x - 2)$

Insert a 0 for the missing x^4-term.

$$\begin{array}{r|rrrrrr} 2 & 1 & 0 & -2 & 3 & -4 & -2 \\ & & 2 & 4 & 4 & 14 & 20 \\ \hline & 1 & 2 & 2 & 7 & 10 & 18 \end{array}$$

Answer:

$$x^4 + 2x^3 + 2x^2 + 7x + 10 + \frac{18}{x - 2}$$

19. $(-3y^5 + 2y^4 - 5y^3 - 6y^2 - 1) \div (y + 2)$

Insert a 0 for the missing y-term.

$$\begin{array}{r|rrrrrr} -2 & -3 & 2 & -5 & -6 & 0 & -1 \\ & & 6 & -16 & 42 & -72 & 144 \\ \hline & -3 & 8 & -21 & 36 & -72 & 143 \end{array}$$

Answer:

$$-3y^4 + 8y^3 - 21y^2 + 36y - 72 + \frac{143}{y + 2}$$

23. $P(x) = 2x^3 - 4x^2 + 5x - 3; k = 2$

To find $P(2)$, divide the polynomial by $x - 2$. $P(2)$ will be the remainder.

$$\begin{array}{r|rrrr} 2 & 2 & -4 & 5 & -3 \\ & & 4 & 0 & 10 \\ \hline & 2 & 0 & 5 & 7 \end{array}$$

Remainder

By the remainder theorem, $P(2) = 7$.

27. $P(y) = 2y^3 - 4y^2 + 5y - 33; k = 3$

$$\begin{array}{r|rrrr} 3 & 2 & -4 & 5 & -33 \\ & & 6 & 6 & 33 \\ \hline & 2 & 2 & 11 & 0 \end{array}$$

Remainder

By the remainder theorem, $P(3) = 0$.

31. $m^4 + 2m^3 - 3m^2 + 8m - 8 = 0; m = -2$

To decide whether -2 is a solution to the given equation, divide the polynomial by $m + 2$.

$$\begin{array}{r|rrrrr} -2 & 1 & 2 & -3 & 8 & -8 \\ & & -2 & 0 & 6 & -28 \\ \hline & 1 & 0 & -3 & 14 & -36 \end{array}$$

Remainder

Since the remainder is not 0, -2 is not a solution to the equation.

5.7 Greatest Common Factors; Factoring by Grouping

5.7 Margin Exercises

1. **(a)** $7k + 28 = 7 \cdot k + 7 \cdot 4 = 7(k + 4)$

   (b) $32m + 24 = 8 \cdot 4m + 8 \cdot 3 = 8(4m + 3)$

   (c) $8a - 9$ cannot be factored because $8a$ and 9 do not have a common factor other than 1.

   (d) $5z + 5 = 5 \cdot z + 5 \cdot 1 = 5(z + 1)$

2. **(a)** $16y^4 + 8y^3$

   The numerical part of the greatest common factor is 8, the largest number that is a factor of both 16 and 8. For the variables, the lowest degree is 3. The greatest common factor is therefore $8y^3$.

$$\begin{aligned} 16y^4 + 8y^3 &= 8y^3 \cdot 2y + 8y^3 \cdot 1 \\ &= 8y^3(2y + 1) \end{aligned}$$

   (b) $14p^2 - 9p^3 + 6p^4$

   The lowest degree is 2. The greatest common factor is thus p^2.

$$= p^2(14 - 9p + 6p^2)$$

   (c) $15z^2 + 45z^5 - 60z^6$
   $= 15z^2(1 + 3z^3 - 4z^4)$

3. **(a)** $12y^5x^2 + 8y^3x^3$

The numerical part of the greatest common factor is 4. The lowest degree of x is 2, and the lowest degree of y is 3. Therefore, the greatest common factor is $4y^3x^2$.

$$= 4y^3x^2(3y^2 + 2x)$$

(b) $5m^4x^3 + 15m^5x^6 - 20m^4x^6$

The numerical part of the greatest common factor is 5. The lowest degree on m is 4, and the lowest degree on x is 3. Therefore, the greatest common factor is $5m^4x^3$.

$$= 5m^4x^3(1 + 3mx^3 - 4x^3)$$

4. **(a)** $(a+2)(a-3) + (a+2)(a+6)$

The greatest common factor is $(a+2)$.

$$\begin{aligned} &= (a+2)[(a-3) + (a+6)] \\ &= (a+2)(a-3+a+6) \\ &= (a+2)(2a+3) \end{aligned}$$

(b) $(y-1)(y+3) - (y-1)(y+4)$

$$\begin{aligned} &= (y-1)[(y+3) - (y+4)] \\ &= (y-1)(y+3-y-4) \\ &= (y-1)(-1) \\ &= -y+1 \end{aligned}$$

(c) $k^2(a+5b) + m^2(a+5b)$

The greatest common factor is $(a+5b)$.

$$= (a+5b)(k^2+m^2)$$

(d) $r^2(y+6) + r^2(y+3)$

$$\begin{aligned} &= r^2[(y+6) + (y+3)] \\ &= r^2(y+6+y+3) \\ &= r^2(2y+9) \end{aligned}$$

5. **(a)** $-k^2 + 3k$

There are two ways to factor this polynomial. One way is to factor out k.

$$-k^2 + 3k = k(-k+3)$$

Another way is to factor out $-k$.

$$-k^2 + 3k = -k(k-3)$$

(b) $-6r^2 + 5r$

Factor out r.

$$-6r^2 + 5r = r(-6r+5)$$

Alternatively, factor out $-r$.

$$-6r^2 + 5r = -r(6r-5)$$

6. **(a)** $mn + 6 + 2n + 3m$

Rearrange the terms so that there is a common factor in the first two terms and a common factor in the last two terms.

$$= mn + 2n + 6 + 3m$$

Group the first two terms and the last two terms.

$$\begin{aligned} &= (mn+2n) + (6+3m) \\ &= n(m+2) + 3(2+m) \\ &= n(m+2) + 3(m+2) \\ &= (m+2)(n+3) \end{aligned}$$

(b) $4y - zx + yx - 4z$

$$\begin{aligned} &= 4y + yx - zx - 4z \\ &= (4y+yx) + (-zx-4z) \\ &= y(4+x) - z(x+4) \\ &= y(x+4) - z(x+4) \\ &= (x+4)(y-z) \end{aligned}$$

7. **(a)** $xy + 2y - 4x - 8$

$$\begin{aligned} &= (xy+2y) - (4x+8) \\ &= y(x+2) - 4(x+2) \\ &= (x+2)(y-4) \end{aligned}$$

(b) $10p^2 + 15p - 12p - 18$

$$\begin{aligned} &= (10p^2+15p) - (12p+18) \\ &= 5p(2p+3) - 6(2p+3) \\ &= (2p+3)(5p-6) \end{aligned}$$

5.7 Section Exercises

3. The greatest common factor of $7z^2(m+n)^4$ and $9z^3(m+n)^5$ is $z^2(m+n)^4$.

7. $6m(r+t)^2$ has the positive factors $1, 2, 3, 6$, and m and the factors of $(r+t)^2$.
$3p(r+t)^4$ has the positive factors $1, 3$, and p and the factors of $(r+t)^4$.
The greatest common factor of the two terms is $1 \cdot 3(r+t)^2$, or $3(r+t)^2$.

11. $8k^3 + 24k = 8k \cdot k^2 + 8k \cdot 3 = 8k(k^2+3)$

15. $-4p^3q^4 - 2p^2q^5 = -2p^2q^4(2p+q)$

The greatest common factor is $2p^2q^4$. Since both terms are negative, factor out -1 also.

19. $15a^2c^3 - 25ac^2 + 5a^2c$
$= 5ac(3ac^2 - 5c + a)$

23. $(m-4)(m+2)+(m-4)(m+3)$

The greatest common factor is $(m-4)$.

$$\begin{aligned}&=(m-4)[(m+2)+(m+3)]\\&=(m-4)(m+2+m+3)\\&=(m-4)(2m+5)\end{aligned}$$

27. $-y^5(r+w)-y^6(z+k)$

The common factor is y^5. Also, factor out the negative factor, -1.

$$\begin{aligned}&=-y^5[(r+w)+y(z+k)]\\&=-y^5(r+w+yz+yk)\end{aligned}$$

31. $42z^3-56z^4$

Factor out $14z^3$.

$$42z^3-56z^4=14z^3(3-4z)$$

Alternatively, factor out $-14z^3$.

$$42z^3-56z^4=-14z^3(-3+4z)$$

35. $2k+2h+jk+jh$

$$\begin{aligned}&=(2k+2h)+(jk+jh)\\&=2(k+h)+j(k+h)\\&=(k+h)(2+j)\end{aligned}$$

39. $z^2-6z-54+9z$

$$\begin{aligned}&=(z^2-6z)+(-54+9z)\\&=z(z-6)+9(-6+z)\\&=z(z-6)+9(z-6)\\&=(z-6)(z+9)\end{aligned}$$

43. $7k+2k^2-6k-21$

$$\begin{aligned}&=(7k+2k^2)+(-6k-21)\\&=k(7+2k)-3(2k+7)\\&=k(2k+7)-3(2k+7)\\&=(2k+7)(k-3)\end{aligned}$$

47. $-16m^3+4m^2p^2-4mp+p^3$

$$\begin{aligned}&=(-16m^3+4m^2p^2)+(-4mp+p^3)\\&=4m^2(-4m+p^2)+p(-4m+p^2)\\&=(-4m+p^2)(4m^2+p)\end{aligned}$$

51. $x^3y^2-3-3y^2+x^3$

$$\begin{aligned}&=(x^3y^2-3y^2)+(-3+x^3)\\&=y^2(x^3-3)+1(-3+x^3)\\&=y^2(x^3-3)+1(x^3-3)\\&=(x^3-3)(y^2+1)\end{aligned}$$

55. $k^{-2}+2k^{-4}$

The smaller exponent is -4. Therefore, factor out k^{-4}.

$$k^{-2}+2k^{-4}=k^{-4}(k^2+2)$$

5.8 Factoring Trinomials

5.8 Margin Exercises

1.

Product	*Sum*
$12\cdot(1)=12;$	$12+1=13$
$6\cdot(2)=12;$	$6+2=8$
$3\cdot(4)=12;$	$3+4=7$
$-12\cdot(-1)=12;$	$-12+(-1)=-13$
$-6\cdot(-2)=12;$	$-6+(-2)=-8$
$-3\cdot(-4)=12;$	$-3+(-4)=-7$

2. **(a)** p^2+6p+5

Look for factors of 5 that add to 6.

Factors	*Sum*	
$5\cdot1$	6	*Yes*

Use 5 and 1.

$$p^2+6p+5=(p+5)(p+1)$$

(b) $a^2+9a+20$

Look for factors of 20 that add to 9.

Factors	*Sums*	
$20\cdot1$	21	*No*
$10\cdot2$	12	*No*
$5\cdot4$	9	*Yes*

Use 5 and 4.

$$a^2+9a+20=(a+5)(a+4)$$

3. **(a)** k^2-k-6

Find two numbers whose product is -6 and whose sum is -1.

Factors	*Sums*	
$-6\cdot1$	-5	*No*
$-3\cdot2$	-1	*Yes*

The numbers are -3 and 2.

$$k^2-k-6=(k-3)(k+2)$$

(b) $b^2-7b+10$

Find two numbers whose product is 10 and whose sum is -7.

Factors	*Sums*	
$-10(-1)$	-11	*No*
$-5(-2)$	-7	*Yes*

Use -5 and -2.

$$b^2 - 7b + 10 = (b-5)(b-2)$$

(c) $y^2 - 8y + 6$

Find two numbers whose product is 6 and whose sum is -8.

Factors	*Sums*	
$-6(-1)$	-7	*No*
$-3(-2)$	-5	*No*
$3(2)$	5	*No*
$6(1)$	7	*No*

None of the pairs of factors have a sum of -8, so $y^2 - 8y + 6$ cannot be factored and is prime.

4. **(a)** $m^2 + 2mn - 8n^2$

We need two expressions whose sum is $2n$ and whose product is $-8n^2$. The quantities $4n$ and $-2n$ have the necessary sum and product.

$$m^2 + 2mn - 8n^2 = (m+4n)(m-2n)$$

(b) $z^2 - 7zx + 9x^2$

We need two expressions whose sum is $-7x$ and whose product is $9x^2$. There are no such quantities. Therefore, the trinomial cannot be factored and is prime.

5. $5m^4 - 5m^3 - 100m^2$

First, factor out the greatest common factor of $5m^2$.

$$5m^4 - 5m^3 - 100m^2 = 5m^2(m^2 - m - 20)$$

Next, look for two integers whose product is -20 and whose sum is -1. The necessary integers are -5 and 4. Remember to write the common factor $5m^2$ as part of the answer.

$$5m^4 - 5m^3 - 100m^2 = 5m^2(m-5)(m+4)$$

6. **(a)** $3y^2 - 11y - 4$

The product ac is $3(-4) = -12$. Look for two integers whose product is -12 and whose sum is -11. The necessary integers are -12 and 1. Write $-11y$ as $-12y + y$.

$$\begin{aligned} 3y^2 - 11y - 4 &= 3y^2 - 12y + y - 4 \\ &= 3y(y-4) + 1(y-4) \\ &= (y-4)(3y+1) \end{aligned}$$

(b) $6k^2 - 19k + 10$

The product ac is $6(10) = 60$. Look for two integers whose product is 60 and whose sum is -19. The necessary integers are -15 and -4. Write $-19k$ as $-15k - 4k$.

$$\begin{aligned} 6k^2 - 19k + 10 &= 6k^2 - 15k - 4k + 10 \\ &= 3k(2k-5) - 2(2k-5) \\ &= (2k-5)(3k-2) \end{aligned}$$

7. **(a)** $2p^2 + 5p - 12$

The factors of $2p^2$ are $2p$ and p. Now find factors of -12 that will produce a middle term with coefficient 5.

Try various possibilities.

$(2p+1)(p-12) = 2p^2 - 23p - 12$
Wrong middle term

$(2p+3)(p-4) = 2p^2 - 5p - 12$
Opposite middle term

Switch the signs on the factors to get the correct answer.

$$2p^2 + 5p - 12 = (2p-3)(p+4)$$

(b) $6k^2 - k - 2$

Try some possibilities.

$(6k+1)(k-2) = 6k^2 - 11k - 2$
Wrong middle term

$(3k+1)(2k-2) = 6k^2 - 4k - 2$
Wrong middle term

$(3k-2)(2k+1) = 6k^2 - k - 2$
Correct middle term

The correct answer is

$$6k^2 - k - 2 = (3k-2)(2k+1).$$

(c) $8m^2 + 18m - 5$

Try some possibilities.

$(8m+1)(m-5) = 8m^2 - 39m - 5$
Wrong middle term

$(8m+5)(m-1) = 8m^2 - 3m - 5$
Wrong middle term

$(4m+1)(2m-5) = 8m^2 - 18m - 5$
Opposite middle term

Exchange the signs on the factors. The correct answer is

$$8m^2 + 18m - 5 = (4m - 1)(2m + 5).$$

(d) $-6r^2 + 13r + 5$

First, factor out -1.

$-6r^2 + 13r + 5 = -(6r^2 - 13r - 5)$

Try some possibilities.

$(6r + 1)(r - 5) = 6r^2 - 29r - 5$
Wrong middle term

$(6r + 5)(r - 1) = 6r^2 - r - 5$
Wrong middle term

$(2r - 5)(3r + 1) = 6r^2 - 13r - 5$
Correct middle term

The correct answer is

$$-6r^2 + 13r + 5 = -(2r - 5)(3r + 1)$$
$$\text{or} \quad (-2r + 5)(3r + 1).$$

8. **(a)** $7p^2 + 15pq + 2q^2$

Try some possibilities.

$(7p + 2q)(p + q) = 7p^2 + 9pq + 2q^2$ *No*

$(7p + q)(p + 2q) = 7p^2 + 15pq + 2q^2$ *Yes*

The correct answer is

$$7p^2 + 15pq + 2q^2 = (7p + q)(p + 2q).$$

(b) $6m^2 + 7mn - 5n^2$

Try some possibilities.

$(3m + n)(2m - 5n) = 6m^2 - 13mn - 5n^2$ *No*

$(3m + 5n)(2m - n) = 6m^2 + 7mn - 5n^2$ *Yes*

The correct answer is

$$6m^2 + 7mn - 5n^2 = (3m + 5n)(2m - n).$$

(c) $12z^2 - 5zy - 2y^2$

Try some possibilities.

$(12z + y)(z - 2y) = 12z^2 - 23yz - 2y^2$ *No*

$(3z - 2y)(4z + y) = 12z^2 - 5yz - 2y^2$ *Yes*

The correct answer is

$$12z^2 - 5yz - 2y^2 = (3z - 2y)(4z + y).$$

9. **(a)** $2m^3 - 4m^2 - 6m$

First, factor out the greatest common factor, $2m$.

$$= 2m(m^2 - 2m - 3)$$

Find two integers whose product is -3 and whose sum is -2. The numbers are 1 and -3.

$$= 2m(m + 1)(m - 3)$$

(b) $12r^4 + 6r^3 - 90r^2$

First, factor out the greatest common factor, $6r^2$.

$$= 6r^2(2r^2 + r - 15)$$

Look for two integers whose product is $2(-15) = -30$ and whose sum is 1. The integers are 6 and -5. Write r as $6r - 5r$.

$$\begin{aligned} &= 6r^2(2r^2 + 6r - 5r - 15) \\ &= 6r^2[2r(r + 3) - 5(r + 3)] \\ &= 6r^2(r + 3)(2r - 5) \end{aligned}$$

(c) $30y^5 - 55y^4 - 50y^3$

First, factor out the greatest common factor, $5y^3$.

$$= 5y^3(6y^2 - 11y - 10)$$

Use one of the above methods to factor the trinomial.

$$= 5y^3(2y - 5)(3y + 2)$$

10. **(a)** $y^4 + y^2 - 6 = (y^2)^2 + (y^2) - 6$
$\qquad = (y^2 - 2)(y^2 + 3)$

(b) $2p^4 + 7p^2 - 15 = 2(p^2)^2 + 7(p^2) - 15$
$\qquad = (2p^2 - 3)(p^2 + 5)$

(c) $6r^4 - 13r^2 + 5 = 6(r^2)^2 - 13(r^2) + 5$
$\qquad = (3r^2 - 5)(2r^2 - 1)$

11. **(a)** $6(a - 1)^2 + (a - 1) - 2$

Let $x = a - 1$.

$$\begin{aligned} &= 6x^2 + x - 2 \\ &= (2x - 1)(3x + 2) \end{aligned}$$

Replace x with $a - 1$. Therefore,

$$\begin{aligned} &6(a - 1)^2 + (a - 1) - 2 \\ &= [2(a - 1) - 1][3(a - 1) + 2] \\ &= (2a - 2 - 1)(3a - 3 + 2) \\ &= (2a - 3)(3a - 1). \end{aligned}$$

(b) $8(z+5)^2 - 2(z+5) - 3$

Let $x = z + 5$.

$$= 8x^2 - 2x - 3$$
$$= (2x+1)(4x-3)$$

Replace x with $z + 5$. Therefore,

$$8(z+5)^2 - 2(z+5) - 3$$
$$= [2(z+5)+1][4(z+5)-3]$$
$$= (2z+10+1)(4z+20-3)$$
$$= (2z+11)(4z+17).$$

(c) $15(m-4)^2 - 11(m-4) + 2$

Let $x = m - 4$.

$$= 15x^2 - 11x + 2$$
$$= (5x-2)(3x-1)$$

Replace x with $m - 4$.

$$= [5(m-4)-2][3(m-4)-1]$$
$$= (5m-20-2)(3m-12-1)$$
$$= (5m-22)(3m-13)$$

5.8 Section Exercises

3. It is not necessary to factor out any common factors when factoring a trinomial. False; begin by factoring out the greatest common factor.

7. $p^2 - p - 56$

Look for the two integers whose product is -56 and whose sum is -1.

Factors	*Sums*	
$-56 \cdot 1$	-55	*No*
$-28 \cdot 2$	-26	*No*
$-14 \cdot 4$	-10	*No*
$-8 \cdot 7$	-1	*Yes*

$$p^2 - p - 56 = (p-8)(p+7)$$

11. $a^2 - 2ab - 35b^2$

Look for two expressions whose product is $-35b^2$ and whose sum is $-2b$. The numbers are $-7b$ and $5b$.

$$a^2 - 2ab - 35b^2 = (a-7b)(a+5b)$$

15. $x^2y^2 + 11xy + 18$

Look for two integers whose product is 18 and whose sum is 11. The integers 9 and 2 satisfy these conditions.

$$x^2y^2 + 11xy + 18 = (xy+9)(xy+2)$$

19. $10x^2 + 3x - 18$

Look for two integers whose product is $10(-18) = -180$ and whose sum is 3. The necessary integers are 15 and -12. Rewrite the polynomial, and factor by grouping. Write $3x$ as $15x - 12x$.

$$10x^2 + 3x - 18$$
$$= 10x^2 + 15x - 12x - 18$$
$$= 5x(2x+3) - 6(2x+3)$$
$$= (2x+3)(5x-6)$$

23. $15a^2 - 22ab + 8b^2$

Find two integers whose product is $15(8) = 120$ and whose sum is -22. The numbers are -10 and -12. Write $-22ab$ as $-10ab - 12ab$.

$$15a^2 - 22ab + 8b^2$$
$$= 15a^2 - 10ab - 12ab + 8b^2$$
$$= 5a(3a-2b) - 4b(3a-2b)$$
$$= (3a-2b)(5a-4b)$$

27. $40x^2 + xy + 6y^2$

Look for two integers whose product is $40(6) = 240$ and whose sum is 1. There are no such integers, so the trinomial cannot be factored and is prime.

31. $24x^2 + 42x + 15$

First factor out the greatest common factor.

$$24x^2 + 42x + 15$$
$$= 3(8x^2 + 14x + 5)$$

Look for two integers whose product is $8(5) = 40$ and whose sum is 14. The numbers are 10 and 4.

$$= 3(8x^2 + 10x + 4x + 5)$$
$$= 3[2x(4x+5) + 1(4x+5)]$$
$$= 3(4x+5)(2x+1)$$

35. $11x^3 - 110x^2 + 264x$

$$= 11x(x^2 - 10x + 24)$$
$$= 11x(x-6)(x-4)$$

39. She lost some credit for factoring

$$4x^2 + 2x - 20$$

as $(4x+10)(x-2)$ because the factor $4x + 10$ can be factored further as $2(2x+5)$.

43. $16x^4 + 16x^2 + 3$

$$= 16(x^2)^2 + 16(x^2) + 3$$
$$= (4x^2+1)(4x^2+3)$$

47. $10(k+1)^2 - 7(k+1) + 1$

Let $x = k+1$.

$$= 10x^2 - 7x + 1$$
$$= (5x-1)(2x-1)$$

Replace x with $k+1$.

$$= [5(k+1)-1][2(k+1)-1]$$
$$= (5k+5-1)(2k+2-1)$$
$$= (5k+4)(2k+1)$$

51. $a^2(a+b)^2 - ab(a+b)^2 - 6b^2(a+b)^2$

The greatest common factor is $(a+b)^2$.

$$= (a+b)^2(a^2 - ab - 6b^2)$$
$$= (a+b)^2(a-3b)(a+2b)$$

53. No, 2 is not a factor of 45 since 45 is an odd number.

54. The positive factors of 45 are 1 and 45, 3 and 15, and 5 and 9. No, 2 is not a factor of any of these factors.

55. No, 5 is not a factor of $10x^2 + 29x + 10$ since it is not a factor of 29.

56. $10x + 29x + 10$

Look for two integers whose product is $10(10) = 100$ and whose sum is 29. The numbers are 4 and 25. Write $29x$ as $4x + 25x$.

$$10x^2 + 29x + 10 = 10x^2 + 4x + 25x + 10$$
$$= 2x(5x+2) + 5(5x+2)$$
$$= (5x+2)(2x+5)$$

No, 5 is not a factor of either of these factors.

57. Since k is odd, 2 is not a factor of $2x^2 + kx + 8$. Because 2 is a factor of $2x+4$, the binomial $2x+4$ cannot be a factor of $2x^2 + kx + 8$.

58. $3y + 15$ cannot be a factor of $12y^2 - 11y - 15$ because 3 is a factor of $3y + 15$, but it is not a factor of $12y^2 - 11y - 15$.

5.9 Special Factoring

5.9 Margin Exercises

1. **(a)** $9a^2 - 16b^2$

This is the difference of two squares.

$$= (3a)^2 - (4b)^2$$
$$= (3a+4b)(3a-4b)$$

(b) $(m+3)^2 - 49z^2$

$$= (m+3)^2 - (7z)^2$$
$$= (m+3+7z)(m+3-7z)$$

2. **(a)** $z^2 + 12z + 36 = z^2 + 12z + 6^2$

If this is a square trinomial the middle term will be

$$2(z)(6) = 12z.$$

Since this is true, $z^2+12z+36$ is a square trinomial.

(b) $2x^2 - 4x + 4$

This is not a square trinomial since $2x^2$ is not a perfect square.

(c) $9a^2 + 12ab + 16b^2 = (3a)^2 + 12ab + (4b)^2$

If this is a square trinomial, the middle term will be

$$2(3a)(4b) = 24ab.$$

Since the middle term is $12ab$, the trinomial is not a square trinomial.

3. **(a)** $49z^2 - 14zk + k^2 = (7z)^2 - 14zk + k^2$
$$= (7z-k)^2$$

Check: $2(7z)(k) = 14zk$, which is the middle term.

$$49z^2 - 14zk + k^2 = (7z-k)^2$$

(b) $9a^2 + 48ab + 64b^2 = (3a)^2 + 48ab + (8b)^2$
$$= (3a+8b)^2$$

Check: $2(3a)(8b) = 48ab$, which is the middle term.

$$9a^2 + 48ab + 64b^2 = (3a+8b)^2$$

(c) $(k+m)^2 - 12(k+m) + 36$
$$= [(k+m)-6]^2$$
$$= (k+m-6)^2$$

Check: $2(k+m)(6) = 12(k+m)$, which is the middle term.

(d) $x^2 - 2x + 1 - y^2$

Group the first three terms.

$$= (x^2 - 2x + 1) - y^2$$

Factor the square trinomial.

$$= (x-1)^2 - y^2$$

This is the difference of two squares.

$$= [(x-1) + y][(x-1) - y]$$
$$= (x - 1 + y)(x - 1 - y)$$

4. **(a)** $x^3 - 1000$
$$= x^3 - 10^3$$

This is the difference of two cubes.

$$= (x - 10)(x^2 + 10x + 10^2)$$
$$= (x - 10)(x^2 + 10x + 100)$$

(b) $8k^3 - y^3$
$$= (2k)^3 - y^3$$
$$= (2k - y)[(2k)^2 + 2ky + y^2]$$
$$= (2k - y)(4k^2 + 2ky + y^2)$$

(c) $27m^3 - 64$
$$= (3m)^3 - 4^3$$
$$= (3m - 4)[(3m)^2 + (3m)(4) + 4^2]$$
$$= (3m - 4)(9m^2 + 12m + 16)$$

5. **(a)** $2x^3 + 2000$

There is a common factor of 2.

$$= 2(x^3 + 1000)$$
$$= 2(x^3 + 10^3)$$

This is the sum of two cubes.

$$= 2(x + 10)(x^2 - 10x + 10^2)$$
$$= 2(x + 10)(x^2 - 10x + 100)$$

(b) $8p^3 + 125$
$$= (2p)^3 + 5^3$$
$$= (2p + 5)[(2p)^2 - (2p)(5) + 5^2]$$
$$= (2p + 5)(4p^2 - 10p + 25)$$

(c) $27m^3 + 125n^3$
$$= (3m)^3 + (5n)^3$$
$$= (3m + 5n)[(3m)^2 - (3m)(5n) + (5n)^2]$$
$$= (3m + 5n)(9m^2 - 15mn + 25n^2)$$

5.9 Section Exercises

3. **(a)** Since $x^2 - 8x - 16$ has a negative third term, it is not a square trinomial.

(b) $4m^2 + 20m + 25 = (2m^2) + 20m + 5^2$
$2(2m)(5) = 20m$, the middle term. Therefore, this trinomial is square.

(c) $9z^4 + 30z^2 + 25 = (3z^2)^2 + 30z^2 + 5^2$
$2(3z^2)(5) = 30z^2$, the middle term. Therefore, this trinomial is square.

(d) $25a^2 - 45a + 81 = (5a)^2 - 45a + (-9)^2$
$2(5a)(-9) = -90a$
This is not the middle term, so the trinomial is not square.

The square trinomials are (b) and (c).

7. $p^2 - 16 = p^2 - 4^2 = (p + 4)(p - 4)$

11. $9a^2 - 49b^2 = (3a)^2 - (7b)^2$
$$= (3a + 7b)(3a - 7b)$$

15. $(y + z)^2 - 81$
$$= (y + z)^2 - 9^2$$
$$= [(y + z) + 9][(y + z) - 9]$$
$$= (y + z + 9)(y + z - 9)$$

19. $(p + q)^2 - (p - q)^2$
$$= [(p + q) + (p - q)][(p + q) - (p - q)]$$
$$= (p + q + p - q)(p + q - p + q)$$
$$= 2p(2q)$$
$$= 4pq$$

23. $4z^2 + 4zw + w^2$
$$= (2z)^2 + 2(2z)(w) + w^2$$
$$= (2z + w)^2$$

27. $4r^2 - 12r + 9 - s^2$

Group the first three terms.

$$= (4r^2 - 12r + 9) - s^2$$
$$= [(2r)^2 - 2(2r)(3) + 3^2] - s^2$$
$$= (2r - 3)^2 - s^2$$
$$= [(2r - 3) + s][(2r - 3) - s]$$
$$= (2r - 3 + s)(2r - 3 - s)$$

31. $98m^2 + 84mn + 18n^2$
$$= 2(49m^2 + 42mn + 9n^2)$$
$$= 2[(7m)^2 + 2(7m)(3n) + (3n)^2]$$
$$= 2(7m + 3n)^2$$

35. $(a - b)^2 + 8(a - b) + 16$
$$= [(a - b) + 4]^2$$
$$= (a - b + 4)^2$$

39. $64g^3 + 27h^3$
$= (4g)^3 + (3h)^3$
$= (4g + 3h)[(4g)^2 - (4g)(3h) + (3h)^2]$
$= (4g + 3h)(16g^2 - 12gh + 9h^2)$

43. $(y + z)^3 - 64$
$= (y + z)^3 - 4^3$
$= [(y + z) - 4][(y + z)^2 + (y + z)(4) + 4^2]$
$= (y + z - 4)$
$\cdot (y^2 + 2yz + z^2 + 4y + 4z + 16)$

47. $(a + b)^3 - (a - b)^3$
$= [(a + b) - (a - b)]$
$\cdot [(a + b)^2 + (a + b)(a - b) + (a - b)^2]$
$= (a + b - a + b)(a^2 + 2ab + b^2$
$+ a^2 - b^2 + a^2 - 2ab + b^2)$
$= 2b(3a^2 + b^2)$

48. $x^6 - y^6$
$= (x^3)^2 - (y^3)^2$
$= (x^3 + y^3)(x^3 - y^3)$
$= [(x + y)(x^2 - xy + y^2)]$
$\cdot [(x - y)(x^2 + xy + y^2)]$
$= (x + y)(x^2 - xy + y^2)(x - y)$
$\cdot (x^2 + xy + y^2)$

49. $x^6 - y^6$
$= (x^2)^3 - (y^2)^3$
$= (x^2 - y^2)[(x^2)^2 + x^2y^2 + (y^2)^2]$
$= (x^2 - y^2)(x^4 + x^2y^2 + y^4)$
$= (x + y)(x - y)(x^4 + x^2y^2 + y^4)$

50. The remaining factors must be equal. To verify this, multiply the trinomial factors in Exercise 48.

$(x^2 + xy + y^2)(x^2 - xy + y^2)$
$= x^2(x^2 - xy + y^2)$
$+ xy(x^2 - xy + y^2)$
$+ y^2(x^2 - xy + y^2)$
$= x^4 - x^3y + x^2y^2 + x^3y - x^2y^2$
$+ xy^3 + x^2y^2 - xy^3 + y^4$
$= x^4 + x^2y^2 + y^4$

This is the trinomial factor in Exercise 49. Therefore, they are equal.

51. Start by factoring as the difference of two squares since doing so resulted in the complete factorization more directly.

Summary Exercises on Factoring Methods

3. $18p^5 - 24p^3 + 12p^6 = 6p^3(3p^2 - 4 + 2p^3)$

7. $49z^2 - 16 = (7z)^2 - 4^2$
$= (7z + 4)(7z - 4)$

11. $k^2 - 6k + 16$

There are no numbers that have a product of 16 and a sum of -6. The trinomial cannot be factored and is prime.

15. $40p^2 - 32p = 8p(5p - 4)$

19. $mn - 2n + 5m - 10$
$= n(m - 2) + 5(m - 2)$
$= (m - 2)(n + 5)$

23. $56k^3 - 875$
$= 7(8k^3 - 125)$
$= 7[(2k)^3 - 5^3]$
$= 7(2k - 5)[(2k)^2 + (2k)(5) + 5^2]$
$= 7(2k - 5)(4k^2 + 10k + 25)$

27. $m^2(m - 2) - 4(m - 2)$
$= (m - 2)(m^2 - 4)$
$= (m - 2)(m + 2)(m - 2)$
$= (m - 2)^2(m + 2)$

31. $125m^6 + 216$
$= (5m^2)^3 + 6^3$
$= (5m^2 + 6)[(5m^2)^2 - (5m^2)(6) + 6^2]$
$= (5m^2 + 6)(25m^4 - 30m^2 + 36)$

35. $4y^2 - 8y = 4y(y - 2)$

39. $256b^2 - 400c^2$
$= 16(16b^2 - 25c^2)$
$= 16[(4b)^2 - (5c)^2]$
$= 16(4b + 5c)(4b - 5c)$

43. $10r^2 + 23rs - 5s^2$

$$ac = 10(-5) = -50$$

The integers 25 and -2 have a product of -50 and a sum of 23.

$10r^2 + 23rs - 5s^2$
$= 10r^2 + 25rs - 2rs - 5s^2$
$= 5r(2r + 5s) - s(2r + 5s)$
$= (2r + 5s)(5r - s)$

47. $14x^2 - 25xq - 25q^2$

$$ac = 14(-25) = -350$$

The integers -35 and 10 have a product of -350 and a sum of -25.

$$\begin{aligned}&14x^2 - 25xq - 25q^2\\ &= 14x^2 - 35xq + 10xq - 25q^2\\ &= 7x(2x - 5q) + 5q(2x - 5q)\\ &= (2x - 5q)(7x + 5q)\end{aligned}$$

51. $2a^3 + 6a^2 - 4a = 2a(a^2 + 3a - 2)$

55. $(x - 2y)^2 - 4$
$$\begin{aligned}&= (x - 2y)^2 - 2^2\\ &= [(x - 2y) + 2][(x - 2y) - 2]\\ &= (x - 2y + 2)(x - 2y - 2)\end{aligned}$$

59. $z^4 - 9z^2 + 20$
$$\begin{aligned}&= (z^2)^2 - 9(z^2) + 20\\ &= (z^2 - 4)(z^2 - 5)\\ &= (z + 2)(z - 2)(z^2 - 5)\end{aligned}$$

5.10 Solving Equations by Factoring

5.10 Margin Exercises

In the margin exercises, check all solutions to the equations by substituting them back in the original equations.

1. **(a)** $(3k + 5)(k + 1) = 0$

$$\begin{aligned}3k + 5 &= 0 \quad \text{or} \quad k + 1 = 0\\ 3k &= -5 \qquad\qquad k = -1\\ k &= -\frac{5}{3}\end{aligned}$$

Solution set: $\{-\frac{5}{3}, -1\}$

(b) $(3r + 11)(5r - 2) = 0$

$$\begin{aligned}3r + 11 = 0 \quad &\text{or} \quad 5r - 2 = 0\\ 3r = -11 \quad & \qquad 5r = 2\\ r = -\frac{11}{3} \quad &\text{or} \quad r = \frac{2}{5}\end{aligned}$$

Solution set: $\{-\frac{11}{3}, \frac{2}{5}\}$

2. **(a)**
$$\begin{aligned}3r^2 - r &= 4\\ 3r^2 - r - 4 &= 0\\ (3r - 4)(r + 1) &= 0\\ 3r - 4 = 0 \quad &\text{or} \quad r + 1 = 0\\ 3r = 4 \quad & \qquad r = -1\\ r = \frac{4}{3}\end{aligned}$$

Solution set: $\{\frac{4}{3}, -1\}$

(b)
$$\begin{aligned}15m^2 + 7m &= 2\\ 15m^2 + 7m - 2 &= 0\\ (5m - 1)(3m + 2) &= 0\\ 5m - 1 = 0 \quad &\text{or} \quad 3m + 2 = 0\\ 5m = 1 \quad & \qquad 3m = -2\\ m = \frac{1}{5} \quad &\text{or} \quad m = -\frac{2}{3}\end{aligned}$$

Solution set: $\{\frac{1}{5}, -\frac{2}{3}\}$

3. **(a)**
$$\begin{aligned}p^2 &= -12p\\ p^2 + 12p &= 0\\ p(p + 12) &= 0\\ p = 0 \quad &\text{or} \quad p + 12 = 0\\ & \qquad p = -12\end{aligned}$$

Solution set: $\{0, -12\}$

(b)
$$\begin{aligned}k^2 - 16 &= 0\\ (k + 4)(k - 4) &= 0\\ k + 4 = 0 \quad &\text{or} \quad k - 4 = 0\\ k = -4 \quad &\text{or} \quad k = 4\end{aligned}$$

Solution set: $\{-4, 4\}$

4.
$$\begin{aligned}(a + 6)(a - 2) &= 2 + a - 10\\ a^2 + 4a - 12 &= -8 + a\\ a^2 + 3a - 4 &= 0\\ (a - 1)(a + 4) &= 0\\ a - 1 = 0 \quad &\text{or} \quad a + 4 = 0\\ a = 1 \quad &\text{or} \quad a = -4\end{aligned}$$

Solution set: $\{1, -4\}$

5. Let $x =$ the length of the deck.
$\frac{1}{2}x - 1 =$ the width.

Use the formula $LW = A$, where $L = x$, $W = \frac{1}{2}x - 1$, and $A = 60$.

$$x\left(\frac{1}{2}x - 1\right) = 60$$
$$\frac{1}{2}x^2 - x = 60$$

Multiply both sides by 2.

$$x^2 - 2x = 120$$
$$x^2 - 2x - 120 = 0$$
$$(x-12)(x+10) = 0$$
$$x - 12 = 0 \quad \text{or} \quad x + 10 = 0$$
$$x = 12 \quad \text{or} \quad x = -10$$

The deck cannot have a negative length, so reject $x = -10$. The length of the deck is 12 m and the width is $\frac{1}{2}(12) - 1 = 5$ m.

5.10 Section Exercises

Check all solutions to the exercises in this section by substituting them back in the original equation.

3. $(x+4)(x-1) = 1$

   The two linear equations lead to $x = -3$ and $x = 2$, but -3 and 2 are not solutions of the equation.

   The correct way to solve the equation is to multiply the factors on the left and subtract 1 from each side, leaving 0 on the right. Then factor the left side, and use the zero-factor property.

7. $(2k-5)(3k+8) = 0$

$$2k - 5 = 0 \quad \text{or} \quad 3k + 8 = 0$$
$$2k = 5 \qquad 3k = -8$$
$$k = \frac{5}{2} \quad \text{or} \quad k = -\frac{8}{3}$$

   Solution set: $\left\{-\frac{8}{3}, \frac{5}{2}\right\}$

11.
$$z^2 + 9z + 18 = 0$$
$$(z+6)(z+3) = 0$$
$$z + 6 = 0 \quad \text{or} \quad z + 3 = 0$$
$$z = -6 \quad \text{or} \quad z = -3$$

   Solution set: $\{-6, -3\}$

15.
$$15k^2 - 7k = 4$$
$$15k^2 - 7k - 4 = 0$$
$$(5k-4)(3k+1) = 0$$
$$5k - 4 = 0 \quad \text{or} \quad 3k + 1 = 0$$
$$5k = 4 \qquad 3k = -1$$
$$k = \frac{4}{5} \quad \text{or} \quad k = -\frac{1}{3}$$

   Solution set: $\left\{-\frac{1}{3}, \frac{4}{5}\right\}$

19.
$$(5y+1)(y+3) = -2(5y+1)$$
$$(5y+1)(y+3) + 2(5y+1) = 0$$
$$5y^2 + 16y + 3 + 10y + 2 = 0$$
$$5y^2 + 26y + 5 = 0$$
$$(5y+1)(y+5) = 0$$
$$5y + 1 = 0 \quad \text{or} \quad y + 5 = 0$$
$$5y = -1 \qquad y = -5$$
$$y = -\frac{1}{5}$$

   Solution set: $\left\{-5, -\frac{1}{5}\right\}$

23.
$$-3m^2 + 27 = 0$$
$$-3(m^2 - 9) = 0$$
$$-3(m+3)(m-3) = 0$$
$$m + 3 = 0 \quad \text{or} \quad m - 3 = 0$$
$$m = -3 \quad \text{or} \quad m = 3$$

   Solution set: $\{-3, 3\}$

27. If the solutions to a quadratic equation are 3 and $\frac{1}{4}$, then the factors of the polynomial must be

$$(x-3)\left(x - \frac{1}{4}\right).$$

   Therefore, the equation is

$$(x-3)\left(x - \frac{1}{4}\right) = 0$$
$$x^2 - \frac{1}{4}x - 3x + \frac{3}{4} = 0$$
$$4x^2 - x - 12x + 3 = 0$$
$$4x^2 - 13x + 3 = 0.$$

31. Let $t =$ the time it takes the rocket to reach the ground.

$$h = -16t^2 + 128t$$

   When the rocket returns to the ground, its height will be 0. Let $h = 0$ in the equation, and solve for t.

$$0 = -16t^2 + 128t$$
$$-16t^2 + 128t = 0$$
$$-16t(t-8) = 0$$
$$-16t = 0 \quad \text{or} \quad t - 8 = 0$$
$$t = 0 \quad \text{or} \quad t = 8$$

   The time $t = 0$ represents the beginning of the launch. Therefore, the rocket will return to the ground 8 sec after launch.

35. Let $w =$ the width of the cardboard.
$w + 6 =$ the length of the cardboard.

If squares that measure 2 in are cut from each corner of the cardboard, then the width becomes $w - 4$ and the length becomes $(w+6) - 4 = w+2$. Use the formula $V = LWH$ and substitute 110 for V, $w + 2$ for L, $w - 4$ for W, and 2 for H.

$$\begin{aligned} V &= LWH \\ 110 &= (w+2)(w-4)2 \\ 110 &= (w^2 - 2w - 8)2 \\ 55 &= w^2 - 2w - 8 \\ 0 &= w^2 - 2w - 63 \\ 0 &= (w-9)(w+7) \\ w - 9 = 0 &\quad \text{or} \quad w + 7 = 0 \\ w = 9 &\quad \text{or} \quad w = -7 \end{aligned}$$

A box cannot have a negative width, so reject -7 as a solution. The only possible solution is 9. The piece of cardboard has width 9 in and length $9 + 6 = 15$ in.

Chapter 5 Review Exercises

1. $4^3 = 4 \cdot 4 \cdot 4 = 64$

2. $\left(\frac{1}{3}\right)^4 = \frac{1}{3} \cdot \frac{1}{3} \cdot \frac{1}{3} \cdot \frac{1}{3} = \frac{1}{81}$

3. $(-.2)^2 = (-.2)(-.2) = .04$

4. $\frac{2}{(-3)^{-2}} = \frac{2}{\frac{1}{(-3)^2}} = 2 \cdot (-3)^2 = 2 \cdot (-3)(-3) = 18$

5. $\left(\frac{2}{3}\right)^{-4} = \left(\frac{3}{2}\right)^4 = \frac{3}{2} \cdot \frac{3}{2} \cdot \frac{3}{2} \cdot \frac{3}{2} = \frac{81}{16}$

6. In $(-6)^0$, the base is -6 and $(-6)^0 = 1$. In -6^0, the base is 6 and $-6^0 = -(6)^0 = -1$.

7. $7^{12} \cdot 7^{-4} = 7^{12+(-4)} = 7^8$

8. $\frac{3^5}{3^{-3}} = 3^{5-(-3)} = 3^{5+3} = 3^8$

9. $m^4 m^6 m = m^4 m^6 m^1 = m^{4+6+1} = m^{11}$

10. $$\begin{aligned} 6(3x^4)(5x^{-7}) &= 6(3)(5)x^4x^{-7} \\ &= 90x^{4+(-7)} \\ &= 90x^{-3} \\ &= \frac{90}{x^3} \end{aligned}$$

11. $\frac{4^{-3}}{4^{-6}} = 4^{-3-(-6)} = 4^3$

12. $\frac{y^{-3}y^{-2}}{y^{-5}y^{-4}} = \frac{y^{-5}}{y^{-9}} = y^{-5-(-9)} = y^4$

13. **(a)** $-3^2 = -9$

$-3^2 = -(3^2) = -9$, so this statement is true.

(b) $-a^4 = (-a)^4$

$(-a)^4 = (-a)(-a)(-a)(-a) = a^4$. This statement is false. The corrected statement is

$$a^4 = (-a)^4.$$

(c) $5^{-2} = -5^2$

$$5^{-2} = \frac{1}{5^2} = \frac{1}{25}$$

$$-5^2 = -(5 \cdot 5) = -25$$

This statement is false. The corrected statement is

$$5^{-2} = \frac{1}{5^2}.$$

(d) $(xy)^3 = x^3y^3$

$(xy)^3 = (xy)(xy)(xy) = x^3y^3$, so this statement is true.

14. $2790 = 2_\wedge 790.$

Place a caret after the first nonzero digit, 2. Count 3 places from the caret to the decimal point. Use a positive exponent on 10 since $2790 > 2.79$.

$$2790 = 2.790 \times 10^3$$

15. $.0000085 = .000008_\wedge 5$
Count 6 places.

Use a negative exponent on 10 since $.0000085 < 8.5$.

$$.0000085 = 8.5 \times 10^{-6}$$

16. $.296 = .2_\wedge 96$
Count 1 place.

Use a negative exponent on 10 since $.296 < 2.96$.

$$.296 = 2.96 \times 10^{-1}$$

17. $$\begin{aligned} 3.6 \times 10^4 &= 3_\wedge 6000. \\ &= 36{,}000 \end{aligned}$$

Since the exponent is positive, move the decimal point 4 places to the right. Attach extra zeros.

18. $$\begin{aligned} 5.71 \times 10^{-3} &= .005_\wedge 71 \\ &= .00571 \end{aligned}$$

Since the exponent is negative, move the decimal point 3 places to the left. Attach an extra zero.

19. $9.04 \times 10^{-2} = .09_{\wedge}04$
$= .0904$

Since the exponent is negative, move the decimal point 2 places to the left.

20. By definition, multiplying by 10^{-a} is equivalent to multiplying by $\frac{1}{10^a}$ which is equivalent to dividing by 10^a.

21. $5449 = 5_{\wedge}449.$
Count 3 places.
$= 5.449 \times 10^3$

22. $$\frac{3.45 \times 10^5}{5.449 \times 10^3} = \frac{3.45}{5.449} \times \frac{10^5}{10^3}$$
$$\approx .63 \times 10^{5-3}$$
$$= .63 \times 10^2$$
$$= 63$$

The area is approximately 63 mi^2.

23. $\frac{5^2}{5^2} = 5^0$
Since $\frac{25}{25} = 1$ and $\frac{25}{25} = 5^0$, it follows that $5^0 = 1$.

24. $(4^{-3})^2 = 4^{-6} = \frac{1}{4^6}$

25. $(k^{-2})^3 k^{-4} = k^{-6}k^{-4} = k^{-10} = \frac{1}{k^{10}}$

26. $\frac{(5z)^3 z^2}{z^{-3}z^{-2}} = \frac{5^3 z^3 z^2}{z^{-3}z^{-2}} = \frac{5^3 z^5}{z^{-5}} = 5^3 z^{10}$

27. $$\left(\frac{7p^{-3}}{p^{-8}}\right)^{-1} \cdot \frac{p^{-3}}{5} = \frac{7^{-1}p^3}{p^8} \cdot \frac{p^{-3}}{5}$$
$$= \frac{7^{-1}}{5p^8}$$
$$= \frac{1}{7 \cdot 5p^8}$$
$$= \frac{1}{35p^8}$$

28. $$\left(\frac{3a^4}{4a^{-2}}\right)^{-3} \left(\frac{9a^{-3}}{8a^{-6}}\right)^4$$
$$= \frac{3^{-3}a^{-12}}{4^{-3}a^6} \cdot \frac{9^4 a^{-12}}{8^4 a^{-24}}$$
$$= \frac{3^{-3}a^{-12} \cdot (3^2)^4 a^{-12}}{(2^2)^{-3}a^6 \cdot (2^3)^4 a^{-24}}$$
$$= \frac{3^{-3}3^8 a^{-24}}{2^{-6}2^{12}a^{-18}}$$
$$= \frac{3^5 a^{-6}}{2^6}$$
$$= \frac{3^5}{2^6 a^6}$$

29. $$\left(\frac{q^{-1}r^{-3}}{5q^3}\right)^{-2} \left(\frac{r^{-3} \cdot 2q^3}{3r^{-4}}\right)^{-1}$$
$$= \frac{q^2 r^6}{5^{-2}q^{-6}} \cdot \frac{r^3 2^{-1} q^{-3}}{3^{-1}r^4}$$
$$= \frac{2^{-1}q^{-1}r^9}{3^{-1}5^{-2}q^{-6}r^4}$$
$$= \frac{3 \cdot 5^2 q^5 r^5}{2}$$

30. $9k + 11k^3 - 3k^2$

(a) In descending powers of k, the polynomial is $11k^3 - 3k^2 + 9k$.

(b) The polynomial is a trinomial since it has three terms.

(c) The degree of the polynomial is 3 since the highest power of k is 3.

31. $14m^6 - 9m^7$

(a) In descending powers of m, the polynomial is $-9m^7 + 14m^6$.

(b) The polynomial is a binomial since it has two terms.

(c) The degree of the polynomial is 7 since the highest power of m is 7.

32. $-5y^4 + 3y^3 + 7y^2 - 2y$

(a) The polynomial is already written in descending powers of y.

(b) The polynomial has four terms, so it is none of the choices.

(c) The degree of the polynomial is 4 since the highest power of y is 4.

33. $P(x) = -6x + 15$

(a) $P(-2) = -6(-2) + 15 = 12 + 15 = 27$

(b) $P(3) = -6(3) + 15 = -18 + 15 = -3$

34. $P(x) = -2x^2 + 5x + 7$

(a) $$P(-2) = -2(-2)^2 + 5(-2) + 7$$
$$= -2(4) - 10 + 7$$
$$= -8 - 10 + 7$$
$$= -11$$

(b) $$P(3) = -2(3)^2 + 5(3) + 7$$
$$= -2(9) + 15 + 7$$
$$= -18 + 15 + 7$$
$$= 4$$

35. Add.

$$\begin{array}{r} 3x^2 - 5x + 6 \\ \underline{-4x^2 + 2x - 5} \\ -1x^2 - 3x + 1 \end{array} \quad \text{or } -x^2 - 3x + 1$$

36. Subtract.

$$\begin{array}{r} 10m - 4 \\ \underline{-8m + 6} \end{array}$$

Change the signs in the second polynomial and add.

$$\begin{array}{r} 10m - 4 \\ \underline{8m - 6} \\ 18m - 10 \end{array}$$

37. Subtract.

$$\begin{array}{rrrr} -5y^3 & & +8y & -3 \\ \hline & 4y^2 & +2y & +9 \end{array}$$

Change the signs and add.

$$\begin{array}{rrrr} -5y^3 & & +8y & -3 \\ & -4y^2 & -2y & -9 \\ \hline -5y^3 & -4y^2 & +6y & -12 \end{array}$$

38. $(4a^3 - 9a + 15) - (-2a^3 + 4a^2 + 7a)$
$= 4a^3 - 9a + 15 + 2a^3 - 4a^2 - 7a$
$= 4a^3 + 2a^3 - 4a^2 - 9a - 7a + 15$
$= 6a^3 - 4a^2 - 16a + 15$

39. $(3y^2 + 2y - 1) + (5y^2 - 11y + 6)$
$= 3y^2 + 5y^2 + 2y - 11y - 1 + 6$
$= 8y^2 - 9y + 5$

40. $-5b(3b^2 + 10) = -5b(3b^2) - 5b(10)$
$= -15b^3 - 50b$

41. $-6k(2k^2 + 7) = -6k(2k^2) - 6k(7)$
$= -12k^3 - 42k$

42. $(3m - 2)(5m + 1)$
$= \underset{\text{F}}{15m^2} + \underset{\text{O}}{3m} - \underset{\text{I}}{10m} - \underset{\text{L}}{2}$
$= 15m^2 - 7m - 2$

43. $(7y - 8)(2y + 3)$
$= 14y^2 + 21y - 16y - 24$
$= 14y^2 + 5y - 24$

44. $(3w - 2t)(2w - 3t)$
$= 6w^2 - 9wt - 4wt + 6t^2$
$= 6w^2 - 13wt + 6t^2$

45. $(2p^2 + 6p)(5p^2 - 4)$
$= 10p^4 - 8p^2 + 30p^3 - 24p$
$= 10p^4 + 30p^3 - 8p^2 - 24p$

46. $(3q^2 + 2q - 4)(q - 5)$
$= 3q^3 + 2q^2 - 4q - 15q^2 - 10q + 20$
$= 3q^3 - 13q^2 - 14q + 20$

47. $(3z^3 - 2z^2 + 4z - 1)(3z - 2)$
$= 9z^4 - 6z^3 + 12z^2 - 3z - 6z^3 + 4z^2$
$- 8z + 2$
$= 9z^4 - 12z^3 + 16z^2 - 11z + 2$

48. $(6r^2 - 1)(6r^2 + 1) = (6r^2)^2 - 1^2$
$= 36r^4 - 1$

49. $\left(z + \frac{3}{5}\right)\left(z - \frac{3}{5}\right) = z^2 - \left(\frac{3}{5}\right)^2$
$= z^2 - \frac{9}{25}$

50. $(4m + 3)^2 = (4m)^2 + 2(4m)(3) + 3^2$
$= 16m^2 + 24m + 9$

51. $(2n - 10)^2 = (2n)^2 - 2(2n)(10) + 10^2$
$= 4n^2 - 40n + 100$

52. $\frac{18x^2 - 32x + 12}{12x} = \frac{18x^2}{12x} - \frac{32x}{12x} + \frac{12}{12x}$
$= \frac{3x}{2} - \frac{8}{3} + \frac{1}{x}$

53. $\frac{4y^3 - 12y^2 + 5y}{4y} = \frac{4y^3}{4y} - \frac{12y^2}{4y} + \frac{5y}{4y}$
$= y^2 - 3y + \frac{5}{4}$

54. $\frac{6x^2 - 17x + 2}{2x - 3}$

$$\begin{array}{r} 3x - 4 \\ 2x - 3\overline{)6x^2 - 17x + 2} \\ \underline{6x^2 - 9x} \\ -8x + 2 \\ \underline{-8x + 12} \\ -10 \end{array}$$

Answer: $3x - 4 + \frac{-10}{2x - 3}$

55. $\dfrac{15k^2+11k-17}{3k-2}$

$$\begin{array}{r} 5k+\ \ 7 \\ 3k-2\overline{)15k^2+11k-17} \\ \underline{15k^2-10k} \\ 21k-17 \\ \underline{21k-14} \\ -3 \end{array}$$

Answer: $5k+7+\dfrac{-3}{3k-2}$

56. $\dfrac{2p^3+9p^2+27}{2p-3}$

$$\begin{array}{r} p^2+\ \ 6p+\ \ 9 \\ 2p-3\overline{)2p^3+\ \ 9p^2+\ \ 0p+27} \\ \underline{2p^3-\ \ 3p^2} \\ 12p^2+\ \ 0p \\ \underline{12p^2-18p} \\ 18p+27 \\ \underline{18p-27} \\ 54 \end{array}$$

Answer: $p^2+6p+9+\dfrac{54}{2p-3}$

57. $\dfrac{12y^4+7y^2-2y+1}{3y^2+1}$

$$\begin{array}{r} 4y^2+1 \\ 3y^2+0y+1\overline{)12y^4+0y^3+7y^2-2y+1} \\ \underline{12y^4+0y^3+4y^2} \\ 3y^2-2y+1 \\ \underline{3y^2+0y+1} \\ -2y \end{array}$$

Answer: $4y^2+1+\dfrac{-2y}{3y^2+1}$

58. $\dfrac{4x^4-28x^3+27x^2-21x+18}{4x^2+3}$

$$\begin{array}{r} x^2-\ \ 7x+\ \ 6 \\ 4x^2+0x+3\overline{)4x^4-28x^3+27x^2-21x+18} \\ \underline{4x^4+\ \ 0x^3+\ \ 3x^2} \\ -28x^3+24x^2-21x \\ \underline{-28x^3-\ \ 0x^2-21x} \\ 24x^2+\ \ 0x+18 \\ \underline{24x^2+\ \ 0x+18} \\ 0 \end{array}$$

Answer: x^2-7x+6

59. $\dfrac{5p^4+15p^3-33p^2-9p+18}{5p^2-3}$

$$\begin{array}{r} p^2+3p-\ \ 6 \\ 5p^2+0p-3\overline{)5p^4+15p^3-33p^2-9p+18} \\ \underline{5p^4+\ \ 0p^3-\ \ 3p^2} \\ 15p^3-30p^2-9p \\ \underline{15p^3+\ \ 0p^2-9p} \\ -30p^2+0p+18 \\ \underline{-30p^2-0p+18} \\ 0 \end{array}$$

Answer: p^2+3p-6

60. $\dfrac{3p^2-p-2}{p-1}$

$$\begin{array}{rrrr} 1) & 3 & -1 & -2 \\ & & 3 & 2 \\ \hline & 3 & 2 & 0 \end{array}$$

Answer: $3p+2$

61. $\dfrac{10k^2-3k-15}{k+2}$

$$\begin{array}{rrrr} -2) & 10 & -3 & -15 \\ & & -20 & 46 \\ \hline & 10 & -23 & 31 \end{array}$$

Answer: $10k-23+\dfrac{31}{k+2}$

62. $(2k^3-5k^2+12)\div(k-3)$

$$\begin{array}{rrrrr} 3) & 2 & -5 & 0 & 12 \\ & & 6 & 3 & 9 \\ \hline & 2 & 1 & 3 & 21 \end{array}$$

Answer: $2k^2+k+3+\dfrac{21}{k-3}$

63. $(-a^4+19a^2+18a+15)\div(a+4)$

$$\begin{array}{rrrrrr} -4) & -1 & 0 & 19 & 18 & 15 \\ & & 4 & -16 & -12 & -24 \\ \hline & -1 & 4 & 3 & 6 & -9 \end{array}$$

Answer: $-a^3+4a^2+3a+6+\dfrac{-9}{a+4}$

64. $2w^3+8w^2-14w-20=0$

Divide the polynomial by $w+5$.

$$\begin{array}{rrrrr} -5) & 2 & 8 & -14 & -20 \\ & & -10 & 10 & 20 \\ \hline & 2 & -2 & -4 & 0 \end{array}$$

Since the remainder is 0, -5 is a solution to the equation.

65. $-3q^4 + 2q^3 + 5q^2 - 9q + 1 = 0$

Divide the polynomial by $q + 5$.

$$\begin{array}{r|rrrrr} -5 & -3 & 2 & 5 & -9 & 1 \\ & & 15 & -85 & 400 & -1955 \\ \hline & -3 & 17 & -80 & 391 & -1954 \end{array}$$

Since the remainder is not 0, -5 is not a solution to the equation.

66. $P(x) = 3x^3 - 5x^2 + 4x - 1; k = -1$

Divide the polynomial by $x + 1$.

$$\begin{array}{r|rrrr} -1 & 3 & -5 & 4 & -1 \\ & & -3 & 8 & -12 \\ \hline & 3 & -8 & 12 & -13 \end{array}$$

By the remainder theorem, $P(-1) = -13$.

67. $P(z) = z^4 - 2z^3 - 9z - 5; k = 3$

Divide the polynomial by $z - 3$.

$$\begin{array}{r|rrrrr} 3 & 1 & -2 & 0 & -9 & -5 \\ & & 3 & 3 & 9 & 0 \\ \hline & 1 & 1 & 3 & 0 & -5 \end{array}$$

By the remainder theorem, $P(3) = -5$.

68. $12p^2 - 6p = 6p(2p - 1)$

69. $21y^2 + 35y = 7y(3y + 5)$

70. $12q^2b + 8qb^2 - 20q^3b^2$
$= 4qb(3q + 2b - 5q^2b)$

71. $6r^3t - 30r^2t^2 + 18rt^3$
$= 6rt(r^2 - 5rt + 3t^2)$

72. $(x + 3)(4x - 1) - (x + 3)(3x + 2)$

The common factor is $(x + 3)$.

$= (x + 3)[(4x - 1) - (3x + 2)]$
$= (x + 3)(4x - 1 - 3x - 2)$
$= (x + 3)(x - 3)$

73. $(z + 1)(z - 4) + (z + 1)(2z + 3)$
$= (z + 1)[(z - 4) + (2z + 3)]$
$= (z + 1)(3z - 1)$

74. $4m + nq + mn + 4q$

Rearrange the terms.

$= (4m + mn) + (nq + 4q)$
$= m(4 + n) + q(n + 4)$
$= m(n + 4) + q(n + 4)$
$= (n + 4)(m + q)$

75. $x^2 + 5y + 5x + xy$
$= (x^2 + 5x) + (5y + xy)$
$= x(x + 5) + y(5 + x)$
$= x(x + 5) + y(x + 5)$
$= (x + 5)(x + y)$

76. $3p^2 - p - 4$

$ac = 3(-4) = -12$

The integers -4 and 3 have a product of -12 and a sum of -1. Write $-p$ as $-4p + 3p$.

$= 3p^2 - 4p + 3p - 4$
$= p(3p - 4) + 1(3p - 4)$
$= (3p - 4)(p + 1)$

77. $6k^2 + 11k - 10$

$ac = 6(-10) = -60$

The integers 15 and -4 have a product of -60 and a sum of 11. Write $11k$ as $15k - 4k$.

$= 6k^2 + 15k - 4k - 10$
$= 3k(2k + 5) - 2(2k + 5)$
$= (2k + 5)(3k - 2)$

78. $12r^2 - 5r - 3$

$ac = 12(-3) = -36$

The integers -9 and 4 have a product of -36 and a sum of -5. Write $-5r$ as $-9r + 4r$.

$= 12r^2 - 9r + 4r - 3$
$= 3r(4r - 3) + 1(4r - 3)$
$= (4r - 3)(3r + 1)$

79. $10m^2 + 37m + 30$

$ac = 10(30) = 300$

The integers 12 and 25 have a product of 300 and a sum of 37. Write $37m$ as $12m + 25m$.

$= 10m^2 + 12m + 25m + 30$
$= 2m(5m + 6) + 5(5m + 6)$
$= (5m + 6)(2m + 5)$

80. $10k^2 - 11kh + 3h^2$

$ac = 10(3) = 30$

The integers -6 and -5 have a product of 30 and a sum of -11. Write $-11kh$ as $-6kh - 5kh$.

$= 10k^2 - 6kh - 5kh + 3h^2$
$= 2k(5k - 3h) - h(5k - 3h)$
$= (5k - 3h)(2k - h)$

81. $9x^2 + 4xy - 2y^2$

$ac = 9(-2) = -18$

There are no integers that have a product of -18 and a sum of 4. Therefore, the trinomial cannot be factored and is prime.

82. $24x - 2x^2 - 2x^3$
$= 2x(12 - x - x^2)$
$= 2x(4 + x)(3 - x)$

83. $6b^3 - 9b^2 - 15b$
$= 3b(2b^2 - 3b - 5)$
$= 3b(2b - 5)(b + 1)$

84. $y^4 + 2y^2 - 8$
$= (y^2)^2 + 2(y^2) - 8$
$= (y^2 + 4)(y^2 - 2)$

85. $2k^4 - 5k^2 - 3$
$= 2(k^2)^2 - 5(k^2) - 3$
$= (2k^2 + 1)(k^2 - 3)$

86. The answer given is a sum, not a product. This is incorrect because the polynomial still has two terms, so it is not factored. The correct answer requires another step to get

$$x^2y^2 - 6x^2 + 5y^2 - 30$$
$$= x^2(y^2 - 6) + 5(y^2 - 6)$$
$$= (y^2 - 6)(x^2 + 5).$$

87. **(a)** $x^2 - y^2 = (x + y)(x - y)$

This is a true statement. It is the difference of two squares.

(b) $x^2 + y^2 = (x + y)(x + y)$

This is a false statement. $x^2 + y^2$ is prime and cannot be factored. Also, $(x+y)(x+y)$ multiplies to $x^2 + 2xy + y^2$.

(c) $x^3 + y^3 = (x + y)^3$

This is a false statement. The sum of two cubes factors as

$$x^3 + y^3 = (x + y)(x^2 - xy + y^2).$$

(d) $x^3 - y^3 = (x - y)^3$

This is a false statement. The difference of two cubes factors as

$$x^3 - y^3 = (x - y)(x^2 + xy + y^2).$$

88. $16p^2 - 9 = (4p)^2 - 3^2$
$= (4p + 3)(4p - 3)$

89. $9z^2 - 100 = (3z)^2 - 10^2$
$= (3z + 10)(3z - 10)$

90. $36(t + 1)^2 - 64(z + 1)^2$
$= 4[9(t + 1)^2 - 16(z + 1)^2]$
$= 4[[3(t + 1)]^2 - [4(z + 1)]^2]$
$= 4[3(t + 1) + 4(z + 1)]$
$\cdot [3(t + 1) - 4(z + 1)]$
$= 4(3t + 3 + 4z + 4)(3t + 3 - 4z - 4)$
$= 4(3t + 4z + 7)(3t - 4z - 1)$

91. $p^2 + 10p + 100 = p^2 + 10p + 10^2$

This polynomial looks like a perfect square trinomial. In this case, however, the middle term would be $2(p)(10) = 20p$, not $10p$. Therefore, the polynomial cannot be factored and is prime.

92. $16x^2 - 40x + 25 = (4x)^2 - 2(4x)(5) + 5^2$
$= (4x - 5)^2$

93. $4z^2 + 12zm + 9m^2$
$= (2z)^2 + 2(2z)(3m) + (3m)^2$
$= (2z + 3m)^2$

94. $1 - y^3 = 1^3 - y^3$
$= (1 - y)(1^2 + 1y + y^2)$
$= (1 - y)(1 + y + y^2)$

95. $125x^3 - 8 = (5x)^3 - 2^3$
$= (5x - 2)[(5x)^2 + (5x)(2) + 2^2]$
$= (5x - 2)(25x^2 + 10x + 4)$

96. $4r^3 + 108 = 4(r^3 + 27)$
$= 4(r^3 + 3^3)$
$= 4(r + 3)(r^2 - 3r + 3^2)$
$= 4(r + 3)(r^2 - 3r + 9)$

97. To use the zero-factor property, one side of the equation must be zero.

$$(x + 2)(x - 4) = -5$$

Multiply on the left.

$$x^2 - 2x - 8 = -5$$

Add 5 to each side.

$$x^2 - 2x - 3 = 0$$

Now factor and apply the zero-factor property.

$$(x-3)(x+1)=0$$
$$x-3=0 \quad \text{or} \quad x+1=0$$
$$x=3 \quad \text{or} \quad x=-1$$

Solution set: $\{-1, 3\}$

98. $$3y^2+10y-8=0$$
$$(3y-2)(y+4)=0$$
$$3y-2=0 \quad \text{or} \quad y+4=0$$
$$3y=2 \qquad y=-4$$
$$y=\frac{2}{3}$$

Solution set: $\{-4, \frac{2}{3}\}$

99. $$16r^2-3=8r$$
$$16r^2-8r-3=0$$
$$(4r+1)(4r-3)=0$$
$$4r+1=0 \quad \text{or} \quad 4r-3=0$$
$$4r=-1 \qquad 4r=3$$
$$r=-\frac{1}{4} \quad \text{or} \quad r=\frac{3}{4}$$

Solution set: $\{-\frac{1}{4}, \frac{3}{4}\}$

100. $$6x^2+2x=20$$
$$6x^2+2x-20=0$$
$$2(3x^2+x-10)=0$$
$$2(3x-5)(x+2)=0$$
$$3x-5=0 \quad \text{or} \quad x+2=0$$
$$3x=5 \qquad x=-2$$
$$x=\frac{5}{3}$$

Solution set: $\{-2, \frac{5}{3}\}$

101. Let $x =$ the length of the shorter side.
$2x+1 =$ the length of the longer side.

The area is 10.5 ft^2. Use the formula for the area of a triangle.

$$\frac{1}{2}bh=A$$
$$\frac{1}{2}(x)(2x+1)=10.5$$

Multiply both sides by 2.

$$x(2x+1)=21$$
$$2x^2+x=21$$
$$2x^2+x-21=0$$
$$(2x+7)(x-3)=0$$
$$2x+7=0 \quad \text{or} \quad x-3=0$$
$$2x=-7 \qquad x=3$$
$$x=-\frac{7}{2}$$

The triangle cannot have a negative length, so reject $x=-\frac{7}{2}$. The length of the shorter side is 3 ft.

102. Let $x =$ the width of the lot.
$x+20 =$ the length of the lot.

The area is 2400 ft^2. Use the formula $LW = A$, where $L = x+20$ and $W = x$.

$$(x+20)x=2400$$
$$x^2+20x=2400$$
$$x^2+20x-2400=0$$
$$(x-40)(x+60)=0$$
$$x-40=0 \quad \text{or} \quad x+60=0$$
$$x=40 \quad \text{or} \quad x=-60$$

The lot cannot have a negative width, so reject $x=-60$. The width of the lot is 40 ft, and the length is $40+20=60$ ft.

103. Let $x =$ the length of the shed.
$x-5 =$ the width of the shed.

The area is 50 m^2. Use the formula $LW = A$, where $L = x$ and $W = x-5$.

$$x(x-5)=50$$
$$x^2-5x=50$$
$$x^2-5x-50=0$$
$$(x-10)(x+5)=0$$
$$x-10=0 \quad \text{or} \quad x+5=0$$
$$x=10 \quad \text{or} \quad x=-5$$

The shed cannot have a negative length, so reject $x=-5$. The length of the shed is 10 m.

104. Let $x =$ the width of the frame.
$x+2 =$ the length of the frame.

The area is 48 in^2.

$$x(x+2)=48$$
$$x^2+2x=48$$
$$x^2+2x-48=0$$
$$(x+8)(x-6)=0$$
$$x+8=0 \quad \text{or} \quad x-6=0$$
$$x=-8 \quad \text{or} \quad x=6$$

The frame cannot have a negative width, so reject $x=-8$. The width of the frame is 6 in.

105. $(2x+1)(4x-3) = 8x^2 - 6x + 4x - 3$
$= 8x^2 - 2x - 3$

106. $\left(\frac{3}{2}\right)^{-2} = \left(\frac{2}{3}\right)^2 = \frac{2^2}{3^2}$

107. $(y^6)^{-5}(2y^{-3})^{-4} = y^{-30}2^{-4}y^{12} = 2^{-4}y^{-18} = \frac{1}{2^4y^{18}}$

108. $(11w^2 - 5w + 6) - (-15 - 8w^2)$
$= 11w^2 - 5w + 6 + 15 + 8w^2$
$= 19w^2 - 5w + 21$

109. $\dfrac{m^2 - 8m + 15}{m - 5}$

$$\begin{array}{r|rrr} 5 & 1 & -8 & 15 \\ & & 5 & -15 \\ \hline & 1 & -3 & 0 \end{array}$$

Answer: $m - 3$

110. $-(-3)^2 = -(-3)(-3) = -9$

111. $\dfrac{(5z^2x^3)^2}{(-10zx^{-3})^{-2}} = \dfrac{5^2z^4x^6}{10^{-2}z^{-2}x^6}$
$= \dfrac{5^2z^6}{10^{-2}}$
$= 5^2 10^2 z^6$
$= 2500z^6$

112. $(2k-1) - (3k^2 - 2k + 6)$
$= 2k - 1 - 3k^2 + 2k - 6$
$= -3k^2 + 4k - 7$

113. $7p^5(3p^4 + p^3 + 2p^2)$
$= 7p^5(3p^4) + 7p^5(p^3) + 7p^5(2p^2)$
$= 21p^9 + 7p^8 + 14p^7$

114. $\dfrac{20y^3x^3 + 15y^4x + 25yx^4}{10yx^2}$
$= \dfrac{20y^3x^3}{10yx^2} + \dfrac{15y^4x}{10yx^2} + \dfrac{25yx^4}{10yx^2}$
$= 2y^2x + \dfrac{3y^3}{2x} + \dfrac{5x^2}{2}$

115. $P(x) = 2x^2 - x + 3$
$P(-2) = 2(-2)^2 - (-2) + 3$
$= 8 + 2 + 3$
$= 13$

116. $10m^2 - m - 3$

$ac = 10(-3) = -30$

The integers -6 and 5 have a product of -30 and a sum of -1. Write $-m$ as $-6m + 5m$.

$= 10m^2 - 6m + 5m - 3$
$= 2m(5m-3) + 1(5m-3)$
$= (5m-3)(2m+1)$

117. $12z - 72z^2 = 12z(1 - 6z)$

118. $64 - p^3 = 4^3 - p^3$
$= (4-p)(4^2 + 4p + p^2)$
$= (4-p)(16 + 4p + p^2)$

119. $2z^2 - 7xz - 4x^2$

$ac = 2(-4) = -8$

The integers -8 and 1 have a product of -8 and a sum of -7. Write $-7xz$ as $-8xz + 1xz$ (or xz).

$= 2z^2 - 8xz + xz - 4x^2$
$= 2z(z - 4x) + x(z - 4x)$
$= (z - 4x)(2z + x)$

120. $15d^4 - 10d^2 = 5d^2(3d^2 - 2)$

121. $25c^2 + 30c + 9 = (5c)^2 + 2(5c)(3) + 3^2$
$= (5c+3)^2$

Chapter 5 Test

1. To show that $(2a)^{-3} \neq \frac{2}{a^3}$, let $a = 4$.

$$(2a)^{-3} = (2\cdot 4)^{-3} = 8^{-3} = \frac{1}{8^3} = \frac{1}{512}$$

$\frac{2}{a^3} = \frac{2}{4^3} = \frac{2}{64} = \frac{1}{32}$

$\frac{1}{512} \neq \frac{1}{32}$

Thus,

$$(2a)^{-3} \neq \frac{2}{a^3}.$$

2. $\left(\frac{3}{2}\right)^{-2}\left(\frac{3}{2}\right)^{5}\left(\frac{3}{2}\right)^{-6} = \left(\frac{3}{2}\right)^{-3}$
$= \left(\frac{2}{3}\right)^3$
$= \frac{2^3}{3^3}$

3. $(-4m^{-2}n^4)^{-1} = -4^{-1}m^2n^{-4} = -\dfrac{m^2}{4n^4}$

4. $\dfrac{5^{-3}a^{-2}}{2(a^3)^{-3}} = \dfrac{5^{-3}a^{-2}}{2a^{-9}} = \dfrac{a^7}{2\cdot 5^3}$

5. $$\begin{aligned}&(4x^3 - 3x^2 + 2x - 5)\\ &\quad - (3x^3 + 11x + 8) + (x^2 - x)\\ &= 4x^3 - 3x^2 + 2x - 5 - 3x^3 - 11x\\ &\quad - 8 + x^2 - x\\ &= x^3 - 2x^2 - 10x - 13\end{aligned}$$

6. $$\begin{aligned}&(5x - 3)(2x + 1)\\ &= 10x^2 + 5x - 6x - 3\\ &= 10x^2 - x - 3\end{aligned}$$

7. $$\begin{aligned}&(2m - 5)(3m^2 + 4m - 5)\\ &= 6m^3 + 8m^2 - 10m - 15m^2 - 20m + 25\\ &= 6m^3 - 7m^2 - 30m + 25\end{aligned}$$

8. $$\begin{aligned}(6x + y)(6x - y) &= (6x)^2 - y^2\\ &= 36x^2 - y^2\end{aligned}$$

9. $$\begin{aligned}(3k + q)^2 &= (3k)^2 + 2(3k)(q) + q^2\\ &= 9k^2 + 6kq + q^2\end{aligned}$$

10. $$\begin{aligned}&\frac{16p^3 - 32p^2 + 24p}{4p^2}\\ &= \frac{16p^3}{4p^2} - \frac{32p^2}{4p^2} + \frac{24p}{4p^2}\\ &= 4p - 8 + \frac{6}{p}\end{aligned}$$

11. $\dfrac{9q^4 - 18q^3 + 11q^2 + 10q - 10}{3q - 2}$

$$\begin{array}{r} 3q^3 - 4q^2 + q + 4 \\ 3q - 2\overline{)9q^4 - 18q^3 + 11q^2 + 10q - 10} \\ \underline{9q^4 - 6q^3} \\ -12q^3 + 11q^2 \\ \underline{-12q^3 + 8q^2} \\ 3q^2 + 10q \\ \underline{3q^2 - 2q} \\ 12q - 10 \\ \underline{12q - 8} \\ -2 \end{array}$$

Answer: $3q^3 - 4q^2 + q + 4 + \dfrac{-2}{3q - 2}$

12. $P(x) = x^4 - 8x^3 + 21x^2 - 14x - 24$

(a) $$\begin{aligned}P(1) &= 1^4 - 8(1)^3 + 21(1)^2 - 14(1) - 24\\ &= 1 - 8 + 21 - 14 - 24\\ &= -24\end{aligned}$$

(b) Divide the polynomial by $x - 4$.

$$\begin{array}{r|rrrrr} 4 & 1 & -8 & 21 & -14 & -24 \\ & & 4 & -16 & 20 & 24 \\ \hline & 1 & -4 & 5 & 6 & 0 \end{array}$$

Since the remainder is 0, 4 is a solution to the equation.

13. $11z^2 - 44z = 11z(z - 4)$

14. $$\begin{aligned}&(h - 1)(3h + 4) - (h - 1)(h + 2)\\ &= (h - 1)[(3h + 4) - (h + 2)]\\ &= (h - 1)(3h + 4 - h - 2)\\ &= (h - 1)(2h + 2)\\ &= (h - 1)2(h + 1)\\ &= 2(h - 1)(h + 1)\end{aligned}$$

15. $$\begin{aligned}&3x + by + bx + 3y\\ &= 3x + bx + by + 3y\\ &= x(3 + b) + y(b + 3)\\ &= x(3 + b) + y(3 + b)\\ &= (3 + b)(x + y)\end{aligned}$$

16. $4p^2 + 3pq - q^2$

$ac = 4(-1) = -4$

Find two integers whose product is -4 and whose sum is 3. The required integers are 4 and -1. Write $3pq$ as $4pq - 1pq$ (or $4pq - pq$).

$$\begin{aligned}&4p^2 + 3pq - q^2\\ &= 4p^2 + 4pq - pq - q^2\\ &= 4p(p + q) - q(p + q)\\ &= (p + q)(4p - q)\end{aligned}$$

17. $$\begin{aligned}&16a^2 + 20ab + 25b^2\\ &= (4a)^2 + 2(4a)(5b) + (5b)^2\end{aligned}$$

To be a perfect square trinomial, the middle term must be $40ab$. Therefore, the trinomial cannot be factored and is prime.

18. $$\begin{aligned}y^3 - 216 &= y^3 - 6^3\\ &= (y - 6)(y^2 + 6y + 6^2)\\ &= (y - 6)(y^2 + 6y + 36)\end{aligned}$$

19. $$\begin{aligned}9k^2 - 121j^2 &= (3k)^2 - (11j)^2\\ &= (3k + 11j)(3k - 11j)\end{aligned}$$

20. $$\begin{aligned}6k^4 - k^2 - 35 &= 6(k^2)^2 - k^2 - 35\\ &= (2k^2 - 5)(3k^2 + 7)\end{aligned}$$

21. **(a)** $$\begin{aligned}&(3 - x)(x + 4)\\ &= 3x + 12 - x^2 - 4x\\ &= -x^2 - x + 12\end{aligned}$$

(b) $-(x-3)(x+4)$
$= -(x^2+4x-3x-12)$
$= -(x^2+x-12)$
$= -x^2-x+12$

(c) $(-x+3)(x+4)$
$= -x^2-4x+3x+12$
$= -x^2-x+12$

(d) $(x-3)(-x+4)$
$= -x^2+4x+3x-12$
$= -x^2+7x-12$

Therefore, only (d) is not a factored form of $-x^2-x+12$.

22. $$3x^2+8x+4=0$$
$$(3x+2)(x+2)=0$$
$$3x+2=0 \quad \text{or} \quad x+2=0$$
$$3x=-2 \qquad x=-2$$
$$x=-\frac{2}{3}$$

Solution set: $\{-2, -\frac{2}{3}\}$

23. $$10x^2 = 17x-3$$
$$10x^2-17x+3=0$$
$$(2x-3)(5x-1)=0$$
$$2x-3=0 \quad \text{or} \quad 5x-1=0$$
$$2x=3 \qquad 5x=1$$
$$x=\frac{3}{2} \quad \text{or} \quad x=\frac{1}{5}$$

Solution set: $\{\frac{1}{5}, \frac{3}{2}\}$

24. $$5m(m-1)=2(1-m)$$
$$5m^2-5m=2-2m$$
$$5m^2-3m-2=0$$
$$(5m+2)(m-1)=0$$
$$5m+2=0 \quad \text{or} \quad m-1=0$$
$$5m=-2 \qquad m=1$$
$$m=-\frac{2}{5}$$

Solution set: $\{-\frac{2}{5}, 1\}$

25. Let $x =$ the height of the triangle.
$x-2 =$ the base of the triangle.

The area is 24 ft^2. Use the formula for the area of a triangle.

$$\frac{1}{2}(x)(x-2)=24$$

Multiply both sides by 2.

$$x(x-2)=48$$
$$x^2-2x=48$$
$$x^2-2x-48=0$$
$$(x-8)(x+6)=0$$
$$x-8=0 \quad \text{or} \quad x+6=0$$
$$x=8 \quad \text{or} \quad x=-6$$

The height cannot be negative, so reject $x=-6$. The height is 8 ft.

Cumulative Review Exercises (Chapters 1-5)

In Exercises 1-3,

$$R=\left\{-5, -\frac{3}{4}, 0, 1.6, \sqrt{7}, 9\right\}.$$

1. The element $\sqrt{7}$ is an irrational number.
2. The elements $-5, -\frac{3}{4}, 0, 1.6$, and 9 are rational numbers.
3. The elements $-5, 0$, and 9 are integers.
4. In interval notation, $\{x \mid -1 < x \le 7\}$ is written $(-1, 7]$. The parenthesis indicates that -1 is not included in the interval. The bracket indicates that 7 is included.
5. $$|2-5|+|-3|-7 = |-3|+|-3|-7$$
$$= 3+3-7$$
$$= 6-7$$
$$= -1$$
6. $$\frac{3\cdot 2^2-4}{-6+2^3\cdot 3} = \frac{3\cdot 4-4}{-6+8\cdot 3}$$
$$= \frac{12-4}{-6+24}$$
$$= \frac{8}{18}$$
$$= \frac{4}{9}$$
7. $$7x-(1+4x)=8+3x$$
$$7x-1-4x=8+3x$$
$$3x-1=8+3x$$
$$-1=8 \quad \textit{False}$$

The equation is a contradiction, and there is no solution.

Solution set: $\emptyset$

8. $|2r-7|=5$

$$\begin{aligned} 2r-7&=5 \quad \text{or} \quad 2r-7=-5 \\ 2r&=12 \qquad\qquad 2r=2 \\ r&=6 \quad \text{or} \qquad\; r=1 \end{aligned}$$

Solution set: $\{1,6\}$

9. Solve $3k-v^2=4p$ for k.

$$\begin{aligned} 3k-v^2&=4p \\ 3k&=4p+v^2 \\ k&=\frac{4p+v^2}{3} \end{aligned}$$

10. $\dfrac{x-6}{4}=1+\dfrac{2x+3}{2}$

Multiply both sides by 4.

$$\begin{aligned} 4\left(\frac{x-6}{4}\right)&=4\left(1+\frac{2x+3}{2}\right) \\ x-6&=4+2(2x+3) \\ x-6&=4+4x+6 \\ x-6&=4x+10 \\ -3x&=16 \\ x&=-\frac{16}{3} \end{aligned}$$

Solution set: $\{-\frac{16}{3}\}$

11. $3-\dfrac{1}{2}m<6+\dfrac{3}{2}m$

Multiply both sides by 2.

$$\begin{aligned} 2\left(3-\frac{1}{2}m\right)&<2\left(6+\frac{3}{2}m\right) \\ 6-m&<12+3m \\ -4m&<6 \end{aligned}$$

Divide by -4; reverse the inequality sign.

$$\begin{aligned} m&>-\frac{6}{4} \\ m&>-\frac{3}{2} \end{aligned}$$

Solution set: $(-\frac{3}{2},\infty)$

12. $3p\le-2$ and $p+1\ge-4$

$$p\le-\frac{2}{3} \qquad\qquad p\ge-5$$

$$-5\le p\le-\frac{2}{3}$$

Solution set: $[-5,-\frac{2}{3}]$

13. $(-1,4)$

This point is located in quadrant II since the x-coordinate is negative and the y-coordinate is positive.

14. $(3,-9)$

This point is located in quadrant IV since the x-coordinate is positive and the y-coordinate is negative.

15. $(-2,-5)$

This point is located in quadrant III since both the x- and y-coordinates are negative.

16. $(0,6)$

This point is located on the y-axis. It is not in any quadrant.

For Exercises 17 and 18, see the answer graphs in the back of the textbook.

17. $f(x)=\dfrac{2}{5}x-2$

Replace $f(x)$ with y to get the linear equation

$$y=\frac{2}{5}x-2$$

with slope $\frac{2}{5}$ and y-intercept $(0,-2)$. Use this information to graph the line.

18. $3x+y<6$

Graph the line $3x+y=6$, which has intercepts $(2,0)$ and $(0,6)$, as a dashed line since the inequality involves "$<$." Test $(0,0)$, which yields $0<6$, a true statement. Shade the region that includes $(0,0)$.

19.
$$\begin{aligned} f(x)&=-x+4 \\ f(-2)&=-(-2)+4=2+4=6 \end{aligned}$$

20.
$$\begin{aligned} f(x)&=\frac{3}{x+1} \\ f(-2)&=\frac{3}{-2+1}=\frac{3}{-1}=-3 \end{aligned}$$

21.
$$\begin{aligned} x+2y&=-1 \quad (1) \\ 3y-\;x&=-4 \quad (2) \\ 4x-2z&=-2 \quad (3) \end{aligned}$$

To eliminate x, add equations (1) and (2).

$$\begin{aligned} x+2y&=-1 \quad (1) \\ -x+3y&=-4 \quad (2) \\ \hline 5y&=-5 \\ y&=-1 \end{aligned}$$

Substitute -1 for y in equation (1) to find x.

$$\begin{aligned} x + 2y &= -1 \quad (1) \\ x + 2(-1) &= -1 \\ x - 2 &= -1 \\ x &= 1 \end{aligned}$$

Substitute 1 for x in equation (3) to find z.

$$\begin{aligned} 4x - 2z &= -2 \quad (3) \\ 4(1) - 2z &= -2 \\ 4 - 2z &= -2 \\ -2z &= -6 \\ z &= 3 \end{aligned}$$

Solution set: $\{(1, -1, 3)\}$

22. Let $x =$ the length of the shorter sides and $y =$ the length of the longer sides.

Since the longer sides are each 2 in longer than the shorter sides, one equation is

$$y = x + 2.$$

Since the distance around or perimeter of the parallelogram, 68, is twice the length plus twice the width, the other equation is

$$2x + 2y = 68.$$

The system is

$$\begin{aligned} y &= x + 2 \\ 2x + 2y &= 68. \end{aligned}$$

23. Rewrite the system as

$$\begin{aligned} x + 2y &= -1 \\ -x + 3y &= -4 \\ 4x - 2z &= -2. \end{aligned}$$

$$D = \begin{vmatrix} 1 & 2 & 0 \\ -1 & 3 & 0 \\ 4 & 0 & -2 \end{vmatrix}$$

Expand about column 3.

$$\begin{aligned} &= 0 - 0 + (-2)\begin{vmatrix} 1 & 2 \\ -1 & 3 \end{vmatrix} \\ &= -2[1(3) - 2(-1)] \\ &= -2(5) = -10 \end{aligned}$$

$$D_x = \begin{vmatrix} -1 & 2 & 0 \\ -4 & 3 & 0 \\ -2 & 0 & -2 \end{vmatrix}$$

Expand about column 3.

$$\begin{aligned} &= 0 - 0 + (-2)\begin{vmatrix} -1 & 2 \\ -4 & 3 \end{vmatrix} \\ &= -2[-1(3) - 2(-4)] \\ &= -2(5) = -10 \end{aligned}$$

$$D_y = \begin{vmatrix} 1 & -1 & 0 \\ -1 & -4 & 0 \\ 4 & -2 & -2 \end{vmatrix}$$

Expand about column 3.

$$\begin{aligned} &= 0 - 0 + (-2)\begin{vmatrix} 1 & -1 \\ -1 & -4 \end{vmatrix} \\ &= -2[1(-4) - (-1)(-1)] \\ &= -2(-5) = 10 \end{aligned}$$

$$D_z = \begin{vmatrix} 1 & 2 & -1 \\ -1 & 3 & -4 \\ 4 & 0 & -2 \end{vmatrix}$$

Expand about row 3.

$$\begin{aligned} &= 4\begin{vmatrix} 2 & -1 \\ 3 & -4 \end{vmatrix} - 0 + (-2)\begin{vmatrix} 1 & 2 \\ -1 & 3 \end{vmatrix} \\ &= 4[2(-4) - (-1)3] + (-2)[1(3) - 2(-1)] \\ &= 4(-5) + (-2)(5) \\ &= -20 + -10 = -30 \end{aligned}$$

$$x = \frac{D_x}{D} = \frac{-10}{-10} = 1$$

$$y = \frac{D_y}{D} = \frac{10}{-10} = -1$$

$$z = \frac{D_z}{D} = \frac{-30}{-10} = 3$$

Solution set: $\{(1, -1, 3)\}$

24. $7^{-3} = \frac{1}{7^3} = \frac{1}{7 \cdot 7 \cdot 7} = \frac{1}{343}$

25. $-3^0 = -(3^0) = -1$

26.
$$\begin{aligned} P(x) &= x^3 - x^2 + x - 5 \\ P(3) &= 3^3 - 3^2 + 3 - 5 \\ &= 27 - 9 + 3 - 5 \\ &= 16 \end{aligned}$$

27.
$$\begin{aligned} &(3x^2y^{-1})^{-2}(2x^{-3}y)^{-1} \\ &= 3^{-2}x^{-4}y^2 \cdot 2^{-1}x^3y^{-1} \\ &= 3^{-2}2^{-1}x^{-1}y \\ &= \frac{y}{3^2 \cdot 2x} \\ &= \frac{y}{18x} \end{aligned}$$

28. $$\frac{5m^{-2}y^3}{3m^{-3}y^{-1}} = \frac{5}{3}m^{-2-(-3)}y^{3-(-1)} = \frac{5}{3}my^4 = \frac{5my^4}{3}$$

29. $(2p+3)(5p^2-4p-8)$
$= 10p^3 - 8p^2 - 16p + 15p^2 - 12p - 24$
$= 10p^3 + 7p^2 - 28p - 24$

30. $(3x^3 + 4x^2 - 7) - (2x^3 - 8x^2 + 3x)$
$= 3x^3 + 4x^2 - 7 - 2x^3 + 8x^2 - 3x$
$= x^3 + 12x^2 - 3x - 7$

31. $16w^2 + 50wz - 21z^2$

$ac = 16(-21) = -336$

Find two integers whose product is -336 and whose sum is 50. The required integers are 56 and -6. Write $50wz$ as $56wz - 6wz$.

$16w^2 + 50wz - 21z^2$
$= 16w^2 + 56wz - 6wz - 21z^2$
$= 8w(2w+7z) - 3z(2w+7z)$
$= (2w+7z)(8w-3z)$

32. $4x^2 - 4x + 1 - y^2$

Group the first three terms.

$= (4x^2 - 4x + 1) - y^2$
$= (2x-1)^2 - y^2$
$= [(2x-1)+y][(2x-1)-y]$
$= (2x-1+y)(2x-1-y)$

33. $4y^2 - 36y + 81 = (2y)^2 - 2(2y)(9) + 9^2$
$= (2y-9)^2$

34. $100x^4 - 81 = (10x^2)^2 - 9^2$
$= (10x^2+9)(10x^2-9)$

35. $8p^3 + 27$
$= (2p)^3 + 3^3$
$= (2p+3)[(2p)^2 - (2p)(3) + 3^2]$
$= (2p+3)(4p^2 - 6p + 9)$

36. $(p-1)(2p+3)(p+4) = 0$
$p - 1 = 0$ or $2p + 3 = 0$ or $p + 4 = 0$
$p = 1$ $\quad 2p = -3$ $\quad p = -4$
$$p = -\frac{3}{2}$$

Solution set: $\left\{-4, -\frac{3}{2}, 1\right\}$

37. $$\begin{aligned} 9q^2 &= 6q - 1 \\ 9q^2 - 6q + 1 &= 0 \\ (3q-1)^2 &= 0 \\ 3q - 1 &= 0 \\ 3q &= 1 \\ q &= \frac{1}{3} \end{aligned}$$

Solution set: $\left\{\frac{1}{3}\right\}$

38. $$\begin{aligned} 5x^2 - 45 &= 0 \\ 5(x^2 - 9) &= 0 \\ 5(x+3)(x-3) &= 0 \end{aligned}$$
$x + 3 = 0$ or $x - 3 = 0$
$x = -3$ or $x = 3$

Solution set: $\{-3, 3\}$

39. Let $p = 3000, I = 384$, and $r = 3.2\%$ or .032 in the interest formula.

$$\begin{aligned} I &= prt \\ 384 &= 3000(.032)t \\ 384 &= 96t \\ 4 &= t \end{aligned}$$

The period of time is 4 yr.

40. Let $\quad x =$ the length of the base.
$x + 3 =$ the height.

The area is 14 m^2. Use the formula for the area of a triangle.

$$\frac{1}{2}(x)(x+3) = 14$$

Multiply both sides by 2.

$$\begin{aligned} x(x+3) &= 28 \\ x^2 + 3x &= 28 \\ x^2 + 3x - 28 &= 0 \\ (x+7)(x-4) &= 0 \end{aligned}$$
$x + 7 = 0$ or $x - 4 = 0$
$x = -7$ or $x = 4$

The length of the base cannot be negative, so reject $x = -7$. The base is 4 m long.

Chapter 6

RATIONAL EXPRESSIONS

6.1 Rational Expressions and Functions; Multiplication and Division

6.1 Margin Exercises

1. (a) $f(x) = \dfrac{x+4}{x-6}$

Set the denominator equal to zero, and solve the equation.

$$x - 6 = 0$$
$$x = 0$$

The number 6 makes the function undefined.

(b) $f(x) = \dfrac{8x}{5}$

Because the denominator is a constant, 5, there is no real number that makes the function undefined.

(c) $f(r) = \dfrac{r+6}{r^2 - r - 6}$

Set the denominator equal to zero, and factor the equation.

$$r^2 - r - 6 = 0$$
$$(r-3)(r+2) = 0$$
$$r - 3 = 0 \quad \text{or} \quad r + 2 = 0$$
$$r = 3 \quad \text{or} \quad r = -2$$

Both 3 and -2 make the function undefined.

(d) $f(k) = \dfrac{k+2}{k^2+1}$

Since for all real numbers $k^2 + 1 > 0$, there is no real number that makes the function undefined.

2. (a) $\dfrac{18m^2}{9m^5} = \dfrac{9m^2 \cdot 2}{9m^2 \cdot m^3} = \dfrac{2}{m^3}$

(b) $\dfrac{6y^2z^2}{12yz^3} = \dfrac{6yz^2 \cdot y}{6yz^2 \cdot 2z} = \dfrac{y}{2z}$

(c) $\dfrac{y^2+2y-3}{y^2-3y+2} = \dfrac{(y+3)(y-1)}{(y-2)(y-1)} = \dfrac{y+3}{y-2}$

(d) $\dfrac{3y+9}{y^2-9} = \dfrac{3(y+3)}{(y-3)(y+3)} = \dfrac{3}{y-3}$

(e) $\dfrac{y+2}{y^2+4}$

This expression cannot be factored and is already in lowest terms.

3. (a) $\dfrac{y-2}{2-y} = \dfrac{y-2}{-1(y-2)} = -1$

(b) $\dfrac{8-b}{8+b}$

Since the expression cannot be simplified further, it is already in lowest terms.

(c) $\dfrac{p-2}{4-p^2} = \dfrac{p-2}{(2-p)(2+p)}$

$$= \frac{p-2}{-1(p-2)(2+p)}$$
$$= \frac{-1}{2+p}$$

(d) $\dfrac{1+p^3}{1+p}$

Factor the sum of cubes.

$$= \frac{(1+p)(1-p+p^2)}{1+p}$$
$$= 1 - p + p^2$$

(e) $\dfrac{3x+3y+rx+ry}{5x+5y-rx-ry}$

$$= \frac{(3x+3y)+(rx+ry)}{(5x+5y)-(rx+ry)}$$
$$= \frac{3(x+y)+r(x+y)}{5(x+y)-r(x+y)}$$

Factor by grouping.

$$= \frac{(3+r)(x+y)}{(5-r)(x+y)}$$
$$= \frac{3+r}{5-r}$$

4. (a) $\dfrac{12r}{7} \cdot \dfrac{14}{r^2} = \dfrac{12 \cdot r \cdot 7 \cdot 2}{7 \cdot r \cdot r} = \dfrac{24}{r}$

(b) $\dfrac{6z^2}{5} \cdot \dfrac{4z}{3} = \dfrac{3 \cdot 2z^2 \cdot 4z}{3 \cdot 5} = \dfrac{8z^3}{5}$

(c) $\dfrac{2r+4}{5r} \cdot \dfrac{3r}{5r+10}$

$$= \frac{2(r+2)}{5r} \cdot \frac{3r}{5(r+2)} = \frac{6}{25}$$

(d) $\dfrac{m^2-16}{m+2} \cdot \dfrac{1}{m+4}$

$$= \frac{(m-4)(m+4)}{m+2} \cdot \frac{1}{m+4}$$
$$= \frac{m-4}{m+2}$$

(e) $\dfrac{c^2+2c}{c^2-4}\cdot\dfrac{c^2-4c+4}{c^2-c}$

$=\dfrac{c(c+2)}{(c-2)(c+2)}\cdot\dfrac{(c-2)(c-2)}{c(c-1)}$

$=\dfrac{c-2}{c-1}$

5. To find the reciprocal, invert the rational expression.

	Expression	*Reciprocal*
(a)	$\dfrac{-3}{r}$	$\dfrac{r}{-3}$
(b)	$\dfrac{7}{y+8}$	$\dfrac{y+8}{7}$
(c)	$\dfrac{a^2+7a}{2a-1}$	$\dfrac{2a-1}{a^2+7a}$
(d)	$\dfrac{0}{-5}$	Undefined; there is no reciprocal for 0.

6. In each case, multiply by the reciprocal of the divisor.

(a) $\dfrac{16k^2}{5}\div\dfrac{3k}{10}=\dfrac{16k^2}{5}\cdot\dfrac{10}{3k}=\dfrac{32k}{3}$

(b) $\dfrac{5p+2}{6}\div\dfrac{15p+6}{5}$

$=\dfrac{5p+2}{6}\cdot\dfrac{5}{15p+6}$

$=\dfrac{5p+2}{6}\cdot\dfrac{5}{3(5p+2)}$

$=\dfrac{5}{18}$

(c) $\dfrac{y^2-2y-3}{y^2+4y+4}\div\dfrac{y^2-1}{y^2+y-2}$

$=\dfrac{y^2-2y-3}{y^2+4y+4}\cdot\dfrac{y^2+y-2}{y^2-1}$

$=\dfrac{(y-3)(y+1)}{(y+2)(y+2)}\cdot\dfrac{(y+2)(y-1)}{(y+1)(y-1)}$

$=\dfrac{y-3}{y+2}$

6.1 Section Exercises

3. $\dfrac{x-3}{x-4}=\dfrac{-1(x-3)}{-1(x-4)}=\dfrac{-x+3}{-x+4}=\dfrac{3-x}{4-x}$

This expression is equivalent to the one in choice F.

7. $f(x)=\dfrac{x+1}{x-7}$

Set the denominator equal to zero, and solve the equation.

$$x-7=0$$
$$x=7$$

The number 7 makes the function undefined.

11. $f(t)=\dfrac{3t+1}{t}$

Set the denominator equal to zero and solve.

$$t=0$$

The number 0 makes the function undefined.

15. $f(x)=\dfrac{x+2}{14}$

The denominator is a constant, so there is no real number that makes the function undefined.

19. (a) $\dfrac{x^2+4x}{x+4}$

The two terms in the numerator are x^2 and $4x$. The two terms in the denominator are x and 4.

(b) To express the rational expression in lowest terms, factor the numerator and denominator and divide both by the common factor $x+4$ to get x.

$$\frac{x^2+4x}{x+4}=\frac{x(x+4)}{x+4}=x$$

23. $\dfrac{(x+4)(x-3)}{(x+5)(x+4)}=\dfrac{x-3}{x+5}$

27. $\dfrac{3x+7}{3x}$

Since the expression cannot be simplified further, it is already in lowest terms.

31. $\dfrac{6m+18}{7m+21}=\dfrac{6(m+3)}{7(m+3)}=\dfrac{6}{7}$

35. $\dfrac{2t+6}{t^2-9}=\dfrac{2(t+3)}{(t-3)(t+3)}=\dfrac{2}{t-3}$

39. $\dfrac{x^2+2x-15}{x^2+6x+5}=\dfrac{(x+5)(x-3)}{(x+5)(x+1)}=\dfrac{x-3}{x+1}$

43. $\dfrac{a^3+b^3}{a+b}=\dfrac{(a+b)(a^2-ab+b^2)}{a+b}=a^2-ab+b^2$

47. $\dfrac{ac-ad+bc-bd}{ac-ad-bc+bd}$

$$=\frac{a(c-d)+b(c-d)}{a(c-d)-b(c-d)}$$

Factor by grouping.

$$=\frac{(c-d)(a+b)}{(c-d)(a-b)}$$
$$=\frac{a+b}{a-b}$$

51. $\dfrac{x^2-y^2}{y-x}=\dfrac{(x+y)(x-y)}{-1(x-y)}=\dfrac{x+y}{-1}=-(x+y)$

55. $\dfrac{5k-10}{20-10k}=\dfrac{5(k-2)}{10(2-k)}$

$$=\frac{5(k-2)}{10\cdot -1(k-2)}$$
$$=-\frac{5}{10}$$
$$=-\frac{1}{2}$$

59. **(a)** $\dfrac{2x+3}{2x-3}$ cannot be simplified further.

(b) $\dfrac{2x-3}{3-2x}=\dfrac{2x-3}{-1(2x-3)}=-1$

(c) $\dfrac{2x+3}{3+2x}=\dfrac{2x+3}{2x+3}=1$

(d) $\dfrac{2x+3}{-2x-3}=\dfrac{2x+3}{-1(2x+3)}=-1$

The expressions in (b) and (d) equal -1.

63. $\dfrac{5a^4b^2}{16a^2b}\div\dfrac{25a^2b}{60a^3b^2}$

Multiply by the reciprocal of the divisor. (Note that each term could be simplified before multiplying by the reciprocal.)

$$=\frac{5a^4b^2}{16a^2b}\cdot\frac{60a^3b^2}{25a^2b}$$
$$=\frac{300a^7b^4}{400a^4b^2}$$
$$=\frac{100a^4b^2\cdot 3a^3b^2}{100a^4b^2\cdot 4}$$
$$=\frac{3a^3b^2}{4}$$

67. $\dfrac{7t+7}{-6}\div\dfrac{4t+4}{15}$

Multiply by the reciprocal of the divisor.

$$=\frac{7t+7}{-6}\cdot\frac{15}{4t+4}$$
$$=\frac{7(t+1)}{-6}\cdot\frac{15}{4(t+1)}$$
$$=-\frac{35}{8}$$

71. $\dfrac{m^2-49}{m+1}\div\dfrac{7-m}{m}$

Multiply by the reciprocal.

$$=\frac{m^2-49}{m+1}\cdot\frac{m}{7-m}$$
$$=\frac{(m-7)(m+7)}{m+1}\cdot\frac{m}{7-m}$$
$$=\frac{(-1)(7-m)(m+7)}{m+1}\cdot\frac{m}{7-m}$$
$$=\frac{-m(m+7)}{m+1}$$

75. $\dfrac{x^2-25}{x^2+x-20}\cdot\dfrac{x^2+7x+12}{x^2-2x-15}$

$$=\frac{(x-5)(x+5)}{(x+5)(x-4)}\cdot\frac{(x+3)(x+4)}{(x-5)(x+3)}$$
$$=\frac{x+4}{x-4}$$

79. $\dfrac{3k^2+17kp+10p^2}{6k^2+13kp-5p^2}\div\dfrac{6k^2+kp-2p^2}{6k^2-5kp+p^2}$

$$=\frac{(3k+2p)(k+5p)}{(3k-p)(2k+5p)}$$
$$\cdot\frac{(3k-p)(2k-p)}{(3k+2p)(2k-p)}$$
$$=\frac{k+5p}{2k+5p}$$

83. $\dfrac{a^2(2a+b)+6a(2a+b)+5(2a+b)}{3a^2(2a+b)-2a(2a+b)+(2a+b)}$

$$\div\frac{a+1}{a-1}$$
$$=\frac{(2a+b)(a^2+6a+5)}{(2a+b)(3a^2-2a+1)}\cdot\frac{a-1}{a+1}$$
$$=\frac{(2a+b)(a+5)(a+1)(a-1)}{(2a+b)(3a^2-2a+1)(a+1)}$$
$$=\frac{(a+5)(a-1)}{3a^2-2a+1}$$

87. From the circle graph, the 1993 investment in New York was \$26 million. The total investment in 1993 was $\$521+\$26+\$31+\$127=\$705$ million. As a ratio, this is written

$$\frac{26}{705}\approx .04.$$

6.2 Addition and Subtraction of Rational Expressions

6.2 Margin Exercises

1. **(a)** $\dfrac{3m}{8}+\dfrac{5n}{8}=\dfrac{3m+5n}{8}$

(b) $\dfrac{7}{3a}+\dfrac{10}{3a}=\dfrac{7+10}{3a}=\dfrac{17}{3a}$

(c) $\dfrac{2}{y^2}-\dfrac{5}{y^2}=\dfrac{2-5}{y^2}=\dfrac{-3}{y^2}=-\dfrac{3}{y^2}$

(d) $\dfrac{a}{a+b}+\dfrac{b}{a+b}=\dfrac{a+b}{a+b}=1$

(e) $\dfrac{2y-1}{y^2+y-2}-\dfrac{y}{y^2+y-2}$

$$\begin{aligned}&=\frac{(2y-1)-y}{y^2+y-2}\\&=\frac{y-1}{y^2+y-2}\\&=\frac{y-1}{(y+2)(y-1)}\\&=\frac{1}{y+2}\end{aligned}$$

2. **(a)** $\dfrac{9}{5k^3s},\dfrac{7}{10ks^4}$

Factor each denominator.

$$\begin{aligned}5k^3s&=5\cdot k^3\cdot s\\10ks^4&=2\cdot 5\cdot k\cdot s^4\end{aligned}$$

The greatest exponent on k is 3; the greatest exponent on s is 4. 5 and 2 are also factors. The LCD is

$$2\cdot 5\cdot k^3\cdot s^4 \text{ or } 10k^3s^4.$$

(b) $\dfrac{10}{3-x},\dfrac{5}{9-x^2}$

The first denominator, $3-x$, is already factored. Factor $9-x^2$.

$$9-x^2=(3+x)(3-x)$$

The LCD is $(3+x)(3-x)$.

(c) $\dfrac{8}{z},\dfrac{8}{z+6}$

Both denominators are already factored. The LCD is $z(z+6)$.

(d) $\dfrac{y-4}{2y^2-3y-2},\dfrac{2y+7}{2y^2+3y+1}$

Factor each denominator.

$$\begin{aligned}2y^2-3y-2&=(2y+1)(y-2)\\2y^2+3y+1&=(2y+1)(y+1)\end{aligned}$$

The LCD is $(2y+1)(y-2)(y+1)$.

3. **(a)** $\dfrac{6}{7}+\dfrac{1}{5}$

The LCD of 7 and 5 is 35. Multiply $\frac{6}{7}$ by $\frac{5}{5}$ and $\frac{1}{5}$ by $\frac{7}{7}$ so that each fraction has denominator 35. Then add the numerators.

$$\begin{aligned}\frac{6}{7}+\frac{1}{5}&=\frac{6\cdot 5}{7\cdot 5}+\frac{1\cdot 7}{5\cdot 7}\\&=\frac{30}{35}+\frac{7}{35}\\&=\frac{30+7}{35}\\&=\frac{37}{35}\end{aligned}$$

(b) $\dfrac{8}{3k}-\dfrac{2}{9k}$

The LCD for $3k$ and $9k$ is $9k$. To write $\frac{8}{3k}$ with a denominator of $9k$, multiply by $\frac{3}{3}$.

$$\begin{aligned}\frac{8}{3k}-\frac{2}{9k}&=\frac{3\cdot 8}{3\cdot 3k}-\frac{2}{9k}\\&=\frac{24}{9k}-\frac{2}{9k}\\&=\frac{24-2}{9k}\\&=\frac{22}{9k}\end{aligned}$$

(c) $\dfrac{2}{y}-\dfrac{1}{y+4}$

The LCD is $y(y+4)$. Rewrite each rational expression with this denominator.

$$\begin{aligned}\frac{2}{y}-\frac{1}{y+4}&=\frac{2(y+4)}{y(y+4)}-\frac{(1)y}{(y+4)y}\\&=\frac{2(y+4)}{y(y+4)}-\frac{y}{y(y+4)}\\&=\frac{2y+8-y}{y(y+4)}\\&=\frac{y+8}{y(y+4)}\end{aligned}$$

4. **(a)** $\dfrac{5x+7}{2x+7} - \dfrac{-x-14}{2x+7}$

The denominator is already the same for both rational expressions, so write them as a single expression. Be careful to apply the subtraction sign to both terms of the second expression.

$$= \frac{5x+7-(-x-14)}{2x+7}$$
$$= \frac{5x+7+x+14}{2x+7}$$
$$= \frac{6x+21}{2x+7}$$
$$= \frac{3(2x+7)}{2x+7}$$
$$= 3$$

(b) $\dfrac{2}{r-2} - \dfrac{r+3}{r-1}$

The LCD is $(r-2)(r-1)$.

$$= \frac{2(r-1)}{(r-2)(r-1)} - \frac{(r+3)(r-2)}{(r-1)(r-2)}$$
$$= \frac{2r-2-(r^2+r-6)}{(r-2)(r-1)}$$
$$= \frac{2r-2-r^2-r+6}{(r-2)(r-1)}$$
$$= \frac{-r^2+r+4}{(r-2)(r-1)}$$

5. **(a)** $\dfrac{8}{x-4} + \dfrac{2}{4-x}$

To get a common denominator of $x-4$, multiply both the numerator and denominator of the second expression by -1.

$$= \frac{8}{x-4} + \frac{2(-1)}{(4-x)(-1)}$$
$$= \frac{8}{x-4} + \frac{-2}{x-4}$$
$$= \frac{8+(-2)}{x-4}$$
$$= \frac{6}{x-4}$$

or

$$= \frac{-1(6)}{-1(x-4)} = \frac{-6}{4-x}$$

(b) $\dfrac{9}{2x-9} - \dfrac{4}{9-2x}$

$$= \frac{9}{2x-9} - \frac{4(-1)}{(9-2x)(-1)}$$
$$= \frac{9}{2x-9} - \frac{-4}{2x-9}$$
$$= \frac{9+4}{2x-9}$$
$$= \frac{13}{2x-9}$$

or

$$= \frac{-1(13)}{-1(2x-9)} = \frac{-13}{9-2x}$$

6. $\dfrac{4}{x-5} + \dfrac{-2}{x} - \dfrac{10}{x^2-5x}$

$$= \frac{4}{x-5} + \frac{-2}{x} - \frac{10}{x(x-5)}$$

The LCD is $x(x-5)$.

$$= \frac{4x}{(x-5)x} + \frac{-2(x-5)}{x(x-5)} - \frac{10}{x(x-5)}$$
$$= \frac{4x}{x(x-5)} + \frac{-2(x-5)}{x(x-5)} - \frac{10}{x(x-5)}$$
$$= \frac{4x+(-2)(x-5)-10}{x(x-5)}$$
$$= \frac{4x-2x+10-10}{x(x-5)}$$
$$= \frac{2x}{x(x-5)}$$
$$= \frac{2}{x-5}$$

7. $\dfrac{-a}{a^2+3a-4} - \dfrac{4a}{a^2+7a+12}$

$$= \frac{-a}{(a+4)(a-1)} - \frac{4a}{(a+3)(a+4)}$$

The LCD is $(a+4)(a-1)(a+3)$.

$$= \frac{-a(a+3)}{(a+4)(a-1)(a+3)}$$
$$- \frac{4a(a-1)}{(a+3)(a+4)(a-1)}$$
$$= \frac{-a(a+3)-4a(a-1)}{(a+4)(a-1)(a+3)}$$
$$= \frac{-a^2-3a-4a^2+4a}{(a+4)(a-1)(a+3)}$$
$$= \frac{-5a^2+a}{(a+4)(a-1)(a+3)}$$

6.2 Section Exercises

3. $\frac{7}{t} + \frac{2}{t} = \frac{7+2}{t} = \frac{9}{t}$

7. $\frac{5x+4}{6x+5} + \frac{x+1}{6x+5}$

$$= \frac{5x+4+x+1}{6x+5}$$
$$= \frac{6x+5}{6x+5} = 1$$

11. $\frac{4}{p^2+7p+12} + \frac{p}{p^2+7p+12}$

$$= \frac{4+p}{p^2+7p+12}$$
$$= \frac{p+4}{(p+3)(p+4)}$$
$$= \frac{1}{p+3}$$

15. $\frac{5}{18x^2y^3}, \frac{-3}{24x^4y^5}$

Factor each denominator.

$$\begin{aligned} 18x^2y^3 &= 2 \cdot 3 \cdot 3 \cdot x^2 \cdot y^3 \\ &= 2 \cdot 3^2 \cdot x^2 \cdot y^2 \\ 24x^4y^5 &= 2 \cdot 2 \cdot 2 \cdot 3 \cdot x^4 \cdot y^5 \\ &= 2^3 \cdot 3 \cdot x^4 \cdot y^5 \end{aligned}$$

Use each factor taken to the greatest exponent.

$$\begin{aligned} \text{LCD} &= 2^3 \cdot 3^2 \cdot x^4 \cdot y^5 \\ &= 8 \cdot 9 \cdot x^4 \cdot y^5 \\ &= 72x^4y^5 \end{aligned}$$

The LCD is $72x^4y^5$.

19. $\frac{9}{2y+8}, \frac{-8}{y+4}$

Factor each denominator.

$$2y + 8 = 2(y+4)$$

The second denominator, $y+4$, is already factored. The LCD is

$$2(y+4).$$

23. $\frac{3m}{m+n}, \frac{2n}{m-n}$

Both $m+n$ and $m-n$ have only 1 and themselves for factors. The LCD is

$$(m+n)(m-n).$$

27. $\frac{t}{2t^2+7t-15}, \frac{t}{t^2+3t-10}$

Factor each denominator.

$$\begin{aligned} 2t^2+7t-15 &= (2t-3)(t+5) \\ t^2+3t-10 &= (t+5)(t-2) \end{aligned}$$

The LCD is $(t+5)(2t-3)(t-2)$.

31. Both answers could be correct since

$$\frac{3}{5-y} = \frac{-1(3)}{-1(5-y)} = \frac{-3}{y-5}.$$

The two expressions are equivalent.

35. $\frac{5}{12x^2y} - \frac{11}{6xy}$

The LCD is $12x^2y$.

$$\begin{aligned} \frac{5}{12x^2y} - \frac{11}{6xy} &= \frac{5}{12x^2y} - \frac{11 \cdot 2x}{6xy \cdot 2x} \\ &= \frac{5}{12x^2y} - \frac{22x}{12x^2y} \\ &= \frac{5-22x}{12x^2y} \end{aligned}$$

39. $\frac{3a}{a+1} + \frac{2a}{a-3}$

The LCD is $(a+1)(a-3)$.

$$\begin{aligned} &\frac{3a}{a+1} + \frac{2a}{a-3} \\ &= \frac{3a(a-3)}{(a+1)(a-3)} + \frac{2a(a+1)}{(a-3)(a+1)} \\ &= \frac{3a(a-3)+2a(a+1)}{(a+1)(a-3)} \\ &= \frac{3a^2-9a+2a^2+2a}{(a+1)(a-3)} \\ &= \frac{5a^2-7a}{(a+1)(a-3)} \end{aligned}$$

43. $\frac{2}{4-x} + \frac{5}{x-4}$

To get a common denominator of $x-4$, multiply both numerator and denominator of the first expression by -1.

$$= \frac{2(-1)}{(4-x)(-1)} + \frac{5}{x-4}$$
$$= \frac{-2}{x-4} + \frac{5}{x-4}$$
$$= \frac{-2+5}{x-4}$$
$$= \frac{3}{x-4}$$

or

$$= \frac{-1(3)}{-1(x-4)} = \frac{-3}{4-x}$$

47. $\frac{5}{12+4x} - \frac{7}{9+3x}$

Factor each denominator to find the LCD.

$$12+4x = 4(3+x)$$
$$9+3x = 3(3+x)$$

The LCD is $12(3+x)$.

$$\frac{5}{12+4x} - \frac{7}{9+3x} = \frac{5}{4(3+x)} - \frac{7}{3(3+x)}$$
$$= \frac{5\cdot 3}{4(3+x)\cdot 3} - \frac{7\cdot 4}{3(3+x)\cdot 4}$$
$$= \frac{15-28}{12(3+x)}$$
$$= \frac{-13}{12(3+x)}$$

51. $\frac{15}{y^2+3y} + \frac{2}{y} + \frac{5}{y+3}$

Factor each denominator to find the LCD.

$$y^2+3y = y(y+3)$$

The other two denominators are already factored. The LCD is $y(y+3)$.

$$\frac{15}{y^2+3y} + \frac{2}{y} + \frac{5}{y+3} = \frac{15}{y(y+3)} + \frac{2(y+3)}{y(y+3)} + \frac{5y}{(y+3)y}$$
$$= \frac{15+2(y+3)+5y}{y(y+3)}$$
$$= \frac{15+2y+6+5y}{y(y+3)}$$
$$= \frac{7y+21}{y(y+3)}$$
$$= \frac{7(y+3)}{y(y+3)}$$
$$= \frac{7}{y}$$

55. $\frac{3x}{x+1} + \frac{4}{x-1} - \frac{6}{x^2-1}$

Factor each denominator to find the LCD. The first two denominators are already factored.

$$x^2-1 = (x+1)(x-1)$$

The LCD is $(x+1)(x-1)$.

$$\frac{3x}{x+1} + \frac{4}{x-1} - \frac{6}{x^2-1}$$
$$= \frac{3x(x-1)}{(x+1)(x-1)} + \frac{4(x+1)}{(x-1)(x+1)}$$
$$- \frac{6}{(x+1)(x-1)}$$
$$= \frac{3x(x-1)+4(x+1)-6}{(x+1)(x-1)}$$
$$= \frac{3x^2-3x+4x+4-6}{(x+1)(x-1)}$$
$$= \frac{3x^2+x-2}{(x+1)(x-1)}$$
$$= \frac{(3x-2)(x+1)}{(x+1)(x-1)}$$
$$= \frac{3x-2}{x-1}$$

59. $\frac{2x+4}{x+3} + \frac{3}{x} - \frac{6}{x^2+3x}$

Factor each denominator to find the LCD. The first two denominators are already factored.

$$x^2+3x = x(x+3)$$

The LCD is $x(x+3)$.

$$\frac{2x+4}{x+3} + \frac{3}{x} - \frac{6}{x^2+3x}$$
$$= \frac{(2x+4)x}{(x+3)x} + \frac{3(x+3)}{x(x+3)} - \frac{6}{x(x+3)}$$
$$= \frac{(2x+4)x+3(x+3)-6}{x(x+3)}$$
$$= \frac{2x^2+4x+3x+9-6}{x(x+3)}$$
$$= \frac{2x^2+7x+3}{x(x+3)}$$
$$= \frac{(2x+1)(x+3)}{x(x+3)}$$
$$= \frac{2x+1}{x}$$

63. $\dfrac{r+s}{3r^2+2rs-s^2}-\dfrac{s-r}{6r^2-5rs+s^2}$

Factor each denominator to find the LCD.

$$3r^2+2rs-s^2=(3r-s)(r+s)$$
$$6r^2-5rs+s^2=(3r-s)(2r-s)$$

The LCD is $(3r-s)(r+s)(2r-s)$.

$$\frac{r+s}{3r^2+2rs-s^2}-\frac{s-r}{6r^2-5rs+s^2}$$
$$=\frac{r+s}{(3r-s)(r+s)}-\frac{s-r}{(3r-s)(2r-s)}$$
$$=\frac{(r+s)(2r-s)}{(3r-s)(r+s)(2r-s)}$$
$$-\frac{(s-r)(r+s)}{(3r-s)(2r-s)(r+s)}$$
$$=\frac{(r+s)(2r-s)-(s-r)(r+s)}{(3r-s)(r+s)(2r-s)}$$
$$=\frac{2r^2+rs-s^2-(s^2-r^2)}{(3r-s)(r+s)(2r-s)}$$
$$=\frac{2r^2+rs-s^2-s^2+r^2}{(3r-s)(r+s)(2r-s)}$$
$$=\frac{3r^2+rs-2s^2}{(3r-s)(r+s)(2r-s)}$$
$$=\frac{(3r-2s)(r+s)}{(3r-s)(r+s)(2r-s)}$$
$$=\frac{3r-2s}{(3r-s)(2r-s)}$$

65. $\dfrac{3}{7}+\dfrac{5}{9}-\dfrac{6}{63}$

The LCD is 63.

$$=\frac{3\cdot 9}{7\cdot 9}+\frac{5\cdot 7}{9\cdot 7}-\frac{6}{63}$$
$$=\frac{27+35-6}{63}$$
$$=\frac{56}{63}$$
$$=\frac{8}{9}$$

66. From Example 6,

$$\frac{3}{x-2}+\frac{5}{x}-\frac{6}{x^2-2x}.$$

Substitute 9 for x.

$$\frac{3}{9-2}+\frac{5}{9}-\frac{6}{9^2-2(9)}=\frac{3}{7}+\frac{5}{9}-\frac{6}{63}$$

The problems in Exercises 65 and 66 are the same.

67. From Exercise 66,

$$\frac{3}{7}+\frac{5}{9}-\frac{6}{63}=\frac{8}{9}.$$

From Example 6, the answer is $\frac{8}{x}$. If we substitute 9 for x, the answer becomes $\frac{8}{9}$. The answers agree.

68. Answers will vary. For example, suppose your last name is Hart. Then $x=4$. The problem is Example 6 becomes

$$\frac{3}{4-2}+\frac{5}{4}-\frac{6}{4^2-2(4)}=\frac{3}{2}+\frac{5}{4}-\frac{6}{8}.$$

The predicted answer is

$$\frac{8}{x}, \text{ or } \frac{8}{4}=2.$$

Perform the operations to verify our prediction.

$$\frac{3}{2}+\frac{5}{4}-\frac{6}{8}=\frac{3\cdot 4}{2\cdot 4}+\frac{5\cdot 2}{4\cdot 2}-\frac{6}{8}$$
$$=\frac{12+10-6}{8}$$
$$=\frac{16}{8}$$
$$=2$$

Our prediction is correct.

69. If $x=2$, then the problem from Example 6,

$$\frac{3}{x-2}+\frac{5}{x}-\frac{6}{x^2-2x},$$

becomes

$$\frac{3}{2-2}+\frac{5}{2}-\frac{6}{2^2-2(2)}=\frac{3}{0}+\frac{5}{2}-\frac{6}{0}.$$

Thus, if $x=2$, then

$$\frac{3}{x-2} \text{ and } \frac{6}{x^2-2x}$$

are undefined.

70. If $x=0$, then

$$\frac{6}{x^2-2x}=\frac{6}{0}$$

which is undefined. Therefore, 0 is not allowed as a value of x.

6.3 Complex Fractions

6.3 Margin Exercises

1. (a) $\dfrac{\frac{a+2}{5a}}{\frac{a-3}{7a}}$

Both the numerator and denominator are already simplified.

$$= \frac{a+2}{5a} \div \frac{a-3}{7a}$$

Multiply by the reciprocal of the divisor.

$$= \frac{a+2}{5a} \cdot \frac{7a}{a-3}$$
$$= \frac{7(a+2)}{5(a-3)}$$

(b) $\dfrac{2+\frac{1}{k}}{2-\frac{1}{k}}$

Simplify the numerator and denominator.

$$= \frac{\frac{2k}{k}+\frac{1}{k}}{\frac{2k}{k}-\frac{1}{k}}$$
$$= \frac{\frac{2k+1}{k}}{\frac{2k-1}{k}}$$

Multiply by the reciprocal of the denominator.

$$= \frac{2k+1}{k} \cdot \frac{k}{2k-1}$$
$$= \frac{2k+1}{2k-1}$$

(c) $\dfrac{\frac{r^2-4}{4}}{1+\frac{2}{r}}$

Simplify the denominator.

$$= \frac{\frac{r^2-4}{4}}{\frac{r+2}{r}}$$

Multiply by the reciprocal of the denominator.

$$= \frac{r^2-4}{4} \cdot \frac{r}{r+2}$$
$$= \frac{(r-2)(r+2)}{4} \cdot \frac{r}{r+2}$$
$$= \frac{r(r-2)}{4}$$

2. (a) $\dfrac{\frac{5}{y}+6}{\frac{8}{3y}-1}$

The LCD of $\frac{5}{y}$ and $\frac{8}{3y}$ is $3y$. Multiply the numerator and denominator by $3y$.

$$= \frac{3y\left(\frac{5}{y}+6\right)}{3y\left(\frac{8}{3y}-1\right)}$$
$$= \frac{3y\left(\frac{5}{y}\right)+3y(6)}{3y\left(\frac{8}{3y}\right)-3y(1)}$$
$$= \frac{15+18y}{8-3y}$$

(b) $\dfrac{\frac{1}{y}+\frac{1}{y-1}}{\frac{1}{y}-\frac{2}{y-1}}$

Multiply the numerator and denominator by $y(y-1)$, the LCD of all the fractions.

$$= \frac{y(y-1)\left(\frac{1}{y}+\frac{1}{y-1}\right)}{y(y-1)\left(\frac{1}{y}-\frac{2}{y-1}\right)}$$
$$= \frac{y(y-1)\left(\frac{1}{y}\right)+y(y-1)\left(\frac{1}{y-1}\right)}{y(y-1)\left(\frac{1}{y}\right)-y(y-1)\left(\frac{2}{y-1}\right)}$$
$$= \frac{(y-1)+y}{(y-1)-2y}$$
$$= \frac{2y-1}{-y-1}$$

or

$$= \frac{(-1)(2y-1)}{(-1)(-y-1)} = \frac{1-2y}{y+1}$$

3. (a) Method 1

$$\frac{\frac{5}{y+2}}{\frac{-3}{y^2-4}} = \frac{\frac{5}{y+2}}{\frac{-3}{(y+2)(y-2)}}$$
$$= \frac{5}{y+2} \div \frac{-3}{(y+2)(y-2)}$$

Multiply by the reciprocal of the divisor.

$$= \frac{5}{y+2} \cdot \frac{(y+2)(y-2)}{-3}$$

$$= \frac{5(y-2)}{-3}$$

Method 2

$$\frac{\frac{5}{y+2}}{\frac{-3}{y^2-4}} = \frac{\frac{5}{y+2}}{\frac{-3}{(y+2)(y-2)}}$$

The LCD of the numerator and denominator is $(y+2)(y-2)$.

$$= \frac{\frac{5}{y+2} \cdot (y+2)(y-2)}{\frac{-3}{(y+2)(y-2)} \cdot (y+2)(y-2)}$$

$$= \frac{5(y-2)}{-3}$$

(b) Method 1

$$\frac{\frac{1}{a} - \frac{1}{b}}{\frac{1}{a^2} - \frac{1}{b^2}}$$

$$= \frac{\frac{b}{ab} - \frac{a}{ab}}{\frac{b^2}{a^2b^2} - \frac{a^2}{a^2b^2}}$$

Simplify the numerator and denominator.

$$= \frac{\frac{b-a}{ab}}{\frac{b^2-a^2}{a^2b^2}}$$

$$= \frac{\frac{b-a}{ab}}{\frac{(b+a)(b-a)}{a^2b^2}}$$

$$= \frac{b-a}{ab} \div \frac{(b+a)(b-a)}{a^2b^2}$$

Multiply by the reciprocal of the divisor.

$$= \frac{b-a}{ab} \cdot \frac{a^2b^2}{(b+a)(b-a)}$$

$$= \frac{ab}{b+a}$$

Method 2

$$\frac{\frac{1}{a} - \frac{1}{b}}{\frac{1}{a^2} - \frac{1}{b^2}}$$

The LCD of the numerator and the denominator is a^2b^2.

$$= \frac{\left(\frac{1}{a} - \frac{1}{b}\right) \cdot a^2b^2}{\left(\frac{1}{a^2} - \frac{1}{b^2}\right) \cdot a^2b^2}$$

$$= \frac{ab^2 - a^2b}{b^2 - a^2}$$

$$= \frac{ab(b-a)}{(b+a)(b-a)}$$

$$= \frac{ab}{b+a}$$

4. **(a)** $\frac{1}{a^{-1} + b^{-1}}$

$$= \frac{1}{\frac{1}{a} + \frac{1}{b}}$$

The LCD is ab. Multiply the numerator and denominator by ab.

$$= \frac{ab(1)}{ab\left(\frac{1}{a} + \frac{1}{b}\right)}$$

$$= \frac{ab}{ab\left(\frac{1}{a}\right) + ab\left(\frac{1}{b}\right)}$$

$$= \frac{ab}{b+a}$$

(b) $\frac{k^{-1}}{3k^{-2} + 1}$

$$= \frac{\frac{1}{k}}{\frac{3}{k^2} + 1}$$

The LCD is k^2. Multiply the numerator and denominator by k^2.

$$= \frac{k^2\left(\frac{1}{k}\right)}{k^2\left(\frac{3}{k^2} + 1\right)}$$

$$= \frac{k}{3 + k^2}$$

6.3 Section Exercises

3. $$\frac{\frac{12}{x-1}}{\frac{6}{x}} = \frac{12}{x-1} \div \frac{6}{x}$$

Multiply by the reciprocal of the divisor.

$$= \frac{12}{x-1} \cdot \frac{x}{6}$$
$$= \frac{2x}{x-1}$$

7. $$\frac{\frac{4z^2x^4}{9}}{\frac{12x^2z^5}{15}} = \frac{\frac{4z^2x^4}{9}}{\frac{4x^2z^5}{5}}$$
$$= \frac{4z^2x^4}{9} \div \frac{4x^2z^5}{5}$$

Multiply by the reciprocal of the divisor.

$$= \frac{4z^2x^4}{9} \cdot \frac{5}{4x^2z^5}$$
$$= \frac{20z^2x^4}{36x^2z^5}$$
$$= \frac{5x^2}{9z^3}$$

11. $$\frac{\frac{3}{x}+\frac{3}{y}}{\frac{3}{x}-\frac{3}{y}}$$

Multiply by the LCD of all the fractions in the numerator and denominator. The LCD is xy.

$$= \frac{\left(\frac{3}{x}+\frac{3}{y}\right)xy}{\left(\frac{3}{x}-\frac{3}{y}\right)xy}$$
$$= \frac{\frac{3}{x}\cdot xy + \frac{3}{y}\cdot xy}{\frac{3}{x}\cdot xy - \frac{3}{y}\cdot xy}$$
$$= \frac{3y+3x}{3y-3x}$$
$$= \frac{3(y+x)}{3(y-x)}$$
$$= \frac{y+x}{y-x}$$

15. $$\frac{\frac{x^2-16y^2}{xy}}{\frac{1}{y}-\frac{4}{x}}$$

Multiply the numerator and denominator by xy, the LCD of all the fractions.

$$= \frac{\left(\frac{x^2-16y^2}{xy}\right)xy}{\left(\frac{1}{y}-\frac{4}{x}\right)xy}$$
$$= \frac{x^2-16y^2}{\frac{1}{y}\cdot xy - \frac{4}{x}\cdot xy}$$
$$= \frac{x^2-16y^2}{x-4y}$$
$$= \frac{(x-4y)(x+4y)}{x-4y}$$
$$= x+4y$$

19. $$\frac{\frac{x+2}{x}+\frac{1}{x+2}}{\frac{5}{x}+\frac{x}{x+2}}$$

Multiply the numerator and denominator by $x(x+2)$, the LCD of all the fractions.

$$= \frac{x(x+2)\left(\frac{x+2}{x}+\frac{1}{x+2}\right)}{x(x+2)\left(\frac{5}{x}+\frac{x}{x+2}\right)}$$
$$= \frac{x(x+2)\left(\frac{x+2}{x}\right)+x(x+2)\left(\frac{1}{x+2}\right)}{x(x+2)\left(\frac{5}{x}\right)+x(x+2)\left(\frac{x}{x+2}\right)}$$
$$= \frac{(x+2)(x+2)+x}{5(x+2)+x^2}$$
$$= \frac{x^2+4x+4+x}{5x+10+x^2}$$
$$= \frac{x^2+5x+4}{x^2+5x+10}$$

21. To add the fractions in the numerator, use the LCD, $m(m-1)$.

$$\frac{4}{m}+\frac{m+2}{m-1} = \frac{4(m-1)}{m(m-1)}+\frac{m(m+2)}{m(m-1)}$$
$$= \frac{4m-4+m^2+2m}{m(m-1)}$$
$$= \frac{m^2+6m-4}{m(m-1)}$$

22. To subtract the fractions in the denominator, use the same LCD, $m(m-1)$.

$$\frac{m+2}{m} - \frac{2}{m-1}$$
$$= \frac{(m+2)(m-1)}{m(m-1)} - \frac{m \cdot 2}{m(m-1)}$$
$$= \frac{m^2+m-2-2m}{m(m-1)}$$
$$= \frac{m^2-m-2}{m(m-1)}$$

23. Exercise 21 answer   Exercise 22 answer

$$\frac{m^2+6m-4}{m(m-1)} \div \frac{m^2-m-2}{m(m-1)}$$

Multiply by the reciprocal.

$$= \frac{m^2+6m-4}{m(m-1)} \cdot \frac{m(m-1)}{m^2-m-2}$$
$$= \frac{m^2+6m-4}{m^2-m-2}$$

24. The LCD of all the denominators in the complex fraction is $m(m-1)$.

25. $$\frac{\left(\frac{4}{m} + \frac{m+2}{m-1}\right) \cdot m(m-1)}{\left(\frac{m+2}{m} - \frac{2}{m-1}\right) \cdot m(m-1)}$$
$$= \frac{4(m-1)+m(m+2)}{(m+2)(m-1)-2m}$$
$$= \frac{4m-4+m^2+2m}{m^2+m-2-2m}$$
$$= \frac{m^2+6m-4}{m^2-m-2}$$

26. Method 1 involves simplifying the numerator and the denominator separately and then dividing. Method 2 involves multiplying the fraction by a form of 1, the identity element for multiplication.

Because of the complicated nature of the numerator and denominator of the complex fraction, Method 1 takes much longer to simplify the complex fraction. Method 2 is a simpler, more direct means of simplifying and, as such, may be the preferred method.

27. $$\frac{1}{x^{-2}+y^{-2}} = \frac{1}{\frac{1}{x^2}+\frac{1}{y^2}}$$
$$= \frac{x^2y^2(1)}{x^2y^2\left(\frac{1}{x^2}+\frac{1}{y^2}\right)}$$
$$= \frac{x^2y^2}{y^2+x^2}$$

31. $$(r^{-1}+s^{-1})^{-1} = \left(\frac{1}{r}+\frac{1}{s}\right)^{-1}$$
$$= \left(\frac{s+r}{rs}\right)^{-1}$$
$$= \left(\frac{rs}{s+r}\right)^{1}$$
$$= \frac{rs}{s+r}$$

6.4 Equations Involving Rational Expressions

6.4 Margin Exercises

1. (a) $\frac{2k}{3} - \frac{k}{6} = -3$

Multiply both sides by the LCD, 6.

$$6\left(\frac{2k}{3} - \frac{k}{6}\right) = 6(-3)$$
$$6\left(\frac{2k}{3}\right) - 6\left(\frac{k}{6}\right) = 6(-3)$$
$$4k - k = -18$$
$$3k = -18$$
$$k = -6$$

Solution set: $\{-6\}$

(b) $\frac{3p}{2} - \frac{p}{4} = 1$

Multiply both sides by the LCD, 4.

$$4\left(\frac{3p}{2} - \frac{p}{4}\right) = 4(1)$$
$$4\left(\frac{3p}{2}\right) - 4\left(\frac{p}{4}\right) = 4$$
$$6p - p = 4$$
$$5p = 4$$
$$p = \frac{4}{5}$$

Solution set: $\left\{\frac{4}{5}\right\}$

(c) $\frac{7k}{12} - \frac{5k}{8} = \frac{1}{4}$

Multiply both sides by the LCD, 24.

$$24\left(\frac{7k}{12} - \frac{5k}{8}\right) = 24\left(\frac{1}{4}\right)$$
$$24\left(\frac{7k}{12}\right) - 24\left(\frac{5k}{8}\right) = 24\left(\frac{1}{4}\right)$$
$$14k - 15k = 6$$
$$-k = 6$$
$$k = -6$$

Solution set: $\{-6\}$

2. **(a)** $\frac{3}{p} - \frac{7}{10} = \frac{8}{5p}$

Multiply both sides by the LCD, $10p$.

$$10p\left(\frac{3}{p} - \frac{7}{10}\right) = 10p\left(\frac{8}{5p}\right)$$
$$10p\left(\frac{3}{p}\right) - 10p\left(\frac{7}{10}\right) = 10p\left(\frac{8}{5p}\right)$$
$$30 - 7p = 16$$
$$-7p = -14$$
$$p = 2$$

Since both sides were multiplied by a term containing a variable, it is especially important to check the potential solution. Substitute 2 for p in the original equation.

$$\frac{3}{p} - \frac{7}{10} = \frac{8}{5p}$$
$$\frac{3}{2} - \frac{7}{10} = \frac{8}{5 \cdot 2} \quad ?$$
$$\frac{15}{10} - \frac{7}{10} = \frac{8}{10} \quad ?$$
$$\frac{8}{10} = \frac{8}{10} \quad \textit{True}$$

Solution set: $\{2\}$

(b) $\frac{3x}{7} + \frac{1}{2} = \frac{5x}{14}$

Multiply both sides by the LCD, 14.

$$14\left(\frac{3x}{7} + \frac{1}{2}\right) = 14\left(\frac{5x}{14}\right)$$
$$14\left(\frac{3x}{7}\right) + 14\left(\frac{1}{2}\right) = 14\left(\frac{5x}{14}\right)$$
$$6x + 7 = 5x$$
$$x = -7$$

Solution set: $\{-7\}$

3. $\frac{3}{a+1} = \frac{1}{a-1} - \frac{2}{a^2-1}$

$$\frac{3}{a+1} = \frac{1}{a-1} - \frac{2}{(a+1)(a-1)}$$

The denominators cannot equal 0. Thus, $a + 1 \neq 0$ or $a - 1 \neq 0$, so $a \neq -1$ or $a \neq 1$. Multiply both sides by the LCD, $(a+1)(a-1)$.

$$(a+1)(a-1) \cdot \frac{3}{a+1}$$
$$= (a+1)(a-1) \cdot \frac{1}{a-1}$$
$$- (a+1)(a-1) \cdot \frac{2}{(a+1)(a-1)}$$
$$3(a-1) = (a+1) - 2$$
$$3a - 3 = a - 1$$
$$2a = 2$$
$$a = 1$$

But $a \neq 1$, or a denominator of 0 will result. The equation has no solution.

Solution set: $\emptyset$

4. **(a)** $\frac{10}{m^2} - \frac{3}{m} = 1$

$m \neq 0$ or a denominator of 0 will result. Multiply both sides by the LCD, m^2.

$$m^2\left(\frac{10}{m^2} - \frac{3}{m}\right) = m^2(1)$$
$$10 - 3m = m^2$$
$$m^2 + 3m - 10 = 0$$
$$(m-2)(m+5) = 0$$
$$m - 2 = 0 \quad \text{or} \quad m + 5 = 0$$
$$m = 2 \quad \text{or} \quad m = -5$$

Solution set: $\{-5, 2\}$

(b) $\frac{x}{x-3} + \frac{2}{x+3} = \frac{-12}{x^2-9}$

$$\frac{x}{x-3} + \frac{2}{x+3} = \frac{-12}{(x-3)(x+3)}$$

$x - 3 \neq 0$ or $x + 3 \neq 0$, so $x \neq 3$ or $x \neq -3$. Multiply both sides by the LCD, $(x-3)(x+3)$.

$$\frac{x}{x-3}(x-3)(x+3) + \frac{2}{x+3}(x-3)(x+3)$$
$$= \frac{-12}{(x-3)(x+3)}(x-3)(x+3)$$
$$x(x+3) + 2(x-3) = -12$$
$$x^2 + 3x + 2x - 6 = -12$$
$$x^2 + 5x + 6 = 0$$
$$(x+3)(x+2) = 0$$

$$x+3=0 \quad \text{or} \quad x+2=0$$
$$x=-3 \quad \text{or} \quad x=-2$$

But $x \neq -3$, or a denominator of 0 will result. The only solution is -2.

Solution set: $\{-2\}$

6.4 Section Exercises

3. $\dfrac{5}{3x+5} - \dfrac{1}{x} = \dfrac{1}{2x+3}$

Set each denominator equal to zero and solve.

$$3x+5=0 \qquad x=0$$
$$3x=-5$$
$$x=-\frac{5}{3}$$
$$2x+3=0$$
$$2x=-3$$
$$x=-\frac{3}{2}$$

So, $-\frac{5}{3}$, 0, and $-\frac{3}{2}$ cannot be solutions.

7. $\dfrac{3x+1}{x-4} = \dfrac{6x+5}{2x-7}$

Set each denominator equal to zero and solve.

$$x-4=0$$
$$x=4$$
$$2x-7=0$$
$$2x=7$$
$$x=\frac{7}{2}$$

So, 4 and $\frac{7}{2}$ cannot be solutions.

11. $\dfrac{x}{4} - \dfrac{x}{6} = \dfrac{2}{3}$

Multiply both sides by the LCD, 12.

$$12\left(\frac{x}{4}-\frac{x}{6}\right)=12\left(\frac{2}{3}\right)$$
$$3x-2x=8$$
$$x=8$$

Solution set: $\{8\}$

15. $\dfrac{x-4}{x+6} = \dfrac{2x+3}{2x-1}$

$x+6 \neq 0$ or $2x-1 \neq 0$, so $x \neq -6$ or $x \neq \frac{1}{2}$.
Multiply both sides by the LCD, $(x+6)(2x-1)$.

$$(x+6)(2x-1)\left(\frac{x-4}{x+6}\right)$$
$$=(x+6)(2x-1)\left(\frac{2x+3}{2x-1}\right)$$
$$(2x-1)(x-4)=(x+6)(2x+3)$$
$$2x^2-9x+4=2x^2+15x+18$$
$$-24x=14$$
$$x=-\frac{14}{24} \text{ or } -\frac{7}{12}$$

Solution set: $\{-\frac{7}{12}\}$

19. $\dfrac{-5}{2x} + \dfrac{3}{4x} = \dfrac{-7}{4}$

If $2x=0$ or $4x=0$, then $x=0$ and a denominator of 0 will result. So $x \neq 0$.
Multiply both sides by the LCD, $4x$.

$$4x\left(\frac{-5}{2x}+\frac{3}{4x}\right)=4x\left(\frac{-7}{4}\right)$$
$$-10+3=-7x$$
$$-7=-7x$$
$$1=x$$

Solution set: $\{1\}$

23. $\dfrac{1}{y-1} + \dfrac{5}{12} = \dfrac{-2}{3y-3}$

$$\frac{1}{y-1}+\frac{5}{12}=\frac{-2}{3(y-1)}$$

$y-1 \neq 0$, so $y \neq 1$.
Multiply both sides by the LCD, $12(y-1)$.

$$12(y-1)\left(\frac{1}{y-1}+\frac{5}{12}\right)=12(y-1)\left(\frac{-2}{3(y-1)}\right)$$
$$12+5(y-1)=-8$$
$$12+5y-5=-8$$
$$5y+7=-8$$
$$5y=-15$$
$$y=-3$$

Solution set: $\{-3\}$

27. $\dfrac{3}{k+2} - \dfrac{2}{k^2-4} = \dfrac{1}{k-2}$

$$\frac{3}{k+2}-\frac{2}{(k-2)(k+2)}=\frac{1}{k-2}$$

$k+2 \neq 0$ or $k-2 \neq 0$, so $k \neq -2$ or $k \neq 2$.
Multiply both sides by the LCD, $(k-2)(k+2)$.

$$(k-2)(k+2)\left(\frac{3}{k+2}-\frac{2}{(k-2)(k+2)}\right)$$
$$=(k-2)(k+2)\left(\frac{1}{k-2}\right)$$
$$\begin{aligned}3(k-2)-2&=k+2\\3k-6-2&=k+2\\3k-8&=k+2\\2k&=10\\k&=5\end{aligned}$$

Solution set: $\{5\}$

31. $\frac{9}{x}+\frac{4}{6x-3}=\frac{2}{6x-3}$

$6x-3\neq 0$, so $x\neq \frac{1}{2}$ or $x\neq 0$.
Multiply both sides by the LCD, $x(6x-3)$.

$$x(6x-3)\left(\frac{9}{x}+\frac{4}{6x-3}\right)$$
$$=x(6x-3)\left(\frac{2}{6x-3}\right)$$
$$\begin{aligned}9(6x-3)+4x&=2x\\54x-27+4x&=2x\\58x-27&=2x\\56x&=27\\x&=\frac{27}{56}\end{aligned}$$

Solution set: $\{\frac{27}{56}\}$

35. $\frac{x}{x-3}+\frac{4}{x+3}=\frac{18}{x^2-9}$

$$\frac{x}{x-3}+\frac{4}{x+3}=\frac{18}{(x-3)(x+3)}$$

$x-3\neq 0$ or $x+3\neq 0$, so $x\neq 3$ or $x\neq -3$.
Multiply both sides by the LCD, $(x-3)(x+3)$.

$$(x-3)(x+3)\left(\frac{x}{x-3}+\frac{4}{x+3}\right)$$
$$=(x-3)(x+3)\left(\frac{18}{(x-3)(x+3)}\right)$$
$$\begin{aligned}x(x+3)+4(x-3)&=18\\x^2+3x+4x-12&=18\\x^2+7x-30&=0\\(x-3)(x+10)&=0\\x-3=0 \quad\text{or}\quad x+10&=0\\x=3 \quad\text{or}\quad x&=-10\end{aligned}$$

But $x\neq 3$, or a denominator of 0 results. The only solution is -10.

Solution set: $\{-10\}$

39. $\frac{2}{4x+7}+\frac{x}{3}=\frac{6}{12x+21}$

$$\frac{2}{4x+7}+\frac{x}{3}=\frac{6}{3(4x+7)}$$

$4x+7\neq 0$, so $x\neq -\frac{7}{4}$.
Multiply by the LCD, $3(4x+7)$.

$$3(4x+7)\left(\frac{2}{4x+7}+\frac{x}{3}\right)=3(4x+7)\left(\frac{6}{3(4x+7)}\right)$$
$$\begin{aligned}2(3)+x(4x+7)&=6\\6+4x^2+7x&=6\\4x^2+7x&=0\\x(4x+7)&=0\\x=0 \quad\text{or}\quad 4x+7&=0\\4x&=-7\\x&=-\frac{7}{4}\end{aligned}$$

But $x\neq -\frac{7}{4}$, or a denominator of 0 will result. The only solution is 0.

Solution set: $\{0\}$

43. $\frac{1}{x+2}+\frac{x}{x-2}=\frac{-4}{(x^2-4)}$

$$\frac{1}{x+2}+\frac{x}{x-2}=\frac{-4}{(x+2)(x-2)}$$

$x+2\neq 0$ or $x-2\neq 0$, so $x\neq -2$ or $x\neq 2$.
Multiply both sides by the LCD, $(x+2)(x-2)$.

$$(x+2)(x-2)\left(\frac{1}{x+2}+\frac{x}{x-2}\right)$$
$$=(x+2)(x-2)\left(\frac{-4}{(x+2)(x-2)}\right)$$
$$\begin{aligned}(x-2)+x(x+2)&=-4\\x-2+x^2+2x&=-4\\x^2+3x+2&=0\\(x+2)(x+1)&=0\\x+2=0 \quad\text{or}\quad x+1&=0\\x=-2 \quad\text{or}\quad x&=-1\end{aligned}$$

But $x\neq -2$, or a denominator of 0 will result. The only solution is -1.

Solution set: $\{-1\}$

44. If $A = 2$ and $B = 1$, then

$$C = -2AB = -2(2)(1) = -4.$$

The equation determined by A, B, and C is

$$\frac{2}{x+1} + \frac{x}{x-1} = \frac{-4}{x^2-1}.$$
$$\frac{2}{x+1} + \frac{x}{x-1} = \frac{-4}{(x+1)(x-1)}$$

$x+1 \neq 0$ or $x-1 \neq 0$, so $x \neq -1$ or $x \neq 1$. Multiply both sides by the LCD, $(x+1)(x-1)$.

$$(x+1)(x-1)\left(\frac{2}{x+1} + \frac{x}{x-1}\right)$$
$$= (x+1)(x-1)\left(\frac{-4}{(x+1)(x-1)}\right)$$
$$2(x-1) + x(x+1) = -4$$
$$2x - 2 + x^2 + x = -4$$
$$x^2 + 3x + 2 = 0$$
$$(x+2)(x+1) = 0$$
$$x+2 = 0 \quad \text{or} \quad x+1 = 0$$
$$x = -2 \quad \text{or} \quad x = -1$$

But $x \neq -1$, or a denominator of 0 will result. The only solution is -2.

Solution set: $\{-2\}$

45. If $A = 4$ and $B = -3$, then

$$C = -2AB = -2(4)(-3) = 24.$$

The equation determined by A, B, and C is

$$\frac{4}{x-3} + \frac{x}{x+3} = \frac{24}{x^2-9}.$$
$$\frac{4}{x-3} + \frac{x}{x+3} = \frac{24}{(x+3)(x-3)}$$

$x-3 \neq 0$ or $x+3 \neq 0$, so $x \neq 3$ or $x \neq -3$. Multiply both sides by the LCD, $(x+3)(x-3)$.

$$(x+3)(x-3)\left(\frac{4}{x-3} + \frac{x}{x+3}\right)$$
$$= (x+3)(x-3)\left(\frac{24}{(x+3)(x-3)}\right)$$
$$4(x+3) + x(x-3) = 24$$
$$4x + 12 + x^2 - 3x = 24$$
$$x^2 + x - 12 = 0$$
$$(x+4)(x-3) = 0$$
$$x+4 = 0 \quad \text{or} \quad x-3 = 0$$
$$x = -4 \quad \text{or} \quad x = 3$$

But $x \neq 3$, or a denominator of 0 will result. The only solution is -4.

Solution set: $\{-4\}$

46. Answers will vary. However, in every case, $-B$ will be the rejected solution and $\{-A\}$ will be the solution set.

47. Answers will vary. For example, start with a solution set of $\{2\}$.

$$x = 2$$

Multiply both sides by 2.

$$2x = 4$$

Rewrite $2x$ as the difference of two terms.

$$6x - 4x = 4$$

Divide both sides by 24.

$$\frac{6x}{24} - \frac{4x}{24} = \frac{4}{24}$$

Simplify to get our equation.

$$\frac{x}{4} - \frac{x}{6} = \frac{1}{6}$$

Now solve this equation. The solution set should be $\{2\}$. Multiply by the LCD, 24.

$$24\left(\frac{x}{4} - \frac{x}{6}\right) = 24\left(\frac{1}{6}\right)$$
$$6x - 4x = 4$$
$$2x = 4$$
$$x = 2$$

As expected, the solution set is $\{2\}$.

Summary Exercises on Operations and Equations Involving Rational Expressions

3. $\dfrac{4x-20}{x^2-25} \cdot \dfrac{(x+5)^2}{10}$ (expression)

$$= \frac{4(x-5)}{(x+5)(x-5)} \cdot \frac{(x+5)(x+5)}{10}$$
$$= \frac{2(x+5)}{5}$$

7. $\frac{x-5}{3}+\frac{1}{3}=\frac{x-2}{5}$ (equation)

Multiply both sides by the LCD, 15.

$$15\left(\frac{x-5}{3}+\frac{1}{3}\right)=15\left(\frac{x-2}{5}\right)$$
$$5(x-5)+5=3(x-2)$$
$$5x-25+5=3x-6$$
$$5x-20=3x-6$$
$$2x=14$$
$$x=7$$

Solution set: $\{7\}$

11. $\frac{8}{r+2}-\frac{7}{4r+8}$ (expression)

$$=\frac{8}{r+2}-\frac{7}{4(r+2)}$$

The LCD is $4(r+2)$.

$$=\frac{4(8)}{4(r+2)}-\frac{7}{4(r+2)}$$
$$=\frac{32}{4(r+2)}-\frac{7}{4(r+2)}$$
$$=\frac{25}{4(r+2)}$$

15. $\frac{a-4}{3}+\frac{11}{6}=\frac{a+1}{2}$ (equation)

Multiply both sides by the LCD, 6.

$$6\left(\frac{a-4}{3}+\frac{11}{6}\right)=6\left(\frac{a+1}{2}\right)$$
$$2(a-4)+11=3(a+1)$$
$$2a-8+11=3a+3$$
$$2a+3=3a+3$$
$$-a=0$$
$$a=0$$

Solution set: $\{0\}$

19. $\frac{6}{t+1}+\frac{4}{5t+5}=\frac{34}{15}$ (equation)

$$\frac{6}{t+1}+\frac{4}{5(t+1)}=\frac{34}{15}$$

$t+1\neq 0$, so $t\neq -1$.

Multiply both sides by the LCD, $15(t+1)$.

$$15(t+1)\left(\frac{6}{t+1}+\frac{4}{5(t+1)}\right)=15(t+1)\left(\frac{34}{15}\right)$$
$$90+12=34(t+1)$$
$$102=34t+34$$
$$68=34t$$
$$2=t$$

Solution set: $\{2\}$

23. $\dfrac{2r^{-1}+5s^{-1}}{\dfrac{4s^2-25r^2}{3rs}}$ (expression)

$$=\frac{\frac{2}{r}+\frac{5}{s}}{\frac{4s^2-25r^2}{3rs}}$$

Multiply the numerator and denominator by $3rs$, the LCD of all the fractions.

$$=\frac{3rs\left(\frac{2}{r}+\frac{5}{s}\right)}{3rs\left(\frac{4s^2-25r^2}{3rs}\right)}$$
$$=\frac{6s+15r}{4s^2-25r^2}$$
$$=\frac{3(2s+5r)}{(2s-5r)(2s+5r)}$$
$$=\frac{3}{2s-5r}$$

27. $\frac{6z^2-5z-6}{6z^2+5z-6}\cdot\frac{12z^2-17z+6}{12z^2-z-6}$

(expression)

$$=\frac{(2z-3)(3z+2)}{(2z+3)(3z-2)}\cdot\frac{(3z-2)(4z-3)}{(3z+2)(4z-3)}$$
$$=\frac{2z-3}{2z+3}$$

31. $\frac{7}{2x^2-8x}+\frac{3}{x^2-16}$ (expression)

$$=\frac{7}{2x(x-4)}+\frac{3}{(x-4)(x+4)}$$

The LCD is $2x(x-4)(x+4)$.

$$=\frac{7(x+4)}{2x(x-4)(x+4)}+\frac{(3)2x}{(x-4)(x+4)2x}$$
$$=\frac{7(x+4)}{2x(x-4)(x+4)}+\frac{6x}{2x(x-4)(x+4)}$$
$$=\frac{7x+28+6x}{2x(x-4)(x+4)}$$
$$=\frac{13x+28}{2x(x-4)(x+4)}$$

6.5 Formulas Involving Rational Expressions

6.5 Margin Exercises

1. **(a)** Find p if $f = 15$ and $q = 25$.

$$\frac{1}{f} = \frac{1}{p} + \frac{1}{q}$$
$$\frac{1}{15} = \frac{1}{p} + \frac{1}{25}$$

Multiply by the LCD, $75p$.

$$75p\left(\frac{1}{15}\right) = 75p\left(\frac{1}{p}\right) + 75p\left(\frac{1}{25}\right)$$
$$5p = 75 + 3p$$
$$2p = 75$$
$$p = \frac{75}{2}$$

(b) Find f if $p = 6$ and $q = 9$.

$$\frac{1}{f} = \frac{1}{p} + \frac{1}{q}$$
$$\frac{1}{f} = \frac{1}{6} + \frac{1}{9}$$

Multiply by the LCD, $18f$.

$$18f\left(\frac{1}{f}\right) = 18f\left(\frac{1}{6}\right) + 18f\left(\frac{1}{9}\right)$$
$$18 = 3f + 2f$$
$$18 = 5f$$
$$\frac{18}{5} = f$$

(c) Find q if $f = 12$ and $p = 16$.

$$\frac{1}{f} = \frac{1}{p} + \frac{1}{q}$$
$$\frac{1}{12} = \frac{1}{16} + \frac{1}{q}$$

Multiply by the LCD, $48q$.

$$48q\left(\frac{1}{12}\right) = 48q\left(\frac{1}{16}\right) + 48q\left(\frac{1}{q}\right)$$
$$4q = 3q + 48$$
$$q = 48$$

2. **(a)** Solve $\frac{1}{f} = \frac{1}{p} + \frac{1}{q}$ for q.

Multiply by the LCD, fpq.

$$fpq\left(\frac{1}{f}\right) = fpq\left(\frac{1}{p}\right) + fpq\left(\frac{1}{q}\right)$$
$$pq = fq + fp$$

Get the terms with q together on one side of the equals sign by subtracting fq.

$$pq - fq = fp$$
$$q(p - f) = fp$$
$$q = \frac{fp}{p - f}$$

(b) Solve $\frac{8}{7x} - \frac{9}{5y} = \frac{2}{z}$ for y.

Multiply by the LCD, $35xyz$.

$$35xyz\left(\frac{8}{7x}\right) - 35xyz\left(\frac{9}{5y}\right) = 35xyz\left(\frac{2}{z}\right)$$
$$40yz - 63xz = 70xy$$

Get the terms with y together on one side of the equals sign.

$$40yz - 70xy = 63xz$$
$$y(40z - 70x) = 63xz$$
$$y = \frac{63xz}{40z - 70x}$$

3. **(a)** Solve for R.

$$A = \frac{Rr}{R + r}$$

Multiply by $R + r$.

$$A(R + r) = \frac{Rr}{R + r}(R + r)$$
$$A(R + r) = Rr$$
$$AR + Ar = Rr$$

Get the terms with R together on one side of the equals sign.

$$AR - Rr = -Ar$$
$$R(A - r) = -Ar$$
$$R = \frac{-Ar}{A - r}$$

or

$$= \frac{-1(-Ar)}{-1(A - r)}$$
$$= \frac{Ar}{r - A}$$

(b) Solve for r.

$$I = \frac{nE}{R+nr}$$

Multiply by $R+nr$.

$$I(R+nr) = \frac{nE}{R+nr}(R+nr)$$
$$I(R+nr) = nE$$
$$IR + Inr = nE$$

Get the term with r on one side of the equation.

$$Inr = nE - IR$$
$$r = \frac{nE - IR}{In}$$

6.5 Section Exercises

3. $A = \frac{1}{2}bh$

To solve for h, multiply both sides by 2.

$$2A = bh$$

Divide both sides by b.

$$\frac{2A}{b} = h \text{ or } h = \frac{2A}{b}$$

The correct answer is (a).

7. $F = \frac{GMm}{d^2}$

To solve for M, multiply both sides by d^2.

$$Fd^2 = \frac{GMm}{d^2} \cdot d^2$$
$$Fd^2 = GMm$$

Divide both sides by Gm to get the term with M on one side of the equation.

$$\frac{Fd^2}{Gm} = M$$

Substitute 10 for F, 6.67×10^{-11} for G, 1 for m, and 3×10^{-6} for d in the equation.

$$M = \frac{Fd^2}{Gm}$$
$$M = \frac{10(3 \times 10^{-6})^2}{(6.67 \times 10^{-11})1}$$
$$M \approx 1.349$$

11. $5x = 4 + 2(x+2)$

(a) $5x = 4 + 2x + 4$

(b) $5x - 2x = 4 + 2x + 4 - 2x$

(c) $(5-2)x = 8$

(d) $3x = 8$

(e) $x = \frac{8}{3}$

12. $bx = c + d(x+2)$

(a) $bx = c + dx + 2d$

(b) $bx - dx = c + dx + 2d - dx$

(c) $(b-d)x = c + 2d$

(d) $x = \frac{c+2d}{b-d}$

13. Only like terms can be combined. In Exercise 11(c), we combined the constant terms and the x-terms. In Exercise 12(c), we combined the x-terms only.

14. The statement is true. There is no difference in the procedures used when solving an equation and when solving a formula.

15. $F = \frac{GMm}{d^2}$

To solve for G, multiply both sides by $\frac{d^2}{Mm}$.

$$\frac{d^2}{Mm}(F) = \frac{d^2}{Mm}\left(\frac{GMm}{d^2}\right)$$
$$\frac{Fd^2}{Mm} = G$$
$$\text{or} \quad G = \frac{Fd^2}{Mm}$$

19. $\frac{PV}{T} = \frac{pv}{t}$

To solve for v, multiply both sides by $\frac{t}{p}$.

$$\frac{t}{p}\left(\frac{PV}{T}\right) = \frac{t}{p}\left(\frac{pv}{t}\right)$$
$$\frac{PVt}{pT} = v$$
$$\text{or} \quad v = \frac{PVt}{pT}$$

23. $A = \frac{1}{2}h(B + b)$

To solve for b, multiply both sides by $\frac{2}{h}$.

$$\frac{2}{h}(A) = \frac{2}{h}\left[\frac{1}{2}h(B + b)\right]$$
$$\frac{2A}{h} = B + b$$
$$\frac{2A}{h} - B = b$$
$$b = \frac{2A}{h} - B$$
$$\text{or} \quad b = \frac{2A - Bh}{h}$$

6.6 Applications Involving Rational Expressions

6.6 Margin Exercises

1. Let $x =$ the average family income.

Set up a proportion.

$$\frac{4296}{x} = \frac{11.1}{100}$$

Multiply by the LCD, $100x$.

$$100x\left(\frac{4296}{x}\right) = 100x\left(\frac{11.1}{100}\right)$$
$$429{,}600 = 11.1x$$
$$x = \frac{429{,}600}{11.1}$$
$$x \approx 38{,}703$$

The average family income was \$38,703.

2. Let $x =$ the amount of tax.

Set up a proportion.

$$\frac{\text{new tax}}{\text{new price}} = \frac{\text{tax}}{\text{price}}$$
$$\frac{x}{3.65} = \frac{.34}{1.22}$$

Multiply by $(3.65)(1.22)$.

$$(3.65)(1.22)\left(\frac{x}{3.65}\right) = (3.65)(1.22)\left(\frac{.34}{1.22}\right)$$
$$1.22x = (3.65)(.34)$$
$$x = \frac{(3.65)(.34)}{1.22}$$
$$x \approx 1.02$$

Since the tax, x, would be 1.02 and the tax was .34, the additional tax would be

$$x - .34 = 1.02 - .34 = .68.$$

The additional tax would be \$.68 per gal.

3. **(a)** Let $x =$ the speed of the plane in still air.

Use $d = rt$, or $t = \frac{d}{r}$, to complete the table.

Direction	Distance	Rate	Time
Against wind	100	$x - 20$	$\frac{100}{x-20}$
With wind	120	$x + 20$	$\frac{120}{x+20}$

(b) Because the time against the wind equals the time with the wind,

$$\frac{100}{x - 20} = \frac{120}{x + 20}.$$

Multiply by the LCD, $(x - 20)(x + 20)$.

$$\frac{100}{x-20}(x - 20)(x + 20) = \frac{120}{x+20}(x - 20)(x + 20)$$
$$100(x + 20) = 120(x - 20)$$
$$100x + 2000 = 120x - 2400$$
$$4400 = 20x$$
$$220 = x$$

The speed of the plane is 220 mph in still air.

4. Let $x =$ the distance at reduced speed.

Use $d = rt$, or $t = \frac{d}{r}$, to complete the table.

	Distance	Rate	Time
Freeway	$300 - x$	55	$\frac{300-x}{55}$
Reduced speed	x	15	$\frac{x}{15}$

Now write an equation.

Time on the freeway	plus	time at reduced speed	equals 6 hr.
$\frac{300-x}{55}$	$+$	$\frac{x}{15}$	$= 6$

Multiply by the LCD, 165.

$$165\left(\frac{300-x}{55}+\frac{x}{15}\right)=165(6)$$
$$3(300-x)+11x=990$$
$$900-3x+11x=990$$
$$8x=90$$
$$x=11\frac{1}{4}$$

She drove $11\frac{1}{4}$ mi at reduced speed.

5. **(a)** Let $t =$ the time it will take them working together.

Make a table.

Worker	Rate	Time together	Part of job done
Stan	$\frac{1}{45}$	t	$\frac{t}{45}$
Bobbie	$\frac{1}{30}$	t	$\frac{t}{30}$

Part done by Stan + part done by Bobbie = 1 whole job.

$$\frac{t}{45} + \frac{t}{30} = 1$$

Multiply by the LCD, 90.

$$90\left(\frac{t}{45}+\frac{t}{30}\right)=90\cdot 1$$
$$2t+3t=90$$
$$5t=90$$
$$t=18$$

It will take them 18 min working together.

(b) Let $t =$ the time for Bobbie to do the job alone.

Make a table.

Worker	Rate	Time together	Part of job done
Stan	$\frac{1}{35}$	15	$\frac{15}{35}$
Bobbie	$\frac{1}{t}$	15	$\frac{15}{t}$

Since together they complete 1 job, the sum of the fractional parts should be 1.

$$\frac{15}{35}+\frac{15}{t}=1$$
$$\frac{3}{7}+\frac{15}{t}=1$$

Multiply by the LCD, $7t$.

$$7t\left(\frac{3}{7}+\frac{15}{t}\right)=7t\cdot 1$$
$$3t+105=7t$$
$$105=4t$$
$$26\frac{1}{4}=t$$

Bobbie can do them in $26\frac{1}{4}$ min.

6.6 Section Exercises

3. In 1993, home video revenue was about $5 billion and box office revenue was about $13 billion, so the ratio was

$$5 \text{ to } 13 \text{ or } \frac{5}{13}.$$

7. Juanita's rate $= \dfrac{1 \text{ job}}{\text{time to complete 1 job}}$

$$= \frac{1 \text{ job}}{2 \text{ hr}}$$
$$= \frac{1}{2} \text{ job per hour}$$

11. Let $x =$ the additional amount of tax.

Write a proportion.

$$\frac{\text{tax}}{\text{income}}=\frac{\text{additional tax}}{\text{additional income}}$$
$$\frac{4500}{30{,}000}=\frac{x}{2000}$$

Multiply by the LCD, 30,000.

$$30{,}000\left(\frac{4500}{30{,}000}\right)=30{,}000\left(\frac{x}{2000}\right)$$
$$4500=15x$$
$$300=x$$

The additional tax is $300.

15. Let $x =$ the speed of the current of the river.

Use $d = rt$, or $t = \frac{d}{r}$, to complete the table.

Direction	Distance	Rate	Time
Downstream	10	$12+x$	$\frac{10}{12+x}$
Upstream	6	$12-x$	$\frac{6}{12-x}$

Because the time upstream equals the time downstream,

$$\frac{6}{12-x} = \frac{10}{12+x}.$$

Multiply by the LCD, $(12-x)(12+x)$.

$$(12-x)(12+x)\left(\frac{6}{12-x}\right)$$
$$= (12-x)(12+x)\left(\frac{10}{12+x}\right)$$
$$6(12+x) = 10(12-x)$$
$$72 + 6x = 120 - 10x$$
$$16x = 48$$
$$x = 3$$

The speed of the current of the river is 3 mph.

19. Let $x =$ the distance on the first part of the trip.

Make a table using the information in the problem and the formula $d = rt$, or $t = \frac{d}{r}$.

	Distance	Rate	Time
First part	x	60	$\frac{x}{60}$
Second part	$x+10$	50	$\frac{x+10}{50}$

From the problem, the equation is stated in words: (Note that 30 min $= \frac{1}{2}$ hr.) Time for the second part $=$ time for the first part $+ \frac{1}{2}$. Use the times given in the table to write the equation.

$$\frac{x+10}{50} = \frac{x}{60} + \frac{1}{2}$$

Multiply by the LCD, 300.

$$300\left(\frac{x+10}{50}\right) = 300\left(\frac{x}{60} + \frac{1}{2}\right)$$
$$6(x+10) = 5x + 150$$
$$6x + 60 = 5x + 150$$
$$x = 90$$

The distance for both parts of the trip is given by

$$90 + (90 + 10) = 190.$$

The distance is 190 mi.

23. Let $x =$ the time it would take them working together.

Make a table.

Worker	Rate	Time together	Part of job done
Lou	$\frac{1}{8}$	x	$\frac{1}{8}x$
Janet	$\frac{1}{5}$	x	$\frac{1}{5}x$

Part done by Lou + part done by Janet = 1 whole job.

$$\frac{1}{8}x + \frac{1}{5}x = 1$$

Multiply by the LCD, 40.

$$40\left(\frac{1}{8}x + \frac{1}{5}x\right) = 40 \cdot 1$$
$$5x + 8x = 40$$
$$13x = 40$$
$$x = \frac{40}{13} = 3\frac{1}{13}$$

Together they can groom Joe's dogs in $\frac{40}{13}$ or $3\frac{1}{13}$ hr.

27. Let $x =$ the time it will take to fill the vat if both pipes are open.

	Rate	Time together	Part of job done
Inlet pipe	$\frac{1}{9}$	x	$\frac{1}{9}x$
Outlet pipe	$-\frac{1}{12}$	x	$-\frac{1}{12}x$

Notice that the rate of the outlet pipe is negative because it will empty the barrel, not fill it.

Part done with the inlet pipe open + part done with the outlet pipe open = 1 whole job.

$$\frac{1}{9}x + \left(-\frac{1}{12}x\right) = 1$$

Multiply by the LCD, 36.

$$36\left(\frac{1}{9}x - \frac{1}{12}x\right) = 36 \cdot 1$$
$$4x - 3x = 36$$
$$x = 36$$

It will take 36 hr to fill the vat.

31. Since $\frac{2}{3} = \frac{4}{6} = \frac{6}{9}$, use the proportion

$$\frac{2}{3} = \frac{2x+1}{2x+5}.$$
$$2(2x+5) = 3(2x+1)$$
$$4x + 10 = 6x + 3$$
$$7 = 2x$$
$$\frac{7}{2} = x$$

If $x = \frac{7}{2}$ or 3.5, then

$$AC = 2x + 1 = 2\left(\frac{7}{2}\right) + 1 = 8$$

and $DF = 2x + 5 = 2\left(\frac{7}{2}\right) + 5 = 12.$

Chapter 6 Review Exercises

1. $f(x) = \dfrac{9}{x+3}$

Set the denominator equal to zero and solve.

$$x + 3 = 0$$
$$x = -3$$

The number -3 makes the function undefined.

2. $f(x) = \dfrac{8}{x^2 - 36}$

Set the denominator equal to zero and solve.

$$x^2 - 36 = 0$$
$$(x+6)(x-6) = 0$$
$$x + 6 = 0 \quad \text{or} \quad x - 6 = 0$$
$$x = -6 \quad \text{or} \quad x = 6$$

The numbers 6 and -6 make the function undefined.

3. $f(x) = \dfrac{x+9}{x^2 - 7x + 10}$

Set the denominator equal to zero and solve.

$$x^2 - 7x + 10 = 0$$
$$(x-5)(x-2) = 0$$
$$x - 5 = 0 \quad \text{or} \quad x - 2 = 0$$
$$x = 5 \quad \text{or} \quad x = 2$$

The numbers 2 and 5 make the function undefined.

4. $\dfrac{55m^4n^3}{10m^5n} = \dfrac{5m^4n \cdot 11n^2}{5m^4n \cdot 2m} = \dfrac{11n^2}{2m}$

5. $\dfrac{12x^2 + 6x}{24x + 12} = \dfrac{6x(2x+1)}{12(2x+1)} = \dfrac{x}{2}$

6. $\dfrac{y^2 + 3y - 10}{y^2 - 5y + 6} = \dfrac{(y-2)(y+5)}{(y-2)(y-3)} = \dfrac{y+5}{y-3}$

7. $$\frac{25m^2 - n^2}{25m^2 - 10mn + n^2} = \frac{(5m-n)(5m+n)}{(5m-n)(5m-n)}$$
$$= \frac{5m+n}{5m-n}$$

8. $\dfrac{r-2}{4-r^2} = \dfrac{r-2}{(2-r)(2+r)} = \dfrac{(-1)(2-r)}{(2-r)(2+r)} = \dfrac{-1}{2+r}$

(There are other ways to write this answer.)

9. $\dfrac{y^5}{6} \cdot \dfrac{9}{y^4} = \dfrac{9y^5}{6y^4} = \dfrac{3y^4 \cdot 3y}{3y^4 \cdot 2} = \dfrac{3y}{2}$

10. $\dfrac{25p^3q^2}{8p^4q} \div \dfrac{15pq^2}{16p^5}$

$$= \frac{25q \cdot p^3q}{8p \cdot p^3q} \div \frac{15q^2p}{16p^4 \cdot p}$$

$$= \frac{25q}{8p} \div \frac{15q^2}{16p^4}$$

Multiply by the reciprocal.

$$= \frac{25q}{8p} \cdot \frac{16p^4}{15q^2}$$

$$= \frac{400p^4q}{120pq^2}$$

$$= \frac{10p^3 \cdot 40pq}{3q \cdot 40pq}$$

$$= \frac{10p^3}{3q}$$

11. $\dfrac{3m+12}{-8} \div \dfrac{5m+20}{4}$

Multiply by the reciprocal.

$$= \frac{3m+12}{-8} \cdot \frac{4}{5m+20}$$
$$= \frac{3(m+4)}{-8} \cdot \frac{4}{5(m+4)}$$
$$= -\frac{3}{10}$$

12. $\dfrac{(2y+3)^2}{5y} \cdot \dfrac{15y^3}{4y^2-9}$

$$= \frac{(2y+3)(2y+3)}{5y} \cdot \frac{15y^3}{(2y+3)(2y-3)}$$
$$= \frac{3y^2(2y+3)}{2y-3}$$

13. $\dfrac{w^2-16}{w} \cdot \dfrac{3}{4-w}$

$$= \frac{(w-4)(w+4)}{w} \cdot \frac{3}{4-w}$$
$$= \frac{(-1)(4-w)(w+4)}{w} \cdot \frac{3}{4-w}$$
$$= \frac{-3(w+4)}{w}$$

14. $\dfrac{y^2+2y}{y^2+y-2} \div \dfrac{y-5}{y^2+4y-5}$

$$= \frac{y(y+2)}{(y+2)(y-1)} \div \frac{y-5}{(y+5)(y-1)}$$

Simplify and multiply by the reciprocal.

$$= \frac{y}{y-1} \cdot \frac{(y+5)(y-1)}{y-5}$$
$$= \frac{y(y+5)}{y-5}$$

15. $\dfrac{z^2-z-6}{z-6} \cdot \dfrac{z^2-6z}{z^2+2z-15}$

$$= \frac{(z-3)(z+2)}{z-6} \cdot \frac{z(z-6)}{(z-3)(z+5)}$$
$$= \frac{z(z+2)}{z+5}$$

16. $\dfrac{2p^2-5p-12}{5p^2-18p-8} \cdot \dfrac{25p^2-4}{30p-12}$

$$= \frac{(2p+3)(p-4)}{(5p+2)(p-4)} \cdot \frac{(5p-2)(5p+2)}{6(5p-2)}$$
$$= \frac{2p+3}{6}$$

17. $\dfrac{m^3-n^3}{m^2-n^2} \div \dfrac{m^2+mn+n^2}{m+n}$

$$= \frac{(m-n)(m^2+mn+n^2)}{(m-n)(m+n)} \div \frac{m^2+mn+n^2}{m+n}$$

Simplify and multiply by the reciprocal.

$$= \frac{m^2+mn+n^2}{m+n} \cdot \frac{m+n}{m^2+mn+n^2}$$
$$= 1$$

18. $\dfrac{x^2+5x}{x+5} \neq \dfrac{x^2}{x} + \dfrac{5x}{5}$

since the terms x and 5 in the denominator cannot be separated into denominators of two fractions. Instead,

$$\frac{x^2+5x}{x+5} = \frac{x(x+5)}{x+5} = x.$$

19. $\dfrac{7x}{12}, \dfrac{8}{15x}$

Factor each denominator.

$$12 = 3 \cdot 2 \cdot 2 = 3 \cdot 2^2$$
$$15x = 3 \cdot 5 \cdot x$$
$$\text{LCD} = 3 \cdot 2^2 \cdot 5 \cdot x = 60x$$

The LCD is $60x$.

20. $\dfrac{5a}{32b^3}, \dfrac{31}{24b^5}$

Factor each denominator.

$$32b^3 = 2 \cdot 2 \cdot 2 \cdot 2 \cdot 2 \cdot b^3 = 2^5 \cdot b^3$$
$$24b^5 = 2 \cdot 2 \cdot 2 \cdot 3 \cdot b^5 = 2^3 \cdot 3 \cdot b^5$$
$$\text{LCD} = 2^5 \cdot 3 \cdot b^5 = 96b^5$$

The LCD is $96b^5$.

21. $\dfrac{17}{9r^2}, \dfrac{5r-3}{3r+1}$

Factor each denominator.

$$9r^2 = 3^2 \cdot r^2$$

The second denominator is already in factored form. The LCD is

$$3^2 \cdot r^2 \cdot (3r+1) \text{ or } 9r^2(3r+1).$$

22. $\dfrac{14}{6k+3}, \dfrac{7k^2+2k+1}{10k+5}, \dfrac{-11k}{18k+9}$

Factor each denominator.

$$6k+3 = 3(2k+1)$$
$$10k+5 = 5(2k+1)$$
$$18k+9 = 9(2k+1)$$
$$= 3^2 \cdot (2k+1)$$

$\text{LCD} = 5 \cdot 3^2(2k+1) = 45(2k+1)$

The LCD is $45(2k+1)$.

23. $\dfrac{4x-9}{6x^2+13x-5}, \dfrac{x+15}{9x^2+9x-4}$

Factor each denominator.

$$6x^2+13x-5 = (3x-1)(2x+5)$$
$$9x^2+9x-4 = (3x-1)(3x+4)$$

The LCD is

$$(3x-1)(2x+5)(3x+4).$$

24. $\dfrac{2}{y} + \dfrac{3}{8y}$

The LCD is $8y$.

$$= \frac{2 \cdot 8}{y \cdot 8} + \frac{3}{8y}$$
$$= \frac{16}{8y} + \frac{3}{8y}$$
$$= \frac{16+3}{8y}$$
$$= \frac{19}{8y}$$

25. $\dfrac{8}{z} - \dfrac{3}{2z^2}$

The LCD is $2z^2$.

$$= \frac{8 \cdot 2z}{z \cdot 2z} - \frac{3}{2z^2}$$
$$= \frac{16z}{2z^2} - \frac{3}{2z^2}$$
$$= \frac{16z-3}{2z^2}$$

26. $\dfrac{3}{t-2} - \dfrac{5}{2-t} = \dfrac{3}{t-2} - \dfrac{(-1)5}{(-1)(2-t)}$

$$= \frac{3}{t-2} - \frac{-5}{t-2}$$
$$= \frac{3+5}{t-2}$$
$$= \frac{8}{t-2}$$

or

$$= \frac{-1(8)}{-1(t-2)}$$
$$= \frac{-8}{2-t}$$

27. $\dfrac{5y+13}{y+1} - \dfrac{1-7y}{y+1}$

$$= \frac{5y+13-(1-7y)}{y+1}$$
$$= \frac{5y+13-1+7y}{y+1}$$
$$= \frac{12y+12}{y+1}$$
$$= \frac{12(y+1)}{y+1}$$
$$= 12$$

28. $\dfrac{6}{5a+10} + \dfrac{7}{6a+12}$

$$= \frac{6}{5(a+2)} + \frac{7}{6(a+2)}$$

The LCD is $30(a+2)$.

$$= \frac{6}{5(a+2)} \cdot \frac{6}{6} + \frac{7}{6(a+2)} \cdot \frac{5}{5}$$
$$= \frac{36}{30(a+2)} + \frac{35}{30(a+2)}$$
$$= \frac{36+35}{30(a+2)}$$
$$= \frac{71}{30(a+2)}$$

29. $\dfrac{3x}{x-5} + \dfrac{20}{x^2-25} + \dfrac{2}{x+5}$

$$= \frac{3x}{x-5} + \frac{20}{(x-5)(x+5)} + \frac{2}{x+5}$$

The LCD is $(x-5)(x+5)$.

$$= \frac{3x(x+5)}{(x-5)(x+5)} + \frac{20}{(x-5)(x+5)}$$
$$+ \frac{2(x-5)}{(x+5)(x-5)}$$
$$= \frac{3x(x+5)+20+2(x-5)}{(x-5)(x+5)}$$
$$= \frac{3x^2+15x+20+2x-10}{(x-5)(x+5)}$$
$$= \frac{3x^2+17x+10}{(x-5)(x+5)}$$
$$= \frac{(3x+2)(x+5)}{(x-5)(x+5)}$$
$$= \frac{3x+2}{x-5}$$

30. $\dfrac{3r}{10r^2 - 3rs - s^2} + \dfrac{2r}{2r^2 + rs - s^2}$

$= \dfrac{3r}{(5r + s)(2r - s)} + \dfrac{2r}{(2r - s)(r + s)}$

The LCD is $(5r + s)(2r - s)(r + s)$.

$= \dfrac{3r(r + s)}{(5r + s)(2r - s)(r + s)}$

$+ \dfrac{2r(5r + s)}{(2r - s)(r + s)(5r + s)}$

$= \dfrac{3r^2 + 3rs + 10r^2 + 2rs}{(5r + s)(2r - s)(r + s)}$

$= \dfrac{13r^2 + 5rs}{(5r + s)(2r - s)(r + s)}$

31. $\dfrac{\frac{3}{t} + 2}{\frac{4}{t} - 7}$

Multiply the numerator and denominator by t, the LCD of all the fractions.

$= \dfrac{t\left(\frac{3}{t} + 2\right)}{t\left(\frac{4}{t} - 7\right)}$

$= \dfrac{3 + 2t}{4 - 7t}$

32. $\dfrac{\frac{4m^5n^6}{mn}}{\frac{8m^7n^3}{m^4n^2}}$

Simplify the numerator and denominator.

$= \dfrac{\frac{4m^4n^5 \cdot mn}{mn}}{\frac{8m^3n \cdot m^4n^2}{m^4n^2}}$

$= \dfrac{4m^4n^5}{8m^3n}$

Simplify again.

$= \dfrac{mn^4}{2}$

33. $\dfrac{\frac{r + 2s}{20}}{\frac{8r + 16s}{5}}$

$= \dfrac{r + 2s}{20} \div \dfrac{8r + 16s}{5}$

Multiply by the reciprocal.

$= \dfrac{r + 2s}{20} \cdot \dfrac{5}{8r + 16s}$

$= \dfrac{r + 2s}{20} \cdot \dfrac{5}{8(r + 2s)}$

$= \dfrac{1}{32}$

34. $\dfrac{3x^{-1} - 5}{6 + x^{-1}}$

$= \dfrac{\frac{3}{x} - 5}{6 + \frac{1}{x}}$

Multiply the numerator and denominator by x, the LCD of all the fractions.

$= \dfrac{x\left(\frac{3}{x} - 5\right)}{x\left(6 + \frac{1}{x}\right)}$

$= \dfrac{3 - 5x}{6x + 1}$

35. $\dfrac{\frac{3}{p} - \frac{2}{q}}{\frac{9q^2 - 4p^2}{qp}}$

Multiply the numerator and denominator by qp, the LCD of all the fractions.

$= \dfrac{qp\left(\frac{3}{p} - \frac{2}{q}\right)}{qp\left(\frac{9q^2 - 4p^2}{qp}\right)}$

$= \dfrac{3q - 2p}{9q^2 - 4p^2}$

$= \dfrac{3q - 2p}{(3q - 2p)(3q + 2p)}$

$= \dfrac{1}{3q + 2p}$

36. $\dfrac{\frac{1}{x}+\frac{2}{x^2}+\frac{1}{x^3}}{\frac{1}{x}-\frac{1}{x^3}}$

Multiply the numerator and denominator by x^3, the LCD of all the fractions.

$$=\frac{x^3\left(\frac{1}{x}+\frac{2}{x^2}+\frac{1}{x^3}\right)}{x^3\left(\frac{1}{x}-\frac{1}{x^3}\right)}$$

$$=\frac{x^2+2x+1}{x^2-1}$$

$$=\frac{(x+1)(x+1)}{(x+1)(x-1)}$$

$$=\frac{x+1}{x-1}$$

37. $\frac{4}{3}-\frac{1}{x}=\frac{1}{3x}$

$3x \neq 0$, so $x \neq 0$.
Multiply both sides by the LCD, $3x$.

$$3x\left(\frac{4}{3}-\frac{1}{x}\right)=3x\left(\frac{1}{3x}\right)$$
$$4x-3=1$$
$$4x=4$$
$$x=1$$

Solution set: $\{1\}$

38. $\frac{1}{t+4}+\frac{1}{2}=\frac{3}{2t+8}$

$$\frac{1}{t+4}+\frac{1}{2}=\frac{3}{2(t+4)}$$

$t+4\neq 0$, so $t \neq -4$.
Multiply both sides by the LCD, $2(t+4)$.

$$2(t+4)\left(\frac{1}{t+4}+\frac{1}{2}\right)=2(t+4)\left(\frac{3}{2(t+4)}\right)$$
$$2+t+4=3$$
$$t+6=3$$
$$t=-3$$

Solution set: $\{-3\}$

39. $\frac{-5m}{m+1}+\frac{m}{3m+3}=\frac{56}{6m+6}$

$$\frac{-5m}{m+1}+\frac{m}{3(m+1)}=\frac{56}{6(m+1)}$$
$$\frac{-5m}{m+1}+\frac{m}{3(m+1)}=\frac{28}{3(m+1)}$$

$m+1\neq 0$, so $m\neq -1$.
Multiply both sides by the LCD, $3(m+1)$.

$$3(m+1)\left(\frac{-5m}{m+1}+\frac{m}{3(m+1)}\right)$$
$$=3(m+1)\left(\frac{28}{3(m+1)}\right)$$
$$-15m+m=28$$
$$-14m=28$$
$$m=-2$$

Solution set: $\{-2\}$

40. $\frac{2}{k-1}-\frac{4k+1}{k^2-1}=\frac{-1}{k+1}$

$$\frac{2}{k-1}-\frac{4k+1}{(k-1)(k+1)}=\frac{-1}{k+1}$$

$k-1\neq 0$ or $k+1\neq 0$, so $k\neq 1$ or $k\neq -1$.
Multiply both sides by the LCD, $(k-1)(k+1)$.

$$(k-1)(k+1)\left(\frac{2}{k-1}-\frac{4k+1}{(k-1)(k+1)}\right)$$
$$=(k-1)(k+1)\left(\frac{-1}{k+1}\right)$$
$$2(k+1)-(4k+1)=-1(k-1)$$
$$2k+2-4k-1=-k+1$$
$$-2k+1=-k+1$$
$$-k=0$$
$$k=0$$

Solution set: $\{0\}$

41. $\frac{x+3}{x^2-5x+4}-\frac{1}{x}=\frac{2}{x^2-4x}$

$$\frac{x+3}{(x-4)(x-1)}-\frac{1}{x}=\frac{2}{x(x-4)}$$

$x-4\neq 0$ or $x-1\neq 0$, so $x\neq 4$ or $x\neq 1$ or $x\neq 0$.
Multiply both sides by the LCD, $x(x-4)(x-1)$.

$$x(x-4)(x-1)\left(\frac{x+3}{(x-4)(x-1)}-\frac{1}{x}\right)$$
$$=x(x-4)(x-1)\left(\frac{2}{x(x-4)}\right)$$
$$x(x+3)-(x-4)(x-1)=2(x-1)$$
$$x^2+3x-(x^2-5x+4)=2x-2$$
$$x^2+3x-x^2+5x-4=2x-2$$
$$8x-4=2x-2$$
$$6x=2$$
$$x=\frac{1}{3}$$

Solution set: $\{\frac{1}{3}\}$

42. $$\frac{5}{x+2}+\frac{3}{x+3}=\frac{x}{x^2+5x+6}$$
$$\frac{5}{x+2}+\frac{3}{x+3}=\frac{x}{(x+2)(x+3)}$$
$x+2\neq 0$ or $x+3\neq 0$, so $x\neq -2$ or $x\neq -3$.
Multiply both sides by the LCD, $(x+2)(x+3)$.
$$(x+2)(x+3)\left(\frac{5}{x+2}+\frac{3}{x+3}\right)$$
$$=(x+2)(x+3)\left(\frac{x}{(x+2)(x+3)}\right)$$
$$\begin{aligned}5(x+3)+3(x+2)&=x\\5x+15+3x+6&=x\\8x+21&=x\\7x&=-21\\x&=-3\end{aligned}$$
But $x\neq -3$, so -3 is not a solution.

Solution set: $\emptyset$

43. The student is wrong, even though she did not make an error in her steps. The number 3 cannot be a solution because it causes division by zero when substituted in the original equation. It must be rejected. The solution set is $\emptyset$.

44. Simplifying the expression means to write the problem as a single rational expression. Solving the equation means to find the values of the variable that make the equation true.

45. $$\frac{1}{A}=\frac{1}{B}+\frac{1}{C}$$
Let $B=30$ and $C=10$.
$$\frac{1}{A}=\frac{1}{30}+\frac{1}{10}$$
To solve for A, multiply both sides by the LCD, $30A$.
$$30A\left(\frac{1}{A}\right)=30A\left(\frac{1}{30}+\frac{1}{10}\right)$$
$$\begin{aligned}30&=A+3A\\30&=4A\\\frac{15}{2}&=A\end{aligned}$$

46. $$\frac{1}{A}=\frac{1}{B}+\frac{1}{C}$$
Let $A=10$ and $C=15$.
$$\frac{1}{10}=\frac{1}{B}+\frac{1}{15}$$
To solve for B, multiply both sides by the LCD, $30B$.
$$30B\left(\frac{1}{10}\right)=30B\left(\frac{1}{B}+\frac{1}{15}\right)$$
$$\begin{aligned}3B&=30+2B\\B&=30\end{aligned}$$

47. $$F=\frac{GMm}{d^2}$$
To solve for m, multiply both sides by d^2.
$$\begin{aligned}Fd^2&=\frac{GMm}{d^2}\cdot d^2\\Fd^2&=GMm\\\frac{Fd^2}{GM}&=m\\\text{or}\quad m&=\frac{Fd^2}{GM}\end{aligned}$$

48. $$S=\frac{n}{2}(a+\ell)$$
To solve for ℓ, multiply both sides by $\frac{2}{n}$.
$$\begin{aligned}\frac{2}{n}\cdot S&=\frac{2}{n}\cdot\frac{n}{2}(a+\ell)\\\frac{2S}{n}&=a+\ell\\\ell&=\frac{2S}{n}-a\\\text{or}\quad \ell&=\frac{2S-na}{n}\end{aligned}$$

49. $$\mu=\frac{Mv}{M+m}$$
To solve for m, multiply both sides by $M+m$.
$$\begin{aligned}\mu(M+m)&=\frac{Mv}{M+m}(M+m)\\\mu(M+m)&=Mv\\\mu M+\mu m&=Mv\\\mu m&=Mv-\mu M\\m&=\frac{Mv-\mu M}{\mu}\\\text{or}\quad m&=\frac{Mv}{\mu}-M\end{aligned}$$

50. $I = \dfrac{nE}{R+nr}$

To solve for R, multiply both sides by $R+nr$.

$$\begin{aligned}(R+nr)I &= (R+nr)\left(\frac{nE}{R+nr}\right)\\ (R+nr)I &= nE\\ RI+nrI &= nE\\ RI &= nE-nrI\\ R &= \frac{nE-Inr}{I}\\ \text{or}\quad R &= \frac{nE}{I}-nr\end{aligned}$$

51. Let $x =$ the cost of 13 gal.

Write a proportion.

$$\frac{x}{13} = \frac{4.86}{3}$$

Multiply by the LCD, 39.

$$\begin{aligned}39\left(\frac{x}{13}\right) &= 39\left(\frac{4.86}{3}\right)\\ 3x &= 13(4.86)\\ 3x &= 63.18\\ x &= 21.06\end{aligned}$$

The cost of 13 gal is \$21.06.

52. Let $x =$ the predicted number of votes for the funding.

Write a proportion.

$$\frac{x}{15{,}000} = \frac{1430}{2000}$$

Multiply by the LCD, 30,000.

$$\begin{aligned}30{,}000\left(\frac{x}{15{,}000}\right) &= 30{,}000\left(\frac{1430}{2000}\right)\\ 2x &= 15(1430)\\ 2x &= 21{,}450\\ x &= 10{,}725\end{aligned}$$

If the survey proves to be accurate, 10,725 voters will vote for funding.

53. Let $x =$ the time it takes to fill the sink with both taps open.

Make a table.

	Rate	Time together	Part of job done
Cold	$\frac{1}{8}$	x	$\frac{x}{8}$
Hot	$\frac{1}{12}$	x	$\frac{x}{12}$

Part done by hot	plus	part done by cold	equals	1 whole job.
$\frac{x}{8}$	$+$	$\frac{x}{12}$	$=$	1

Multiply by the LCD, 24.

$$\begin{aligned}24\left(\frac{x}{8}+\frac{x}{12}\right) &= 24\cdot 1\\ 3x+2x &= 24\\ 5x &= 24\\ x &= \frac{24}{5} \text{ or } 4\frac{4}{5}\end{aligned}$$

The sink will be filled in $\frac{24}{5}$ or $4\frac{4}{5}$ min.

54. Let $x =$ the time to do the job working together.

Make a table.

Worker	Rate	Time together	Part of job done
Jane	$\frac{1}{9}$	x	$\frac{x}{9}$
Greg	$\frac{1}{6}$	x	$\frac{x}{6}$

Part done by Jane	plus	part done by Greg	equals	1 whole job.
$\frac{x}{9}$	$+$	$\frac{x}{6}$	$=$	1

Multiply by the LCD, 36.

$$36\left(\frac{x}{9}+\frac{x}{6}\right)=36\cdot 1$$
$$4x+6x=36$$
$$10x=36$$
$$x=\frac{36}{10}$$
$$x=\frac{18}{5} \text{ or } 3\frac{3}{5}$$

Working together, they can do the job in $\frac{18}{5}$ or $3\frac{3}{5}$ hr.

55. Let $x =$ the speed of the bus and $x+10 =$ the speed of the train.

Use $d=rt$, or $t=\frac{d}{r}$, to make a table.

	Distance	Rate	Time
Bus	80	x	$\frac{80}{x}$
Train	96	$x+10$	$\frac{96}{x+10}$

Because the times are equal,

$$\frac{80}{x}=\frac{96}{x+10}.$$

Multiply by the LCD, $x(x+10)$.

$$x(x+10)\left(\frac{80}{x}\right)=x(x+10)\left(\frac{96}{x+10}\right)$$
$$80(x+10)=96x$$
$$80x+800=96x$$
$$-16x=-800$$
$$x=50$$

So,

$$x+10=50+10=60.$$

The speed of the bus is 50 mph. The speed of the train is 60 mph.

56. Let $x =$ the speed of the boat in still water.

Use $d=rt$, or $t=\frac{d}{r}$, to make a table.

Direction	Distance	Rate	Time
Downstream	40	$x+4$	$\frac{40}{x+4}$
Upstream	24	$x-4$	$\frac{24}{x-4}$

Because the times are equal,

$$\frac{40}{x+4}=\frac{24}{x-4}.$$

Multiply by the LCD, $(x+4)(x-4)$.

$$(x+4)(x-4)\left(\frac{40}{x+4}\right)$$
$$=(x+4)(x-4)\left(\frac{24}{x-4}\right)$$

$$40(x-4)=24(x+4)$$
$$40x-160=24x+96$$
$$16x=256$$
$$x=16$$

The speed of the boat in still water is 16 km/hr.

57. $\frac{2}{m}+\frac{5}{3m^2}$

The LCD is $3m^2$.

$$=\frac{2\cdot 3m}{m\cdot 3m}+\frac{5}{3m^2}$$
$$=\frac{6m}{3m^2}+\frac{5}{3m^2}$$
$$=\frac{6m+5}{3m^2}$$

58. $\frac{k^2-6k+9}{1-216k^3}\cdot\frac{6k^2+17k-3}{9-k^2}$

$$=\frac{(k-3)(k-3)}{(1-6k)(1+6k+36k^2)}$$
$$\cdot\frac{(6k-1)(k+3)}{(3-k)(3+k)}$$
$$=\frac{(k-3)(k-3)}{(-1)(6k-1)(1+6k+36k^2)}$$
$$\cdot\frac{(6k-1)(k+3)}{(-1)(k-3)(k+3)}$$
$$=\frac{k-3}{1+6k+36k^2} \text{ or } \frac{k-3}{36k^2+6k+1}$$

59. $\dfrac{\frac{-3}{x}+\frac{x}{2}}{1+\frac{x+1}{x}}$

Multiply the numerator and denominator by $2x$, the LCD of all the fractions.

$$=\frac{2x\left(\frac{-3}{x}+\frac{x}{2}\right)}{2x\left(1+\frac{x+1}{x}\right)}$$

$$=\frac{-6+x^2}{2x+2(x+1)}$$

$$=\frac{x^2-6}{2x+2x+2}$$

$$=\frac{x^2-6}{4x+2}$$

$$=\frac{x^2-6}{2(2x+1)}$$

60. $\dfrac{9x^2+46x+5}{3x^2-2x-1}\div\dfrac{x^2+11x+30}{x^3+5x^2-6x}$

Multiply by the reciprocal.

$$=\frac{9x^2+46x+5}{3x^2-2x-1}\cdot\frac{x^3+5x^2-6x}{x^2+11x+30}$$

$$=\frac{(9x+1)(x+5)}{(3x+1)(x-1)}\cdot\frac{x(x+6)(x-1)}{(x+6)(x+5)}$$

$$=\frac{x(9x+1)}{3x+1}$$

61. $\dfrac{9}{3-x}-\dfrac{2}{x-3}=\dfrac{9}{3-x}-\dfrac{-1(2)}{-1(x-3)}$

$$=\frac{9-(-2)}{3-x}$$

$$=\frac{11}{3-x}$$

or

$$=\frac{-1(11)}{-1(3-x)}$$

$$=\frac{-11}{x-3}$$

62. $\dfrac{4y+16}{30}\div\dfrac{2y+8}{5}$

Multiply by the reciprocal.

$$=\frac{4y+16}{30}\cdot\frac{5}{2y+8}$$

$$=\frac{4(y+4)}{30}\cdot\frac{5}{2(y+4)}$$

$$=\frac{1}{3}$$

63. $\dfrac{t^{-2}+s^{-2}}{t^{-1}-s^{-1}}$

$$=\frac{\frac{1}{t^2}+\frac{1}{s^2}}{\frac{1}{t}-\frac{1}{s}}$$

Multiply the numerator and denominator by t^2s^2, the LCD of all the fractions.

$$=\frac{t^2s^2\left(\frac{1}{t^2}+\frac{1}{s^2}\right)}{t^2s^2\left(\frac{1}{t}-\frac{1}{s}\right)}$$

$$=\frac{s^2+t^2}{ts^2-t^2s}$$

$$=\frac{s^2+t^2}{st(s-t)}$$

64. $\dfrac{4a}{a^2-ab-2b^2}-\dfrac{6b}{a^2+4ab+3b^2}$

$$=\frac{4a}{(a-2b)(a+b)}-\frac{6b}{(a+3b)(a+b)}$$

The LCD is $(a+3b)(a-2b)(a+b)$.

$$=\frac{4a(a+3b)}{(a-2b)(a+b)(a+3b)}$$

$$-\frac{6b(a-2b)}{(a+3b)(a+b)(a-2b)}$$

$$=\frac{4a(a+3b)-6b(a-2b)}{(a+3b)(a+b)(a-2b)}$$

$$=\frac{4a^2+12ab-6ab+12b^2}{(a+3b)(a+b)(a-2b)}$$

$$=\frac{4a^2+6ab+12b^2}{(a+3b)(a+b)(a-2b)}$$

$$=\frac{2(2a^2+3ab+6b^2)}{(a+3b)(a+b)(a-2b)}$$

65. $A = \dfrac{Rr}{R+r}$

To solve for r, multiply both sides by $R + r$.

$$A(R+r) = \frac{Rr}{R+r}(R+r)$$
$$A(R+r) = Rr$$
$$AR + Ar = Rr$$
$$AR = Rr - Ar$$
$$AR = (R-A)r$$
$$\frac{AR}{R-A} = r$$
$$r = \frac{AR}{R-A}$$

or

$$r = \frac{-1(AR)}{-1(R-A)}$$
$$r = \frac{-AR}{A-R}$$

66. $1 - \dfrac{5}{r} = \dfrac{-4}{r^2}$

$r \neq 0$, or a denominator of 0 will result.
Multiply by the LCD, r^2.

$$r^2\left(1 - \frac{5}{r}\right) = r^2\left(\frac{-4}{r^2}\right)$$
$$r^2 - 5r = -4$$
$$r^2 - 5r + 4 = 0$$
$$(r-4)(r-1) = 0$$
$$r - 4 = 0 \quad \text{or} \quad r - 1 = 0$$
$$r = 4 \quad \text{or} \quad r = 1$$

Solution set: $\{1, 4\}$

67. $\dfrac{3x}{x-4} + \dfrac{2}{x} = \dfrac{48}{x^2 - 4x}$

$$\frac{3x}{x-4} + \frac{2}{x} = \frac{48}{x(x-4)}$$

$x - 4 \neq 0$, so $x \neq 4$ or $x \neq 0$.
Multiply by the LCD, $x(x-4)$.

$$x(x-4)\left(\frac{3x}{x-4} + \frac{2}{x}\right) = x(x-4)\left(\frac{48}{x(x-4)}\right)$$
$$3x^2 + 2(x-4) = 48$$
$$3x^2 + 2x - 8 = 48$$
$$3x^2 + 2x - 56 = 0$$
$$(3x+14)(x-4) = 0$$
$$3x + 14 = 0 \quad \text{or} \quad x - 4 = 0$$
$$3x = -14$$
$$x = -\frac{14}{3} \quad \text{or} \quad x = 4$$

But $x \neq 4$, so $-\frac{14}{3}$ is the only solution.

Solution set: $\{-\frac{14}{3}\}$

68. Let $x =$ the time to fill the tub working together.

Make a table.

	Rate	Time together	Part of job done
Hot	$\frac{1}{20}$	x	$\frac{x}{20}$
Cold	$\frac{1}{15}$	x	$\frac{x}{15}$

Part done by hot	plus	part done by cold	equals	1 whole job.
$\frac{x}{20}$	$+$	$\frac{x}{15}$	$=$	1

Multiply by the LCD, 60.

$$60\left(\frac{x}{20} + \frac{x}{15}\right) = 60 \cdot 1$$
$$3x + 4x = 60$$
$$7x = 60$$
$$x = \frac{60}{7} \text{ or } 8\frac{4}{7}$$

Working together, the tub can be filled in $\frac{60}{7}$ or $8\frac{4}{7}$ min.

Chapter 6 Test

1. $f(x) = \dfrac{2x-1}{3x^2 + 2x - 8}$

Set the denominator equal to zero and solve.

$$3x^2 + 2x - 8 = 0$$
$$(3x-4)(x+2) = 0$$
$$3x - 4 = 0 \quad \text{or} \quad x + 2 = 0$$
$$3x = 4$$
$$x = \frac{4}{3} \quad \text{or} \quad x = -2$$

The numbers -2 and $\frac{4}{3}$ make the function undefined.

2. $$\frac{6x^2 - 13x - 5}{9x^3 - x} = \frac{(3x+1)(2x-5)}{x(3x-1)(3x+1)}$$
$$= \frac{2x-5}{x(3x-1)}$$

3. $\dfrac{4x^2y^5}{7xy^8} \div \dfrac{8xy^6}{21xy}$

Simplify the numerator and denominator.

$$= \frac{4x}{7y^3} \div \frac{8y^5}{21}$$

Multiply by the reciprocal.

$$= \frac{4x}{7y^3} \cdot \frac{21}{8y^5}$$

$$= \frac{84x}{56y^8}$$

$$= \frac{3x}{2y^8}$$

4. $\dfrac{y^2-16}{y^2-25} \cdot \dfrac{y^2+2y-15}{y^2-7y+12}$

$$= \frac{(y-4)(y+4)}{(y-5)(y+5)} \cdot \frac{(y+5)(y-3)}{(y-4)(y-3)}$$

$$= \frac{y+4}{y-5}$$

5. $\dfrac{x^2-9}{x^3+3x^2} \div \dfrac{x^2+x-12}{x^3+9x^2+20x}$

Multiply by the reciprocal.

$$= \frac{x^2-9}{x^3+3x^2} \cdot \frac{x^3+9x^2+20x}{x^2+x-12}$$

$$= \frac{(x-3)(x+3)}{x^2(x+3)} \cdot \frac{x(x+5)(x+4)}{(x+4)(x-3)}$$

$$= \frac{x+5}{x}$$

6. $\dfrac{t}{t^2+t-6}, \dfrac{17t^2}{t^2+3t}, \dfrac{-1}{t^2}$

Factor each denominator.

$$t^2+t-6 = (t+3)(t-2)$$
$$t^2+3t = t(t+3)$$

The third denominator is already in factored form. The LCD is

$$t^2(t+3)(t-2).$$

7. $\dfrac{7}{6t^2} - \dfrac{1}{3t}$

The LCD is $6t^2$.

$$= \frac{7}{6t^2} - \frac{1 \cdot 2t}{3t \cdot 2t}$$

$$= \frac{7}{6t^2} - \frac{2t}{6t^2}$$

$$= \frac{7-2t}{6t^2}$$

8. $\dfrac{9}{x-7} + \dfrac{4}{x+7}$

The LCD is $(x-7)(x+7)$.

$$= \frac{9(x+7)}{(x-7)(x+7)} + \frac{4(x-7)}{(x+7)(x-7)}$$

$$= \frac{9(x+7)+4(x-7)}{(x-7)(x+7)}$$

$$= \frac{9x+63+4x-28}{(x-7)(x+7)}$$

$$= \frac{13x+35}{(x-7)(x+7)}$$

9. $\dfrac{6}{x+4} + \dfrac{1}{x+2} - \dfrac{3x}{x^2+6x+8}$

$$= \frac{6}{x+4} + \frac{1}{x+2} - \frac{3x}{(x+4)(x+2)}$$

The LCD is $(x+4)(x+2)$.

$$= \frac{6(x+2)}{(x+4)(x+2)} + \frac{1(x+4)}{(x+2)(x+4)} - \frac{3x}{(x+4)(x+2)}$$

$$= \frac{6(x+2)+x+4-3x}{(x+4)(x+2)}$$

$$= \frac{6x+12+x+4-3x}{(x+4)(x+2)}$$

$$= \frac{4x+16}{(x+4)(x+2)}$$

$$= \frac{4(x+4)}{(x+4)(x+2)}$$

$$= \frac{4}{x+2}$$

10. $\dfrac{\frac{12}{r+4}}{\frac{11}{6r+24}}$

$$= \frac{12}{r+4} \div \frac{11}{6r+24}$$

Multiply by the reciprocal.

$$= \frac{12}{r+4} \cdot \frac{6r+24}{11}$$

$$= \frac{12}{r+4} \cdot \frac{6(r+4)}{11}$$

$$= \frac{72}{11}$$

11. $\dfrac{\frac{1}{a}-\frac{1}{b}}{\frac{a}{b}-\frac{b}{a}}$

Multiply numerator and denominator by ab, the LCD of all the fractions.

$$=\frac{ab\left(\frac{1}{a}-\frac{1}{b}\right)}{ab\left(\frac{a}{b}-\frac{b}{a}\right)}$$

$$=\frac{b-a}{a^2-b^2}$$

$$=\frac{b-a}{(a-b)(a+b)}$$

$$=\frac{(-1)(a-b)}{(a-b)(a+b)}$$

$$=\frac{-1}{a+b}$$

12. **(a)** Simplify this expression.

$\frac{2x}{3}+\frac{x}{4}-\frac{11}{2}$

The LCD is 12.

$$=\frac{2x\cdot 4}{3\cdot 4}+\frac{x\cdot 3}{4\cdot 3}-\frac{11\cdot 6}{2\cdot 6}$$

$$=\frac{8x}{12}+\frac{3x}{12}-\frac{66}{12}$$

$$=\frac{8x+3x-66}{12}$$

$$=\frac{11x-66}{12}$$

$$=\frac{11(x-6)}{12}$$

(b) Solve this equation.

$$\frac{2x}{3}+\frac{x}{4}=\frac{11}{2}$$

Multiply by the LCD, 12.

$$12\left(\frac{2x}{3}+\frac{x}{4}\right)=12\left(\frac{11}{2}\right)$$

$$8x+3x=66$$
$$11x=66$$
$$x=6$$

Solution set: $\{6\}$

13. $1-\frac{3}{y}=\frac{1}{2}$

$y\neq 0$, or a denominator of 0 will result. Multiply by the LCD, $2y$.

$$2y\left(1-\frac{3}{y}\right)=2y\left(\frac{1}{2}\right)$$

$$2y-6=y$$
$$y=6$$

Solution set: $\{6\}$

14. $\frac{1}{x}-\frac{4}{3x}=\frac{1}{x-2}$

$3x\neq 0$ or $x-2\neq 0$, so $x\neq 0$ or $x\neq 2$. Multiply by the LCD, $3x(x-2)$.

$$3x(x-2)\left(\frac{1}{x}-\frac{4}{3x}\right)=3x(x-2)\left(\frac{1}{x-2}\right)$$

$$3(x-2)-4(x-2)=3x$$
$$3x-6-4x+8=3x$$
$$-x+2=3x$$
$$-4x=-2$$
$$x=\frac{1}{2}$$

Solution set: $\{\frac{1}{2}\}$

15. $\frac{y}{y+2}-\frac{1}{y-2}=\frac{8}{y^2-4}$

$$\frac{y}{y+2}-\frac{1}{y-2}=\frac{8}{(y+2)(y-2)}$$

$y+2\neq 0$ or $y-2\neq 0$, so $y\neq -2$ or $y\neq 2$. Multiply by the LCD, $(y+2)(y-2)$.

$$(y+2)(y-2)\left(\frac{y}{y+2}-\frac{1}{y-2}\right)$$

$$=(y+2)(y-2)\left(\frac{8}{(y+2)(y-2)}\right)$$

$$y(y-2)-(y+2)=8$$
$$y^2-2y-y-2=8$$
$$y^2-3y-10=0$$
$$(y-5)(y+2)=0$$
$$y-5=0 \quad\text{or}\quad y+2=0$$
$$y=5 \quad\text{or}\quad y=-2$$

But $y\neq -2$, so 5 is the only solution.

Solution set: $\{5\}$

16. $I = \dfrac{En}{Rn + r}$

To solve for r, multiply both sides by $Rn + r$.

$$I(Rn + r) = \left(\frac{En}{Rn + r}\right)(Rn + r)$$
$$IRn + Ir = En$$
$$Ir = En - IRn$$
$$r = \frac{En - IRn}{I}$$
$$\text{or} \quad r = \frac{En}{I} - Rn$$

17. $P = \dfrac{W(R - r)}{2R}$

To solve for R, multiply both sides by $2R$.

$$P \cdot 2R = \frac{W(R - r)}{2R} \cdot 2R$$
$$2RP = WR - Wr$$
$$2RP - WR = -Wr$$
$$R(2P - W) = -Wr$$
$$R = \frac{-Wr}{2P - W}$$

Let $W = 120$, $r = 8$, and $P = 10$.

$$R = \frac{-(120)(8)}{2(10) - 120}$$
$$R = \frac{-960}{20 - 120}$$
$$R = \frac{-960}{-100}$$
$$R = \frac{48}{5}$$

18. Let $x =$ the time to do the job working together.

Make a table.

Worker	Rate	Time together	Part of job done
Wayne	$\frac{1}{9}$	x	$\frac{x}{9}$
Susan	$\frac{1}{5}$	x	$\frac{x}{5}$

Part done by Wayne	plus	part done by Susan	equals	1 whole job.
$\frac{x}{9}$	$+$	$\frac{x}{5}$	$=$	1

Multiply by the LCD, 45.

$$45\left(\frac{x}{9} + \frac{x}{5}\right) = 45 \cdot 1$$
$$5x + 9x = 45$$
$$14x = 45$$
$$x = \frac{45}{14} \text{ or } 3\frac{3}{14}$$

Working together, they can do the job in $\frac{45}{14}$ or $3\frac{3}{14}$ hr.

19. Let $x =$ the speed of the boat in still water.

Use $d = rt$, or $t = \frac{d}{r}$, to make a table.

Direction	Distance	Rate	Time
Downstream	36	$x + 3$	$\frac{36}{x+3}$
Upstream	24	$x - 3$	$\frac{24}{x-3}$

Because the times are equal,

$$\frac{36}{x + 3} = \frac{24}{x - 3}.$$

Multiply by the LCD, $(x + 3)(x - 3)$.

$$\left(\frac{36}{x + 3}\right)(x + 3)(x - 3)$$
$$= \left(\frac{24}{x - 3}\right)(x + 3)(x - 3)$$
$$36(x - 3) = 24(x + 3)$$
$$36x - 108 = 24x + 72$$
$$12x = 180$$
$$x = 15$$

The speed of the boat in still water is 15 mph.

20. Let $x =$ the number of fish in Lake Linda.

Write a proportion.

$$\frac{x}{600} = \frac{800}{10}$$

Multiply by the LCD, 600.

$$600\left(\frac{x}{600}\right) = 600\left(\frac{800}{10}\right)$$
$$x = 60 \cdot 800$$
$$x = 48{,}000$$

There are about 48,000 fish in the lake.

Cumulative Review Exercises (Chapters 1-6)

1. $$\begin{aligned} 3x + 2(x - 4) &= 5x - 8 \\ 3x + 2x - 8 &= 5x - 8 \\ 5x - 8 &= 5x - 8 \quad \textit{True} \end{aligned}$$

Solution set: $(-\infty, \infty)$

2. $$\begin{aligned} -3 \le \frac{2}{3}x - 1 \le 1 \\ -2 \le \frac{2}{3}x \le 2 \\ -3 \le x \le 3 \end{aligned}$$

Solution set: $[-3, 3]$

See the answer graph in the back of the textbook.

3. Let $x =$ the amount of money invested at 4%
and $2x =$ the amount of money invested at 3%.

Since $I = prt$ and the time, t, is 1 yr,

$$\begin{aligned} .04x + .03(2x) &= 400 \\ .04x + .06x &= 400 \\ .10x &= 400. \end{aligned}$$

Multiply by 10 to clear the decimal.

$$x = 4000$$

Then $2x = 2(4000) = 8000$.
He invested \$4000 at 4% and \$8000 at 3%.

4. Let $x =$ his score on the fourth test.

Since his average, the sum of the scores divided by the number of scores, must be at least 80,

$$\begin{aligned} \frac{79 + 75 + 88 + x}{4} &\ge 80 \\ \frac{242 + x}{4} &\ge 80 \\ 242 + x &\ge 320 \\ x &\ge 78. \end{aligned}$$

He must score 78 or greater on the fourth test.

5. $$\begin{aligned} |6t - 4| + 8 &= 3 \\ |6t - 4| &= -5 \end{aligned}$$

Since the absolute value of an expression must be nonnegative, the equation has no solution.

Solution set: $\emptyset$

6. $-4^2 + (-4)^2 = -(4)^2 + (-4)^2 = -16 + 16 = 0$

7. $3x - 5 \ge 1$ or $2x + 7 \le 9$

Solve each inequality.

$$\begin{aligned} 3x - 5 \ge 1 \quad &\text{or} \quad 2x + 7 \le 9 \\ 3x \ge 6 \quad & \quad 2x \le 2 \\ x \ge 2 \quad &\text{or} \quad x \le 1 \end{aligned}$$

Solution set: $(-\infty, 1] \cup [2, \infty)$

8. $.04t + .06(t - 1) = 1.04$

Multiply by 100 to clear the decimals.

$$\begin{aligned} 4t + 6(t - 1) &= 104 \\ 4t + 6t - 6 &= 104 \\ 10t - 6 &= 104 \\ 10t &= 110 \\ t &= 11 \end{aligned}$$

Solution set: $\{11\}$

For Exercises 9 and 10, see the answers graphs in the back of the textbook.

9. $$\begin{aligned} 4 - 7(q + 4) &< -3q \\ 4 - 7q - 28 &< -3q \\ -7q - 24 &< -3q \\ -4q &< 24 \\ q &> -6 \end{aligned}$$

Solution set: $(-6, \infty)$

10. $$\begin{aligned} 3x - 2 < 10 \quad &\text{and} \quad -2x < 10 \\ 3x < 12 \quad & \quad x > \frac{10}{-2} \\ x < 4 \quad &\text{and} \quad x > -5 \end{aligned}$$

Since the inequality is true whenever $x > -5$ and $x < 4$, the solution consists of all numbers between -5 and 4, not including -5 or 4.

Solution set: $(-5, 4)$

11. $-4x + 5y = 20$

To find the x-intercept, let $y = 0$.

$$\begin{aligned} -4x + 5(0) &= 20 \\ -4x &= 20 \\ x &= -5 \end{aligned}$$

The x-intercept is $(-5, 0)$.

To find the y-intercept, let $x = 0$.

$$\begin{aligned} -4(0) + 5y &= 20 \\ 5y &= 20 \\ y &= 4 \end{aligned}$$

The y-intercept is $(0, 4)$.

12. $y = -4$

There is no x-intercept since the equation $y = -4$ represents a horizontal line. The y-intercept is $(0, -4)$.

13. $x = 3y$

When $x = 0$ in this equation, $y = 0$ as well. Therefore, the origin, $(0, 0)$, is the x-intercept and the y-intercept.

14. $5x + 2y = 16$ *(1)*
$3x - 3y = 18$ *(2)*

To eliminate y, multiply equation (1) by 3 and equation (2) by 2 and add the results.

$$\begin{aligned} 15x + 6y &= 48 \\ 6x - 6y &= 36 \\ \hline 21x \quad &= 84 \\ x &= 4 \end{aligned}$$

To find y, substitute 4 for x in equation (2).

$$\begin{aligned} 3x - 3y &= 18 \quad (2) \\ 3(4) - 3y &= 18 \\ 12 - 3y &= 18 \\ -3y &= 6 \\ y &= -2 \end{aligned}$$

Solution set: $\{(4, -2)\}$

15. $f(x) = \dfrac{x+1}{3-x}$

$f(2) = \dfrac{2+1}{3-2} = \dfrac{3}{1} = 3$

16. $.000076 = 7.6 \times 10^{-5}$

The decimal point was moved 5 places to the right.

17. $5.6 \times 10^9 = 5{,}600{,}000{,}000$

Because of the positive exponent, the decimal point was moved 9 places to the right.

18. $\dfrac{3x^{-4}y^3}{7^{-1}x^5y^{-4}} = \dfrac{7 \cdot 3y^3y^4}{x^4x^5} = \dfrac{21y^7}{x^9}$

19. $$\begin{aligned} \left(\frac{a^{-3}b^4}{a^2b^{-1}}\right)^{-2} &= \left(\frac{b^4b^1}{a^2a^3}\right)^{-2} \\ &= \left(\frac{b^5}{a^5}\right)^{-2} \\ &= \left(\frac{a^5}{b^5}\right)^{2} \\ &= \frac{a^{10}}{b^{10}} \end{aligned}$$

20. $$\begin{aligned} \left(\frac{m^{-4}n^2}{m^2n^{-3}}\right) \cdot \left(\frac{m^5n^{-1}}{m^{-2}n^5}\right) &= \frac{n^2n^3}{m^2m^4} \cdot \frac{m^5m^2}{n^1n^5} \\ &= \frac{n^5}{m^6} \cdot \frac{m^7}{n^6} \\ &= \frac{n^5m^7}{n^6m^6} \\ &= \frac{m}{n} \end{aligned}$$

21. $$\begin{aligned} &9(-3x^3 - 4x + 12) + 2(x^3 - x^2 + 3) \\ &= -27x^3 - 36x + 108 + 2x^3 - 2x^2 + 6 \\ &= -25x^3 - 2x^2 - 36x + 114 \end{aligned}$$

22. $$\begin{aligned} (4f + 3)(3f - 1) &= 12f^2 - 4f + 9f - 3 \\ &= 12f^2 + 5f - 3 \end{aligned}$$

23. $$\begin{aligned} &(x + y)(x^2 - xy + y^2) \\ &= x^3 - x^2y + xy^2 + x^2y - xy^2 + y^3 \\ &= x^3 + y^3 \end{aligned}$$

24. $(7t^3 + 8)(7t^3 - 8)$

This is the product of the sum and difference of two terms.

$$\begin{aligned} &= (7t^3)^2 - 8^2 \\ &= 49t^6 - 64 \end{aligned}$$

25. $\left(\dfrac{1}{4}x + 5\right)^2$

Use the formula for the square of a binomial, $(a + b)^2 = a^2 + 2ab + b^2$.

$$\begin{aligned} &= \left(\frac{1}{4}x\right)^2 + 2\left(\frac{1}{4}x\right)(5) + 5^2 \\ &= \frac{1}{16}x^2 + \frac{5}{2}x + 25 \end{aligned}$$

26. $$\begin{aligned} P(x) &= -4x^3 + 2x - 8 \\ P(-2) &= -4(-2)^3 + 2(-2) - 8 \\ &= -4(-8) - 4 - 8 \\ &= 32 - 4 - 8 = 20 \end{aligned}$$

27. $(2x^4 + 3x^3 - 8x^2 + x + 2) \div (x - 1)$

$$\begin{array}{r|rrrrr} 1 & 2 & 3 & -8 & 1 & 2 \\ & & 2 & 5 & -3 & -2 \\ \hline & 2 & 5 & -3 & -2 & 0 \end{array}$$

Answer: $2x^3 + 5x^2 - 3x - 2$

28. $2x^2 - 13x - 45 = (2x + 5)(x - 9)$

29. $$\begin{aligned} 100t^4 - 25 &= 25(4t^4 - 1) \\ &= 25(2t^2 + 1)(2t^2 - 1) \end{aligned}$$

30. $$3x^2 + 4x = 7$$
$$3x^2 + 4x - 7 = 0$$
$$(3x + 7)(x - 1) = 0$$
$$3x + 7 = 0 \quad \text{or} \quad x - 1 = 0$$
$$3x = -7$$
$$x = -\frac{7}{3} \quad \text{or} \quad x = 1$$

Solution set: $\left\{-\frac{7}{3}, 1\right\}$

31. $f(x) = \dfrac{x+8}{x^2 - 3x - 4}$

Set the denominator equal to zero and factor the equation.

$$x^2 - 3x - 4 = 0$$
$$(x - 4)(x + 1) = 0$$
$$x - 4 = 0 \quad \text{or} \quad x + 1 = 0$$
$$x = 4 \quad \text{or} \quad x = -1$$

Both -1 and 4 make the function undefined.

32. $$\frac{8x^2 - 18}{8x^2 + 4x - 12} = \frac{2(4x^2 - 9)}{4(2x^2 + x - 3)}$$
$$= \frac{2(2x - 3)(2x + 3)}{4(2x + 3)(x - 1)}$$
$$= \frac{2x - 3}{2(x - 1)}$$

33. $$\frac{x+4}{x-2} + \frac{2x-10}{x-2} = \frac{x + 4 + 2x - 10}{x - 2}$$
$$= \frac{3x - 6}{x - 2}$$
$$= \frac{3(x - 2)}{x - 2}$$
$$= 3$$

34. $\dfrac{2}{a+b} - \dfrac{3}{a-b}$

The LCD is $(a + b)(a - b)$.

$$= \frac{2(a - b)}{(a + b)(a - b)} - \frac{3(a + b)}{(a - b)(a + b)}$$
$$= \frac{2(a - b) - 3(a + b)}{(a + b)(a - b)}$$
$$= \frac{2a - 2b - 3a - 3b}{(a + b)(a - b)}$$
$$= \frac{-a - 5b}{(a + b)(a - b)}$$

35. $\dfrac{2x}{2x-1} + \dfrac{4}{2x+1} + \dfrac{8}{4x^2 - 1}$

$$= \frac{2x}{2x-1} + \frac{4}{2x+1} + \frac{8}{(2x - 1)(2x + 1)}$$

The LCD is $(2x - 1)(2x + 1)$.

$$= \frac{2x(2x + 1)}{(2x - 1)(2x + 1)} + \frac{4(2x - 1)}{(2x + 1)(2x - 1)}$$
$$+ \frac{8}{(2x - 1)(2x + 1)}$$
$$= \frac{2x(2x + 1) + 4(2x - 1) + 8}{(2x - 1)(2x + 1)}$$
$$= \frac{4x^2 + 2x + 8x - 4 + 8}{(2x - 1)(2x + 1)}$$
$$= \frac{4x^2 + 10x + 4}{(2x - 1)(2x + 1)}$$
$$= \frac{2(2x^2 + 5x + 2)}{(2x - 1)(2x + 1)}$$
$$= \frac{2(2x + 1)(x + 2)}{(2x - 1)(2x + 1)}$$
$$= \frac{2(x + 2)}{2x - 1}$$

36. $\dfrac{5}{x^3 - y^3} + \dfrac{3}{x^2 + xy + y^2}$

$$= \frac{5}{(x - y)(x^2 + xy + y^2)}$$
$$+ \frac{3}{x^2 + xy + y^2}$$

The LCD is $(x - y)(x^2 + xy + y^2)$.

$$= \frac{5}{(x - y)(x^2 + xy + y^2)}$$
$$+ \frac{3(x - y)}{(x^2 + xy + y^2)(x - y)}$$
$$= \frac{5 + 3(x - y)}{(x - y)(x^2 + xy + y^2)}$$
$$= \frac{5 + 3x - 3y}{(x - y)(x^2 + xy + y^2)}$$

37. $$\frac{-3x}{x+1} + \frac{4x+1}{x} = \frac{-3}{x^2 + x}$$
$$\frac{-3x}{x+1} + \frac{4x+1}{x} = \frac{-3}{x(x+1)}$$

$x + 1 \neq 0$, so $x \neq -1$ or $x \neq 0$.
Multiply by the LCD, $x(x + 1)$.

$$x(x + 1)\left(\frac{-3x}{x+1} + \frac{4x+1}{x}\right)$$
$$= x(x + 1)\left(\frac{-3}{x(x+1)}\right)$$

$$\begin{aligned} x(-3x)+(x+1)(4x+1)&=-3\\ -3x^2+4x^2+x+4x+1&=-3\\ x^2+5x+1&=-3\\ x^2+5x+4&=0\\ (x+4)(x+1)&=0 \end{aligned}$$

$$x+4=0 \quad \text{or} \quad x+1=0$$
$$x=-4 \quad \text{or} \quad x=-1$$

But $x \neq -1$, so the only solution is -4.

Solution set: $\{-4\}$

38. $\frac{1}{f}=\frac{1}{p}+\frac{1}{q}$

To solve for q, multiply by the LCD, fpq.

$$fpq\left(\frac{1}{f}\right)=fpq\left(\frac{1}{p}+\frac{1}{q}\right)$$

$$\begin{aligned} pq&=fq+fp\\ pq-fq&=fp\\ q(p-f)&=fp\\ q&=\frac{fp}{p-f}\\ \text{or}\quad q&=\frac{-fp}{f-p} \end{aligned}$$

39. Let $x=$ the speed of the plane in still air.

Use $d=rt$, or $t=\frac{d}{r}$, to make a table.

Direction	Distance	Rate	Time
Against the wind	200	$x-30$	$\frac{200}{x-30}$
With the wind	300	$x+30$	$\frac{300}{x+30}$

The times are the same, so

$$\frac{200}{x-30}=\frac{300}{x+30}.$$

Multiply by the LCD, $(x-30)(x+30)$.

$$(x-30)(x+30)\left(\frac{200}{x-30}\right)$$
$$=(x-30)(x+30)\left(\frac{300}{x+30}\right)$$

$$\begin{aligned} 200(x+30)&=300(x-30)\\ 200x+6000&=300x-9000\\ -100x&=-15,000\\ x&=150 \end{aligned}$$

The speed of the plane in still air is 150 mph.

40. Let $x=$ the time to do the job working together.

Make a table.

	Rate	Time together	Part of job done
Machine A	$\frac{1}{2}$	x	$\frac{x}{2}$
Machine B	$\frac{1}{3}$	x	$\frac{x}{3}$

Part done by A	plus	Part done by B	equals	1 whole job.
$\frac{x}{2}$	$+$	$\frac{x}{3}$	$=$	1

Multiply both sides by the LCD, 6.

$$\begin{aligned} 6\left(\frac{x}{2}+\frac{x}{3}\right)&=6\cdot 1\\ 3x+2x&=6\\ 5x&=6\\ x&=\frac{6}{5}\text{ or }1\frac{1}{5} \end{aligned}$$

Working together, the machines can do the job in $\frac{6}{5}$ or $1\frac{1}{5}$ hr.

Chapter 7

ROOTS AND RADICALS

7.1 Radicals

7.1 Margin Exercises

1. A square root of a number a is a number which, when squared, gives a.

   (a) Square roots of 64 are 8 and -8, since $8^2 = 64$ and $(-8)^2 = 64$.

   (b) Both 13 and -13 are square roots of 169, since $13^2 = 169$ and $(-13)^2 = 169$.

   (c) There is no real number that when squared equals -9, so -9 has no real number square roots.

2. **(a)** $\sqrt[3]{8} = 2$, since $2^3 = 8$.

   (b) $\sqrt[3]{1000} = 10$, since $10^3 = 1000$.

   (c) $\sqrt[3]{-1} = -1$, since $(-1)^3 = -1$.

   (d) $\sqrt[4]{81} = 3$, since $3^4 = 81$.

   (e) $\sqrt[4]{-1}$ is not a real number.

   (f) $\sqrt[6]{64} = 2$, since $2^6 = 64$.

3. **(a)** $\sqrt{4} = 2$, since $2^2 = 4$.

   (b) $\sqrt[3]{27} = 3$, since $3^3 = 27$.

   (c) $-\sqrt{36} = -6$, since $6^2 = 36$.

   The negative sign outside the radical makes the answer negative.

   (d) $\sqrt[4]{625} = 5$, since $5^4 = 625$.

   (e) $\sqrt[5]{-32} = -2$, since $(-2)^5 = -32$.

4. Since the symbol $\sqrt{a}$ means the nonnegative square root of a, the symbol $-\sqrt{a}$ means the negative square root of a. Whenever $\sqrt{a}$ is used, only one answer is required.

   (a) $\sqrt{49} = 7$, since 7 is the only *nonnegative* number whose square is 49.

   (b) $-\sqrt{\dfrac{36}{25}} = -\dfrac{6}{5}$

   (c) $\sqrt{(-6)^2} = |-6| = 6$

   (d) $\sqrt{r^2} = |r|$

5. **(a)** $\sqrt[6]{64} = 2$, since $2^6 = 64$.

   (b) $-\sqrt[4]{16} = -2$, since $2^4 = 16$ and we want the negative fourth root.

   (c) $\sqrt[3]{\dfrac{216}{125}} = \dfrac{6}{5}$, since $\left(\dfrac{6}{5}\right)^3 = \dfrac{216}{125}$.

   (d) $\sqrt[5]{-243} = -3$, since $(-3)^5 = -243$.

   (e) $\sqrt[6]{(-p)^6} = |-p| = |p|$

   (f) $-\sqrt[6]{y^{24}} = -\left|y^4\right| = -y^4$

6. Use a calculator, and round to three decimal places.

   (a) $\sqrt{10} \approx 3.162$

   (b) $\sqrt{51} \approx 7.141$

   (c) $-\sqrt{99} \approx -9.950$

   (d) $\sqrt{950} \approx 30.822$

   (e) $-\sqrt{670} \approx -25.884$

7.1 Section Exercises

3. The principal cube root of 27 is *3*.

7. Both 2 and -2 are square roots of 4, since $2^2 = 4$ and $(-2)^2 = 4$.

11. Use a calculator. Both 42 and -42 are square roots of 1764, since $42^2 = 1764$ and $(-42)^2 = 1764$.

15. $\sqrt{123.5}$ is between 11 and 12, and closer to 11, since $11^2 = 121$ and $12^2 = 144$. The answer is (c).

19. $\sqrt{36} = 6$, since $6^2 = 36$.

23. $\sqrt{-81}$ is not a real number, since the square of a real number is nonnegative.

27. $-\sqrt{-169}$ is not a real number, since the square of a real number is nonnegative.

31. $\sqrt{81} = 9$, since $9^2 = 81$. So, $-\sqrt{81} = -9$.

35. $\sqrt[3]{-64} = -4$, since $(-4)^3 = -64$.

39. Use a calculator and take the square root twice.

$$\sqrt[4]{1296} = 6, \text{ since } 6^4 = 1296.$$

43. $\sqrt[4]{-16}$ is not a real number, since the fourth power of a real number is nonnegative.

47. Since 5 is odd, $\sqrt[5]{a^5} = a$, so

$$\sqrt[5]{(-9)^5} = -9.$$

51. Since the index is even, $\sqrt{x^2} = |x|$.

55. $\sqrt[3]{x^{15}} = x^5$, since $(x^5)^3 = x^{15}$.

59. Use a calculator, and round to three decimal places.

$$\sqrt{284.361} \approx 16.863$$

61. $\sqrt{16} = 4$ and $\sqrt{16} = -4$, since $4^2 = 16$ and $(-4)^2 = 16$.

62. The principal square root is the positive root. The principal square root of $\sqrt{16}$ is 4.

63. $\sqrt{16} = 4$ and $-\sqrt{16} = -(4) = -4$. The principal square root is 4.

64. Since $4^2 = 16$ and $(-4)^2 = 16$, the solution set of $x^2 = 16$ is $\{-4, 4\}$.

65. $\sqrt[4]{81} = 3$ and $\sqrt[4]{81} = -3$, since $3^4 = 81$ and $(-3)^4 = 81$.

66. The principal fourth root is the positive root. The principal fourth root is 3.

67. $\sqrt[4]{81} = 3$ and $-\sqrt[4]{81} = -(3) = -3$. The principal fourth root is 3.

68. Since $3^4 = 81$ and $(-3)^4 = 81$, the solution set of $x^4 = 81$ is $\{-3, 3\}$.

69. $\pm\sqrt{25}$ means both the positive (principal) and negative square roots of 25, that is, $\sqrt{25} = 5$ and $-\sqrt{25} = -5$.

70. $\sqrt[4]{x^4}$ represents the *nonnegative* fourth root. $\sqrt[4]{x^4}$ is simplified as $|x|$ because the index is even and x could be a negative number. For example, if $x = -2$, then

$$\sqrt[4]{x^4} = \sqrt[4]{(-2)^4} = \sqrt[4]{16} = 2 \neq x.$$

However,

$$x = |-2| = 2.$$

$\sqrt[3]{x^3}$ could be a positive or negative number. $\sqrt[3]{x^3}$ is simplified as x because the index is odd. If x is negative, so is $\sqrt[3]{x^3}$.

7.2 Rational Exponents

7.2 Margin Exercises

1. **(a)** $8^{1/3} = \sqrt[3]{8} = 2$

   (b) $9^{1/2} = \sqrt{9} = 3$

   (c) $-81^{1/4} = -\sqrt[4]{81} = -3$

   (d) $(-16)^{1/4}$ is not a real number, since the fourth power of a real number is nonnegative.

   (e) $a^{1/7} = \sqrt[7]{a}$

2. **(a)** $64^{2/3} = (64^{1/3})^2 = 4^2 = 16$

   (b) $100^{3/2} = (100^{1/2})^3 = 10^3 = 1000$

   (c) $-16^{3/4} = -(16^{1/4})^3 = -(2)^3 = -8$

   (d) $(-16)^{3/4}$ is not a real number, since $(-16)^{1/4}$ is not a real number.

3. **(a)** $36^{-3/2} = \dfrac{1}{36^{3/2}} = \dfrac{1}{(\sqrt{36})^3} = \dfrac{1}{6^3} = \dfrac{1}{216}$

   (b) $32^{-4/5} = \dfrac{1}{32^{4/5}} = \dfrac{1}{(\sqrt[5]{32})^4} = \dfrac{1}{2^4} = \dfrac{1}{16}$

   (c) $$\left(\frac{4}{9}\right)^{-5/2} = \frac{1}{\left(\frac{4}{9}\right)^{5/2}} = \frac{1}{\left(\sqrt{\frac{4}{9}}\right)^5}$$
   $$= \frac{1}{\left(\frac{2}{3}\right)^5} = \frac{1}{\frac{32}{243}} = \frac{243}{32}$$

4. **(a)** $11^{3/4} \cdot 11^{5/4} = 11^{3/4+5/4} = 11^{8/4} = 11^2 = 121$

   (b) $\dfrac{7^{3/4}}{7^{7/4}} = 7^{3/4-7/4} = 7^{-4/4} = 7^{-1} = \dfrac{1}{7}$

   (c) $\dfrac{9^{2/3}}{9^{-1/3}} = 9^{2/3-(-1/3)} = 9^{2/3+1/3} = 9^1 = 9$

   (d) $(x^{3/2})^4 = x^{(3/2)4} = x^6 \quad (x > 0)$

5. **(a)** $\sqrt{y^{10}} = (y^{10})^{1/2} = y^5$

   (b) $$\begin{aligned}\sqrt[6]{27y^9} &= (27y^9)^{1/6} = (3^3y^9)^{1/6} \\ &= 3^{3/6}y^{9/6} = 3^{1/2}y^{3/2} \\ &= 3^{1/2}y^{1+1/2} = y \cdot 3^{1/2}y^{1/2} \\ &= y \cdot (3y)^{1/2} = y\sqrt{3y}\end{aligned}$$

   (c) $$\begin{aligned}\sqrt[5]{y} \cdot \sqrt[10]{y^7} &= y^{1/5} \cdot y^{7/10} \\ &= y^{2/10} \cdot y^{7/10} \\ &= y^{9/10} = \sqrt[10]{y^9}\end{aligned}$$

   (d) $$\frac{\sqrt[3]{m^2}}{\sqrt[5]{m}} = \frac{m^{2/3}}{m^{1/5}} = m^{2/3-1/5}$$
   $$= m^{7/15} = \sqrt[15]{m^7}$$

7.2 Section Exercises

3. The statement is false.

$$\sqrt{3^2+4^2} = \sqrt{9+16} = \sqrt{25} = 5,$$

not $3+4=7$.

7. $169^{1/2} = \sqrt{169} = 13$

11. $16^{1/4} = \sqrt[4]{16} = 2$

15. $(-27)^{1/3} = \sqrt[3]{-27} = -3$

19. $-144^{1/2} = -\sqrt{144} = -12$

23. $64^{-3/2} = \frac{1}{64^{3/2}} = \frac{1}{(\sqrt{64})^3} = \frac{1}{8^3} = \frac{1}{512}$

27. $$\left(-\frac{8}{27}\right)^{-2/3} = \frac{1}{\left(-\frac{8}{27}\right)^{2/3}} = \frac{1}{\left(\sqrt[3]{-\frac{8}{27}}\right)^2}$$
$$= \frac{1}{\left(-\frac{2}{3}\right)^2} = \frac{1}{\frac{4}{9}} = \frac{9}{4}$$

In Exercises 31 and 35, keystrokes will vary depending on the calculator used.

31. To find $169^{1/2}$ on one calculator, the keystrokes are:

[1] [6] [9] [y^x] [(] [1] [÷] [2] [)] [=].

Or, since $\frac{1}{2} = .5$, the following keystrokes might be used.

[1] [6] [9] [y^x] [.] [5] [=]

The display is 13, which is the same as the answer to Exercise 7.

35. To find $125^{2/3}$ on one calculator, the keystrokes are:

[1] [2] [5] [y^x] [(] [2] [÷] [3] [)] [=].

The display is 25, which is the same as the answer to Example 2(b).

39. $\frac{64^{5/3}}{64^{4/3}} = 64^{(5/3-4/3)} = 64^{1/3} = \sqrt[3]{64} = 4$

43. $\frac{k^{1/3}}{k^{2/3}\cdot k^{-1}} = \frac{k^{1/3}}{k^{-1/3}} = k^{1/3-(-1/3)} = k^{2/3}$

47. $\frac{(x^{2/3})^2}{(x^2)^{7/3}} = \frac{x^{4/3}}{x^{14/3}} = x^{4/3-14/3} = x^{-10/3} = \frac{1}{x^{10/3}}$

51. $$\left(\frac{b^{-3/2}}{c^{-5/3}}\right)^2 \cdot (b^{-1/4}c^{-1/3})^{-1}$$
$$= \left(\frac{c^{5/3}}{b^{3/2}}\right)^2 \cdot (b^{1/4}c^{1/3})$$
$$= \frac{c^{10/3}}{b^3}(b^{1/4}c^{1/3})$$
$$= \frac{c^{10/3}b^{1/4}c^{1/3}}{b^3}$$
$$= c^{10/3+1/3}b^{1/4-3}$$
$$= c^{11/3}b^{-11/4}$$
$$= \frac{c^{11/3}}{b^{11/4}}$$

55. $\sqrt{2^{12}} = 2^{12/2} = 2^6 = 64$

59. $\sqrt{x^{20}} = x^{20/2} = x^{10}$

63. $$\sqrt[4]{49y^6} = (49y^6)^{1/4} = (7^2y^6)^{1/4}$$
$$= 7^{2/4}y^{6/4} = 7^{1/2}y^{3/2}$$
$$= 7^{1/2}y^{1+1/2} = y \cdot 7^{1/2}y^{1/2}$$
$$= y\sqrt{7y}$$

67. In $3x^{-1/2} - 4x^{1/2}$, factor out the variable with the smallest exponent, $x^{-1/2}$.

68. In $m^3 - 3m^{5/2}, m^3 = m^{6/2}$ and $m^{5/2} < m^{6/2}$. Factor out $m^{5/2}$.

69. In $9k^{-3/4} + 2k^{-1/4}, k^{-3/4} < k^{-1/4}$, so factor out $k^{-3/4}$.

70. $3x^{-1/2} - 4x^{1/2}$

From Exercise 67, factor out $x^{-1/2}$. Rewrite the second expression to allow this: $x^{1/2} = x^{1-1/2}$.

$$= 3x^{-1/2} - 4x^1x^{-1/2}$$
$$= x^{-1/2}(3-4x)$$

71. $m^3 - 3m^{5/2}$

From Exercise 68, factor out $m^{5/2}$. Rewrite the first expression to allow this: $m^3 = m^{5/2+1/2}$.

$$= m^{5/2}m^{1/2} - 3m^{5/2}$$
$$= m^{5/2}(m^{1/2}-3)$$

72. $9k^{-3/4} + 2k^{-1/4}$

From Exercise 69, factor out $k^{-3/4}$.

$$= 9k^{-3/4} + 2k^{1/2}k^{-3/4}$$
$$= k^{-3/4}(9+2k^{1/2})$$

73. $\dfrac{4}{\sqrt{t}}+7\sqrt{t^3}$

$= 4t^{-1/2}+7t^{3/2}$

Factor out $t^{-1/2}$, the smallest exponent.

$$\begin{aligned}&= 4t^{-1/2}+7t^{4/2-1/2}\\&= 4t^{-1/2}+7t^{4/2}t^{-1/2}\\&= t^{-1/2}(4+7t^{4/2})\\&= t^{-1/2}(4+7t^2)\end{aligned}$$

74. $8\sqrt[3]{x^2}-\dfrac{5}{\sqrt[3]{x}}$

$= 8x^{2/3}-5x^{-1/3}$

Factor out $x^{-1/3}$.

$$\begin{aligned}&= 8x^{3/3-1/3}-5x^{-1/3}\\&= 8x^{3/3}x^{-1/3}-5x^{-1/3}\\&= x^{-1/3}(8x^{3/3}-5)\\&= x^{-1/3}(8x^1-5)\\&= x^{-1/3}(8x-5)\end{aligned}$$

75. Use $T(D)=.07D^{3/2}$ with $D=16$.

$$\begin{aligned}T(16)&=.07(16)^{3/2}=.07(16^{1/2})^3\\&=.07(4)^3=.07(64)=4.48\end{aligned}$$

To the nearest tenth of an hour, time T is 4.5 hr.

7.3 Simplifying Radicals

7.3 Margin Exercises

1. **(a)** $\sqrt{5}\cdot\sqrt{13}=\sqrt{5\cdot 13}=\sqrt{65}$

   (b) $\sqrt{10y}\cdot\sqrt{3k}=\sqrt{10y\cdot 3k}=\sqrt{30yk}$

   (c) $\sqrt{\dfrac{5}{a}}\cdot\sqrt{\dfrac{11}{z}}=\sqrt{\dfrac{5}{a}\cdot\dfrac{11}{z}}=\sqrt{\dfrac{55}{az}}$

2. **(a)** $\sqrt[3]{2}\cdot\sqrt[3]{7}=\sqrt[3]{2\cdot 7}=\sqrt[3]{14}$

   (b) $\sqrt[6]{8r^2}\cdot\sqrt[6]{2r^3}=\sqrt[6]{8r^2\cdot 2r^3}=\sqrt[6]{16r^5}$

   (c) $\sqrt[5]{9y^2x}\cdot\sqrt[5]{8xy^2}=\sqrt[5]{9y^2x\cdot 8xy^2}=\sqrt[5]{72y^4x^2}$

   (d) $\sqrt{7}\cdot\sqrt[3]{5}$ cannot be simplified by the product rule because the two indexes (2 and 3) are different.

3. **(a)** $\sqrt{\dfrac{100}{81}}=\dfrac{\sqrt{100}}{\sqrt{81}}=\dfrac{10}{9}$

   (b) $\sqrt{\dfrac{11}{25}}=\dfrac{\sqrt{11}}{\sqrt{25}}=\dfrac{\sqrt{11}}{5}$

   (c) $\sqrt[3]{\dfrac{18}{125}}=\dfrac{\sqrt[3]{18}}{\sqrt[3]{125}}=\dfrac{\sqrt[3]{18}}{5}$

   (d) $\sqrt{\dfrac{y^8}{16}}=\dfrac{\sqrt{y^8}}{\sqrt{16}}=\dfrac{y^4}{4}$

   (e) $\sqrt[3]{\dfrac{x^2}{r^{12}}}=\dfrac{\sqrt[3]{x^2}}{\sqrt[3]{r^{12}}}=\dfrac{\sqrt[3]{x^2}}{r^4}$

4. **(a)** $\sqrt{32}=\sqrt{16\cdot 2}=\sqrt{16}\cdot\sqrt{2}=4\sqrt{2}$

   (b) $\sqrt{45}=\sqrt{9\cdot 5}=\sqrt{9}\cdot\sqrt{5}=3\sqrt{5}$

   (c) $\sqrt{300}=\sqrt{100\cdot 3}=\sqrt{100}\cdot\sqrt{3}=10\sqrt{3}$

   (d) $\sqrt{35}=\sqrt{5\cdot 7}$

   Since 35 has no perfect square factor, $\sqrt{35}$ cannot be further simplified.

   (e) $\sqrt[3]{54}=\sqrt[3]{27\cdot 2}=\sqrt[3]{27}\cdot\sqrt[3]{2}=3\sqrt[3]{2}$

   (f) $\sqrt[4]{243}=\sqrt[4]{81\cdot 3}=\sqrt[4]{81}\cdot\sqrt[4]{3}=3\sqrt[4]{3}$

5. **(a)** $\sqrt{25p^7}=\sqrt{25p^6\cdot p}=\sqrt{25p^6}\cdot\sqrt{p}=5p^3\sqrt{p}$

   (b) $\begin{aligned}[t]\sqrt{72y^3x}&=\sqrt{36y^2\cdot 2yx}=\sqrt{36y^2}\cdot\sqrt{2yx}\\&=6y\sqrt{2yx}\end{aligned}$

   (c) $\begin{aligned}[t]\sqrt[3]{y^7x^5z^6}&=\sqrt[3]{y^6x^3z^6\cdot yx^2}=\sqrt[3]{y^6x^3z^6}\cdot\sqrt[3]{yx^2}\\&=y^2xz^2\sqrt[3]{yx^2}\end{aligned}$

   (d) $\begin{aligned}[t]\sqrt[4]{32a^5b^6}&=\sqrt[4]{16a^4b^4\cdot 2ab^2}\\&=\sqrt[4]{16a^4b^4}\cdot\sqrt[4]{2ab^2}\\&=2ab\sqrt[4]{2ab^2}\end{aligned}$

6. **(a)** $\sqrt[12]{2^3}=2^{3/12}=2^{1/4}=\sqrt[4]{2}$

   (b) $\sqrt[6]{t^2}=t^{2/6}=t^{1/3}=\sqrt[3]{t}$

7. $\sqrt{5}\cdot\sqrt[3]{4}$

   The least common index of 2 and 3 is 6. Write each radical as a sixth root.

   $$\begin{aligned}\sqrt{5}&=5^{1/2}=5^{3/6}=\sqrt[6]{5^3}\\\sqrt[3]{4}&=4^{1/3}=4^{2/6}=\sqrt[6]{4^2}\end{aligned}$$

   So,

   $$\begin{aligned}\sqrt{5}\cdot\sqrt[3]{4}&=\sqrt[6]{5^3}\cdot\sqrt[6]{4^2}=\sqrt[6]{5^3\cdot 4^2}\\&=\sqrt[6]{125\cdot 16}=\sqrt[6]{2000}.\end{aligned}$$

8. **(a)** Substitute 14 for a and 8 for b in the Pythagorean formula to find c.

   $$\begin{aligned}c^2&=a^2+b^2\\c&=\sqrt{a^2+b^2}=\sqrt{14^2+8^2}=\sqrt{196+64}\\&=\sqrt{260}=\sqrt{4}\cdot\sqrt{65}=2\sqrt{65}\end{aligned}$$

   The length of the hypotenuse is $2\sqrt{65}$.

(b) Substitute 4 for a and 6 for c to find b.

$$\begin{aligned} a^2 + b^2 &= c^2 \\ b &= \sqrt{c^2 - a^2} = \sqrt{6^2 - 4^2} \\ &= \sqrt{36 - 16} = \sqrt{20} = \sqrt{4} \cdot \sqrt{5} = 2\sqrt{5} \end{aligned}$$

The length of the unknown side is $2\sqrt{5}$.

9. Use the distance formula

$$d = \sqrt{(x_2 - x_1)^2 + (y_2 - y_1)^2}.$$

(a) The distance between $(2, -1)$ and $(5, 3)$ is

$$\begin{aligned} d &= \sqrt{(2-5)^2 + (-1-3)^2} \\ &= \sqrt{(-3)^2 + (-4)^2} \\ &= \sqrt{9 + 16} \\ &= 25 \\ &= 5. \end{aligned}$$

(b) The distance between $(-3, 2)$ and $(0, -4)$ is

$$\begin{aligned} d &= \sqrt{(-3-0)^2 + [2-(-4)]^2} \\ &= \sqrt{(-3)^2 + 6^2} \\ &= \sqrt{9 + 36} \\ &= \sqrt{45} \\ &= \sqrt{9} \cdot \sqrt{5} \\ &= 3\sqrt{5}. \end{aligned}$$

7.3 Section Exercises

3. A simplified radical cannot have a *radical* in its *denominator.*

7. $\sqrt{5} \cdot \sqrt{6} = \sqrt{5 \cdot 6} = \sqrt{30}$

11. $\sqrt[4]{12} \cdot \sqrt[4]{3} = \sqrt[4]{12 \cdot 3} = \sqrt[4]{36}$

15. $\sqrt{\dfrac{3}{25}} = \dfrac{\sqrt{3}}{\sqrt{25}} = \dfrac{\sqrt{3}}{5}$

19. $\sqrt{\dfrac{p^6}{81}} = \dfrac{\sqrt{p^6}}{\sqrt{81}} = \dfrac{p^3}{9}$

23. $\sqrt[3]{-\dfrac{r^2}{8}} = \dfrac{\sqrt[3]{r^2}}{\sqrt[3]{-8}} = \dfrac{\sqrt[3]{r^2}}{-2} = -\dfrac{\sqrt[3]{r^2}}{2}$

27. $\sqrt{288} = \sqrt{144 \cdot 2} = \sqrt{144} \cdot \sqrt{2} = 12\sqrt{2}$

31. $-\sqrt{28} = -\sqrt{4 \cdot 7} = -\sqrt{4} \cdot \sqrt{7} = -2\sqrt{7}$

35. $\sqrt[3]{128} = \sqrt[3]{64 \cdot 2} = \sqrt[3]{64} \cdot \sqrt[3]{2} = 4\sqrt[3]{2}$

39. $\sqrt[3]{40} = \sqrt[3]{8 \cdot 5} = \sqrt[3]{8} \cdot \sqrt[3]{5} = 2\sqrt[3]{5}$

43. $\sqrt[5]{64} = \sqrt[5]{2^5 \cdot 2} = \sqrt[5]{2^5} \cdot \sqrt[5]{2} = 2\sqrt[5]{2}$

47. $\sqrt{72k^2} = \sqrt{36k^2 \cdot 2} = \sqrt{36k^2} \cdot \sqrt{2} = 6k\sqrt{2}$

51. $\sqrt{121x^6} = \sqrt{(11x^3)^2} = 11x^3$

55. $-\sqrt{100m^8z^4} = -\sqrt{(10m^4z^2)^2} = -10m^4z^2$

59. $\sqrt[4]{\dfrac{1}{16}r^8t^{20}} = \sqrt[4]{\left(\dfrac{1}{2}r^2t^5\right)^4} = \dfrac{1}{2}r^2t^5$

63. $$\begin{aligned} -\sqrt{500r^{11}} &= -\sqrt{100r^{10} \cdot 5r} = -\sqrt{100r^{10}} \cdot \sqrt{5r} \\ &= -10r^5\sqrt{5r} \end{aligned}$$

67. $\sqrt[3]{8z^6w^9} = \sqrt[3]{(2z^2w^3)^3} = 2z^2w^3$

71. $\sqrt[4]{81x^{12}y^{16}} = \sqrt[4]{(3x^3y^4)^4} = 3x^3y^4$

75. $\sqrt{\dfrac{y^{11}}{36}} = \dfrac{\sqrt{y^{11}}}{\sqrt{36}} = \dfrac{\sqrt{y^{10}y}}{6} = \dfrac{y^5\sqrt{y}}{6}$

79. $$\begin{aligned} \sqrt[4]{48^2} &= 48^{2/4} = 48^{1/2} = \sqrt{48} = \sqrt{16 \cdot 3} \\ &= \sqrt{16} \cdot \sqrt{3} = 4\sqrt{3} \end{aligned}$$

83. $\sqrt[3]{4} \cdot \sqrt{3}$

The least common index of 3 and 2 is 6. Write each radical as a sixth root.

$$\begin{aligned} \sqrt[3]{4} &= \sqrt[3]{2^2} = 2^{2/3} = 2^{4/6} = \sqrt[6]{2^4} \\ \sqrt{3} &= 3^{1/2} = 3^{3/6} = \sqrt[6]{3^3} \end{aligned}$$

So,

$$\begin{aligned} \sqrt[3]{4} \cdot \sqrt{3} &= \sqrt[6]{2^4} \cdot \sqrt[6]{3^3} = \sqrt[6]{2^4 \cdot 3^3} \\ &= \sqrt[6]{16 \cdot 27} = \sqrt[6]{432}. \end{aligned}$$

87. Substitute 3 for a and 4 for b in the Pythagorean formula to find c.

$$\begin{aligned} c^2 &= a^2 + b^2 \\ c &= \sqrt{a^2 + b^2} = \sqrt{3^2 + 4^2} \\ &= \sqrt{9 + 16} = \sqrt{25} = 5 \end{aligned}$$

The length of the hypotenuse is 5.

In Exercises 91 and 95, use the distance formula

$$d = \sqrt{(x_2 - x_1)^2 + (y_2 - y_1)^2}.$$

91. The distance between $(5, 3)$ and $(-1, 2)$ is

$$\begin{aligned} d &= \sqrt{[5-(-1)]^2 + (3-2)^2} \\ &= \sqrt{6^2 + 1^2} \\ &= \sqrt{36 + 1} \\ &= \sqrt{37}. \end{aligned}$$

95. The distance between $(\sqrt{2}, \sqrt{6})$ and $(-2\sqrt{2}, 4\sqrt{6})$ is

$$\begin{aligned} d &= \sqrt{(-2\sqrt{2}-\sqrt{2})^2+(4\sqrt{6}-\sqrt{6})^2} \\ &= \sqrt{(-3\sqrt{2})^2+(3\sqrt{6})^2} \\ &= \sqrt{9\cdot 2+9\cdot 6} \\ &= \sqrt{72} \\ &= \sqrt{36}\cdot\sqrt{2} \\ &= 6\sqrt{2}. \end{aligned}$$

99. Substitute 640 for k and 2 for I in the equation to find d.

$$\begin{aligned} d &= \sqrt{\frac{k}{I}} = \sqrt{\frac{640}{2}} = \sqrt{320} = \sqrt{64\cdot 5} \\ &= \sqrt{64}\cdot\sqrt{5} = 8\sqrt{5} \approx 17.9 \end{aligned}$$

The illumination will be 2 footcandles at $8\sqrt{5}$ ft or about 17.9 ft from the light source.

7.4 Addition and Subtraction of Radical Expressions

7.4 Margin Exercises

1. (a) $3\sqrt{5}+7\sqrt{5}=(3+7)\sqrt{5}=10\sqrt{5}$

(b) $$\begin{aligned} &2\sqrt{11}-\sqrt{11}+3\sqrt{44} \\ &= \sqrt{11}+3\sqrt{44}=\sqrt{11}+3\sqrt{4\cdot 11} \\ &= \sqrt{11}+3\cdot 2\sqrt{11}=\sqrt{11}+6\sqrt{11} \\ &= 7\sqrt{11} \end{aligned}$$

(c) $$\begin{aligned} &5\sqrt{12y}+6\sqrt{75y},\quad y\geq 0 \\ &= 5\sqrt{4\cdot 3y}+6\sqrt{25\cdot 3y} \\ &= 5\cdot 2\sqrt{3y}+6\cdot 5\sqrt{3y} \\ &= 10\sqrt{3y}+30\sqrt{3y} \\ &= 40\sqrt{3y} \end{aligned}$$

(d) $$\begin{aligned} &3\sqrt{8}-6\sqrt{50}+2\sqrt{200} \\ &= 3\sqrt{4\cdot 2}-6\sqrt{25\cdot 2}+2\sqrt{100\cdot 2} \\ &= 3\cdot 2\sqrt{2}-6\cdot 5\sqrt{2}+2\cdot 10\sqrt{2} \\ &= 6\sqrt{2}-30\sqrt{2}+20\sqrt{2} \\ &= -4\sqrt{2} \end{aligned}$$

(e) $9\sqrt{5}-4\sqrt{10}$

Each radical differs and is already simplified, so this expression cannot be simplified further.

2. (a) $$\begin{aligned} 7\sqrt[3]{81}+3\sqrt[3]{24} &= 7\sqrt[3]{27\cdot 3}+3\sqrt[3]{8\cdot 3} \\ &= 7\sqrt[3]{27}\cdot\sqrt[3]{3}+3\sqrt[3]{8}\cdot\sqrt[3]{3} \\ &= 7\cdot 3\sqrt[3]{3}+3\cdot 2\sqrt[3]{3} \\ &= 21\sqrt[3]{3}+6\sqrt[3]{3}=27\sqrt[3]{3} \end{aligned}$$

(b) $$\begin{aligned} &-2\sqrt[4]{32}-7\sqrt[4]{162} \\ &= -2\sqrt[4]{16\cdot 2}-7\sqrt[4]{81\cdot 2} \\ &= -2\cdot 2\sqrt[4]{2}-7\cdot 3\sqrt[4]{2} \\ &= -4\sqrt[4]{2}-21\sqrt[4]{2}=-25\sqrt[4]{2} \end{aligned}$$

(c) $$\begin{aligned} &\sqrt[3]{p^4q^7}-\sqrt[3]{64pq} \\ &= \sqrt[3]{p^3q^6\cdot pq}-\sqrt[3]{64\cdot pq} \\ &= pq^2\sqrt[3]{pq}-4\sqrt[3]{pq}=(pq^2-4)\sqrt[3]{pq} \end{aligned}$$

3. The formula for the area of a trapezoid is $A=.5(B+b)h$. Substitute $\sqrt{45}$ and $\sqrt{80}$ for B and b and $\sqrt{18}$ for h to find A.

$$\begin{aligned} \text{Area} &= .5(B+b)h \\ &= .5(\sqrt{45}+\sqrt{80})\sqrt{18} \\ &= .5(\sqrt{9\cdot 5}+\sqrt{16\cdot 5})\sqrt{9\cdot 2} \\ &= .5(3\sqrt{5}+4\sqrt{5})3\sqrt{2} \\ &= .5(7\sqrt{5})3\sqrt{2} \\ &= .5(21\sqrt{10}) \\ &= 10.5\sqrt{10} \end{aligned}$$

The area is $10.5\sqrt{10}$ in^2.

7.4 Section Exercises

3. $$\begin{aligned} \sqrt{64}+\sqrt[3]{125}+\sqrt[4]{16} &= \sqrt{8^2}+\sqrt[3]{5^3}+\sqrt[4]{2^4} \\ &= 8+5+2=15 \end{aligned}$$

This sum can be found easily since each radicand is a whole number power corresponding to the index of the radical. In other words, each term is a perfect root.

7. $$\begin{aligned} -2\sqrt{48}+3\sqrt{75} &= -2\sqrt{16\cdot 3}+3\sqrt{25\cdot 3} \\ &= -2\cdot 4\sqrt{3}+3\cdot 5\sqrt{3} \\ &= -8\sqrt{3}+15\sqrt{3}=7\sqrt{3} \end{aligned}$$

11. $$\begin{aligned} &-2\sqrt{63}+2\sqrt{28}+2\sqrt{7} \\ &= -2\sqrt{9\cdot 7}+2\sqrt{4\cdot 7}+2\sqrt{7} \\ &= -2\cdot 3\sqrt{7}+2\cdot 2\sqrt{7}+2\sqrt{7} \\ &= -6\sqrt{7}+4\sqrt{7}+2\sqrt{7}=0 \end{aligned}$$

15. $$\begin{aligned} &8\sqrt{2x}-\sqrt{8x}+\sqrt{72x} \\ &= 8\sqrt{2x}-\sqrt{4\cdot 2x}+\sqrt{36\cdot 2x} \\ &= 8\sqrt{2x}-2\sqrt{2x}+6\sqrt{2x}=12\sqrt{2x} \end{aligned}$$

19. $$\begin{aligned} -\sqrt[3]{54}+2\sqrt[3]{16} &= -\sqrt[3]{27\cdot 2}+2\sqrt[3]{8\cdot 2} \\ &= -3\sqrt[3]{2}+2\cdot 2\sqrt[3]{2} \\ &= -3\sqrt[3]{2}+4\sqrt[3]{2}=\sqrt[3]{2} \end{aligned}$$

23. $$\begin{aligned} 5\sqrt[4]{32}+3\sqrt[4]{162} &= 5\sqrt[4]{16\cdot 2}+3\sqrt[4]{81\cdot 2} \\ &= 5\cdot 2\sqrt[4]{2}+3\cdot 3\sqrt[4]{2} \\ &= 10\sqrt[4]{2}+9\sqrt[4]{2}=19\sqrt[4]{2} \end{aligned}$$

27. $\sqrt[3]{64xy^2} + \sqrt[3]{27x^4y^5}$
$= \sqrt[3]{64 \cdot xy^2} + \sqrt[3]{27x^3y^3 \cdot xy^2}$
$= 4\sqrt[3]{xy^2} + 3xy\sqrt[3]{xy^2} = (4 + 3xy)\sqrt[3]{xy^2}$

31. $\frac{\sqrt{32}}{3} + \frac{2\sqrt{2}}{3} - \frac{\sqrt{2}}{\sqrt{9}}$
$= \frac{\sqrt{16}\sqrt{2}}{3} + \frac{2\sqrt{2}}{3} - \frac{\sqrt{2}}{3}$
$= \frac{4\sqrt{2} + 2\sqrt{2} - \sqrt{2}}{3}$
$= \frac{5\sqrt{2}}{3}$

35. Does $2\sqrt{40} + 6\sqrt{90} - 3\sqrt{160} = 10\sqrt{10}$?

Find a calculator approximation of each term.

$12.64911064 + 56.92099788 - 37.94733192$
$= 31.6227766$
$31.6227766 = 31.6227766$

The calculator approximations are the same. The statement is true.

39. The perimeter, P, of a triangle is the sum of the measures of the sides.

$P = 3\sqrt{20} + 2\sqrt{45} + \sqrt{75}$
$= 3\sqrt{4 \cdot 5} + 2\sqrt{9 \cdot 5} + \sqrt{25 \cdot 3}$
$= 3 \cdot 2\sqrt{5} + 2 \cdot 3\sqrt{5} + 5\sqrt{3}$
$= 6\sqrt{5} + 6\sqrt{5} + 5\sqrt{3}$
$= 12\sqrt{5} + 5\sqrt{3}$

The perimeter is $12\sqrt{5} + 5\sqrt{3}$ in.

7.5 Multiplication and Division of Radical Expressions

7.5 Margin Exercises

1. Use the FOIL method and multiply as two binomials.

(a) $(2 + \sqrt{3})(1 + \sqrt{5})$
$= 2 \cdot 1 + 2\sqrt{5} + \sqrt{3} \cdot 1 + \sqrt{3} \cdot \sqrt{5}$
$= 2 + 2\sqrt{5} + \sqrt{3} + \sqrt{15}$

(b) $(2\sqrt{3} + \sqrt{5})(\sqrt{6} - 3\sqrt{5})$
$= 2\sqrt{3} \cdot \sqrt{6} - 2\sqrt{3} \cdot 3\sqrt{5}$
$+ \sqrt{5} \cdot \sqrt{6} - \sqrt{5} \cdot 3\sqrt{5}$
$= 2\sqrt{18} - 6\sqrt{15} + \sqrt{30} - 3(5)$
$= 2\sqrt{9 \cdot 2} - 6\sqrt{15} + \sqrt{30} - 15$
$= 6\sqrt{2} - 6\sqrt{15} + \sqrt{30} - 15$

(c) $(4 + \sqrt{3})(4 - \sqrt{3})$
$= 4 \cdot 4 - 4\sqrt{3} + 4\sqrt{3} - \sqrt{3} \cdot \sqrt{3}$
$= 16 - 3$
$= 13$

(d) $(\sqrt{6} - \sqrt{5})^2$
$= (\sqrt{6} - \sqrt{5})(\sqrt{6} - \sqrt{5})$
$= \sqrt{6} \cdot \sqrt{6} - \sqrt{6} \cdot \sqrt{5} - \sqrt{5} \cdot \sqrt{6}$
$+ \sqrt{5} \cdot \sqrt{5}$
$= 6 - \sqrt{30} - \sqrt{30} + 5$
$= 11 - 2\sqrt{30}$

(e) $(4 + \sqrt[3]{7})(4 - \sqrt[3]{7})$
$= 4 \cdot 4 - 4\sqrt[3]{7} + 4\sqrt[3]{7} - \sqrt[3]{7} \cdot \sqrt[3]{7}$
$= 16 - \sqrt[3]{49}$

2. **(a)** $\frac{8}{\sqrt{3}} = \frac{8}{\sqrt{3}} \cdot \frac{\sqrt{3}}{\sqrt{3}} = \frac{8\sqrt{3}}{3}$

(b) $\frac{9}{\sqrt{6}} = \frac{9}{\sqrt{6}} \cdot \frac{\sqrt{6}}{\sqrt{6}} = \frac{9\sqrt{6}}{6} = \frac{3\sqrt{6}}{2}$

(c) $\frac{\sqrt{3}}{\sqrt{7}} = \frac{\sqrt{3}}{\sqrt{7}} \cdot \frac{\sqrt{7}}{\sqrt{7}} = \frac{\sqrt{21}}{7}$

3. **(a)** $\frac{3}{\sqrt{48}} = \frac{3}{\sqrt{16 \cdot 3}} = \frac{3}{4\sqrt{3}} = \frac{3}{4\sqrt{3}} \cdot \frac{\sqrt{3}}{\sqrt{3}}$
$= \frac{3\sqrt{3}}{4 \cdot 3} = \frac{\sqrt{3}}{4}$

(b) $\frac{9}{\sqrt{20}} = \frac{9}{\sqrt{4 \cdot 5}} = \frac{9}{2\sqrt{5}} = \frac{9}{2\sqrt{5}} \cdot \frac{\sqrt{5}}{\sqrt{5}}$
$= \frac{9\sqrt{5}}{2 \cdot 5} = \frac{9\sqrt{5}}{10}$

(c) $\frac{-16}{\sqrt{32}} = \frac{-16}{\sqrt{16 \cdot 2}} = \frac{-16}{4\sqrt{2}} = \frac{-4}{\sqrt{2}}$
$= \frac{-4}{\sqrt{2}} \cdot \frac{\sqrt{2}}{\sqrt{2}} = \frac{-4\sqrt{2}}{2}$
$= -2\sqrt{2}$

4. **(a)** $\sqrt{\frac{8}{45}} = \frac{\sqrt{8}}{\sqrt{45}} = \frac{\sqrt{4 \cdot 2}}{\sqrt{9 \cdot 5}} = \frac{2\sqrt{2}}{3\sqrt{5}}$
$= \frac{2\sqrt{2} \cdot \sqrt{5}}{3\sqrt{5} \cdot \sqrt{5}} = \frac{2\sqrt{10}}{3 \cdot 5} = \frac{2\sqrt{10}}{15}$

(b) $\sqrt{\frac{72}{y}} = \frac{\sqrt{72}}{\sqrt{y}} \cdot \frac{\sqrt{y}}{\sqrt{y}} = \frac{\sqrt{72y}}{y} = \frac{\sqrt{36 \cdot 2y}}{y}$
$= \frac{6\sqrt{2y}}{y}$

(c) $\sqrt{\dfrac{200k^6}{y^7}} = \dfrac{\sqrt{100k^6 \cdot 2}}{\sqrt{y^6 \cdot y}} = \dfrac{10k^3\sqrt{2}}{y^3\sqrt{y}}$

$= \dfrac{10k^3\sqrt{2}\cdot\sqrt{y}}{y^3\sqrt{y}\cdot\sqrt{y}} = \dfrac{10k^3\sqrt{2y}}{y^3 \cdot y}$

$= \dfrac{10k^3\sqrt{2y}}{y^4}$

5. (a) $\sqrt[3]{\dfrac{15}{32}} = \dfrac{\sqrt[3]{15}}{\sqrt[3]{32}} = \dfrac{\sqrt[3]{15}}{\sqrt[3]{2^5}} = \dfrac{\sqrt[3]{15}\cdot\sqrt[3]{2}}{\sqrt[3]{2^5}\cdot\sqrt[3]{2}} = \dfrac{\sqrt[3]{30}}{\sqrt[3]{2^6}}$

$= \dfrac{\sqrt[3]{30}}{2^2} = \dfrac{\sqrt[3]{30}}{4}$

(b) $\sqrt[3]{\dfrac{m^{12}}{n}} = \dfrac{\sqrt[3]{m^{12}}}{\sqrt[3]{n}} = \dfrac{m^4\cdot\sqrt[3]{n^2}}{\sqrt[3]{n}\cdot\sqrt[3]{n^2}}$

$= \dfrac{m^4\sqrt[3]{n^2}}{n}\ (n \neq 0)$

6. (a) $\dfrac{-4}{\sqrt{5}+2}$

Multiply numerator and denominator by $\sqrt{5}-2$.

$= \dfrac{-4(\sqrt{5}-2)}{(\sqrt{5}+2)(\sqrt{5}-2)}$

$= \dfrac{-4(\sqrt{5}-2)}{5-4}$

$= -4(\sqrt{5}-2)$

(b) $\dfrac{18}{9+\sqrt{18}}$

Multiply numerator and denominator by $9-\sqrt{18}$.

$= \dfrac{18(9-\sqrt{18})}{(9+\sqrt{18})(9-\sqrt{18})}$

$= \dfrac{18(9-\sqrt{18})}{81-18} = \dfrac{18(9-\sqrt{18})}{63}$

$= \dfrac{2(9-3\sqrt{2})}{7} = \dfrac{18-6\sqrt{2}}{7}$

$= \dfrac{6(3-\sqrt{2})}{7}$

7. (a) $\dfrac{15}{\sqrt{7}+\sqrt{2}} = \dfrac{15(\sqrt{7}-\sqrt{2})}{(\sqrt{7}+\sqrt{2})(\sqrt{7}-\sqrt{2})}$

$= \dfrac{15(\sqrt{7}-\sqrt{2})}{7-2}$

$= \dfrac{15(\sqrt{7}-\sqrt{2})}{5}$

$= 3(\sqrt{7}-\sqrt{2})$

(b) $\dfrac{\sqrt{3}+\sqrt{5}}{\sqrt{2}-\sqrt{7}} = \dfrac{(\sqrt{3}+\sqrt{5})(\sqrt{2}+\sqrt{7})}{(\sqrt{2}-\sqrt{7})(\sqrt{2}+\sqrt{7})}$

$= \dfrac{\sqrt{6}+\sqrt{21}+\sqrt{10}+\sqrt{35}}{2-7}$

$= \dfrac{-(\sqrt{6}+\sqrt{21}+\sqrt{10}+\sqrt{35})}{5}$

(c) $\dfrac{2}{\sqrt{k}+\sqrt{z}} = \dfrac{2(\sqrt{k}-\sqrt{z})}{(\sqrt{k}+\sqrt{z})(\sqrt{k}-\sqrt{z})}$

$= \dfrac{2(\sqrt{k}-\sqrt{z})}{k-z}$

$(k \neq z, k > 0, z > 0)$

8. (a) $\dfrac{15-5\sqrt{3}}{5} = \dfrac{5(3-\sqrt{3})}{5} = 3-\sqrt{3}$

(b) $\dfrac{24-36\sqrt{7}}{16} = \dfrac{4(6-9\sqrt{7})}{16} = \dfrac{6-9\sqrt{7}}{4}$

7.5 Section Exercises

1. $\sqrt{a}\cdot\sqrt{b} = \sqrt{ab}$

2. $(x+y)(x-y) = x^2-y^2$

3. $(x+\sqrt{y})(x-\sqrt{y}) = x^2-y$

This exercise is an example of the product of the sum and difference of two terms. Note that $(\sqrt{y})^2 = \sqrt{y^2} = y$.

4. $(\sqrt{x}+\sqrt{y})(\sqrt{x}-\sqrt{y}) = (\sqrt{x})^2-(\sqrt{y})^2$
$= x-y$

This is another example of the product of the sum and difference of two terms.

5. $(x+y)^2 = x^2+2xy+y^2$

This exercise is an example of the square of a binomial.

6. $(\sqrt{x}+\sqrt{y})^2 = (\sqrt{x})^2+2\sqrt{x}\sqrt{y}+(\sqrt{y})^2$
$= x+2\sqrt{xy}+y$

This is another example of the square of a binomial.

7. $\sqrt{3}(\sqrt{12}-4) = \sqrt{36}-4\sqrt{3} = 6-4\sqrt{3}$

11. $(\sqrt{6}+2)(\sqrt{6}-2)$
$= \sqrt{6}\cdot\sqrt{6}-2\sqrt{6}+2\sqrt{6}-4$
$= 6-4 = 2$

15. $(\sqrt{3}+2)(\sqrt{6}-5)$
$= \sqrt{3}\cdot\sqrt{6}-5\sqrt{3}+2\sqrt{6}-10$
$= \sqrt{18}-5\sqrt{3}+2\sqrt{6}-10$
$= \sqrt{9\cdot 2}-5\sqrt{3}+2\sqrt{6}-10$
$= 3\sqrt{2}-5\sqrt{3}+2\sqrt{6}-10$

19. $(2\sqrt{x}+\sqrt{y})(2\sqrt{x}-\sqrt{y})$
$= 2\sqrt{x}\cdot 2\sqrt{x} - 2\sqrt{xy} + 2\sqrt{xy} - \sqrt{y}\cdot\sqrt{y}$
$= 4x - y$

23. $(9-\sqrt[3]{2})(9+\sqrt[3]{2})$
$= 9\cdot 9 + 9\sqrt[3]{2} - 9\sqrt[3]{2} - \sqrt[3]{2}\cdot\sqrt[3]{2}$
$= 81 - \sqrt[3]{4}$

27. $\frac{7}{\sqrt{7}} = \frac{7\cdot\sqrt{7}}{\sqrt{7}\cdot\sqrt{7}} = \frac{7\sqrt{7}}{7} = \sqrt{7}$

31. $\frac{\sqrt{3}}{\sqrt{2}} = \frac{\sqrt{3}\cdot\sqrt{2}}{\sqrt{2}\cdot\sqrt{2}} = \frac{\sqrt{6}}{2}$

35. $\frac{-6}{\sqrt{18}} = \frac{-6}{\sqrt{9\cdot 2}} = \frac{-6}{3\sqrt{2}} = \frac{-2}{\sqrt{2}} = \frac{-2\cdot\sqrt{2}}{\sqrt{2}\cdot\sqrt{2}}$
$= \frac{-2\sqrt{2}}{2} = -\sqrt{2}$

39. $\frac{6\sqrt{3y}}{\sqrt{y^3}} = \frac{6\sqrt{3y}}{\sqrt{y^2\cdot y}} = \frac{6\sqrt{3y}}{y\sqrt{y}} = \frac{6\sqrt{3y}\cdot\sqrt{y}}{y\sqrt{y}\cdot\sqrt{y}}$
$= \frac{6\sqrt{3y^2}}{y\cdot y} = \frac{6y\sqrt{3}}{y^2} = \frac{6\sqrt{3}}{y}$

43. $\sqrt{\frac{7}{2}} = \frac{\sqrt{7}}{\sqrt{2}} = \frac{\sqrt{7}\cdot\sqrt{2}}{\sqrt{2}\cdot\sqrt{2}} = \frac{\sqrt{14}}{2}$

47. $\sqrt{\frac{24}{x}} = \frac{\sqrt{24}}{\sqrt{x}} = \frac{\sqrt{4\cdot 6}}{\sqrt{x}} = \frac{2\sqrt{6}}{\sqrt{x}}$
$= \frac{2\sqrt{6}\cdot\sqrt{x}}{\sqrt{x}\cdot\sqrt{x}} = \frac{2\sqrt{6x}}{x}$

51. $\sqrt{\frac{288x^7}{y^9}} = \frac{\sqrt{288x^7}}{\sqrt{y^9}} = \frac{\sqrt{144x^6\cdot 2x}}{\sqrt{y^8\cdot y}}$
$= \frac{12x^3\sqrt{2x}}{y^4\sqrt{y}} = \frac{12x^3\sqrt{2x}\cdot\sqrt{y}}{y^4\sqrt{y}\cdot\sqrt{y}}$
$= \frac{12x^3\sqrt{2xy}}{y^4\cdot y} = \frac{12x^3\sqrt{2xy}}{y^5}$

55. $\sqrt[3]{\frac{4}{9}} = \frac{\sqrt[3]{4}}{\sqrt[3]{9}} = \frac{\sqrt[3]{4}}{\sqrt[3]{3^2}} = \frac{\sqrt[3]{4}\cdot\sqrt[3]{3}}{\sqrt[3]{3^2}\cdot\sqrt[3]{3}} = \frac{\sqrt[3]{12}}{\sqrt[3]{3^3}} = \frac{\sqrt[3]{12}}{3}$

59. $\sqrt[4]{\frac{16}{x}} = \frac{\sqrt[4]{16}}{\sqrt[4]{x}} = \frac{2}{\sqrt[4]{x}} = \frac{2\cdot\sqrt[4]{x^3}}{\sqrt[4]{x}\cdot\sqrt[4]{x^3}}$
$= \frac{2\sqrt[4]{x^3}}{\sqrt[4]{x^4}} = \frac{2\sqrt[4]{x^3}}{x}$

63. $\frac{2}{4+\sqrt{3}}$

Multiply by the conjugate of the denominator, $4-\sqrt{3}$.

$= \frac{2(4-\sqrt{3})}{(4+\sqrt{3})(4-\sqrt{3})}$
$= \frac{2(4-\sqrt{3})}{16-3} = \frac{2(4-\sqrt{3})}{13}$

67. $\frac{-4}{\sqrt{3}-\sqrt{7}}$

Multiply by the conjugate of the denominator, $\sqrt{3}+\sqrt{7}$.

$= \frac{-4(\sqrt{3}+\sqrt{7})}{(\sqrt{3}-\sqrt{7})(\sqrt{3}+\sqrt{7})}$
$= \frac{-4(\sqrt{3}+\sqrt{7})}{3-7}$
$= \frac{-4(\sqrt{3}+\sqrt{7})}{-4} = \sqrt{3}+\sqrt{7}$

71. $\frac{4\sqrt{x}}{\sqrt{x}-2\sqrt{y}}$

Multiply by the conjugate of the denominator, $\sqrt{x}+2\sqrt{y}$.

$= \frac{4\sqrt{x}(\sqrt{x}+2\sqrt{y})}{(\sqrt{x}-2\sqrt{y})(\sqrt{x}+2\sqrt{y})}$
$= \frac{4\sqrt{x}(\sqrt{x}+2\sqrt{y})}{x-4y}$

75. Let $a = \frac{\sqrt{6}-\sqrt{2}}{4}$. Let $b = \frac{\sqrt{2-\sqrt{3}}}{2}$.

Since the numerators and denominators are all positive, a and b are both positive.

$$a^2 = \left(\frac{\sqrt{6}-\sqrt{2}}{4}\right)^2 = \frac{(\sqrt{6}-\sqrt{2})(\sqrt{6}-\sqrt{2})}{16}$$
$$= \frac{6-2\sqrt{12}+2}{16} = \frac{8-2\sqrt{12}}{16}$$
$$= \frac{8-2\sqrt{4\cdot 3}}{16} = \frac{8-4\sqrt{3}}{16}$$
$$= \frac{4(2-\sqrt{3})}{16} = \frac{2-\sqrt{3}}{4}$$
$$b^2 = \left(\frac{\sqrt{2-\sqrt{3}}}{2}\right)^2 = \frac{\sqrt{2-\sqrt{3}}\cdot\sqrt{2-\sqrt{3}}}{4}$$
$$= \frac{2-\sqrt{3}}{4}$$

Since $a^2 = b^2$ and a and b are both positive, $a = b$.

79. $\frac{16+4\sqrt{8}}{12} = \frac{16+4\sqrt{4\cdot 2}}{12} = \frac{16+8\sqrt{2}}{12}$
$= \frac{4(4+2\sqrt{2})}{12} = \frac{4+2\sqrt{2}}{3}$

83. $$\frac{1}{\sqrt{2}}\cdot\frac{\sqrt{3}}{2}-\frac{1}{\sqrt{2}}\cdot\frac{1}{2}=\frac{\sqrt{3}}{2\sqrt{2}}-\frac{1}{2\sqrt{2}}=\frac{\sqrt{3}-1}{2\sqrt{2}}$$
$$=\frac{(\sqrt{3}-1)\sqrt{2}}{(2\sqrt{2})\sqrt{2}}=\frac{\sqrt{6}-\sqrt{2}}{2\cdot 2}$$
$$=\frac{\sqrt{6}-\sqrt{2}}{4}$$

Using a calculator,

$$\frac{1}{\sqrt{2}}\cdot\frac{\sqrt{3}}{2}-\frac{1}{\sqrt{2}}\cdot\frac{1}{2}\approx .2588190451 \text{ and}$$
$$\frac{\sqrt{6}-\sqrt{2}}{4}\approx .2588190451.$$

85. $$\frac{8\sqrt{5}-1}{6}=\frac{(8\sqrt{5}-1)(8\sqrt{5}+1)}{6(8\sqrt{5}+1)}$$
$$=\frac{64(5)-1}{6(8\sqrt{5}+1)}$$
$$=\frac{319}{6(8\sqrt{5}+1)}$$

86. $$\frac{3\sqrt{a}+\sqrt{b}}{\sqrt{b}-\sqrt{a}}=\frac{(3\sqrt{a}+\sqrt{b})(3\sqrt{a}-\sqrt{b})}{(\sqrt{b}-\sqrt{a})(3\sqrt{a}-\sqrt{b})}$$
$$=\frac{9a-b}{(\sqrt{b}-\sqrt{a})(3\sqrt{a}-\sqrt{b})}$$

87. $$\frac{3\sqrt{a}+\sqrt{b}}{\sqrt{b}-\sqrt{a}}=\frac{(3\sqrt{a}+\sqrt{b})(\sqrt{b}+\sqrt{a})}{(\sqrt{b}-\sqrt{a})(\sqrt{b}+\sqrt{a})}$$
$$=\frac{(3\sqrt{a}+\sqrt{b})(\sqrt{b}+\sqrt{a})}{b-a}$$

88. In Exercise 86, we multiplied by the conjugate of the numerator, while in Exercise 87, we multiplied by the conjugate of the denominator.

7.6 Equations with Radicals

7.6 Margin Exercises

In Exercises 1-6, check each solution in the original equation.

1. **(a)** $\sqrt{r}=3$

Square both sides.

$$(\sqrt{r})^2=3^2$$
$$r=9$$

Check 9 in the original equation.

$$\sqrt{9}=3 \quad ?$$
$$3=3 \qquad \textit{True}$$

Solution set: $\{9\}$

(b) $\sqrt{5x+1}=4$

Square both sides.

$$(\sqrt{5x+1})^2=4^2$$
$$5x+1=16$$
$$5x=15$$
$$x=3$$

Check 3 in the original equation.

$$\sqrt{5\cdot 3+1}=4 \quad ?$$
$$\sqrt{16}=4 \quad ?$$
$$4=4 \qquad \textit{True}$$

Solution set: $\{3\}$

2. **(a)** $\sqrt{k}+4=-3$

Isolate the radical on one side of the equation.

$$\sqrt{k}=-7$$

The equation has no solution because the square root of a real number must be nonnegative.

Solution set: $\emptyset$

(b) $\sqrt{x-9}-3=0$

Isolate the radical on one side.

$$\sqrt{x-9}=3$$

Square both sides.

$$(\sqrt{x-9})^2=3^2$$
$$x-9=9$$
$$x=18$$

Check 18 in the original equation.

$$\sqrt{18-9}-3=0 \quad ?$$
$$\sqrt{9}-3=0 \quad ?$$
$$3-3=0 \quad ?$$
$$0=0 \qquad \textit{True}$$

The potential solution, 18, checks.

Solution set: $\{18\}$

3. $\sqrt{3z-5} = z-1$

Square both sides.

$$\begin{aligned}(\sqrt{3z-5})^2 &= (z-1)^2\\ 3z-5 &= z^2-2z+1\\ 0 &= z^2-5z+6\\ 0 &= (z-3)(z-2)\end{aligned}$$

$$z-3=0 \quad \text{or} \quad z-2=0$$
$$z=3 \quad \text{or} \quad z=2$$

Check 3 in the original equation.

$$\begin{aligned}\sqrt{3(3)-5} &= 3-1 \quad ?\\ \sqrt{9-5} &= 2 \quad ?\\ \sqrt{4} &= 2 \quad ?\\ 2 &= 2 \quad \textit{True}\end{aligned}$$

Now check 2.

$$\begin{aligned}\sqrt{3(2)-5} &= 2-1 \quad ?\\ \sqrt{6-5} &= 1 \quad ?\\ \sqrt{1} &= 1 \quad ?\\ 1 &= 1 \quad \textit{True}\end{aligned}$$

Solution set: $\{2,3\}$

4. $\sqrt{4a^2+2a-3} = 2a+7$

Square both sides.

$$\begin{aligned}(\sqrt{4a^2+2a-3})^2 &= (2a+7)^2\\ 4a^2+2a-3 &= 4a^2+28a+49\\ -52 &= 26a\\ -2 &= a\end{aligned}$$

Check -2 in the original equation.

$$\begin{aligned}\sqrt{4(-2)^2+2(-2)-3} &= 2(-2)+7 \quad ?\\ \sqrt{16-4-3} &= -4+7 \quad ?\\ \sqrt{9} &= 3 \quad ?\\ 3 &= 3 \quad \textit{True}\end{aligned}$$

Solution set: $\{-2\}$

5. $\sqrt{p+1} - \sqrt{p-4} = 1$

Get one radical on each side of the equals sign.

$$\sqrt{p+1} = 1 + \sqrt{p-4}$$

Square both sides.

$$\begin{aligned}(\sqrt{p+1})^2 &= (1+\sqrt{p-4})^2\\ p+1 &= 1+2\sqrt{p-4}+(p-4)\end{aligned}$$

Isolate the remaining radical.

$$\begin{aligned}p+1 &= 2\sqrt{p-4}+p-3\\ 4 &= 2\sqrt{p-4}\\ 2 &= \sqrt{p-4}\end{aligned}$$

Square both sides again.

$$\begin{aligned}2^2 &= (\sqrt{p-4})^2\\ 4 &= p-4\\ 8 &= p\end{aligned}$$

Check 8 in the original equation.

$$\begin{aligned}\sqrt{8+1}-\sqrt{8-4} &= 1 \quad ?\\ \sqrt{9}-\sqrt{4} &= 1 \quad ?\\ 3-2 &= 1 \quad ?\\ 1 &= 1 \quad \textit{True}\end{aligned}$$

Solution set: $\{8\}$

6. **(a)** $\sqrt[3]{p^2+3p+12} = \sqrt[3]{p^2}$

Cube both sides.

$$\begin{aligned}(\sqrt[3]{p^2+3p+12})^3 &= (\sqrt[3]{p^2})^3\\ p^2+3p+12 &= p^2\\ 3p+12 &= 0\\ 3p &= -12\\ p &= -4\end{aligned}$$

Check -4.

$$\begin{aligned}\sqrt[3]{(-4)^2+3(-4)+12} &= \sqrt[3]{(-4)^2} \quad ?\\ \sqrt[3]{16-12+12} &= \sqrt[3]{16} \quad ?\\ \sqrt[3]{16} &= \sqrt[3]{16} \quad \textit{True}\end{aligned}$$

Solution set: $\{-4\}$

(b) $\sqrt[4]{2k+5}+1=0$

$$\sqrt[4]{2k+5} = -1$$

Since no real number has a principal fourth root that is negative, the equation has no solution.

Solution set: $\emptyset$

7.6 Section Exercises

3. $\sqrt{x-3} = 4$

Square both sides.

$$\begin{aligned}(\sqrt{x-3})^2 &= 4^2\\ x-3 &= 16\\ x &= 19\end{aligned}$$

Check 19 in the original equation.

$$\begin{aligned} \sqrt{19-3} &= 4 \quad ? \\ \sqrt{16} &= 4 \quad ? \\ 4 &= 4 \quad \textit{True} \end{aligned}$$

Solution set: $\{19\}$

7. $\sqrt{x}+9=0$

$$\sqrt{x}=-9$$

No real number has a negative principal square root. There is no solution.

Solution set: $\emptyset$

11. $\sqrt{6x+2}-\sqrt{5x+3}=0$

Get one radical on each side of the equals sign.

$$\sqrt{6x+2}=\sqrt{5x+3}$$

Square both sides.

$$\begin{aligned} (\sqrt{6x+2})^2 &= (\sqrt{5x+3})^2 \\ 6x+2 &= 5x+3 \\ x &= 1 \end{aligned}$$

Check 1 in the original equation.

$$\begin{aligned} \sqrt{6(1)+2}-\sqrt{5(1)+3} &= 0 \quad ? \\ \sqrt{8}-\sqrt{8} &= 0 \quad ? \\ 0 &= 0 \quad \textit{True} \end{aligned}$$

Solution set: $\{1\}$

15. $(8-x)^2$ equals $64-16x+x^2$, not $64+x^2$. The first step should be

$$3x+4=64-16x+x^2.$$

19. $\sqrt{13+4t}=t+4$

Square both sides.

$$\begin{aligned} (\sqrt{13+4t})^2 &= (t+4)^2 \\ 13+4t &= t^2+8t+16 \\ 0 &= t^2+4t+3 \\ 0 &= (t+3)(t+1) \end{aligned}$$

$$t+3=0 \quad \text{or} \quad t+1=0$$

$$t=-3 \quad \text{or} \quad t=-1$$

Check -3 in the original equation.

$$\begin{aligned} \sqrt{13+4(-3)} &= -3+4 \quad ? \\ \sqrt{1} &= 1 \quad ? \\ 1 &= 1 \quad \textit{True} \end{aligned}$$

Now check -1.

$$\begin{aligned} \sqrt{13+4(-1)} &= -1+4 \quad ? \\ \sqrt{9} &= 3 \quad ? \\ 3 &= 3 \quad \textit{True} \end{aligned}$$

Solution set: $\{-3,-1\}$

23. $\sqrt{r+4}-\sqrt{r-4}=2$

Get one radical on each side of the equals sign.

$$\sqrt{r+4}=\sqrt{r-4}+2$$

Square both sides.

$$\begin{aligned} (\sqrt{r+4})^2 &= (\sqrt{r-4}+2)^2 \\ r+4 &= r-4+4\sqrt{r-4}+4 \end{aligned}$$

Isolate the remaining radical.

$$\begin{aligned} r+4 &= r+4\sqrt{r-4} \\ 4 &= 4\sqrt{r-4} \\ 1 &= \sqrt{r-4} \end{aligned}$$

Square both sides again.

$$\begin{aligned} 1^2 &= (\sqrt{r-4})^2 \\ 1 &= r-4 \\ 5 &= r \end{aligned}$$

Check 5 in the original equation.

$$\begin{aligned} \sqrt{5+4}-\sqrt{5-4} &= 2 \quad ? \\ \sqrt{9}-\sqrt{1} &= 2 \quad ? \\ 3-1 &= 2 \quad ? \\ 2 &= 2 \quad \textit{True} \end{aligned}$$

Solution set: $\{5\}$

27. $\sqrt{3-3p}-\sqrt{3p+2}=3$

Get one radical on each side of the equals sign.

$$\sqrt{3-3p}=\sqrt{3p+2}+3$$

Square both sides.

$$\begin{aligned} (\sqrt{3-3p})^2 &= (\sqrt{3p+2}+3)^2 \\ 3-3p &= 3p+2+6\sqrt{3p+2}+9 \end{aligned}$$

Isolate the remaining radical.

$$\begin{aligned} 3-3p &= 3p+11+6\sqrt{3p+2} \\ -6p-8 &= 6\sqrt{3p+2} \\ -3p-4 &= 3\sqrt{3p+2} \end{aligned}$$

Square both sides again.

$$\begin{aligned}(-3p-4)^2 &= (3\sqrt{3p+2})^2\\ 9p^2+24p+16 &= 9(3p+2)\\ 9p^2+24p+16 &= 27p+18\\ 9p^2-3p-2 &= 0\\ (3p+1)(3p-2) &= 0\\ 3p+1=0 \quad &\text{or} \quad 3p-2=0\\ 3p=-1 \quad & \quad\quad 3p=2\\ p=-\frac{1}{3} \quad &\text{or} \quad p=\frac{2}{3}\end{aligned}$$

Check $-\frac{1}{3}$.

$$\begin{aligned}\sqrt{3-3\left(-\frac{1}{3}\right)}-\sqrt{3\left(-\frac{1}{3}\right)+2} &= 3 \quad ?\\ \sqrt{3+1}-\sqrt{-1+2} &= 3 \quad ?\\ \sqrt{4}-\sqrt{1} &= 3 \quad ?\\ 2-1 &= 3 \quad ?\\ 1 &= 3 \quad \textit{False}\end{aligned}$$

Check $\frac{2}{3}$.

$$\begin{aligned}\sqrt{3-3\left(\frac{2}{3}\right)}-\sqrt{3\left(\frac{2}{3}\right)+2} &= 3 \quad ?\\ \sqrt{3-2}-\sqrt{2+2} &= 3 \quad ?\\ \sqrt{1}-\sqrt{4} &= 3 \quad ?\\ 1-2 &= 3 \quad ?\\ -1 &= 3 \quad \textit{False}\end{aligned}$$

Solution set: $\emptyset$

31. $\sqrt[3]{2x^2+3x-7}=\sqrt[3]{2x^2+4x+6}$

Cube both sides.

$$\begin{aligned}(\sqrt[3]{2x^2+3x-7})^3 &= (\sqrt[3]{2x^2+4x+6})^3\\ 2x^2+3x-7 &= 2x^2+4x+6\\ -x &= 13\\ x &= -13\end{aligned}$$

Check -13.

$$\begin{aligned}\sqrt[3]{2(-13)^2+3(-13)-7} &= \sqrt[3]{2(-13)^2+4(-13)+6}\\ \sqrt[3]{338-39-7} &= \sqrt[3]{338-52+6} \quad ?\\ \sqrt[3]{292} &= \sqrt[3]{292} \quad \textit{True}\end{aligned}$$

Solution set: $\{-13\}$

35. $\sqrt[4]{x-1}+2=0$

$$\sqrt[4]{x-1}=-2$$

No real number has a negative principal fourth root. There is no solution.

Solution set: $\emptyset$

39. $(5r-6)^{1/2}=2+(3r-6)^{1/2}$

$$\sqrt{5r-6}=2+\sqrt{3r-6}$$

Square both sides.

$$\begin{aligned}(\sqrt{5r-6})^2 &= (2+\sqrt{3r-6})^2\\ 5r-6 &= 4+4\sqrt{3r-6}+3r-6\\ 5r-6 &= 3r-2+4\sqrt{3r-6}\\ 2r-4 &= 4\sqrt{3r-6}\\ r-2 &= 2\sqrt{3r-6}\end{aligned}$$

Square both sides again.

$$\begin{aligned}(r-2)^2 &= (2\sqrt{3r-6})^2\\ r^2-4r+4 &= 4(3r-6)\\ r^2-4r+4 &= 12r-24\\ r^2-16r+28 &= 0\\ (r-2)(r-14) &= 0\\ r-2=0 \quad &\text{or} \quad r-14=0\\ r=2 \quad &\text{or} \quad r=14\end{aligned}$$

Check 2.

$$\begin{aligned}[5(2)-6]^{1/2} &= 2+[3(2)-6]^{1/2} \quad ?\\ 4^{1/2} &= 2+0 \quad ?\\ 2 &= 2 \quad \textit{True}\end{aligned}$$

Check 14.

$$\begin{aligned}[5(14)-6]^{1/2} &= 2+[3(14)-6]^{1/2} \quad ?\\ 64^{1/2} &= 2+36^{1/2} \quad ?\\ 8 &= 2+6 \quad ?\\ 8 &= 8 \quad \textit{True}\end{aligned}$$

Solution set: $\{2, 14\}$

43. From the screen, the solution is 4, the value of x when $y=0$. Check 4 in the equation.

$$\begin{aligned}\sqrt{x^2-4x+9} &= x-1\\ \sqrt{4^2-4(4)+9} &= 4-1 \quad ?\\ \sqrt{9} &= 3 \quad ?\\ 3 &= 3 \quad \textit{True}\end{aligned}$$

Solution set: $\{4\}$

47. Substitute 2.5 for P in the formula.

$$\begin{aligned}P &= 2\pi\sqrt{\frac{L}{32}}\\ 2.5 &= 2\pi\sqrt{\frac{L}{32}}\\ \frac{2.5}{2\pi} &= \sqrt{\frac{L}{32}}\end{aligned}$$

Square both sides.

$$\left(\frac{2.5}{2\pi}\right)^2 = \left(\sqrt{\frac{L}{32}}\right)^2$$
$$\frac{6.25}{4\pi^2} = \frac{L}{32}$$
$$\frac{50}{\pi^2} = L$$
$$5.1 \approx L$$

The pendulum is about 5.1 ft long.

7.7 Complex Numbers

7.7 Margin Exercises

1. **(a)** $\sqrt{-16} = i\sqrt{16} = 4i$

   (b) $-\sqrt{-81} = -i\sqrt{81} = -9i$

   (c) $\sqrt{-7} = i\sqrt{7}$

2. **(a)** $\sqrt{-7}\cdot\sqrt{-7} = i\sqrt{7}\cdot i\sqrt{7} = i^2(\sqrt{7})^2$
   $= -1(7) = -7$

   (b) $\sqrt{-5}\cdot\sqrt{-10} = i\sqrt{5}\cdot i\sqrt{10} = i^2\sqrt{50}$
   $= i^2\sqrt{25\cdot 2} = -1(5\sqrt{2})$
   $= -5\sqrt{2}$

   (c) $\sqrt{-15}\cdot\sqrt{2} = i\sqrt{15}\cdot\sqrt{2} = i\sqrt{30}$

3. **(a)** $\frac{\sqrt{-32}}{\sqrt{-2}} = \frac{i\sqrt{32}}{i\sqrt{2}} = \sqrt{\frac{32}{2}} = \sqrt{16} = 4$

   (b) $\frac{\sqrt{-27}}{\sqrt{-3}} = \frac{i\sqrt{27}}{i\sqrt{3}} = \sqrt{\frac{27}{3}} = \sqrt{9} = 3$

   (c) $\frac{\sqrt{-40}}{\sqrt{10}} = \frac{i\sqrt{4\cdot 10}}{\sqrt{10}} = \frac{2i\sqrt{10}}{\sqrt{10}} = 2i$

4. To add complex numbers, add the real parts and add the imaginary parts.

   (a) $(4+6i)+(-3+5i)$
   $= [4+(-3)]+(6+5)i$
   $= 1+11i$

   (b) $(-1+8i)+(9-3i)$
   $= (-1+9)+(8-3)i$
   $= 8+5i$

5. To subtract complex numbers, subtract the real parts and subtract the imaginary parts.

   (a) $(7+3i)-(4+2i) = (7-4)+(3-2)i = 3+i$

   (b) $(-6-i)-(-5-4i) = (-6+5)+(-1+4)i$
   $= -1+3i$

   (c) $8-(3-2i) = (8-3)+2i = 5+2i$

6. **(a)** $6i(4+3i) = 24i+18i^2$
   $= 24i+18(-1)$
   $= 24i-18$
   $= -18+24i$

   (b) $(6-4i)(2+4i)$

   Use the FOIL method.

   $$= \underset{\mathbf{F}}{6(2)} + \underset{\mathbf{O}}{6(4i)} - \underset{\mathbf{I}}{2(4i)} - \underset{\mathbf{L}}{4i(4i)}$$
   $$= 12+24i-8i-16i^2$$
   $$= 12+16i-16(-1)$$
   $$= 12+16i+16$$
   $$= 28+16i$$

   (c) $(3-2i)(3+2i)$

   This is the product of conjugates, so $(a+bi)\cdot(a-bi) = a^2-b^2$.

   $$= 3^2-(2i)^2$$
   $$= 9-4i^2$$
   $$= 9-4(-1)$$
   $$= 9+4 = 13$$

7. In each case, multiply the numerator and denominator by the conjugate of the denominator.

   (a) $\frac{2+i}{3-i} = \frac{(2+i)(3+i)}{(3-i)(3+i)}$
   $$= \frac{6+2i+3i+i^2}{3^2-i^2}$$
   $$= \frac{6+2i+3i-1}{9-(-1)}$$
   $$= \frac{5+5i}{10}$$
   $$= \frac{5(1+i)}{10}$$
   $$= \frac{1+i}{2} = \frac{1}{2}+\frac{1}{2}i$$

   (b) $\frac{6+2i}{4-3i} = \frac{(6+2i)(4+3i)}{(4-3i)(4+3i)}$
   $$= \frac{24+18i+8i+6i^2}{4^2-(3i)^2}$$
   $$= \frac{24+26i+6(-1)}{16-9(-1)}$$
   $$= \frac{24+26i-6}{16+9}$$
   $$= \frac{18+26i}{25} = \frac{18}{25}+\frac{26}{25}i$$

(c) $\frac{5}{3-2i} = \frac{5(3+2i)}{(3-2i)(3+2i)}$

$= \frac{5(3+2i)}{3^2-(2i)^2}$

$= \frac{5(3+2i)}{9-4(-1)}$

$= \frac{5(3+2i)}{13}$

$= \frac{15+10i}{13} = \frac{15}{13} + \frac{10}{13}i$

(d) $\frac{5-i}{i}$

The conjugate of i is $-i$.

$= \frac{(5-i)(-i)}{i(-i)}$

$= \frac{(5-i)(-i)}{-i^2}$

$= \frac{-5i+i^2}{-(-1)}$

$= \frac{-5i-1}{1}$

$= -1-5i$

8. **(a)** $i^{21} = i^{20} \cdot i = (i^4)^5 \cdot i = 1^5 \cdot i = i$

(b) $i^{36} = (i^4)^9 = 1^9 = 1$

(c) $i^{50} = i^{48} \cdot i^2 = (i^4)^{12} \cdot i^2 = 1^{12} \cdot (-1) = -1$

(d) $i^{-9} = \frac{1}{i^9} = \frac{1}{i^8 \cdot i} = \frac{1}{(i^4)^2 \cdot i} = \frac{1}{1^2 \cdot i} = \frac{1}{i}$

Multiply by $-i$, the conjugate of i.

$= \frac{1(-i)}{i(-i)} = \frac{-i}{-i^2} = \frac{-i}{-(-1)} = -i$

7.7 Section Exercises

3. $\sqrt{-b} = i\sqrt{b}$, for $b > 0$.

7. $\sqrt{-169} = i\sqrt{169} = 13i$

11. $\sqrt{-5} = i\sqrt{5}$

15. $\sqrt{-15} \cdot \sqrt{-15} = i\sqrt{15} \cdot i\sqrt{15} = i^2(\sqrt{15})^2$
$= -1(15) = -15$

19. $\frac{\sqrt{-300}}{\sqrt{-100}} = \frac{i\sqrt{300}}{i\sqrt{100}} = \sqrt{\frac{300}{100}} = \sqrt{3}$

23. $(3+2i) + (-4+5i)$
$= [3+(-4)] + (2+5)i$
$= -1+7i$

27. $(4+i) - (-3-2i)$
$= [4-(-3)] + [1-(-2)]i$
$= 7+3i$

31. $(-4+11i) + (-2-4i) + (7+6i)$
$= (-4-2+7) + (11-4+6)i$
$= 1+13i$

35. If $a - c = b$, then $b + c = a$. So, $(4+2i) - (3+i)$ $= 1+i$ implies that

$$(1+i) + (3+i) = 4+2i.$$

39. $(-8i)(-2i) = 16i^2 = 16(-1) = -16$

43. $(4+3i)(1-2i) = 4 - 8i + 3i - 6i^2$
$= 4 - 5i - 6(-1)$
$= 10 - 5i$

47. $(12+3i)(12-3i) = 12^2 - (3i)^2$
$= 144 - 9i^2$
$= 144 - 9(-1)$
$= 144 + 9$
$= 153$

51. $\frac{2}{1-i}$

Multiply by the conjugate of the denominator, $1+i$.

$= \frac{2(1+i)}{(1-i)(1+i)}$

$= \frac{2(1+i)}{1^2-i^2}$

$= \frac{2(1+i)}{1-(-1)}$

$= \frac{2(1+i)}{2}$

$= 1+i$

55. $\frac{8i}{2+2i}$

Write in lowest terms.

$= \frac{8i}{2(1+i)}$

$= \frac{4i}{1+i}$

Multiply by the conjugate of the denominator, $1-i$.

$$\begin{aligned}&=\frac{4i(1-i)}{(1+i)(1-i)}\\&=\frac{4i(1-i)}{1-i^2}\\&=\frac{4i(1-i)}{1-(-1)}\\&=\frac{4i(1-i)}{2}\\&=2i(1-i)\\&=2i-2i^2\\&=2i-2(-1)\\&=2+2i\end{aligned}$$

59. **(a)** $(x+2)+(3x-1)$
$=(1+3)x+(2-1)$
$=4x+1$

(b) $(1+2i)+(3-i)$
$=(1+3)+(2-1)i$
$=4+i$

60. **(a)** $(x+2)-(3x-1)$
$=(x-3x)+[2-(-1)]$
$=(1-3)x+(2+1)$
$=-2x+3$

(b) $(1+2i)-(3-i)$
$=(1-3)+[2-(-1)]i$
$=-2+3i$

61. **(a)** $(x+2)(3x-1)$
$=3x^2-x+6x-2$
$=3x^2+5x-2$

(b) $(1+2i)(3-i)$
$=3-i+6i-2i^2$
$=3+5i-2(-1)$
$=3+5i+2$
$=5+5i$

62. **(a)** $$\begin{aligned}\frac{\sqrt{3}-1}{1+\sqrt{2}}&=\frac{(\sqrt{3}-1)(1-\sqrt{2})}{(1+\sqrt{2})(1-\sqrt{2})}\\&=\frac{\sqrt{3}-\sqrt{6}-1+\sqrt{2}}{1-2}\\&=-\sqrt{3}+\sqrt{6}+1-\sqrt{2}\end{aligned}$$

(b) $$\begin{aligned}\frac{3-i}{1+2i}&=\frac{(3-i)(1-2i)}{(1+2i)(1-2i)}\\&=\frac{3-6i-i+2i^2}{1-(2i)^2}\\&=\frac{3-7i+2(-1)}{1-4(-1)}\\&=\frac{3-7i-2}{1+4}\\&=\frac{1-7i}{5}\\&=\frac{1}{5}-\frac{7}{5}i\end{aligned}$$

63. The first step of parts (a) and (b) in Exercise 61 are comparable. After that we begin to simplify (b) since $i^2=-1$, and the answers no longer correspond.

64. Again, the first step of parts (a) and (b) in Exercise 62 are comparable. After that we begin to simplify (b) since $i^2=-1$, and the answers no longer correspond.

67. $i^{18}=i^{16}\cdot i^2=(i^4)^4\cdot i^2=1^4\cdot(-1)=-1$

71. $i^{96}=(i^4)^{24}=1^{24}=1$

75. $i^{-18}=i^{-18}\cdot i^{20}$, since
$i^{20}=(i^4)^5=1^5=1.$

The student multiplied by 1, which is justified by the identity property of multiplication.

79. To check that $1+5i$ is a solution to the equation, substitute $1+5i$ for x.

$$\begin{aligned}x^2-2x+26&=0\\(1+5i)^2-2(1+5i)+26&=0\quad ?\\(1+10i+25i^2)-2-10i+26&=0\quad ?\\1+10i-25-2-10i+26&=0\quad ?\\(1-25-2+26)+(10-10)i&=0\quad ?\\0&=0\qquad \textit{True}\end{aligned}$$

So, $1+5i$ is a solution of

$$x^2-2x+26=0.$$

Chapter 7 Review Exercises

1. Use a calculator. The square root of 1764 is 42, since $42^2=1764$.

2. Use a calculator. The square root of 289 is 17, since $17^2=289$. So,

$$-\sqrt{289}=-17.$$

3. $-\sqrt{-841}$ is not a real number, since the square of a real number is nonnegative.

4. $\sqrt[3]{216} = 6$, since $6^3 = 216$.

5. $\sqrt[3]{-125} = -5$, since $(-5)^3 = -125$.

6. $-\sqrt[3]{27z^{12}} = -\sqrt[3]{(3z^4)^3} = -3z^4$

7. $\sqrt[5]{-32} = -2$, since $(-2)^5 = -32$.

8. $\sqrt[n]{a}$ is not a real number if n is even and a is negative.

In Exercises 9-11, use a calculator, and round to three decimal places.

9. $\sqrt{40} \approx 6.325$

10. $\sqrt{77} \approx 8.775$

11. $\sqrt{310} \approx 17.607$

12. $16^{5/4} = (\sqrt[4]{16})^5 = 2^5 = 32$

13. $-8^{2/3} = -(\sqrt[3]{8})^2 = -(2)^2 = -4$

14. $-\left(\frac{36}{25}\right)^{3/2} = -\left(\sqrt{\frac{36}{25}}\right)^3 = -\left(\frac{6}{5}\right)^3 = -\frac{216}{125}$

15. $$\left(-\frac{1}{8}\right)^{-5/3} = (-8)^{5/3} = (\sqrt[3]{-8})^5$$
$$= (-2)^5 = -32$$

16. $$\left(\frac{81}{10,000}\right)^{-3/4} = \left(\frac{10,000}{81}\right)^{3/4}$$
$$= \left(\sqrt[4]{\frac{10,000}{81}}\right)^3$$
$$= \left(\frac{10}{3}\right)^3 = \frac{1000}{27}$$

17. $7^{1/3} \cdot 7^{5/3} = 7^{1/3+5/3} = 7^{6/3} = 7^2 = 49$

18. $\frac{96^{2/3}}{96^{-1/3}} = 96^{2/3-(-1/3)} = 96^1 = 96$

19. $$\frac{k^{2/3}k^{-1/2}k^{3/4}}{2(k^2)^{-1/4}} = \frac{k^{2/3-1/2+3/4}}{2k^{-2/4}}$$
$$= \frac{k^{8/12-6/12+9/12}}{2k^{-1/2}}$$
$$= \frac{k^{11/12}}{2k^{-6/12}} = \frac{k^{17/12}}{2}$$

20. The expression with fractional exponents, $a^{m/n}$, is equivalent to the radical expression, $\sqrt[n]{a^m}$. The denominator of the exponent is the index of the radical.

21. $\sqrt{3^{18}} = 3^{18/2} = 3^9$

22. $$\sqrt{7^9} = 7^{9/2} = 7^{8/2+1/2} = 7^{4+1/2}$$
$$= 7^4 \cdot 7^{1/2} = 7^4\sqrt{7}$$

23. $$\sqrt[3]{m^5} \cdot \sqrt[3]{m^8} = m^{5/3} \cdot m^{8/3} = m^{5/3+8/3}$$
$$= m^{13/3} = m^{4+1/3}$$
$$= m^4 \cdot m^{1/3} = m^4\sqrt[3]{m}$$

24. $$\sqrt[4]{k^2} \cdot \sqrt[4]{k^7} = k^{2/4} \cdot k^{7/4} = k^{2/4+7/4}$$
$$= k^{9/4} = k^{2+1/4} = k^2 \cdot k^{1/4}$$
$$= k^2\sqrt[4]{k}$$

25. $\sqrt{6} \cdot \sqrt{11} = \sqrt{6 \cdot 11} = \sqrt{66}$

26. $\sqrt{5} \cdot \sqrt{r} = \sqrt{5 \cdot r} = \sqrt{5r}$

27. $\sqrt[3]{6} \cdot \sqrt[3]{5} = \sqrt[3]{6 \cdot 5} = \sqrt[3]{30}$

28. $\sqrt[4]{7} \cdot \sqrt[4]{3} = \sqrt[4]{7 \cdot 3} = \sqrt[4]{21}$

29. $\sqrt{20} = \sqrt{4 \cdot 5} = 2\sqrt{5}$

30. $\sqrt{75} = \sqrt{25 \cdot 3} = 5\sqrt{3}$

31. $-\sqrt{125} = -\sqrt{25 \cdot 5} = -5\sqrt{5}$

32. $\sqrt[3]{-108} = \sqrt[3]{-27 \cdot 4} = -3\sqrt[3]{4}$

33. $\sqrt{100y^7} = \sqrt{100y^6 \cdot y} = 10y^3\sqrt{y}$

34. $\sqrt[3]{64p^4q^6} = \sqrt[3]{64p^3q^6 \cdot p} = 4pq^2\sqrt[3]{p}$

35. $\sqrt{\frac{49}{81}} = \frac{\sqrt{49}}{\sqrt{81}} = \frac{7}{9}$

36. $\sqrt{\frac{y^3}{144}} = \frac{\sqrt{y^3}}{\sqrt{144}} = \frac{\sqrt{y^2 \cdot y}}{12} = \frac{y\sqrt{y}}{12}$

37. $\sqrt[3]{\frac{m^{15}}{27}} = \frac{\sqrt[3]{m^{15}}}{\sqrt[3]{27}} = \frac{m^5}{3}$

38. $\sqrt[3]{\frac{r^2}{8}} = \frac{\sqrt[3]{r^2}}{\sqrt[3]{8}} = \frac{\sqrt[3]{r^2}}{2}$

In Exercises 39 and 40, use the distance formula

$$d = \sqrt{(x_2 - x_1)^2 + (y_2 - y_1)^2}.$$

39. The distance between $(2, 7)$ and $(-1, -4)$ is

$$\begin{aligned} d &= \sqrt{[2-(-1)]^2 + [7-(-4)]^2} \\ &= \sqrt{3^2 + 11^2} \\ &= \sqrt{9 + 121} \\ &= \sqrt{130}. \end{aligned}$$

40. The distance between $(-3, -5)$ and $(4, -3)$ is

$$\begin{aligned} d &= \sqrt{(-3-4)^2 + [-5-(-3)]^2} \\ &= \sqrt{(-7)^2 + (-2)^2} \\ &= \sqrt{49 + 4} \\ &= \sqrt{53}. \end{aligned}$$

41. $$\begin{aligned} 2\sqrt{8} - 3\sqrt{50} &= 2\sqrt{4 \cdot 2} - 3\sqrt{25 \cdot 2} \\ &= 2 \cdot 2\sqrt{2} - 3 \cdot 5\sqrt{2} \\ &= 4\sqrt{2} - 15\sqrt{2} = -11\sqrt{2} \end{aligned}$$

42. $$\begin{aligned} 8\sqrt{80} - 3\sqrt{45} &= 8\sqrt{16 \cdot 5} - 3\sqrt{9 \cdot 5} \\ &= 8 \cdot 4\sqrt{5} - 3 \cdot 3\sqrt{5} \\ &= 32\sqrt{5} - 9\sqrt{5} = 23\sqrt{5} \end{aligned}$$

43. $$\begin{aligned} -\sqrt{27y} + 2\sqrt{75y} &= -\sqrt{9 \cdot 3y} + 2\sqrt{25 \cdot 3y} \\ &= -3\sqrt{3y} + 2 \cdot 5\sqrt{3y} \\ &= -3\sqrt{3y} + 10\sqrt{3y} \\ &= 7\sqrt{3y} \end{aligned}$$

44. $$\begin{aligned} &2\sqrt{54m^3} + 5\sqrt{96m^3} \\ &= 2\sqrt{9m^2 \cdot 6m} + 5\sqrt{16m^2 \cdot 6m} \\ &= 2 \cdot 3m\sqrt{6m} + 5 \cdot 4m\sqrt{6m} \\ &= 6m\sqrt{6m} + 20m\sqrt{6m} \\ &= 26m\sqrt{6m} \end{aligned}$$

45. $$\begin{aligned} 3\sqrt[3]{54} + 5\sqrt[3]{16} &= 3\sqrt[3]{27 \cdot 2} + 5\sqrt[3]{8 \cdot 2} \\ &= 3 \cdot 3\sqrt[3]{2} + 5 \cdot 2\sqrt[3]{2} \\ &= 9\sqrt[3]{2} + 10\sqrt[3]{2} = 19\sqrt[3]{2} \end{aligned}$$

46. $$\begin{aligned} -6\sqrt[4]{32} + \sqrt[4]{512} &= -6\sqrt[4]{16 \cdot 2} + \sqrt[4]{256 \cdot 2} \\ &= -6 \cdot 2\sqrt[4]{2} + 4\sqrt[4]{2} \\ &= -12\sqrt[4]{2} + 4\sqrt[4]{2} \\ &= -8\sqrt[4]{2} \end{aligned}$$

47. $$\begin{aligned} (\sqrt{3} + 1)(\sqrt{3} - 2) &= 3 - 2\sqrt{3} + \sqrt{3} - 2 \\ &= 1 - \sqrt{3} \end{aligned}$$

48. $$\begin{aligned} &(\sqrt{7} + \sqrt{5})(\sqrt{7} - \sqrt{5}) \\ &= 7 - \sqrt{35} + \sqrt{35} - 5 \\ &= 2 \end{aligned}$$

49. $$\begin{aligned} &(3\sqrt{2} + 1)(2\sqrt{2} - 3) \\ &= 6 \cdot 2 - 9\sqrt{2} + 2\sqrt{2} - 3 \\ &= 12 - 7\sqrt{2} - 3 = 9 - 7\sqrt{2} \end{aligned}$$

50. $$\begin{aligned} &(\sqrt{11} + 3\sqrt{5})(\sqrt{11} + 5\sqrt{5}) \\ &= 11 + 5\sqrt{55} + 3\sqrt{55} + 15 \cdot 5 \\ &= 11 + 8\sqrt{55} + 75 \\ &= 86 + 8\sqrt{55} \end{aligned}$$

51. $$\begin{aligned} (\sqrt{13} - \sqrt{2})^2 &= (\sqrt{13} - \sqrt{2})(\sqrt{13} - \sqrt{2}) \\ &= 13 - \sqrt{26} - \sqrt{26} + 2 \\ &= 15 - 2\sqrt{26} \end{aligned}$$

52. $$\begin{aligned} (\sqrt{5} - \sqrt{7})^2 &= (\sqrt{5} - \sqrt{7})(\sqrt{5} - \sqrt{7}) \\ &= 5 - \sqrt{35} - \sqrt{35} + 7 \\ &= 12 - 2\sqrt{35} \end{aligned}$$

53. $12 - 2\sqrt{35} \approx .1678404338$
$10\sqrt{35} \approx 59.16079783$

So, $12 - 2\sqrt{35} \neq 10\sqrt{35}$.

54. $$\frac{\sqrt{6}}{\sqrt{5}} = \frac{\sqrt{6} \cdot \sqrt{5}}{\sqrt{5} \cdot \sqrt{5}} = \frac{\sqrt{30}}{5}$$

55. $$\frac{-6\sqrt{3}}{\sqrt{2}} = \frac{-6\sqrt{3} \cdot \sqrt{2}}{\sqrt{2} \cdot \sqrt{2}} = \frac{-6\sqrt{6}}{2} = -3\sqrt{6}$$

56. $$\frac{3\sqrt{7p}}{\sqrt{y}} = \frac{3\sqrt{7p} \cdot \sqrt{y}}{\sqrt{y} \cdot \sqrt{y}} = \frac{3\sqrt{7py}}{y}$$

57. $$\begin{aligned} \sqrt{\frac{11}{8}} &= \frac{\sqrt{11}}{\sqrt{8}} = \frac{\sqrt{11}}{\sqrt{4 \cdot 2}} = \frac{\sqrt{11}}{2\sqrt{2}} = \frac{\sqrt{11} \cdot \sqrt{2}}{2\sqrt{2} \cdot \sqrt{2}} \\ &= \frac{\sqrt{22}}{2 \cdot 2} = \frac{\sqrt{22}}{4} \end{aligned}$$

58. $$\begin{aligned} -\sqrt[3]{\frac{9}{25}} &= -\frac{\sqrt[3]{9}}{\sqrt[3]{5^2}} = -\frac{\sqrt[3]{9} \cdot \sqrt[3]{5}}{\sqrt[3]{5^2} \cdot \sqrt[3]{5}} = -\frac{\sqrt[3]{45}}{\sqrt[3]{5^3}} \\ &= -\frac{\sqrt[3]{45}}{5} \end{aligned}$$

59. $$\begin{aligned} \sqrt[3]{\frac{108m^3}{n^5}} &= \frac{\sqrt[3]{108m^3}}{\sqrt[3]{n^5}} = \frac{\sqrt[3]{27m^3 \cdot 4}}{\sqrt[3]{n^3} \cdot \sqrt[3]{n^2}} = \frac{3m\sqrt[3]{4}}{n\sqrt[3]{n^2}} \\ &= \frac{3m\sqrt[3]{4} \cdot \sqrt[3]{n}}{n\sqrt[3]{n^2} \cdot \sqrt[3]{n}} = \frac{3m\sqrt[3]{4n}}{n \cdot n} = \frac{3m\sqrt[3]{4n}}{n^2} \end{aligned}$$

60. $$\frac{1}{\sqrt{2} + \sqrt{7}}$$

Multiply by the conjugate of the denominator, $\sqrt{2} - \sqrt{7}$.

$$\begin{aligned} &= \frac{1(\sqrt{2} - \sqrt{7})}{(\sqrt{2} + \sqrt{7})(\sqrt{2} - \sqrt{7})} \\ &= \frac{\sqrt{2} - \sqrt{7}}{2 - 7} = \frac{\sqrt{2} - \sqrt{7}}{-5} \end{aligned}$$

61. $\dfrac{-5}{\sqrt{6}-\sqrt{3}}$

Multiply by the conjugate of the denominator, $\sqrt{6}+\sqrt{3}$.

$$= \frac{-5(\sqrt{6}+\sqrt{3})}{(\sqrt{6}-\sqrt{3})(\sqrt{6}+\sqrt{3})}$$
$$= \frac{-5(\sqrt{6}+\sqrt{3})}{6-3}$$
$$= \frac{-5(\sqrt{6}+\sqrt{3})}{3}$$

62. $\sqrt{8y+9}=5$

Square both sides.

$$(\sqrt{8y+9})^2 = 5^2$$
$$8y+9=25$$
$$8y=16$$
$$y=2$$

Check 2 in the original equation.

$$\sqrt{8(2)+9}=5 \quad ?$$
$$\sqrt{16+9}=5 \quad ?$$
$$\sqrt{25}=5 \quad ?$$
$$5=5 \qquad \textit{True}$$

Solution set: $\{2\}$

63. $\sqrt{2z-3}-3=0$
$\sqrt{2z-3}=3$

Square both sides.

$$(\sqrt{2z-3})^2=3^2$$
$$2z-3=9$$
$$2z=12$$
$$z=6$$

Check 6 in the original equation.

$$\sqrt{2(6)-3}-3=0 \quad ?$$
$$\sqrt{9}-3=0 \quad ?$$
$$3-3=0 \quad ?$$
$$0=0 \qquad \textit{True}$$

Solution set: $\{6\}$

64. $\sqrt{3m+1}=-1$

No real number has a negative principal square root.

Solution set: $\emptyset$

65. $\sqrt{7z+1}=z+1$

Square both sides.

$$(\sqrt{7z+1})^2=(z+1)^2$$
$$7z+1=z^2+2z+1$$
$$0=z^2-5z$$
$$0=z(z-5)$$
$$z=0 \quad \text{or} \quad z-5=0$$
$$z=5$$

Check 0.

$$\sqrt{7(0)+1}=0+1 \quad ?$$
$$\sqrt{1}=1 \quad ?$$
$$1=1 \qquad \textit{True}$$

Check 5.

$$\sqrt{7(5)+1}=5+1 \quad ?$$
$$\sqrt{36}=6 \quad ?$$
$$6=6 \qquad \textit{True}$$

Solution set: $\{0,5\}$

66. $3\sqrt{m}=\sqrt{10m-9}$

Square both sides.

$$(3\sqrt{m})^2=(\sqrt{10m-9})^2$$
$$9m=10m-9$$
$$-m=-9$$
$$m=9$$

Check 9.

$$3\sqrt{9}=\sqrt{10(9)-9} \quad ?$$
$$3\cdot 3=\sqrt{90-9} \quad ?$$
$$9=\sqrt{81} \quad ?$$
$$9=9 \qquad \textit{True}$$

Solution set: $\{9\}$

67. $\sqrt{p^2+3p+7}=p+2$

Square both sides.

$$(\sqrt{p^2+3p+7})^2=(p+2)^2$$
$$p^2+3p+7=p^2+4p+4$$
$$-p=-3$$
$$p=3$$

Check 3.

$$\sqrt{3^2+3(3)+7}=3+2 \quad ?$$
$$\sqrt{9+9+7}=5 \quad ?$$
$$\sqrt{25}=5 \quad ?$$
$$5=5 \qquad \textit{True}$$

Solution set: $\{3\}$

68. $\sqrt{a+2} - \sqrt{a-3} = 1$

Get one radical on each side of the equals sign.

$$\sqrt{a+2} = 1 + \sqrt{a-3}$$

Square both sides.

$$\begin{aligned}(\sqrt{a+2})^2 &= (1+\sqrt{a-3})^2 \\ a+2 &= 1 + 2\sqrt{a-3} + a - 3 \\ a+2 &= a - 2 + 2\sqrt{a-3} \\ 4 &= 2\sqrt{a-3} \\ 2 &= \sqrt{a-3}\end{aligned}$$

Square both sides again.

$$\begin{aligned}2^2 &= (\sqrt{a-3})^2 \\ 4 &= a-3 \\ 7 &= a\end{aligned}$$

Check 7.

$$\begin{aligned}\sqrt{7+2} - \sqrt{7-3} &= 1 \quad ? \\ \sqrt{9} - \sqrt{4} &= 1 \quad ? \\ 3 - 2 &= 1 \quad ? \\ 1 &= 1 \qquad \textit{True}\end{aligned}$$

Solution set: $\{7\}$

69. $\sqrt[3]{5m-1} = \sqrt[3]{3m-2}$

Cube both sides.

$$\begin{aligned}(\sqrt[3]{5m-1})^3 &= (\sqrt[3]{3m-2})^3 \\ 5m - 1 &= 3m - 2 \\ 2m &= -1 \\ m &= -\frac{1}{2}\end{aligned}$$

Check $-\frac{1}{2}$.

$$\begin{aligned}\sqrt[3]{5\left(-\frac{1}{2}\right) - 1} &= \sqrt[3]{3\left(-\frac{1}{2}\right) - 2} \quad ? \\ \sqrt[3]{-\frac{5}{2} - \frac{2}{2}} &= \sqrt[3]{-\frac{3}{2} - \frac{4}{2}} \quad ? \\ \sqrt[3]{-\frac{7}{2}} &= \sqrt[3]{-\frac{7}{2}} \qquad \textit{True}\end{aligned}$$

Solution set: $\{-\frac{1}{2}\}$

70. $\sqrt[4]{b+6} = \sqrt[4]{2b}$

Raise both sides to the fourth power.

$$\begin{aligned}(\sqrt[4]{b+6})^4 &= (\sqrt[4]{2b})^4 \\ b+6 &= 2b \\ -b &= -6 \\ b &= 6\end{aligned}$$

Check 6.

$$\begin{aligned}\sqrt[4]{6+6} &= \sqrt[4]{2(6)} \quad ? \\ \sqrt[4]{12} &= \sqrt[4]{12} \qquad \textit{True}\end{aligned}$$

Solution set: $\{6\}$

71. $\sqrt{-25} = i\sqrt{25} = 5i$

72. $\sqrt{-200} = i\sqrt{100 \cdot 2} = 10i\sqrt{2}$

73. If a is a positive real number, then $-a$ is negative. So, $\sqrt{-a}$ is not a real number. Therefore, $-\sqrt{-a}$ is not a real number either.

74. $$\begin{aligned}&(-2+5i) + (-8-7i) \\ &= [-2 + (-8)] + [5 + (-7)]i \\ &= -10 - 2i\end{aligned}$$

75. $$\begin{aligned}&(5+4i) - (-9-3i) \\ &= [5 - (-9)] + [4 - (-3)]i \\ &= 14 + 7i\end{aligned}$$

76. $$\begin{aligned}\sqrt{-5} \cdot \sqrt{-7} &= i\sqrt{5} \cdot i\sqrt{7} = i^2\sqrt{35} \\ &= -1(\sqrt{35}) = -\sqrt{35}\end{aligned}$$

77. $\sqrt{-25} \cdot \sqrt{-81} = 5i \cdot 9i = 45i^2 = 45(-1) = -45$

78. $\dfrac{\sqrt{-72}}{\sqrt{-8}} = \dfrac{i\sqrt{72}}{i\sqrt{8}} = \sqrt{\dfrac{72}{8}} = \sqrt{9} = 3$

79. $$\begin{aligned}(2+3i)(1-i) &= 2 - 2i + 3i - 3i^2 \\ &= 2 + i - 3(-1) \\ &= 2 + i + 3 \\ &= 5 + i\end{aligned}$$

80. $$\begin{aligned}(6-2i)^2 &= (6-2i)(6-2i) \\ &= 36 - 12i - 12i + 4i^2 \\ &= 36 - 24i + 4(-1) \\ &= 36 - 24i - 4 \\ &= 32 - 24i\end{aligned}$$

81. $\dfrac{3-i}{2+i}$

Multiply by the conjugate of the denominator, $2-i$.

$$= \frac{(3-i)(2-i)}{(2+i)(2-i)}$$
$$= \frac{6-3i-2i+i^2}{4-i^2}$$
$$= \frac{6-5i-1}{4-(-1)} = \frac{5-5i}{5}$$
$$= \frac{5(1-i)}{5} = 1-i$$

82. $\dfrac{5+14i}{2+3i}$

Multiply by the conjugate of the denominator, $2-3i$.

$$= \frac{(5+14i)(2-3i)}{(2+3i)(2-3i)}$$
$$= \frac{10-15i+28i-42i^2}{4-9i^2}$$
$$= \frac{10+13i-42(-1)}{4-9(-1)} = \frac{52+13i}{13}$$
$$= \frac{13(4+i)}{13} = 4+i$$

83. $i^{11} = i^8 \cdot i^3 = (i^4)^2 \cdot i^3 = 1^2 \cdot (-i) = -i$

84. $i^{52} = (i^4)^{13} = 1^{13} = 1$

85. $i^{-13} = i^{-13} \cdot i^{16}$, since $i^{16} = (i^4)^4 = 1^4 = 1$.

So,
$$i^{-13} = i^{-13} \cdot i^{16} = i^3 = -i.$$

86. $-\sqrt[4]{256} = -\sqrt[4]{4^4} = -4$

87. $-\sqrt{169a^2b^4} = -\sqrt{(13ab^2)^2} = -13ab^2$

88. $1000^{-2/3} = \dfrac{1}{1000^{2/3}} = \dfrac{1}{(\sqrt[3]{1000})^2}$
$$= \frac{1}{10^2} = \frac{1}{100}$$

89. $\dfrac{y^{-1/3} \cdot y^{5/6}}{y} = \dfrac{y^{-2/6+5/6}}{y} = \dfrac{y^{3/6}}{y}$
$$= \frac{y^{1/2}}{y} = y^{1/2-1}$$
$$= y^{-1/2} = \frac{1}{y^{1/2}}$$

90. $\dfrac{z^{-1/4}x^{1/2}}{z^{1/2}x^{-1/4}}$
$$= z^{-1/4-1/2} \cdot x^{1/2-(-1/4)}$$
$$= z^{-1/4-2/4} \cdot x^{2/4-(-1/4)}$$
$$= z^{-3/4} \cdot x^{3/4} = \frac{x^{3/4}}{z^{3/4}}$$

91. $\sqrt[4]{k^{24}} = k^{24/4} = k^6$

92. $\sqrt[3]{54z^9t^8} = \sqrt[3]{27z^9t^6 \cdot 2t^2} = 3z^3t^2\sqrt[3]{2t^2}$

93. $-5\sqrt{18} + 12\sqrt{72} = -5\sqrt{9\cdot 2} + 12\sqrt{36\cdot 2}$
$$= -5\cdot 3\sqrt{2} + 12\cdot 6\sqrt{2}$$
$$= -15\sqrt{2} + 72\sqrt{2} = 57\sqrt{2}$$

94. $8\sqrt[3]{x^3y^2} - 2x\sqrt[3]{y^2} = 8x\sqrt[3]{y^2} - 2x\sqrt[3]{y^2} = 6x\sqrt[3]{y^2}$

95. $(\sqrt{5}-\sqrt{3})(\sqrt{7}+\sqrt{3}) = \sqrt{35}+\sqrt{15}-\sqrt{21}-3$

96. $\dfrac{-1}{\sqrt{12}} = \dfrac{-1}{\sqrt{4\cdot 3}} = \dfrac{-1}{2\sqrt{3}} = \dfrac{-1\cdot\sqrt{3}}{2\sqrt{3}\cdot\sqrt{3}}$
$$= \frac{-\sqrt{3}}{2\cdot 3} = -\frac{\sqrt{3}}{6}$$

97. $\sqrt[3]{\dfrac{12}{25}} = \dfrac{\sqrt[3]{12}}{\sqrt[3]{25}} = \dfrac{\sqrt[3]{12}}{\sqrt[3]{5^2}} = \dfrac{\sqrt[3]{12}\cdot\sqrt[3]{5}}{\sqrt[3]{5^2}\cdot\sqrt[3]{5}}$
$$= \frac{\sqrt[3]{60}}{\sqrt[3]{5^3}} = \frac{\sqrt[3]{60}}{5}$$

98. $\dfrac{2\sqrt{z}}{\sqrt{z}-2}$

Multiply by the conjugate of the denominator, $\sqrt{z}+2$.

$$= \frac{2\sqrt{z}(\sqrt{z}+2)}{(\sqrt{z}-2)(\sqrt{z}+2)}$$
$$= \frac{2\sqrt{z}(\sqrt{z}+2)}{z-4}$$

99. $\sqrt{-49} = i\sqrt{49} = 7i$

100. $(4-9i)+(-1+2i)$
$$= (4-1)+(-9+2)i$$
$$= 3-7i$$

101. $\dfrac{\sqrt{50}}{\sqrt{-2}} = \dfrac{\sqrt{25\cdot 2}}{i\sqrt{2}} = \dfrac{5\sqrt{2}}{i\sqrt{2}} = \dfrac{5}{i}$

The conjugate of i is $-i$.

$$\frac{5(-i)}{i(-i)} = \frac{-5i}{-i^2} = \frac{-5i}{-1(-1)} = -5i$$

102. $i^{-1000} = (i^4)^{-250} = (1)^{-250} = 1$

103. $\sqrt{x+4} = x - 2$

Square both sides.

$$\begin{aligned}(\sqrt{x+4})^2 &= (x-2)^2\\ x+4 &= x^2 - 4x + 4\\ 0 &= x^2 - 5x\\ 0 &= x(x-5)\\ x = 0 \quad \text{or} \quad x - 5 &= 0\\ x &= 5\end{aligned}$$

Check 0.

$$\begin{aligned}\sqrt{0+4} &= 0 - 2 \quad ?\\ \sqrt{4} &= -2 \quad ?\\ 2 &= -2 \qquad \textit{False}\end{aligned}$$

Check 5.

$$\begin{aligned}\sqrt{5+4} &= 5 - 2 \quad ?\\ \sqrt{9} &= 3 \quad ?\\ 3 &= 3 \qquad \textit{True}\end{aligned}$$

Solution set: $\{5\}$

104. $\sqrt{6+2y} - 1 = \sqrt{7-2y}$

Square both sides.

$$\begin{aligned}(\sqrt{6+2y} - 1)^2 &= (\sqrt{7-2y})^2\\ 6 + 2y - 2\sqrt{6+2y} + 1 &= 7 - 2y\\ 7 + 2y - 2\sqrt{6+2y} &= 7 - 2y\\ 4y &= 2\sqrt{6+2y}\\ 2y &= \sqrt{6+2y}\end{aligned}$$

Square both sides again.

$$\begin{aligned}(2y)^2 &= (\sqrt{6+2y})^2\\ 4y^2 &= 6 + 2y\\ 4y^2 - 2y - 6 &= 0\\ 2y^2 - y - 3 &= 0\\ (2y-3)(y+1) &= 0\\ 2y - 3 = 0 \quad &\text{or} \quad y + 1 = 0\\ 2y &= 3\\ y = \frac{3}{2} \quad &\text{or} \quad y = -1\end{aligned}$$

Check $\frac{3}{2}$.

$$\begin{aligned}\sqrt{6 + 2\left(\frac{3}{2}\right)} - 1 &= \sqrt{7 - 2\left(\frac{3}{2}\right)} \quad ?\\ \sqrt{6+3} - 1 &= \sqrt{7-3} \quad ?\\ \sqrt{9} - 1 &= \sqrt{4} \quad ?\\ 3 - 1 &= 2 \quad ?\\ 2 &= 2 \qquad \textit{True}\end{aligned}$$

Check -1.

$$\begin{aligned}\sqrt{6 + 2(-1)} - 1 &= \sqrt{7 - 2(-1)} \quad ?\\ \sqrt{6-2} - 1 &= \sqrt{7+2} \quad ?\\ \sqrt{4} - 1 &= \sqrt{9} \quad ?\\ 2 - 1 &= 3 \quad ?\\ 1 &= 3 \qquad \textit{False}\end{aligned}$$

Solution set: $\{\frac{3}{2}\}$

105. The distance between $(6, -2)$ and $(5, 8)$ is

$$\begin{aligned}d &= \sqrt{(x_2 - x_1)^2 + (y_2 - y_1)^2}\\ &= \sqrt{(6-5)^2 + (-2-8)^2}\\ &= \sqrt{1^2 + (-10)^2}\\ &= \sqrt{1 + 100}\\ &= \sqrt{101}.\end{aligned}$$

106. Substitute 12 for L and 9 for W in the formula.

$$\begin{aligned}L &= \sqrt{H^2 + W^2}\\ 12 &= \sqrt{H^2 + 9^2}\end{aligned}$$

Square both sides.

$$\begin{aligned}12^2 &= (\sqrt{H^2 + 9^2})^2\\ 144 &= H^2 + 81\\ 63 &= H^2\\ H &= \pm\sqrt{63} \approx \pm 7.9\end{aligned}$$

Since H must be positive, -7.9 cannot be a solution. The height is about 7.9 ft.

107. Since $H = 28 + 6 = 34$ ft, substitute 34 for H in the function.

$$D(H) = (2H)^{1/2} = \sqrt{2 \cdot 34} = \sqrt{68} \approx 8.2$$

She will be able to see about 8.2 mi.

Chapter 7 Test

1. Use a calculator. The square root of 841 is 29, since $29^2 = 841$. So, $-\sqrt{841} = -29$.

2. Use a calculator.

$$\sqrt[3]{3375} = 15$$

3. Since $12^2 = 144$, and $13^2 = 169$, the closest estimate for $\sqrt{146.25}$ is 12. The answer is (c).

4. Use a calculator, and round to the nearest hundredth.

$$\sqrt{146.25} \approx 12.09$$

5. $$(-64)^{-4/3} = \frac{1}{(-64)^{4/3}} = \frac{1}{(\sqrt[3]{-64})^4}$$
$$= \frac{1}{(-4)^4} = \frac{1}{256}$$

6. $$\frac{3^{2/5}x^{-1/4}y^{2/5}}{3^{-8/5}x^{7/4}y^{1/10}}$$
$$= 3^{2/5-(-8/5)}x^{-1/4-7/4}y^{2/5-1/10}$$
$$= 3^{10/5}x^{-8/4}y^{4/10-1/10}$$
$$= 3^2x^{-2}y^{3/10} = \frac{9y^{3/10}}{x^2}$$

7. $\sqrt{54x^5y^6} = \sqrt{9x^4y^6 \cdot 6x} = 3x^2y^3\sqrt{6x}$

8. $$\sqrt[4]{32a^7b^{13}} = \sqrt[4]{16a^4b^{12} \cdot 2a^3b}$$
$$= 2ab^3\sqrt[4]{2a^3b}$$

9. $$\sqrt{2} \cdot \sqrt[3]{5} = 2^{1/2} \cdot 5^{1/3} = 2^{3/6} \cdot 5^{2/6}$$
$$= \sqrt[6]{2^3 \cdot 5^2} = \sqrt[6]{8 \cdot 25} = \sqrt[6]{200}$$

10. $$3\sqrt{20} - 5\sqrt{80} + 4\sqrt{500}$$
$$= 3\sqrt{4 \cdot 5} - 5\sqrt{16 \cdot 5} + 4\sqrt{100 \cdot 5}$$
$$= 3 \cdot 2\sqrt{5} - 5 \cdot 4\sqrt{5} + 4 \cdot 10\sqrt{5}$$
$$= 6\sqrt{5} - 20\sqrt{5} + 40\sqrt{5} = 26\sqrt{5}$$

11. $$(7\sqrt{5} + 4)(2\sqrt{5} - 1)$$
$$= 14 \cdot 5 - 7\sqrt{5} + 8\sqrt{5} - 4$$
$$= 70 + \sqrt{5} - 4 = 66 + \sqrt{5}$$

12. $$\frac{-4}{\sqrt{7} + \sqrt{5}}$$

Multiply by the conjugate of the denominator, $\sqrt{7} - \sqrt{5}$.

$$= \frac{-4(\sqrt{7} - \sqrt{5})}{(\sqrt{7} + \sqrt{5})(\sqrt{7} - \sqrt{5})}$$
$$= \frac{-4(\sqrt{7} - \sqrt{5})}{7 - 5}$$
$$= \frac{-4(\sqrt{7} - \sqrt{5})}{2}$$
$$= -2(\sqrt{7} - \sqrt{5})$$

13. $$\frac{-5}{\sqrt{40}} = \frac{-5}{\sqrt{4 \cdot 10}} = \frac{-5}{2\sqrt{10}} = \frac{-5 \cdot \sqrt{10}}{2\sqrt{10} \cdot \sqrt{10}}$$
$$= \frac{-5\sqrt{10}}{2 \cdot 10} = \frac{-5\sqrt{10}}{20} = \frac{-\sqrt{10}}{4}$$

14. $$\frac{2}{\sqrt[3]{5}} = \frac{2}{\sqrt[3]{5}} \cdot \frac{\sqrt[3]{25}}{\sqrt[3]{25}} = \frac{2\sqrt[3]{25}}{5}$$

15. The distance between $(-3, 8)$ and $(2, 7)$ is

$$d = \sqrt{(x_2 - x_1)^2 + (y_2 - y_1)^2}$$
$$= \sqrt{(-3 - 2)^2 + (8 - 7)^2}$$
$$= \sqrt{(-5)^2 + 1^2}$$
$$= \sqrt{26}.$$

16. Substitute 50 for V_0, .01 for k, and 30 for T in the function.

$$V(T) = \frac{V_0}{\sqrt{1 - kT}}$$
$$V(30) = \frac{50}{\sqrt{1 - .01(30)}}$$
$$= \frac{50}{\sqrt{1 - .3}}$$
$$= \frac{50}{\sqrt{.7}} \approx 59.8$$

The velocity is about 59.8.

17. $x - 5 = \sqrt{7 - x}$

Square both sides.

$$(x - 5)^2 = (\sqrt{7 - x})^2$$
$$x^2 - 10x + 25 = 7 - x$$
$$x^2 - 9x + 18 = 0$$
$$(x - 3)(x - 6) = 0$$
$$x - 3 = 0 \quad \text{or} \quad x - 6 = 0$$
$$x = 3 \quad \text{or} \quad x = 6$$

Check 3.

$$3 - 5 = \sqrt{7 - 3} \quad ?$$
$$-2 = \sqrt{4} \quad ?$$
$$-2 = 2 \quad \textit{False}$$

Check 6.

$$6 - 5 = \sqrt{7 - 6} \quad ?$$
$$1 = \sqrt{1} \quad ?$$
$$1 = 1 \quad \textit{True}$$

Solution set: $\{6\}$

18. $$(-2 + 5i) - (3 + 6i) - 7i$$
$$= (-2 - 3) + (5 - 6 - 7)i$$
$$= -5 - 8i$$

19. $\dfrac{7+i}{1-i}$

Multiply by the conjugate of the denominator, $1+i$.

$$= \frac{(7+i)(1+i)}{(1-i)(1+i)}$$
$$= \frac{7+7i+i+i^2}{1-i^2}$$
$$= \frac{7+8i-1}{1-(-1)}$$
$$= \frac{6+8i}{2}$$
$$= \frac{2(3+4i)}{2} = 3+4i$$

20. $i^{35} = i^{32} \cdot i^3 = (i^4)^8 \cdot i^3 = 1^8 \cdot i^3$
$= 1 \cdot i^3 = -i$

Cumulative Review Exercises (Chapters 1-7)

1. $7-(4+3t)+2t = -6(t-2)-5$
$$\begin{aligned} 7-4-3t+2t &= -6t+12-5 \\ 3-t &= -6t+7 \\ 5t &= 4 \\ t &= \frac{4}{5} \end{aligned}$$

Solution set: $\{\frac{4}{5}\}$

2. $|6x-9| = |-4x+2|$

$6x-9 = -4x+2$ or $6x-9 = -(-4x+2)$
$6x-9 = 4x-2$
$10x = 11$ $\quad$ $2x = 7$
$x = \frac{11}{10}$ or $x = \frac{7}{2}$

Solution set: $\{\frac{11}{10}, \frac{7}{2}\}$

3. $-5-3(m-2) < 11-2(m+2)$
$$\begin{aligned} -5-3m+6 &< 11-2m-4 \\ 1-3m &< 7-2m \\ -m &< 6 \end{aligned}$$

Multiply by -1; reverse the inequality.

$$m > -6$$

Solution set: $(-6, \infty)$

4. $1+4x > 5$ and $-2x > -6$
$4x > 4$
$x > 1$ and $x < 3$

Solution set: $(1,3)$

5. $-2 < 1-3y < 7$
$-3 < -3y < 6$
$1 > y > -2$ or $-2 < y < 1$

Solution set: $(-2,1)$

6. To write an equation of the line through $(-4,6)$ and $(7,-6)$, first find the slope.

$$m = \frac{-6-6}{7-(-4)} = \frac{-12}{11} = -\frac{12}{11}$$

Use the point-slope form with $(x_1, y_1) = (-4,6)$ and $m = -\frac{12}{11}$.

$$\begin{aligned} y-y_1 &= m(x-x_1) \\ y-6 &= -\frac{12}{11}[x-(-4)] \end{aligned}$$

Multiply by 11.

$$\begin{aligned} 11y-66 &= -12(x+4) \\ 11y-66 &= -12x-48 \\ 12x+11y &= 18 \end{aligned}$$

7. $2x+3y = 8$ and $6y = 4x+16$

Find the slope of each line by writing them in slope-intercept form.

$$\begin{aligned} 2x+3y &= 8 \\ 3y &= -2x+8 \\ y &= -\frac{2}{3}x+\frac{8}{3} \end{aligned}$$

The slope of this line is $-\frac{2}{3}$.

$$\begin{aligned} 6y &= 4x+16 \\ y &= \frac{4}{6}x+\frac{16}{6} \\ \text{or} \quad y &= \frac{2}{3}x+\frac{8}{3} \end{aligned}$$

The slope of this line is $\frac{2}{3}$.

The slopes are not the same, so the lines are not parallel. The slopes are not negative reciprocals, so the lines are not perpendicular. Therefore, the answer is (c) neither.

8. $f(x) = -3x + 6$

Replace $f(x)$ with y to get an equation in slope-intercept form.

$$y = -3x + 6$$

(a) From the equation, the y-intercept is $(0, 6)$.

(b) Let $y = 0$ to find the x-intercept.

$$0 = -3x + 6$$
$$3x = 6$$
$$x = 2$$

The x-intercept is $(2, 0)$.

9. Let $c =$ cost per item and
$n =$ number of items manufactured.

c varies inversely as n, so $c = \frac{k}{n}$ for some constant k. Substitute $c = 200$ and $n = 1500$ in the equation and solve for k.

$$\begin{aligned} c &= \frac{k}{n} \\ 200 &= \frac{k}{1500} \\ 300,000 &= k \end{aligned}$$

So, $c = \frac{300,000}{n}$. When $n = 2500$,

$$\begin{aligned} c &= \frac{300,000}{2500} \\ c &= 120. \end{aligned}$$

It will cost \$120 per item.

10. To graph $-2x + y < -6$, graph the line $-2x + y = -6$, which has intercepts $(3, 0)$ and $(0, -6)$, as a dashed line since the inequality involves " $<$." Test $(0, 0)$, which yields $0 < -6$, a false statement. Shade the region that does not include $(0, 0)$. See the answer graph in the back of the textbook.

11. The two angles have the same measure, so

$$10x - 70 = 7x - 25$$
$$3x = 45$$
$$x = 15.$$

Then,

$$10x - 70 = 10(15) - 70 = 150 - 70 = 80$$

and

$$7x - 25 = 7(15) - 25 = 105 - 25 = 80.$$

Each angle measures 80°.

12. $$\begin{aligned} 5x + 2y &= 7 \quad (1) \\ 10x + 4y &= 12 \quad (2) \end{aligned}$$

Multiplying equation (1) by -2 and adding the result to equation (2) yields a false statement. The system is inconsistent.

Solution set: $\emptyset$

13. $$\begin{aligned} 2x + y - z &= 5 \quad (1) \\ 3x + 2y + z &= 8 \quad (2) \\ 4x + 2y - 2z &= 10 \quad (3) \end{aligned}$$

Multiplying equation (1) by -2 and adding the result to equation (3) yields a true statement, $0 = 0$. Equations (1) and (3) are dependent. Since the *two* equations that are left include *three* variables, it is not possible to find a unique solution. Therefore, there is an infinite number of solutions to the system.

14. $\begin{vmatrix} 1 & 5 & 2 \\ 2 & 7 & 4 \\ 3 & -3 & 6 \end{vmatrix}$ Expand about row 1.

$$= 1\begin{vmatrix} 7 & 4 \\ -3 & 6 \end{vmatrix} - 5\begin{vmatrix} 2 & 4 \\ 3 & 6 \end{vmatrix} + 2\begin{vmatrix} 2 & 7 \\ 3 & -3 \end{vmatrix}$$

$$\begin{aligned} &= 1[42 - (-12)] - 5(12 - 12) + 2(-6 - 21) \\ &= 54 - 0 - 54 = 0 \end{aligned}$$

15. $\begin{vmatrix} 5 & 6 \\ 7 & 3 \end{vmatrix} = 5(3) - 6(7) = 15 - 42 = -27$

16. $$\begin{aligned} &(3k^3 - 5k^2 + 8k - 2) - (4k^3 + 11k + 7) \\ &\quad + (2k^2 - 5k) \\ &= 3k^3 - 4k^3 - 5k^2 + 2k^2 + 8k - 11k \\ &\quad - 5k - 2 - 7 \\ &= -k^3 - 3k^2 - 8k - 9 \end{aligned}$$

17. $$\begin{aligned} (8x - 7)(x + 3) &= 8x^2 + 24x - 7x - 21 \\ &= 8x^2 + 17x - 21 \end{aligned}$$

18. $$\begin{aligned} \frac{8z^3 - 16z^2 + 24z}{8z^2} &= \frac{8z^3}{8z^2} - \frac{16z^2}{8z^2} + \frac{24z}{8z^2} \\ &= z - 2 + \frac{3}{z} \end{aligned}$$

19.
$$\begin{array}{r} 3y^3 - 3y^2 + 4y + 1 \\ 2y+1\overline{)6y^4 - 3y^3 + 5y^2 + 6y - 9} \\ \underline{6y^4 + 3y^3} \\ -6y^3 + 5y^2 \\ \underline{-6y^3 - 3y^2} \\ 8y^2 + 6y \\ \underline{8y^2 + 4y} \\ 2y - 9 \\ \underline{2y + 1} \\ -10 \end{array}$$

Answer: $3y^3 - 3y^2 + 4y + 1 + \dfrac{-10}{2y+1}$

20. $2p^2 - 5pq + 3q^2 = (2p - 3q)(p - q)$

21. $18k^4 + 9k^2 - 20 = (3k^2 + 4)(6k^2 - 5)$

22. $x^3 + 512 = x^3 + 8^3 = (x + 8)(x^2 - 8x + 64)$

23. $$\frac{y^2 + y - 12}{y^3 + 9y^2 + 20y} \div \frac{y^2 - 9}{y^3 + 3y^2}$$
$$= \frac{y^2 + y - 12}{y^3 + 9y^2 + 20y} \cdot \frac{y^3 + 3y^2}{y^2 - 9}$$
$$= \frac{(y+4)(y-3)}{y(y+4)(y+5)} \cdot \frac{y^2(y+3)}{(y-3)(y+3)}$$
$$= \frac{y}{y+5}$$

24. $$\frac{1}{x+y} + \frac{3}{x-y}$$

The LCD is $(x+y)(x-y)$.

$$= \frac{1(x-y)}{(x+y)(x-y)} + \frac{3(x+y)}{(x-y)(x+y)}$$
$$= \frac{(x-y) + 3(x+y)}{(x+y)(x-y)}$$
$$= \frac{x - y + 3x + 3y}{(x+y)(x-y)}$$
$$= \frac{4x + 2y}{(x+y)(x-y)}$$

25. $$\frac{\frac{-6}{x-2}}{\frac{8}{3x-6}} = \frac{-6}{x-2} \div \frac{8}{3x-6}$$
$$= \frac{-6}{x-2} \cdot \frac{3x-6}{8}$$
$$= \frac{-6}{x-2} \cdot \frac{3(x-2)}{8} = -\frac{9}{4}$$

26. $$\frac{\frac{1}{a} - \frac{1}{b}}{\frac{a}{b} - \frac{b}{a}}$$

The LCD in both numerator and denominator is ab.

$$= \frac{\frac{b-a}{ab}}{\frac{a^2 - b^2}{ab}}$$
$$= \frac{b-a}{ab} \div \frac{a^2 - b^2}{ab}$$
$$= \frac{b-a}{ab} \cdot \frac{ab}{a^2 - b^2}$$
$$= \frac{b-a}{a^2 - b^2}$$
$$= \frac{-(a-b)}{(a-b)(a+b)}$$
$$= \frac{-1}{a+b}$$

27.
$$\begin{aligned} 2x^2 + 11x + 15 &= 0 \\ (2x+5)(x+3) &= 0 \\ 2x + 5 = 0 \quad &\text{or} \quad x + 3 = 0 \\ 2x = -5 \quad & \\ x = -\frac{5}{2} \quad &\text{or} \quad x = -3 \end{aligned}$$

Solution set: $\{-3, -\frac{5}{2}\}$

28.
$$\begin{aligned} 5t(t-1) &= 2(1-t) \\ 5t^2 - 5t &= 2 - 2t \\ 5t^2 - 3t - 2 &= 0 \\ (5t+2)(t-1) &= 0 \\ 5t + 2 = 0 \quad &\text{or} \quad t - 1 = 0 \\ 5t = -2 \quad & \\ t = -\frac{2}{5} \quad &\text{or} \quad t = 1 \end{aligned}$$

Solution set: $\{-\frac{2}{5}, 1\}$

29. $27^{-5/3} = \dfrac{1}{27^{5/3}} = \dfrac{1}{(27^{1/3})^5} = \dfrac{1}{3^5} = \dfrac{1}{243}$

30. $\dfrac{x^{-2/3}}{x^{-3/4}} = x^{-2/3-(-3/4)} = x^{-2/3+3/4} = x^{1/12}$

$(x \neq 0)$

31.
$$\begin{aligned} &8\sqrt{20} + 3\sqrt{80} - 2\sqrt{500} \\ &= 8\sqrt{4 \cdot 5} + 3\sqrt{16 \cdot 5} - 2\sqrt{100 \cdot 5} \\ &= 8 \cdot 2\sqrt{5} + 3 \cdot 4\sqrt{5} - 2 \cdot 10\sqrt{5} \\ &= 16\sqrt{5} + 12\sqrt{5} - 20\sqrt{5} = 8\sqrt{5} \end{aligned}$$

32. $$\frac{-9}{\sqrt{80}} = \frac{-9}{\sqrt{16 \cdot 5}} = \frac{-9}{4\sqrt{5}}$$
$$= \frac{-9}{4\sqrt{5}} \cdot \frac{\sqrt{5}}{\sqrt{5}} = \frac{-9\sqrt{5}}{20}$$

33. $$\frac{4}{\sqrt{6}-\sqrt{5}} = \frac{4}{\sqrt{6}-\sqrt{5}} \cdot \frac{\sqrt{6}+\sqrt{5}}{\sqrt{6}+\sqrt{5}}$$
$$= \frac{4(\sqrt{6}+\sqrt{5})}{6-5} = 4(\sqrt{6}+\sqrt{5})$$

34. $$\frac{12}{\sqrt[3]{2}} = \frac{12}{\sqrt[3]{2}} \cdot \frac{\sqrt[3]{4}}{\sqrt[3]{4}} = \frac{12\sqrt[3]{4}}{\sqrt[3]{8}} = \frac{12\sqrt[3]{4}}{2} = 6\sqrt[3]{4}$$

35. The distance between $(-4, 4)$ and $(-2, 9)$ is
$$\begin{aligned} d &= \sqrt{(x_2-x_1)^2 + (y_2-y_1)^2} \\ &= \sqrt{[-4-(-2)]^2 + (4-9)^2} \\ &= \sqrt{(-2)^2 + (-5)^2} \\ &= \sqrt{4+25} \\ &= \sqrt{29}. \end{aligned}$$

36. $$\begin{aligned} \sqrt{8x-4} - \sqrt{7x+2} &= 0 \\ \sqrt{8x-4} &= \sqrt{7x+2} \end{aligned}$$

Square both sides.
$$\begin{aligned} 8x-4 &= 7x+2 \\ x &= 6 \end{aligned}$$

Check 6.
$$\begin{aligned} \sqrt{8(6)-4} - \sqrt{7(6)+2} &= 0 \quad ? \\ \sqrt{44} - \sqrt{44} &= 0 \quad ? \\ 0 &= 0 \quad \textit{True} \end{aligned}$$

Solution set: $\{6\}$

37. Let $x =$ the speed of the boat in still water.

Direction	Distance	Rate	Time
Downstream	36	$x+3$	$\frac{36}{x+3}$
Upstream	24	$x-3$	$\frac{24}{x-3}$

Both times are the same, so
$$\frac{36}{x+3} = \frac{24}{x-3}.$$

Multiply by $(x-3)(x+3)$.
$$\begin{aligned} 36(x-3) &= 24(x+3) \\ 36x - 108 &= 24x + 72 \\ 12x &= 180 \\ x &= 15 \end{aligned}$$

The speed of the boat is 15 mph.

38. Let $x =$ Chuck's speed and $x + 4 =$ Brenda's speed.

Use $d = rt$, or $t = \frac{d}{r}$.

	Distance	Rate	Time
Chuck	24	x	$\frac{24}{x}$
Brenda	48	$x+4$	$\frac{48}{x+4}$

Since the times are the same,
$$\frac{24}{x} = \frac{48}{x+4}.$$

Multiply by the LCD, $x(x+4)$.
$$\begin{aligned} x(x+4)\left(\frac{24}{x}\right) &= x(x+4)\left(\frac{48}{x+4}\right) \\ 24(x+4) &= 48x \\ 24x + 96 &= 48x \\ -24x &= -96 \\ x &= 4 \end{aligned}$$

Chuck's speed is 4 mph; Brenda's speed is $x + 4 = 4 + 4 = 8$ mph.

39. Let $x =$ the number of dimes. $29 - x =$ the number of quarters.

The total value of the coins is \$4.70, so
$$.10x + .25(29-x) = 4.70.$$

Multiply by 100 to clear the decimals.
$$\begin{aligned} 10x + 25(29-x) &= 470 \\ 10x + 725 - 25x &= 470 \\ -15x &= -255 \\ x &= 17 \\ 29 - x = 29 - 17 &= 12 \end{aligned}$$

There are 17 dimes and 12 quarters.

40. Let $x =$ the amount of pure alcohol.

Solution	Strength	Amount	Pure Alcohol
#1	100%	x	$1 \cdot x = x$
#2	18%	40	$.18(40) = 7.2$
Mixture	22%	$x+40$	$.22(x+40)$

$$x + 7.2 = .22(x + 40)$$
$$x + 7.2 = .22x + 8.8$$
$$.78x = 1.6$$
$$x = \frac{1.6}{.78} = \frac{160}{78} = \frac{80}{39} \text{ or } 2\frac{2}{39}$$

The required amount is $\frac{80}{39}$ or $2\frac{2}{39}$ L.

Chapter 8

QUADRATIC EQUATIONS AND INEQUALITIES

8.1 The Square Root Property and Completing the Square

8.1 Margin Exercises

1. **(a)** $m^2 = 64$
$m^2 = 8^2$

By the square root property,

$$m = 8 \quad \text{or} \quad m = -8.$$

Solution set: $\{-8, 8\}$

(b) $p^2 = 7$
$p = \sqrt{7}$ or $p = -\sqrt{7}$

Solution set: $\{-\sqrt{7}, \sqrt{7}\}$

2. **(a)** $(a-3)^2 = 25$

By the square root property,

$$\begin{aligned} a - 3 &= 5 \quad \text{or} \quad a - 3 = -5 \\ a &= 8 \quad \text{or} \quad a = -2. \end{aligned}$$

Solution set: $\{-2, 8\}$

(b) $(3k+1)^2 = 2$

$$\begin{aligned} 3k + 1 &= \sqrt{2} \quad &\text{or} \quad 3k + 1 &= -\sqrt{2} \\ 3k &= -1 + \sqrt{2} & 3k &= -1 - \sqrt{2} \\ k &= \frac{-1+\sqrt{2}}{3} \quad &\text{or} \quad k &= \frac{-1-\sqrt{2}}{3} \end{aligned}$$

Solution set: $\left\{\frac{-1+\sqrt{2}}{3}, \frac{-1-\sqrt{2}}{3}\right\}$

(c) $(2r+3)^2 = 8$

$$\begin{aligned} 2r + 3 &= \sqrt{8} \quad &\text{or} \quad 2r + 3 &= -\sqrt{8} \\ 2r + 3 &= 2\sqrt{2} & 2r + 3 &= -2\sqrt{2} \\ 2r &= -3 + 2\sqrt{2} & 2r &= -3 - 2\sqrt{2} \\ r &= \frac{-3+2\sqrt{2}}{2} \quad &\text{or} \quad r &= \frac{-3-2\sqrt{2}}{2} \end{aligned}$$

Solution set: $\left\{\frac{-3+2\sqrt{2}}{2}, \frac{-3-2\sqrt{2}}{2}\right\}$

3. $(k+5)^2 = -100$

$$\begin{aligned} k + 5 &= \sqrt{-100} \quad &\text{or} \quad k + 5 &= -\sqrt{-100} \\ k + 5 &= 10i & k + 5 &= -10i \\ k &= -5 + 10i \quad &\text{or} \quad k &= -5 - 10i \end{aligned}$$

Solution set: $\{-5 + 10i, -5 - 10i\}$

4. **(a)** $x^2 + 6x + \_\_\_$

Take half of 6, the coefficient of x, and square the result.

$$\left(\frac{1}{2} \cdot 6\right)^2 = 3^2 = 9$$

Adding 9 will complete the perfect square trinomial.

(b) $m^2 - 14m + \_\_\_$

Take half of -14 and square the result.

$$\left[\frac{1}{2}(-14)\right]^2 = (-7)^2 = 49$$

Adding 49 will complete the perfect square trinomial.

(c) $r^3 + 3r + \_\_\_$

Take half of 3 and square the result.

$$\left(\frac{1}{2} \cdot 3\right)^2 = \left(\frac{3}{2}\right)^2 = \frac{9}{4}$$

Adding $\frac{9}{4}$ will complete the perfect square trinomial.

5. **(a)** $y^2 + 6y + 4 = 0$
$y^2 + 6y = -4$

Complete the square. Take half of 6, the coefficient of y, and square the result.

$$\left(\frac{1}{2} \cdot 6\right)^2 = 3^2 = 9$$

Add 9 to both sides of the equation.

$$\begin{aligned} y^2 + 6y + 9 &= -4 + 9 \\ (y+3)^2 &= 5 \end{aligned}$$

Use the square root property.

$$\begin{aligned} y + 3 &= \sqrt{5} \quad &\text{or} \quad y + 3 &= -\sqrt{5} \\ y &= -3 + \sqrt{5} \quad &\text{or} \quad y &= -3 - \sqrt{5} \end{aligned}$$

Solution set: $\{-3 + \sqrt{5}, -3 - \sqrt{5}\}$

(b) $x^2+2x-10=0$
$x^2+2x=10$

Complete the square.

$$\left(\frac{1}{2}\cdot 2\right)^2=1^2=1$$

Add 1 to both sides.

$$x^2+2x+1=10+1$$
$$(x+1)^2=11$$

Use the square root property.

$x+1=\sqrt{11}$ or $x-1=-\sqrt{11}$
$x=-1+\sqrt{11}$ or $x=-1-\sqrt{11}$

Solution set: $\{-1+\sqrt{11},-1-\sqrt{11}\}$

6. **(a)** $2r^2-4r+1=0$
$2r^2-4r=-1$
$r^2-2r=-\frac{1}{2}$

Complete the square.

$$\left[\frac{1}{2}(-2)\right]^2=(-1)^2=1$$

Add 1 to both sides.

$$r^2-2r+1=-\frac{1}{2}+1$$
$$(r-1)^2=\frac{1}{2}$$

$r-1=\sqrt{\frac{1}{2}}$ or $r-1=-\sqrt{\frac{1}{2}}$
$r-1=\frac{\sqrt{2}}{2}$ $\quad r-1=\frac{-\sqrt{2}}{2}$
$r=1+\frac{\sqrt{2}}{2}$ $\quad r=1-\frac{\sqrt{2}}{2}$
$r=\frac{2+\sqrt{2}}{2}$ or $r=\frac{2-\sqrt{2}}{2}$

Solution set: $\left\{\frac{2+\sqrt{2}}{2},\frac{2-\sqrt{2}}{2}\right\}$

(b) $5t^2-15t+12=0$
$5t^2-15t=-12$
$t^2-3t=-\frac{12}{5}$

Complete the square.

$$\left[\frac{1}{2}(-3)\right]^2=\left(-\frac{3}{2}\right)^2=\frac{9}{4}$$

Add $\frac{9}{4}$ to both sides.

$$t^2-3t+\frac{9}{4}=-\frac{12}{5}+\frac{9}{4}$$
$$\left(t-\frac{3}{2}\right)^2=-\frac{3}{20}$$

$t-\frac{3}{2}=\sqrt{-\frac{3}{20}}$ or $t-\frac{3}{2}=-\sqrt{-\frac{3}{20}}$
$t-\frac{3}{2}=\frac{i\sqrt{3}}{\sqrt{20}}\cdot\frac{\sqrt{5}}{\sqrt{5}}$ $\quad t-\frac{3}{2}=\frac{-i\sqrt{3}}{\sqrt{20}}\cdot\frac{\sqrt{5}}{\sqrt{5}}$
$t-\frac{3}{2}=\frac{i\sqrt{15}}{10}$ $\quad t-\frac{3}{2}=\frac{-i\sqrt{15}}{10}$
$t=\frac{3}{2}+\frac{i\sqrt{15}}{10}$ $\quad t=\frac{3}{2}-\frac{i\sqrt{15}}{10}$
$t=\frac{15+i\sqrt{15}}{10}$ or $t=\frac{15-i\sqrt{15}}{10}$

Solution set: $\left\{\frac{15+i\sqrt{15}}{10},\frac{15-i\sqrt{15}}{10}\right\}$

8.1 Section Exercises

3. To solve $(2x+1)^2=5$, we find that $2x+1=\sqrt{5}$ or $2x+1=-\sqrt{5}$ using the square root property. Then we solve for x in both equations.

To solve $x^2+4x=12$, we find that

$$x^2+4x+4=12+4$$
$$(x+2)^2=16$$

by completing the square.

7. $t^2=17$
$t=\sqrt{17}$ or $t=-\sqrt{17}$
Solution set: $\{-\sqrt{17},\sqrt{17}\}$

11. $(1-3k)^2=7$
$1-3k=\sqrt{7}$ or $1-3k=-\sqrt{7}$
$-3k=-1+\sqrt{7}$ $\quad -3k=-1-\sqrt{7}$
$k=\frac{-1+\sqrt{7}}{-3}$ $\quad k=\frac{-1-\sqrt{7}}{-3}$
$k=\frac{1-\sqrt{7}}{3}$ or $k=\frac{1+\sqrt{7}}{3}$
Solution set: $\left\{\frac{1+\sqrt{7}}{3},\frac{1-\sqrt{7}}{3}\right\}$

15. $x^2=-12$
$x=\sqrt{-12}$ or $x=-\sqrt{-12}$
$x=i\sqrt{12}$ $\quad x=-i\sqrt{12}$
$x=2i\sqrt{3}$ or $x=-2i\sqrt{3}$
Solution set: $\{-2i\sqrt{3},2i\sqrt{3}\}$

19. $(6k-1)^2 = -8$

$$6k-1=\sqrt{-8} \quad \text{or} \quad 6k-1=-\sqrt{-8}$$
$$6k-1=i\sqrt{8} \qquad 6k-1=-i\sqrt{8}$$
$$6k-1=2i\sqrt{2} \qquad 6k-1=-2i\sqrt{2}$$
$$6k=1+2i\sqrt{2} \qquad 6k=1-2i\sqrt{2}$$
$$k=\frac{1+2i\sqrt{2}}{6} \quad \text{or} \quad k=\frac{1-2i\sqrt{2}}{6}$$

Solution set: $\left\{\frac{1+2i\sqrt{2}}{6}, \frac{1-2i\sqrt{2}}{6}\right\}$

23. $3y^2+y=24$

$$y^2+\frac{1}{3}y=8$$

Complete the square by taking half of $\frac{1}{3}$, the coefficient of y, and squaring the result.

$$\left(\frac{1}{2}\cdot\frac{1}{3}\right)^2=\left(\frac{1}{6}\right)^2=\frac{1}{36}$$

Add $\frac{1}{36}$ to both sides.

$$y^2+\frac{1}{3}y+\frac{1}{36}=8+\frac{1}{36}$$
$$\left(y+\frac{1}{6}\right)^2=\frac{288}{36}+\frac{1}{36}$$
$$\left(y+\frac{1}{6}\right)^2=\frac{289}{36}$$

$$y+\frac{1}{6}=\sqrt{\frac{289}{36}} \quad \text{or} \quad y+\frac{1}{6}=-\sqrt{\frac{289}{36}}$$
$$y=-\frac{1}{6}+\frac{\sqrt{289}}{\sqrt{36}} \qquad y=-\frac{1}{6}-\frac{\sqrt{289}}{\sqrt{36}}$$
$$y=-\frac{1}{6}+\frac{17}{6} \qquad y=-\frac{1}{6}-\frac{17}{6}$$
$$y=\frac{16}{6} \qquad y=-\frac{18}{6}$$
$$y=\frac{8}{3} \quad \text{or} \quad y=-3$$

Solution set: $\left\{-3, \frac{8}{3}\right\}$

27. $m^2+4m+13=0$

$$m^2+4m=-13$$

Complete the square.

$$\left(\frac{1}{2}\cdot 4\right)^2=2^2=4$$

Add 4 to both sides.

$$m^2+4m+4=-13+4$$
$$(m+2)^2=-9$$

$$m+2=\sqrt{-9} \quad \text{or} \quad m+2=-\sqrt{-9}$$
$$m=-2+3i \quad \text{or} \quad m=-2-3i$$

Solution set: $\{-2+3i, -2-3i\}$

31. $z^2-\frac{4}{3}z=-\frac{1}{9}$

Complete the square.

$$\left[\frac{1}{2}\left(-\frac{4}{3}\right)\right]^2=\left(-\frac{2}{3}\right)^2=\frac{4}{9}$$

Add $\frac{4}{9}$ to both sides.

$$z^2-\frac{4}{3}z+\frac{4}{9}=-\frac{1}{9}+\frac{4}{9}$$
$$\left(z-\frac{2}{3}\right)^2=\frac{3}{9}$$

$$z-\frac{2}{3}=\sqrt{\frac{3}{9}} \quad \text{or} \quad z-\frac{2}{3}=-\sqrt{\frac{3}{9}}$$
$$z=\frac{2}{3}+\frac{\sqrt{3}}{3} \qquad z=\frac{2}{3}-\frac{\sqrt{3}}{3}$$
$$z=\frac{2+\sqrt{3}}{3} \quad \text{or} \quad z=\frac{2-\sqrt{3}}{3}$$

Solution set: $\left\{\frac{2+\sqrt{3}}{3}, \frac{2-\sqrt{3}}{3}\right\}$

35. $.1x^2-.2x-.1=0$

Multiply both sides by 10 to clear the decimals.

$$x^2-2x-1=0$$
$$x^2-2x=1$$

Complete the square.

$$\left[\frac{1}{2}(-2)\right]^2=(-1)^2=1$$

Add 1 to both sides.

$$x^2-2x+1=1+1$$
$$(x-1)^2=2$$

$$x-1=\sqrt{2} \quad \text{or} \quad x-1=-\sqrt{2}$$
$$x=1+\sqrt{2} \quad \text{or} \quad x=1-\sqrt{2}$$

Solution set: $\{1+\sqrt{2}, 1-\sqrt{2}\}$

39. The area of the original square is $x\cdot x$, or x^2.

40. Each rectangular strip has length x and width 1, so each strip has an area of $x\cdot 1$, or x.

41. From Exercise 40, the area of a rectangular strip is x. The area of 6 rectangular strips is $6x$.

42. These are 1 by 1 squares, so each has an area of $1 \cdot 1$, or 1.

43. There are 9 small squares, each with area 1 (from Exercise 42), so the total area is $9 \cdot 1$, or 9.

44. The area of the new larger square is $(x+3)^2$. Using the results from Exercises 39-43,

$$(x+3)^2 = x^2 + 6x + 9.$$

8.2 The Quadratic Formula

8.2 Margin Exercises

1. **(a)** $-3q^2 + 9q - 4 = 0$

Here $a = -3, b = 9$, and $c = -4$.

(b)
$$3y^2 = 6y + 2$$
$$3y^2 - 6y - 2 = 0$$

So, $a = 3$, $b = -6$, and $c = -2$.

2. $6x^2 + 4x - 1 = 0$

Here $a = 6, b = 4$, and $c = -1$.

$$x = \frac{-b \pm \sqrt{b^2 - 4ac}}{2a}$$
$$x = \frac{-4 \pm \sqrt{4^2 - 4(6)(-1)}}{2(6)}$$
$$= \frac{-4 \pm \sqrt{16 + 24}}{12}$$
$$= \frac{-4 \pm \sqrt{40}}{12}$$
$$= \frac{-4 \pm 2\sqrt{10}}{12}$$
$$= \frac{2(-2 \pm \sqrt{10})}{12}$$
$$x = \frac{-2 \pm \sqrt{10}}{6}$$

Solution set: $\left\{\frac{-2+\sqrt{10}}{6}, \frac{-2-\sqrt{10}}{6}\right\}$

3. **(a)**
$$2k^2 + 19 = 14k$$
$$2k^2 - 14k + 19 = 0$$

Here $a = 2, b = -14$, and $c = 19$.

$$k = \frac{-b \pm \sqrt{b^2 - 4ac}}{2a}$$
$$k = \frac{-(-14) \pm \sqrt{(-14)^2 - 4(2)(19)}}{2(2)}$$
$$= \frac{14 \pm \sqrt{196 - 152}}{4}$$
$$= \frac{14 \pm \sqrt{44}}{4}$$
$$= \frac{14 \pm 2\sqrt{11}}{4}$$
$$= \frac{2(7 \pm \sqrt{11})}{4}$$
$$k = \frac{7 \pm \sqrt{11}}{2}$$

Solution set: $\left\{\frac{7+\sqrt{11}}{2}, \frac{7-\sqrt{11}}{2}\right\}$

(b)
$$z^2 = 4z - 5$$
$$z^2 - 4z + 5 = 0$$

Here $a = 1, b = -4$, and $c = 5$.

$$z = \frac{-b \pm \sqrt{b^2 - 4ac}}{2a}$$
$$z = \frac{-(-4) \pm \sqrt{(-4)^2 - 4(1)(5)}}{2(1)}$$
$$= \frac{4 \pm \sqrt{16 - 20}}{2}$$
$$= \frac{4 \pm \sqrt{-4}}{2}$$
$$= \frac{4 \pm 2i}{2}$$
$$= \frac{2(2 \pm i)}{2}$$
$$z = 2 \pm i$$

Solution set: $\{2 + i, 2 - i\}$

4. When the ball is 32 ft from the ground,

$$s(t) = -16t^2 + 64t$$

can be written as

$$32 = -16t^2 + 64t.$$

Solve the equation for time, t.

$$16t^2 - 64t + 32 = 0$$
$$t^2 - 4t + 2 = 0$$

Use the quadratic formula with $a = 1$, $b = -4$, and $c = 2$.

$$t = \frac{-b \pm \sqrt{b^2 - 4ac}}{2a}$$

$$t = \frac{4 \pm \sqrt{16 - 8}}{2}$$

$$t = \frac{4 \pm \sqrt{8}}{2}$$

A calculator gives

$$\sqrt{8} \approx 2.83, \text{ so}$$

$$t \approx \frac{4 + 2.83}{2} \quad \text{or} \quad t \approx \frac{4 - 2.83}{2}$$

$$t \approx 3.4 \quad \text{or} \quad t \approx .6.$$

At .6 sec and 3.4 sec the ball will be 32 ft from the ground.

5. The discriminant is the expression under the square root sign in the quadratic formula: $b^2 - 4ac$.

(a) $6m^2 - 13m - 28 = 0$

Here $a = 6, b = -13$, and $c = -28$, so the discriminant is

$$\begin{aligned} b^2 - 4ac &= (-13)^2 - 4(6)(-28) \\ &= 169 + 672 \\ &= 841 \text{ or } 29^2. \end{aligned}$$

The discriminant, 841, is a perfect square.

(b) $4y^2 + 2y + 1 = 0$

Here $a = 4, b = 2$, and $c = 1$, so the discriminant is

$$\begin{aligned} b^2 - 4ac &= 2^2 - 4(4)(1) \\ &= 4 - 16 \\ &= -12. \end{aligned}$$

The discriminant, -12, is not a perfect square.

(c)

$$\begin{aligned} 15k^2 + 11k &= 14 \\ 15k^2 + 11k - 14 &= 0 \end{aligned}$$

Here $a = 15, b = 11$, and $c = -14$, so the discriminant is

$$\begin{aligned} b^2 - 4ac &= 11^2 - 4(15)(-14) \\ &= 121 + 840 \\ &= 961 \text{ or } 31^2. \end{aligned}$$

The discriminant, 961, is a perfect square.

6. **(a)**

$$\begin{aligned} 2x^2 + 3x &= 4 \\ 2x^2 + 3x - 4 &= 0 \end{aligned}$$

Here $a = 2, b = 3$, and $c = -4$, so the discriminant is

$$\begin{aligned} b^2 - 4ac &= 3^2 - 4(2)(-4) \\ &= 9 + 32 \\ &= 41. \end{aligned}$$

The discriminant is positive but not a perfect square, so there are two different irrational solutions.

(b) $2x^2 + 3x + 4 = 0$

Here $a = 2, b = 3$, and $c = 4$, so the discriminant is

$$\begin{aligned} b^2 - 4ac &= 3^2 - 4(2)(4) \\ &= 9 - 32 \\ &= -23. \end{aligned}$$

The discriminant is negative, so there are two different imaginary solutions.

(c) $x^2 + 20x + 100 = 0$

Here $a = 1, b = 20$, and $c = 100$, so the discriminant is

$$\begin{aligned} b^2 - 4ac &= 20^2 - 4(1)(100) \\ &= 400 - 400 \\ &= 0. \end{aligned}$$

The discriminant is zero, so there is one rational solution.

(d) $3x^2 + 7x = 0$

Here $a = 3, b = 7$, and $c = 0$, so the discriminant is

$$\begin{aligned} b^2 - 4ac &= 7^2 - 4(3)(0) \\ &= 49 \text{ or } 7^2. \end{aligned}$$

The discriminant is a perfect square, so there are two different rational solutions.

7. **(a)** $2y^2 + 13y - 7$

The corresponding quadratic equation is

$$2y^2 + 13y - 7 = 0.$$

Here $a = 2, b = 13$, and $c = -7$, so the discriminant is

$$\begin{aligned} b^2 - 4ac &= 13^2 - 4(2)(-7) \\ &= 169 + 56 \\ &= 225 \text{ or } 15^2. \end{aligned}$$

Since 225 is a perfect square, the trinomial can be factored.

$$2y^2 + 13y - 7 = (2y - 1)(y + 7)$$

(b) $6z^2 - 11z + 18$

The corresponding quadratic equation is

$$6z^2 - 11z + 18 = 0.$$

Here $a = 6$, $b = -11$, and $c = 18$. Evaluate the discriminant.

$$\begin{aligned} b^2 - 4ac &= (-11)^2 - 4(6)(18) \\ &= 121 - 432 \\ &= -311 \end{aligned}$$

Since the discriminant is negative, the trinomial cannot be factored.

8.2 Section Exercises

3. If we multiply

$$-x^2 - 3x + 4 = 0$$

by -1, we get

$$x^2 + 3x - 4 = 0.$$

Thus, the two equations are the same and will have the same solution set. The statement is true.

7. $m^2 - 8m + 15 = 0$

Here $a = 1, b = -8$, and $c = 15$.

$$m = \frac{-b \pm \sqrt{b^2 - 4ac}}{2a}$$

$$m = \frac{-(-8) \pm \sqrt{(-8)^2 - 4(1)(15)}}{2(1)}$$

$$= \frac{8 \pm \sqrt{64 - 60}}{2}$$

$$= \frac{8 \pm \sqrt{4}}{2}$$

$$= \frac{8 \pm 2}{2}$$

$$m = \frac{8 + 2}{2} = \frac{10}{2} = 5 \quad \text{or}$$

$$m = \frac{8 - 2}{2} = \frac{6}{2} = 3$$

Solution set: $\{3, 5\}$

11.
$$\begin{aligned} 2x^2 - 2x &= 1 \\ 2x^2 - 2x - 1 &= 0 \end{aligned}$$

Here $a = 2, b = -2$, and $c = -1$.

$$x = \frac{-b \pm \sqrt{b^2 - 4ac}}{2a}$$

$$x = \frac{-(-2) \pm \sqrt{(-2)^2 - 4(2)(-1)}}{2(2)}$$

$$= \frac{2 \pm \sqrt{4 + 8}}{4}$$

$$= \frac{2 \pm \sqrt{12}}{4}$$

$$= \frac{2 \pm 2\sqrt{3}}{4}$$

$$= \frac{2(1 \pm \sqrt{3})}{4}$$

$$x = \frac{1 \pm \sqrt{3}}{2}$$

Solution set: $\left\{\frac{1+\sqrt{3}}{2}, \frac{1-\sqrt{3}}{2}\right\}$

15.
$$\begin{aligned} -2t(t + 2) &= -3 \\ -2t^2 - 4t &= -3 \\ -2t^2 - 4t + 3 &= 0 \end{aligned}$$

$$a = -2, b = -4, c = 3$$

$$t = \frac{-b \pm \sqrt{b^2 - 4ac}}{2a}$$

$$t = \frac{-(-4) \pm \sqrt{(-4)^2 - 4(-2)(3)}}{2(-2)}$$

$$= \frac{4 \pm \sqrt{16 + 24}}{-4}$$

$$= \frac{4 \pm \sqrt{40}}{-4}$$

$$= \frac{4 \pm 2\sqrt{10}}{-4}$$

$$= \frac{2(2 \pm \sqrt{10})}{-4}$$

$$= \frac{2 \pm \sqrt{10}}{-2}$$

$$t = \frac{-2 \pm \sqrt{10}}{2}$$

Solution set: $\left\{\frac{-2+\sqrt{10}}{2}, \frac{-2-\sqrt{10}}{2}\right\}$

19. $k^2 + 47 = 0$

$a = 1, b = 0, c = 47$

$$k = \frac{-b \pm \sqrt{b^2 - 4ac}}{2a}$$

$$k = \frac{-(0) \pm \sqrt{0^2 - 4(1)(47)}}{2(1)}$$

$$= \frac{0 \pm \sqrt{0 - 188}}{2}$$

$$= \frac{\pm\sqrt{-188}}{2}$$

$$= \frac{\pm 2i\sqrt{47}}{2}$$

$$k = \pm i\sqrt{47}$$

Solution set: $\{i\sqrt{47}, -i\sqrt{47}\}$

23. $4x^2 - 4x = -7$

$$4x^2 - 4x + 7 = 0$$

$a = 4, b = -4, c = 7$

$$x = \frac{-b \pm \sqrt{b^2 - 4ac}}{2a}$$

$$x = \frac{-(-4) \pm \sqrt{(-4)^2 - 4(4)(7)}}{2(4)}$$

$$= \frac{4 \pm \sqrt{16 - 112}}{8}$$

$$= \frac{4 \pm \sqrt{-96}}{8}$$

$$= \frac{4 \pm 4i\sqrt{6}}{8}$$

$$= \frac{4(1 \pm i\sqrt{6})}{8}$$

$$x = \frac{1 \pm i\sqrt{6}}{2}$$

Solution set: $\left\{\frac{1+i\sqrt{6}}{2}, \frac{1-i\sqrt{6}}{2}\right\}$

27. $\dfrac{x+5}{2x-1} = \dfrac{x-4}{x-6}$

Multiply by the LCD, $(2x-1)(x-6)$.

$$(2x-1)(x-6)\left(\frac{x+5}{2x-1}\right) = (2x-1)(x-6)\left(\frac{x-4}{x-6}\right)$$

$$(x-6)(x+5) = (2x-1)(x-4)$$

$$x^2 - x - 30 = 2x^2 - 9x + 4$$

$$-x^2 + 8x - 34 = 0$$

$$x^2 - 8x + 34 = 0$$

$a = 1, b = -8, c = 34$

$$x = \frac{-b \pm \sqrt{b^2 - 4ac}}{2a}$$

$$x = \frac{-(-8) \pm \sqrt{(-8)^2 - 4(1)(34)}}{2(1)}$$

$$= \frac{8 \pm \sqrt{64 - 136}}{2}$$

$$= \frac{8 \pm \sqrt{-72}}{2}$$

$$= \frac{8 \pm 6i\sqrt{2}}{2}$$

$$= \frac{2(4 \pm 3i\sqrt{2})}{2}$$

$$x = 4 \pm 3i\sqrt{2}$$

Solution set: $\{4 + 3i\sqrt{2}, 4 - 3i\sqrt{2}\}$

31. Let $s(t) = 240$ in the function. Solve for t.

$$s(t) = -16t^2 + 128t$$

$$240 = -16t^2 + 128t$$

$$0 = -16t^2 + 128t - 240$$

Divide by -16.

$$0 = t^2 - 8t + 15$$

$$0 = (t-5)(t-3)$$

$$t - 5 = 0 \quad \text{or} \quad t - 3 = 0$$

$$t = 5 \quad \text{or} \quad t = 3$$

The ball will be 240 ft above the ground at 3 sec (on the way up) and again at 5 sec (on the way down).

In Exercises 35-40, use the quadratic function

$$f(x) = -6.5x^2 + 132.5x + 2117.$$

35. For 1988, $x = 0$. If $x = 0$, then

$$f(0) = -6.5(0)^2 + 132.5(0) + 2117$$

$$= 2117.$$

There were 2117 prisoners in 1988.

36. For 1989, $x = 1$. If $x = 1$, then

$$f(1) = -6.5(1)^2 + 132.5(1) + 2117$$

$$= 2243.$$

There were 2243 prisoners in 1989.

37. For 1990, $x = 2$. If $x = 2$, then

$$\begin{aligned} f(2) &= -6.5(2)^2 + 132.5(2) + 2117 \\ &= 2356. \end{aligned}$$

There were 2356 prisoners in 1990.

38. For 1991, let $x = 3$. Then evaluate $f(x)$ for $x = 3$, that is, find $f(3)$.

39. Substitute 2543 for $f(x)$ in the quadratic function.

$$\begin{aligned} -6.5x^2 + 132.5x + 2117 &= 2543 \\ -6.5x^2 + 132.5x - 426 &= 0 \end{aligned}$$

$$\begin{aligned} x &= \frac{-b \pm \sqrt{b^2 - 4ac}}{2a} \\ x &= \frac{-132.5 \pm \sqrt{(132.5)^2 - 4(-6.5)(-426)}}{2(-6.5)} \\ &= \frac{-132.5 \pm \sqrt{6480.25}}{-13} \\ &= \frac{-132.5 \pm 80.5}{-13} \\ x &= \frac{-132.5 + 80.5}{-13} = 4 \quad \text{or} \\ x &= \frac{-132.5 - 80.5}{-13} \approx 16 \end{aligned}$$

$x = 4$ corresponds to $1988 + 4 = 1992$, so the number of prisoners was 2543 in 1992.

(Note that the quadratic function would not necessarily be accurate for $x \approx 16$, which would correspond to the year $1988 + 16 = 2004$, since the function is based on data collected only from 1988-1990.)

40. Substitute 700 for $f(x)$ in the given function.

$$\begin{aligned} 5x^2 + 45x + 350 &= 700 \\ 5x^2 + 45x - 350 &= 0 \end{aligned}$$

Divide by 5.

$$x^2 + 9x - 70 = 0$$

This equation can be solved by factoring.

$$\begin{aligned} (x + 14)(x - 5) &= 0 \\ x + 14 = 0 \quad &\text{or} \quad x - 5 = 0 \\ x = -14 \quad &\text{or} \quad x = 5 \end{aligned}$$

Discard the negative solution. $x = 5$ corresponds to $1994 + 5 = 1999$. The population would be 700 in the year 1999.

43. $x^2 + 4x + 2 = 0$

Here $a = 1, b = 4$, and $c = 2$, so the discriminant is

$$\begin{aligned} b^2 - 4ac &= 4^2 - 4(1)(2) \\ &= 16 - 8 \\ &= 8. \end{aligned}$$

Since the discriminant is positive, but not a perfect square, there are two distinct irrational number solutions. The answer is (c).

47. $3y^2 - 10y + 15 = 0$

Here $a = 3$, $b = -10$, and $c = 15$, so the discriminant is

$$\begin{aligned} b^2 - 4ac &= (-10)^2 - 4(3)(15) \\ &= 100 - 180 \\ &= -80. \end{aligned}$$

Since the discriminant is negative, there are two distinct imaginary number solutions. The answer is (d).

51. $36x^2 + 21x - 24$

The corresponding quadratic equation is

$$36x^2 + 21x - 24 = 0.$$

Since $a = 36$, $b = 21$, and $c = -24$, the discriminant is

$$\begin{aligned} b^2 - 4ac &= 21^2 - 4(36)(-24) \\ &= 441 + 3456 \\ &= 3897. \end{aligned}$$

The discriminant is not a perfect square, so the trinomial $36x^2 + 21x - 24$ cannot be factored.

55. $2x^2 + kx + 2 = 0$

In order to have exactly one rational number solution, the discriminant $b^2 - 4ac$ must equal 0. Here, $a = 2, b = k$, and $c = 2$, so

$$\begin{aligned} b^2 - 4ac = k^2 - 4(2)(2) &= 0 \\ k^2 - 16 &= 0 \\ k^2 &= 16 \\ k &= \pm 4. \end{aligned}$$

Therefore, $k = 4$ or $k = -4$.

8.3 Equations Quadratic in Form

8.3 Margin Exercises

1. **(a)** $\frac{5}{m}+\frac{12}{m^2}=2$

To clear the fractions, multiply each term by the common denominator, m^2.

$$m^2\left(\frac{5}{m}\right)+m^2\left(\frac{12}{m^2}\right)=m^2(2)$$
$$5m+12=2m^2$$
$$-2m^2+5m+12=0$$
$$2m^2-5m-12=0$$
$$(2m+3)(m-4)=0$$
$$2m+3=0 \quad \text{or} \quad m-4=0$$
$$m=-\frac{3}{2} \quad \text{or} \quad m=4$$

Both potential solutions check.

Solution set: $\{-\frac{3}{2},4\}$

(b) $\frac{2}{x}+\frac{1}{x-2}=\frac{5}{3}$

Multiply by $3x(x-2)$, the common denominator.

$$3x(x-2)\left(\frac{2}{x}+\frac{1}{x-2}\right)=3x(x-2)\left(\frac{5}{3}\right)$$
$$6(x-2)+3x=5x(x-2)$$
$$6x-12+3x=5x^2-10x$$
$$-5x^2+19x-12=0$$
$$5x^2-19x+12=0$$
$$(5x-4)(x-3)=0$$
$$5x-4=0 \quad \text{or} \quad x-3=0$$
$$x=\frac{4}{5} \quad \text{or} \quad x=3$$

Both potential solutions check.

Solution set: $\{\frac{4}{5},3\}$

(c) $\frac{4}{m-1}+9=-\frac{7}{m}$

Multiply by the common denominator, $m(m-1)$.

$$m(m-1)\left(\frac{4}{m-1}+9\right)=m(m-1)\left(-\frac{7}{m}\right)$$
$$4m+9m(m-1)=-7(m-1)$$
$$4m+9m^2-9m=-7m+7$$
$$9m^2+2m-7=0$$
$$(9m-7)(m+1)=0$$
$$9m-7=0 \quad \text{or} \quad m+1=0$$
$$m=\frac{7}{9} \quad \text{or} \quad m=-1$$

Both potential solutions check.

Solution set: $\{-1,\frac{7}{9}\}$

2. **(a)** Let $x=$ the speed of the boat in still water.

Kerrie went 15 mi each way. Use $d=rt$, or $t=\frac{d}{r}$.

	d	r	t
Up	15	$x-5$	$\frac{15}{x-5}$
Down	15	$x+5$	$\frac{15}{x+5}$

(b) To solve, use the fact that she can make the entire trip in 4 hr; that is, the time going upstream added to the time going downstream is 4.

$$\frac{15}{x-5}+\frac{15}{x+5}=4$$

Multiply both sides by $(x-5)(x+5)$.

$$\frac{15}{x-5}(x-5)(x+5)+\frac{15}{x+5}(x-5)(x+5)$$
$$=4(x-5)(x+5)$$
$$15(x+5)+15(x-5)=4(x^2-25)$$
$$15x+75+15x-75=4x^2-100$$
$$30x=4x^2-100$$
$$0=4x^2-30x-100$$
$$0=2x^2-15x-50$$
$$0=(2x+5)(x-10)$$
$$2x+5=0 \quad \text{or} \quad x-10=0$$
$$x=-\frac{5}{2} \quad \text{or} \quad x=10$$

The speed of the boat must be nonnegative, so $-\frac{5}{2}$ is not a solution. The speed is 10 mph.

(c) Let x be the speed Ken can row.

Make a table.

	d	r	t
Up	5	$x-3$	$\frac{5}{x-3}$
Down	5	$x+3$	$\frac{5}{x+3}$

In this problem, the time going upstream added to the time going downstream is $1\frac{3}{4}$ or $\frac{7}{4}$ hr.

$$\frac{5}{x-3}+\frac{5}{x+3}=\frac{7}{4}$$

Multiply both sides by $4(x-3)(x+3)$.

$$4(x-3)(x+3)\frac{5}{x-3}+4(x-3)(x+3)\frac{5}{x+3}$$
$$=4(x-3)(x+3)\left(\frac{7}{4}\right)$$

$$20(x+3)+20(x-3)=7(x-3)(x+3)$$
$$20x+60+20x-60=7x^2-63$$
$$40x=7x^2-63$$
$$0=7x^2-40x-63$$
$$0=(7x+9)(x-7)$$
$$7x+9=0 \quad \text{or} \quad x-7=0$$
$$x=-\frac{9}{7} \quad \text{or} \quad x=7$$

Speed can't be negative, so Ken rows at the speed of 7 mph.

3. Let x = Jaime's time alone and $x-2$ = Carlos' time alone.

Make a table.

Worker	Rate	Time together	Part of job done
Jaime	$\frac{1}{x}$	2	$\frac{2}{x}$
Carlos	$\frac{1}{x-2}$	2	$\frac{2}{x-2}$

Since together they complete 1 whole job,

$$\frac{2}{x}+\frac{2}{x-2}=1.$$

Multiply both sides by $x(x-2)$.

$$x(x-2)\left(\frac{2}{x}+\frac{2}{x-2}\right)=x(x-2)(1)$$
$$2(x-2)+2x=x(x-2)$$
$$2x-4+2x=x^2-2x$$
$$0=x^2-6x+4$$

Now use the quadratic formula.
$a=1, b=-6, c=4$

$$x=\frac{-b\pm\sqrt{b^2-4ac}}{2a}$$
$$x=\frac{-(-6)\pm\sqrt{(-6)^2-4(1)(4)}}{2(1)}$$
$$=\frac{6\pm\sqrt{36-16}}{2}$$
$$=\frac{6\pm\sqrt{20}}{2}$$
$$=\frac{6\pm2\sqrt{5}}{2}$$
$$=\frac{2(3\pm\sqrt{5})}{2}$$
$$x=3+\sqrt{5}\approx5.2 \text{ or}$$
$$x=3-\sqrt{5}\approx.8$$

Jaime's time cannot be .8, since Carlos' time would then be .8 − 2 or −1.2. So, Jaime's time alone is 5.2 hr, and Carlos' time alone is 3.2 hr.

4. **(a)** $x=\sqrt{7x-10}$

Square both sides.

$$x^2=(\sqrt{7x-10})^2$$
$$x^2=7x-10$$
$$x^2-7x+10=0$$
$$(x-2)(x-5)=0$$
$$x-2=0 \quad \text{or} \quad x-5=0$$
$$x=2 \quad \text{or} \quad x=5$$

Check each potential solution in the original equation.

Solution set: $\{2,5\}$

(b) $2x=\sqrt{x}+1$
$2x-1=\sqrt{x}$

Square both sides.

$$(2x-1)^2=(\sqrt{x})^2$$
$$4x^2-4x+1=x$$
$$4x^2-5x+1=0$$
$$(4x-1)(x-1)=0$$
$$4x-1=0 \quad \text{or} \quad x-1=0$$
$$x=\frac{1}{4} \quad \text{or} \quad x=1$$

Check each potential solution in the original solution.

Let $x=\frac{1}{4}$.

$$2\left(\frac{1}{4}\right)=\sqrt{\frac{1}{4}}+1 \quad ?$$
$$\frac{1}{2}=\frac{1}{2}+1 \quad ?$$
$$\frac{1}{2}=\frac{3}{2} \qquad \textit{False}$$

$\frac{1}{4}$ is not a solution.

Let $x = 1$.

$$\begin{aligned} 2(1) &= \sqrt{1} + 1 \quad ? \\ 2 &= 2 \qquad \textit{True} \end{aligned}$$

1 is a solution.

Solution set: $\{1\}$

5. **(a)** $m^4 - 10m^2 + 9 = 0$

Let $y = m^2$. Then $y^2 = (m^2)^2 = m^4$. The equation becomes

$$\begin{aligned} y^2 - 10y + 9 &= 0 \\ (y-9)(y-1) &= 0 \\ y - 9 = 0 \quad &\text{or} \quad y - 1 = 0 \\ y = 9 \quad &\text{or} \quad y = 1. \end{aligned}$$

To find m, substitute m^2 for y.

$$\begin{aligned} m^2 = 9 \quad &\text{or} \quad m^2 = 1 \\ m = \pm 3 \quad &\text{or} \quad m = \pm 1 \end{aligned}$$

The potential solutions all check.

Solution set: $\{-3, -1, 1, 3\}$

(b) $9k^4 - 37k^2 + 4 = 0$

Let $y = k^2$, so $y^2 = (k^2)^2 = k^4$. The equation becomes

$$\begin{aligned} 9y^2 - 37y + 4 &= 0 \\ (y-4)(9y-1) &= 0 \\ y - 4 = 0 \quad &\text{or} \quad 9y - 1 = 0 \\ y = 4 \quad &\text{or} \quad y = \frac{1}{9}. \end{aligned}$$

To find k, substitute k^2 for y.

$$\begin{aligned} k^2 = 4 \quad &\text{or} \quad k^2 = \frac{1}{9} \\ k = \pm 2 \quad &\text{or} \quad k = \pm\frac{1}{3} \end{aligned}$$

The potential solutions all check.

Solution set: $\{-2, -\frac{1}{3}, \frac{1}{3}, 2\}$

6. **(a)** $5(r+3)^2 + 9(r+3) = 2$

Let $x = r + 3$. The equation becomes

$$\begin{aligned} 5x^2 + 9x &= 2 \\ 5x^2 + 9x - 2 &= 0 \\ (5x-1)(x+2) &= 0 \\ 5x - 1 = 0 \quad &\text{or} \quad x + 2 = 0 \\ x = \frac{1}{5} \quad &\text{or} \quad x = -2. \end{aligned}$$

To find r, substitute $r + 3$ for x.

$$\begin{aligned} r + 3 = \frac{1}{5} \quad &\text{or} \quad r + 3 = -2 \\ r = -\frac{14}{5} \quad &\text{or} \quad r = -5 \end{aligned}$$

Both potential solutions check.

Solution set: $\{-5, -\frac{14}{5}\}$

(b) $4m^{2/3} = 3m^{1/3} + 1$

Let $x = m^{1/3}$ and $x^2 = m^{2/3}$. The equation becomes

$$\begin{aligned} 4x^2 &= 3x + 1 \\ 4x^2 - 3x - 1 &= 0 \\ (4x+1)(x-1) &= 0 \\ 4x + 1 = 0 \quad &\text{or} \quad x - 1 = 0 \\ x = -\frac{1}{4} \quad &\text{or} \quad x = 1. \end{aligned}$$

To find m, substitute $m^{1/3}$ for x.

$$m^{1/3} = -\frac{1}{4} \quad \text{or} \quad m^{1/3} = 1$$

Cube both sides of each equation.

$$\begin{aligned} (m^{1/3})^3 = \left(-\frac{1}{4}\right)^3 \quad &\text{or} \quad (m^{1/3})^3 = 1^3 \\ m = -\frac{1}{64} \quad &\text{or} \quad m = 1 \end{aligned}$$

Both potential solutions check.

Solution set: $\{-\frac{1}{64}, 1\}$

8.3 Section Exercises

3. $(r^2 + r)^2 - 8(r^2 + r) + 12 = 0$

To solve this equation, make a substitution for $r^2 + r$.

7. $3 - \frac{1}{t} = \frac{2}{t^2}$

Multiply each term by the common denominator, t^2.

$$\begin{aligned} t^2(3) - t^2\left(\frac{1}{t}\right) &= t^2\left(\frac{2}{t^2}\right) \\ 3t^2 - t &= 2 \\ 3t^2 - t - 2 &= 0 \\ (3t+2)(t-1) &= 0 \\ 3t + 2 = 0 \quad &\text{or} \quad t - 1 = 0 \\ 3t &= -2 \\ t = -\frac{2}{3} \quad &\text{or} \quad t = 1 \end{aligned}$$

Both potential solutions check.

Solution set: $\{-\frac{2}{3}, 1\}$

11. $\frac{2}{x+1} + \frac{3}{x+2} = \frac{7}{2}$

Multiply by the common denominator, $2(x+1)(x+2)$.

$$2(x+1)(x+2)\left(\frac{2}{x+1} + \frac{3}{x+2}\right)$$
$$= 2(x+1)(x+2)\left(\frac{7}{2}\right)$$

$$\begin{aligned} 2(x+2)(2) + 2(x+1)(3) &\\ = (x+1)(x+2)(7) &\\ 4x + 8 + 6x + 6 &= (x^2 + 3x + 2)(7) \\ 10x + 14 &= 7x^2 + 21x + 14 \\ 7x^2 + 11x &= 0 \\ x(7x + 11) &= 0 \\ x = 0 \quad \text{or} \quad 7x + 11 &= 0 \\ x &= -\frac{11}{7} \end{aligned}$$

Both potential solutions check.

Solution set: $\{-\frac{11}{7}, 0\}$

15. The grader's rate $= \dfrac{1 \text{ job}}{\text{time to finish}}$

$= \dfrac{1}{m}$ job per hour.

19. Let $x =$ Steve's speed.

Make a table. Use $d = rt$, or $t = \frac{d}{r}$.

	Distance	Rate	Time
Steve	300	x	$\frac{300}{x}$
Paula	300	$x - 20$	$\frac{300}{x-20}$

Steve's time = Paula's time − $1\frac{1}{4}$

$$\frac{300}{x} = \frac{300}{x-20} - \frac{5}{4}$$

Multiply by the common denominator, $4x(x-20)$.

$$4x(x-20)\left(\frac{300}{x}\right) = 4x(x-20)\left(\frac{300}{x-20} - \frac{5}{4}\right)$$

$$\begin{aligned} 4(x-20)(300) &= 4x(300) - 5x(x-20) \\ 1200x - 24{,}000 &= 1200x - 5x^2 + 100x \\ 5x^2 - 100x - 24{,}000 &= 0 \\ x^2 - 20x - 4800 &= 0 \\ (x-80)(x+60) &= 0 \\ x - 80 = 0 \quad \text{or} \quad x + 60 &= 0 \\ x = 80 \quad \text{or} \quad x &= -60 \end{aligned}$$

Speed can't be negative, so Steve's speed is 80 mph.

23. $2x = \sqrt{11x+3}$

Square both sides.

$$\begin{aligned} (2x)^2 &= (\sqrt{11x+3})^2 \\ 4x^2 &= 11x + 3 \\ 4x^2 - 11x - 3 &= 0 \\ (4x+1)(x-3) &= 0 \\ 4x + 1 = 0 \quad &\text{or} \quad x - 3 = 0 \\ x = -\frac{1}{4} \quad &\text{or} \quad x = 3 \end{aligned}$$

Since the square root must be nonnegative, $-\frac{1}{4}$ is not a solution. The potential solution, 3, checks.

Solution set: $\{3\}$

27. $p - 2\sqrt{p} = 8$

$p - 8 = 2\sqrt{p}$

Square both sides.

$$\begin{aligned} (p-8)^2 &= (2\sqrt{p})^2 \\ p^2 - 16p + 64 &= 4p \\ p^2 - 20p + 64 &= 0 \\ (p-4)(p-16) &= 0 \\ p - 4 = 0 \quad &\text{or} \quad p - 16 = 0 \\ p = 4 \quad &\text{or} \quad p = 16 \end{aligned}$$

Check each potential solution.

Let $p = 4$.

$$\begin{aligned} 4 - 2\sqrt{4} &= 8 \quad ? \\ 4 - 2(2) &= 8 \quad ? \\ 0 &= 8 \quad \textit{False} \end{aligned}$$

4 is not a solution.

Let $p = 16$.

$$\begin{aligned} 16 - 2\sqrt{16} &= 8 \quad ? \\ 16 - 2(4) &= 8 \quad ? \\ 16 - 8 &= 8 \quad ? \\ 8 &= 8 \quad \textit{True} \end{aligned}$$

16 is a solution.

Solution set: $\{16\}$

31. $t^4 - 18t^2 + 81 = 0$

Let $u = t^2$, so $u^2 = t^4$. The equation becomes

$$\begin{aligned} u^2 - 18u + 81 &= 0 \\ (u-9)^2 &= 0 \\ u - 9 &= 0 \\ u &= 9. \end{aligned}$$

To find t, substitute t^2 for u.

$$\begin{aligned} t^2 &= 9 \\ t = 3 \quad &\text{or} \quad t = -3 \end{aligned}$$

Both potential solutions check.

Solution set: $\{-3, 3\}$

35. $(x+3)^2 + 5(x+3) + 6 = 0$

Let $u = x + 3$. The equation becomes

$$\begin{aligned} u^2 + 5u + 6 &= 0 \\ (u+3)(u+2) &= 0 \\ u + 3 = 0 \quad \text{or} \quad u + 2 &= 0 \\ u = -3 \quad \text{or} \quad u &= -2. \end{aligned}$$

To find x, substitute $x + 3$ for u.

$$\begin{aligned} x + 3 = -3 \quad \text{or} \quad x + 3 &= -2 \\ x = -6 \quad \text{or} \quad x &= -5 \end{aligned}$$

Both potential solutions check.

Solution set: $\{-6, -5\}$

39. $2 + \dfrac{5}{3k-1} = \dfrac{-2}{(3k-1)^2}$

Let $u = 3k - 1$. The equation becomes

$$2 + \frac{5}{u} = -\frac{2}{u^2}.$$

Multiply both sides by u^2.

$$\begin{aligned} u^2\left(2 + \frac{5}{u}\right) &= u^2\left(-\frac{2}{u^2}\right) \\ 2u^2 + 5u &= -2 \\ 2u^2 + 5u + 2 &= 0 \\ (2u+1)(u+2) &= 0 \\ 2u + 1 = 0 \quad \text{or} \quad u + 2 &= 0 \\ u = -\frac{1}{2} \quad \text{or} \quad u &= -2 \end{aligned}$$

To find k, substitute $3k - 1$ for u.

$$\begin{aligned} 3k - 1 = -\frac{1}{2} \quad &\text{or} \quad 3k - 1 = -2 \\ 3k = \frac{1}{2} \quad &\phantom{\text{or}} \quad 3k = -1 \\ k = \frac{1}{6} \quad &\text{or} \quad k = -\frac{1}{3} \end{aligned}$$

Both potential solutions check.

Solution set: $\{-\frac{1}{3}, \frac{1}{6}\}$

43. $x^{2/3} + x^{1/3} - 2 = 0$

Let $u = x^{1/3}$. The equation becomes

$$\begin{aligned} u^2 + u - 2 &= 0 \\ (u+2)(u-1) &= 0 \\ u + 2 = 0 \quad \text{or} \quad u - 1 &= 0 \\ u = -2 \quad \text{or} \quad u &= 1. \end{aligned}$$

To find x, substitute $x^{1/3}$ for u.

$$x^{1/3} = -2 \quad \text{or} \quad x^{1/3} = 1$$

Cube both sides of each equation.

$$\begin{aligned} (x^{1/3})^3 = (-2)^3 \quad &\text{or} \quad (x^{1/3})^3 = 1^3 \\ x = -8 \quad &\text{or} \quad x = 1 \end{aligned}$$

Both potential solutions check.

Solution set: $\{-8, 1\}$

For Exercises 47-51, use the equation

$$\frac{x^2}{(x-3)^2} + \frac{3x}{x-3} - 4 = 0.$$

47. The number 3 cannot possibly be a solution because it would make the denominators zero. Division by zero is undefined.

48. $(x-3)^2\left(\dfrac{x^2}{(x-3)^2}\right) + (x-3)^2\left(\dfrac{3x}{(x-3)}\right)$

$$\begin{aligned} -(x-3)^2 \cdot 4 &= (x-3)^2 \cdot 0 \\ x^2 + 3x(x-3) - 4(x^2 - 6x + 9) &= 0 \\ x^2 + 3x^2 - 9x - 4x^2 + 24x - 36 &= 0 \\ 15x - 36 &= 0 \\ 15x &= 36 \\ x &= \frac{36}{15} \\ x &= \frac{12}{5} \end{aligned}$$

The solution is $\frac{12}{5}$.

49. $$\begin{aligned} \frac{x^2}{(x-3)^2} + \frac{3x}{x-3} - 4 &= 0 \\ \left(\frac{x}{x-3}\right)^2 + 3\left(\frac{x}{x-3}\right) - 4 &= 0 \end{aligned}$$

50. The expression $\frac{x}{x-3}$ cannot equal 1 because there is no value of x for which $x = x - 3$, that is, for which the numerator equals the denominator which equals 1.

51. Let $t = \frac{x}{x+3}$. The equation becomes

$$\begin{aligned} t^2 + 3t - 4 &= 0 \\ (t-1)(t+4) &= 0 \\ t - 1 = 0 \quad &\text{or} \quad t + 4 = 0 \\ t = 1 \quad &\text{or} \quad t = -4. \end{aligned}$$

To find x, substitute $\frac{x}{x-3}$ for t.

$$\frac{x}{x-3} = 1 \quad \text{or} \quad \frac{x}{x-3} = -4$$

The equation $\frac{x}{x-3} = 1$ has no solution since there is no value of x for which $x = x - 3$. (See Exercise 50.)

$$\begin{aligned} \frac{x}{x-3} &= -4 \\ x &= -4(x-3) \\ x &= -4x + 12 \\ 5x &= 12 \\ x &= \frac{12}{5} \end{aligned}$$

Solution set: $\{\frac{12}{5}\}$

52. $x^2(x-3)^{-2} + 3x(x-3)^{-1} - 4 = 0$

Let $s = (x-3)^{-1}$. Thus, $s = \frac{1}{x-3}$.

$$\begin{aligned} x^2s^2 + 3xs - 4 &= 0 \\ (xs+4)(xs-1) &= 0 \\ xs + 4 = 0 \quad &\text{or} \quad xs - 1 = 0 \\ s = -\frac{4}{x} \quad &\text{or} \quad s = \frac{1}{x} \end{aligned}$$

To find x, substitute $\frac{1}{x-3}$ for s.

$$\frac{1}{x-3} = -\frac{4}{x} \quad \text{or} \quad \frac{1}{x-3} = \frac{1}{x}$$

$$\begin{aligned} x &= -4(x-3) & x &= x - 3 \\ x &= -4x + 12 & 0 &= -3 \quad \textit{False} \\ 5x &= 12 \\ x &= \frac{12}{5} \end{aligned}$$

Solution set: $\{\frac{12}{5}\}$

8.4 Formulas and Applications Involving Quadratic Equations

8.4 Margin Exercises

1.
$$\begin{aligned} 5y^2 &= z^2 + 9x^2 \\ 5y^2 - z^2 &= 9x^2 \\ \frac{5y^2 - z^2}{9} &= x^2 \\ x &= \pm\sqrt{\frac{5y^2 - z^2}{9}} \\ x &= \frac{\pm\sqrt{5y^2 - z^2}}{3} \end{aligned}$$

Since x represents a length, it must be positive. The only solution is

$$x = \frac{\sqrt{5y^2 - z^2}}{3}.$$

2. $2t^2 - 5t + k = 0$

Solve for t. Use the quadratic formula with $a = 2, b = -5$, and $c = k$.

$$t = \frac{-b \pm \sqrt{b^2 - 4ac}}{2a}$$

$$t = \frac{-(-5) \pm \sqrt{(-5)^2 - 4(2)k}}{2(2)}$$

$$t = \frac{5 \pm \sqrt{25 - 8k}}{4}$$

The solutions are

$$t = \frac{5 + \sqrt{25 - 8k}}{4} \quad \text{and}$$

$$t = \frac{5 - \sqrt{25 - 8k}}{4}.$$

3. Let $x =$ the distance to the top of the ladder from the ground.

Then $x - 7 =$ the distance to the bottom of the ladder from the house.

The wall of the house is perpendicular to the ground, so this is a right triangle. Use the Pythagorean theorem.

$$\begin{aligned} a^2 + b^2 &= c^2 \\ x^2 + (x-7)^2 &= 13^2 \\ x^2 + x^2 - 14x + 49 &= 169 \\ 2x^2 - 14x - 120 &= 0 \\ x^2 - 7x - 60 &= 0 \\ (x-12)(x+5) &= 0 \end{aligned}$$

$$x - 12 = 0 \quad \text{or} \quad x + 5 = 0$$
$$x = 12 \quad \text{or} \quad x = -5$$

Since x represents a length, it must be positive, so -5 is not a solution. If $x = 12$, then $x - 7 = 12 - 7 = 5$. The bottom of the ladder is 5 ft from the house.

4. Let $x =$ the width of the grass strip.

The width of the large rectangle is $20 + 2x$, and the length is $40 + 2x$.

Area of big rectangle	$-$	area of pool	$=$	area of strip.

$$(20 + 2x)(40 + 2x) - 20(40) = 700$$
$$800 + 120x + 4x^2 - 800 = 700$$
$$4x^2 + 120x - 700 = 0$$
$$x^2 + 30x - 175 = 0$$
$$(x - 5)(x + 35) = 0$$
$$x - 5 = 0 \quad \text{or} \quad x + 35 = 0$$
$$x = 5 \quad \text{or} \quad x = -35$$

The width cannot be -35, so the strip should be 5 ft wide.

5. Let $x =$ the time in hours for the experienced worker,

$x + 1 =$ time in hours for the new worker, and

$5 =$ time in hours for both working together.

Then $\frac{1}{x} =$ fraction of the job done by the experienced worker,

$\frac{1}{x+1} =$ fraction of the job done by the new worker, and

$\frac{1}{5} =$ fraction of the job done together in one hour.

Solve the equation.

$$\frac{1}{x} + \frac{1}{x+1} = \frac{1}{5}$$

Multiply both sides by $5x(x + 1)$.

$$5x(x+1)\left(\frac{1}{x} + \frac{1}{x+1}\right) = 5x(x+1)\left(\frac{1}{5}\right)$$
$$5(x + 1) + 5x = x(x + 1)$$
$$5x + 5 + 5x = x^2 + x$$
$$0 = x^2 - 9x - 5$$

Use the quadratic formula with $a = 1, b = -9$, and $c = -5$.

$$x = \frac{-b \pm \sqrt{b^2 - 4ac}}{2a}$$
$$x = \frac{-(-9) \pm \sqrt{(-9)^2 - 4(1)(-5)}}{2(1)}$$
$$= \frac{9 \pm \sqrt{81 + 20}}{2}$$
$$= \frac{9 \pm \sqrt{101}}{2}$$
$$x = \frac{9 + \sqrt{101}}{2} \approx 9.5$$
$$\text{or} \quad x = \frac{9 - \sqrt{101}}{2} \approx -.5$$

Time cannot be negative, so the experienced worker requires 9.5 hr to clean the building alone.

6. Use a calculator.

$$\frac{1.22 \pm \sqrt{(-1.22)^2 - 4(.011)(23.625)}}{2(.011)}$$
$$\approx \frac{1.22 \pm .67}{.022}$$
$$\frac{1.22 + .67}{.022} \approx 85.9$$
$$\frac{1.22 - .67}{.022} = 25$$

The smaller positive solution is 25.

8.4 Section Exercises

3. $d = kt^2$

Solve for t.

$$kt^2 = d$$

Divide both sides by k.

$$t^2 = \frac{d}{k}$$
$$t = \pm\sqrt{\frac{d}{k}}$$

Rationalize the denominator.

$$t = \frac{\pm\sqrt{d}}{\sqrt{k}} \cdot \frac{\sqrt{k}}{\sqrt{k}}$$
$$t = \frac{\pm\sqrt{dk}}{k}$$

7. $F = \dfrac{kA}{v^2}$

Solve for v. Multiply both sides by v^2.

$$v^2F = v^2\left(\frac{kA}{v^2}\right)$$
$$v^2F = kA$$

Divide both sides by F.

$$v^2 = \frac{kA}{F}$$
$$v = \pm\sqrt{\frac{kA}{F}}$$

Rationalize the denominator.

$$v = \frac{\pm\sqrt{kA}}{\sqrt{F}} \cdot \frac{\sqrt{F}}{\sqrt{F}}$$
$$v = \frac{\pm\sqrt{kAF}}{F}$$

11. $At^2 + Bt = -C$

Solve for t.

$$At^2 + Bt + C = 0$$

Use the quadratic formula.

$$t = \frac{-B \pm \sqrt{B^2 - 4AC}}{2A}$$

15. $p = \sqrt{\dfrac{k\ell}{g}}$

Solve for ℓ. Square both sides.

$$p^2 = \left(\sqrt{\frac{k\ell}{g}}\right)^2$$
$$p^2 = \frac{k\ell}{g}$$
$$p^2g = k\ell$$
$$\frac{p^2g}{k} = \ell$$

19. Let $x =$ the distance traveled by the eastbound ship and
$x + 70 =$ the distance traveled by the southbound ship.

Since the ships are traveling at right angles to one another, the distance, d, between them can be found using the Pythagorean theorem.

$$c^2 = a^2 + b^2$$
$$d^2 = x^2 + (x + 70)^2$$

Let $d = 170$, and solve for x.

$$170^2 = x^2 + (x + 70)^2$$
$$28,900 = x^2 + x^2 + 140x + 4900$$
$$0 = 2x^2 + 140x - 24,000$$
$$0 = x^2 + 70x - 12,000$$
$$0 = (x + 150)(x - 80)$$
$$x + 150 = 0 \quad \text{or} \quad x - 80 = 0$$
$$x = -150 \quad \text{or} \quad x = 80$$

Distance cannot be negative, so reject -150. If $x = 80$, then $x + 70 = 80 + 70 = 150$. The eastbound ship traveled 80 mi, and the southbound ship traveled 150 mi.

23. Let $x =$ the amount removed from one dimension.

The area of the square is 256 cm^2, so the length of one side is $\sqrt{256}$ or 16 cm. The dimensions of the new rectangle are $16 + x$ and $16 - x$ cm. The area of the new rectangle is 16 cm^2 less than the area of the square. So,

$$(16 + x)(16 - x) = 256 - 16$$
$$256 - x^2 = 240$$
$$-x^2 = -16$$
$$x^2 = 16$$
$$x = 4 \quad \text{or} \quad x = -4.$$

Length cannot be negative, so reject -4. If $x = 4$, then $16 + x = 20$, and $16 - x = 12$. The dimensions are 20 cm by 12 cm.

27. Let $x =$ time for Carmen working alone.

Make a table.

Worker	Rate	Time together	Part of job done
Carmen	$\frac{1}{x}$	2	$\frac{2}{x}$
Paul	$\frac{1}{x - \frac{1}{2}}$	2	$\frac{2}{x - \frac{1}{2}}$

Part done by Carmen	+	part done by Paul	=	1 whole job.
$\frac{2}{x}$	+	$\frac{2}{x - \frac{1}{2}}$	=	1

Multiply both sides by $x\left(x-\frac{1}{2}\right)$.

$$x\left(x-\frac{1}{2}\right)\left(\frac{2}{x}+\frac{2}{x-\frac{1}{2}}\right)=x\left(x-\frac{1}{2}\right)\cdot 1$$
$$2\left(x-\frac{1}{2}\right)+2x=x^2-\frac{1}{2}x$$
$$2x-1+2x=x^2-\frac{1}{2}x$$
$$4x-1=x^2-\frac{1}{2}x$$

Multiply both sides by 2.

$$8x-2=2x^2-x$$
$$0=2x^2-9x+2$$

Use the quadratic formula.
$a=2, b=-9, c=2$

$$x=\frac{-b\pm\sqrt{b^2-4ac}}{2a}$$
$$x=\frac{-(-9)\pm\sqrt{(-9)^2-4(2)(2)}}{2(2)}$$
$$=\frac{9\pm\sqrt{65}}{4}$$
$$x=\frac{9+\sqrt{65}}{4}\approx 4.3 \quad \text{or}$$
$$x=\frac{9-\sqrt{65}}{4}\approx .2$$

Reject about .2 for Carmen's time, because that would make Paul's time negative. So, Carmen's time is about 4.3 hr.

31. $x=0$ corresponds to 1984. Let $x=0$ in the quadratic model.

$$f(x)=18.7x^2+105.3x+4814.1$$
$$f(0)=18.7(0)^2+105.3(0)+4814.1$$
$$=4814.1$$

The threshold was \$4814.10 in 1984.

35. $D(t)=13t^2-100t$

Let $D(t)=180$, and solve for t.

$$180=13t^2-100t$$
$$13t^2-100t-180=0$$

Use the quadratic formula.
$a=13, b=-100, c=-180$

$$t=\frac{-b\pm\sqrt{b^2-4ac}}{2a}$$
$$t=\frac{100\pm\sqrt{(-100)^2-4(13)(-180)}}{2(13)}$$
$$=\frac{100\pm\sqrt{10,000+9360}}{26}$$
$$=\frac{100\pm\sqrt{19,360}}{26}$$
$$\approx\frac{100\pm 139.1}{26}$$
$$t\approx\frac{100+139.1}{26}\approx 9.2 \quad \text{or}$$
$$t\approx\frac{100-139.1}{26}\approx -1.5$$

Since time cannot be negative, the time is 9.2 sec.

39. Let r = the interest rate.

$$A=P(1+r)^2$$

Let $A=2142.25$ and $P=2000$. Solve for r.

$$2142.25=2000(1+r)^2$$
$$1.071125=(1+r)^2$$

Use the square root property.

$$1+r\approx 1.035 \quad \text{or} \quad 1+r\approx -1.035$$
$$r\approx .035 \quad \text{or} \quad r\approx -2.035$$

Since interest cannot be negative, reject -2.035. The interest rate is 3.5%.

43. Write a proportion.

$$\frac{x-4}{3x-19}=\frac{4}{x-3}$$

Multiply by the LCD, $(3x-19)(x-3)$.

$$(3x-19)(x-3)\left(\frac{x-4}{3x-19}\right)$$
$$=(3x-19)(x-3)\left(\frac{4}{x-3}\right)$$
$$(x-3)(x-4)=(3x-19)4$$
$$x^2-7x+12=12x-76$$
$$x^2-19x+88=0$$
$$(x-8)(x-11)=0$$
$$x-8=0 \quad \text{or} \quad x-11=0$$
$$x=8 \quad \text{or} \quad x=11$$

If $x=8$, then

$$3x-19=3(8)-19=5.$$

If $x = 11$, then

$$3x - 19 = 3(11) - 19 = 14.$$

Thus, $AC = 5$ or $AC = 14$.

8.5 Nonlinear and Fractional Inequalities

8.5 Margin Exercises

1. $x^2 - x - 12 > 0$

Let $x = 5$ in the inequality.

$$\begin{aligned} 5^2 - 5 - 12 &> 0 \quad ? \\ 25 - 5 - 12 &> 0 \quad ? \\ 8 &> 0 \quad \textit{True} \end{aligned}$$

Yes, 5 satisfies $x^2 - x - 12 > 0$.

For Exercises 2-5, see the answer graphs for the margin exercises in the textbook.

2. **(a)** $x^2 + x - 6 > 0$

Use factoring to solve the quadratic equation

$$\begin{aligned} x^2 + x - 6 &= 0 \\ (x+3)(x-2) &= 0 \\ x + 3 = 0 \quad &\text{or} \quad x - 2 = 0 \\ x = -3 \quad &\text{or} \quad x = 2. \end{aligned}$$

Locate the numbers -3 and 2 that divide the number line into three regions A, B, and C.

A B C
-3 2

Choose a number from each region to substitute in the inequality

$$x^2 + x - 6 > 0.$$

Region A: Let $x = -4$.

$$\begin{aligned} (-4)^2 + (-4) - 6 &> 0 \quad ? \\ 16 - 4 - 6 &> 0 \quad ? \\ 6 &> 0 \quad \textit{True} \end{aligned}$$

Region B: Let $x = 0$.

$$\begin{aligned} 0^2 + 0 - 6 &> 0 \quad ? \\ -6 &> 0 \quad \textit{False} \end{aligned}$$

Region C: Let $x = 3$.

$$\begin{aligned} 3^2 + 3 - 6 &> 0 \quad ? \\ 9 + 3 - 6 &> 0 \quad ? \\ 6 &> 0 \quad \textit{True} \end{aligned}$$

The numbers in Regions A and C are solutions. The numbers -3 and 2 are not included because of "$>$."

Solution set: $(-\infty, -3) \cup (2, \infty)$

(b) $3m^2 - 13m - 10 \leq 0$

Solve the quadratic equation

$$\begin{aligned} 3m^2 - 13m - 10 &= 0 \\ (3m+2)(m-5) &= 0 \\ 3m + 2 = 0 \quad &\text{or} \quad m - 5 = 0 \\ m = -\frac{2}{3} \quad &\text{or} \quad m = 5. \end{aligned}$$

Locate these numbers on a number line.

A B C
-2/3 5

Choose a number from each region to substitute in the inequality

$$3m^2 - 13m - 10 \leq 0.$$

Region A: Let $m = -1$.

$$\begin{aligned} 3(-1)^2 - 13(-1) - 10 &\leq 0 \quad ? \\ 3 + 13 - 10 &\leq 0 \quad ? \\ 6 &\leq 0 \quad \textit{False} \end{aligned}$$

Region B: Let $m = 0$.

$$\begin{aligned} 3(0)^2 - 13(0) - 10 &\leq 0 \quad ? \\ -10 &\leq 0 \quad \textit{True} \end{aligned}$$

Region C: Let $m = 6$.

$$\begin{aligned} 3(6)^2 - 13(6) - 10 &\leq 0 \quad ? \\ 108 - 78 - 10 &\leq 0 \quad ? \\ 20 &\leq 0 \quad \textit{False} \end{aligned}$$

The numbers in region B, including $-\frac{2}{3}$ and 5 because of "$\leq$," are solutions.

Solution set: $\left[-\frac{2}{3}, 5\right]$

3. **(a)** $(y-3)(y+2)(y+1) > 0$

Solve the equation

$$(y-3)(y+2)(y+1) = 0$$

$$y-3=0 \quad \text{or} \quad y+2=0 \quad \text{or} \quad y+1=0$$
$$y=3 \quad \text{or} \quad y=-2 \quad \text{or} \quad y=-1.$$

Locate these numbers and the regions A, B, C, and D on a number line.

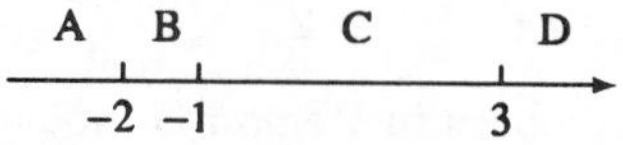

Test a number from each region in the inequality

$$(y-3)(y+2)(y+1)>0.$$

Region A: Let $y=-3$.

$$(-3-3)(-3+2)(-3+1)>0 \quad ?$$
$$-6(-1)(-2)>0 \quad ?$$
$$-12>0 \qquad \textit{False}$$

Region B: Let $y=-\frac{3}{2}$.

$$\left(-\frac{3}{2}-3\right)\left(-\frac{3}{2}+2\right)\left(-\frac{3}{2}+1\right)>0 \quad ?$$
$$-\frac{9}{2}\left(\frac{1}{2}\right)\left(-\frac{1}{2}\right)>0 \quad ?$$
$$\frac{9}{8}>0 \qquad \textit{True}$$

Region C: Let $y=0$.

$$(0-3)(0+2)(0+1)>0 \quad ?$$
$$-3(2)(1)>0 \quad ?$$
$$-6>0 \qquad \textit{False}$$

Region D: Let $y=4$.

$$(4-3)(4+2)(4+1)>0 \quad ?$$
$$1(6)(5)>0 \quad ?$$
$$30>0 \qquad \textit{True}$$

The numbers in Regions B and D, not including $-2, -1$, or 3 because of "$>$," are solutions.

Solution set: $(-2,-1)\cup(3,\infty)$

(b) $(k-5)(k+1)(k-3)\le 0$

Solve the equation

$$(k-5)(k+1)(k-3)=0$$

$$k-5=0 \quad \text{or} \quad k+1=0 \quad \text{or} \quad k-3=0$$
$$k=5 \quad \text{or} \quad k=-1 \quad \text{or} \quad k=3.$$

Locate these numbers on a number line.

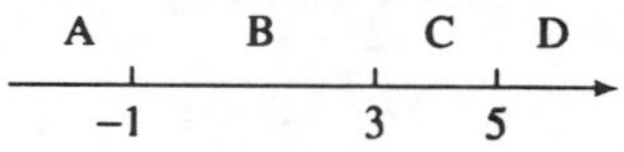

Test a number from each region in the inequality

$$(k-5)(k+1)(k-3)\le 0.$$

Region A: Let $k=-2$.

$$(-2-5)(-2+1)(-2-3)\le 0 \quad ?$$
$$-7(-1)(-5)\le 0 \quad ?$$
$$-35\le 0 \qquad \textit{True}$$

Region B: Let $k=0$.

$$(0-5)(0+1)(0-3)\le 0 \quad ?$$
$$-5(1)(-3)\le 0 \quad ?$$
$$15\le 0 \qquad \textit{False}$$

Region C: Let $k=4$.

$$(4-5)(4+1)(4-3)\le 0 \quad ?$$
$$-1(5)(1)\le 0 \quad ?$$
$$-5\le 0 \qquad \textit{True}$$

Region D: Let $k=6$.

$$(6-5)(6+1)(6-3)\le 0 \quad ?$$
$$1(7)(3)\le 0 \quad ?$$
$$21\le 0 \qquad \textit{False}$$

The numbers in Regions A and C, including $-1, 3$, and 5 because of "$\le$," are solutions.

Solution set: $(-\infty,-1]\cup[3,5]$

4. **(a)** $\dfrac{2}{x-4}<3$

Solve the equation

$$\frac{2}{x-4}=3$$
$$2=3(x-4)$$
$$2=3x-12$$
$$\frac{14}{3}=x.$$

If $x-4=0$, then $x=4$ makes the denominator 0. Locate $\frac{14}{3}$ and 4 to determine Regions A, B, and C on a number line.

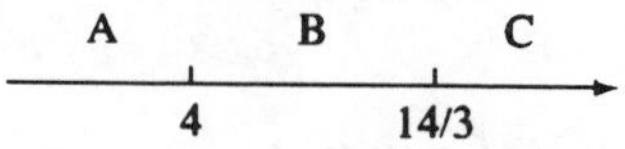

Test a number from each region in the inequality

$$\frac{2}{x-4}<3.$$

Region A: Let $x = 0$.

$$\frac{2}{0-4} < 3 \quad ?$$
$$-\frac{1}{2} < 3 \qquad \textit{True}$$

Region B: Let $x = \frac{13}{3}$.

$$\frac{2}{\frac{13}{3}-4} < 3 \quad ?$$
$$6 < 3 \qquad \textit{False}$$

Region C: Let $x = 5$.

$$\frac{2}{5-4} < 3 \quad ?$$
$$2 < 3 \qquad \textit{True}$$

The numbers in Regions A and C, not including 4 or $\frac{14}{3}$, are solutions.

Solution set: $(-\infty, 4) \cup (\frac{14}{3}, \infty)$

(b) $\dfrac{5}{y+1} > 4$

Solve the equation

$$\frac{5}{y+1} = 4$$
$$5 = 4(y+1)$$
$$5 = 4y+4$$
$$\frac{1}{4} = y.$$

If $y+1=0$, then $y=-1$ makes the denominator 0. Locate $\frac{1}{4}$, -1, and Regions A, B, and C on a number line.

A B C
-1 1/4

Test a number from each region in the inequality

$$\frac{5}{y+1} > 4.$$

Region A: Let $y = -2$.

$$\frac{5}{-2+1} > 4 \quad ?$$
$$-5 > 4 \qquad \textit{False}$$

Region B: Let $y = 0$.

$$\frac{5}{0+1} > 4 \quad ?$$
$$5 > 4 \qquad \textit{True}$$

Region C: Let $y = 1$.

$$\frac{5}{1+1} > 4 \quad ?$$
$$\frac{5}{2} > 4 \qquad \textit{False}$$

The numbers in Region B, not including -1 or $\frac{1}{4}$, are solutions.

Solution set: $(-1, \frac{1}{4})$

5. $\dfrac{k+2}{k-1} \leq 5$

Solve the equation

$$\frac{k+2}{k-1} = 5$$
$$k+2 = 5(k-1)$$
$$k+2 = 5k-5$$
$$7 = 4k$$
$$k = \frac{7}{4}.$$

If $k-1=0$, then $k=1$ makes the denominator 0. The numbers $\frac{7}{4}$ and 1 determine three regions on a number line.

A B C
1 7/4

Test a number from each region in the inequality

$$\frac{k+2}{k-1} \leq 5.$$

Region A: Let $k = 0$.

$$\frac{0+2}{0-1} \leq 5 \quad ?$$
$$-2 \leq 5 \qquad \textit{True}$$

Region B: Let $k = \frac{3}{2}$.

$$\frac{\frac{3}{2}+2}{\frac{3}{2}-1} \leq 5 \quad ?$$
$$\frac{\frac{7}{2}}{\frac{1}{2}} \leq 5 \quad ?$$
$$7 \leq 5 \qquad \textit{False}$$

Region C: Let $k = 2$.

$$\frac{2+2}{2-1} \leq 5 \quad ?$$
$$4 \leq 5 \qquad \textit{True}$$

The numbers in Regions A and C are solutions. 1 is not in the solution set (since it makes the denominator 0), but $\frac{7}{4}$ is.

Solution set: $(-\infty, 1) \cup \left[\frac{7}{4}, \infty\right)$

6. $\dfrac{m-2}{m+2} \le 2$

Subtract 2 so one side of the inequality is 0.

$$\frac{m-2}{m+2} - 2 \le 0$$

Use $m+2$ as the common denominator.

$$\frac{m-2}{m+2} - \frac{2(m+2)}{m+2} \le 0$$
$$\frac{m-2-2m-4}{m+2} \le 0$$
$$\frac{-m-6}{m+2} \le 0$$

The number -6 makes the numerator 0, and -2 makes the denominator 0. Use these numbers to divide a number line into regions.

A B C

-6 -2

Test a number from each region in the inequality

$$\frac{m-2}{m+2} \le 2.$$

Region A: Let $m = -7$.

$$\frac{-7-2}{-7+2} \le 2 \quad ?$$
$$\frac{-9}{-5} \le 2 \quad ?$$
$$\frac{9}{5} \le 2 \quad \textit{True}$$

Region B: Let $m = -4$.

$$\frac{-4-2}{-4+2} \le 2 \quad ?$$
$$\frac{-6}{-2} \le 2 \quad ?$$
$$3 \le 2 \quad \textit{False}$$

Region C: Let $m = 0$.

$$\frac{0-2}{0+2} \le 2 \quad ?$$
$$-1 \le 2 \quad \textit{True}$$

The numbers in Regions A and C, including -6 but not -2 (since it makes the denominator 0), are solutions.

Solution set: $(-\infty, -6] \cup (-2, \infty)$

7. **(a)** $(3k-2)^2 > -2$

The square of any real number is always greater than or equal to 0, so any real number satisfies this inequality.

Solution set: $(-\infty, \infty)$

(b) $(5z+3)^2 < -3$

The square of a real number is never negative.

Solution set: $\emptyset$

8.5 Section Exercises

3. The solution set of the inequality $x^2 + x - 12 \ge 0$ is outside the solution of the inequality $x^2+x-12 < 0$. So, the solution set is $(-\infty, -4] \cup [3, \infty)$.

For Exercises 7-31, see the answer graphs in the back of the textbook.

7. $(r+4)(r-6) < 0$

Solve the equation

$$(r+4)(r-6) = 0$$
$$r+4 = 0 \quad \text{or} \quad r-6 = 0$$
$$r = -4 \quad \text{or} \quad r = 6.$$

These numbers divide the number line into three regions.

A B C

-4 6

Test a number from each region in the inequality

$$(r+4)(r-6) < 0.$$

Region A: Let $r = -5$.

$$(-5+4)(-5-6) < 0 \quad ?$$
$$-1(-11) < 0 \quad ?$$
$$11 < 0 \quad \textit{False}$$

Region B: Let $r = 0$.

$$(0+4)(0-6) < 0 \quad ?$$
$$4(-6) < 0 \quad ?$$
$$-24 < 0 \quad \textit{True}$$

Region C: Let $r = 7$.

$$(7+4)(7-6) < 0 \quad ?$$
$$11(1) < 0 \quad ?$$
$$11 < 0 \quad \textit{False}$$

The numbers in Region B, not including -4 or 6 because of " <," are solutions.

Solution set: $(-4, 6)$

11. $$10a^2 + 9a \geq 9$$
$$10a^2 + 9a - 9 \geq 0$$

Solve the equation

$$10a^2 + 9a - 9 = 0$$
$$(2a + 3)(5a - 3) = 0$$
$$2a + 3 = 0 \quad \text{or} \quad 5a - 3 = 0$$
$$a = -\frac{3}{2} \quad \text{or} \quad a = \frac{3}{5}.$$

These numbers divide the number line into three regions.

A B C
−3/2 3/5

Test a number from each region in the inequality

$$10a^2 + 9a \geq 9.$$

Region A: Let $a = -2$.

$$10(-2)^2 + 9(-2) \geq 9 \quad ?$$
$$40 - 18 \geq 9 \quad ?$$
$$22 \geq 9 \quad \textit{True}$$

Region B: Let $a = 0$.

$$10(0)^2 + 9(0) \geq 9 \quad ?$$
$$0 \geq 9 \quad \textit{False}$$

Region C: Let $a = 1$.

$$10(1)^2 + 9(1) \geq 9 \quad ?$$
$$10 + 9 \geq 9 \quad ?$$
$$19 \geq 9 \quad \textit{True}$$

The numbers in Regions A and C, including $-\frac{3}{2}$ and $\frac{3}{5}$ because of "$\geq$," are solutions.

Solution set: $\left(-\infty, -\frac{3}{2}\right] \cup \left[\frac{3}{5}, \infty\right)$

15. $$6x^2 + x \geq 1$$
$$6x^2 + x - 1 \geq 0$$

Solve the equation

$$6x^2 + x - 1 = 0$$
$$(2x + 1)(3x - 1) = 0$$
$$2x + 1 = 0 \quad \text{or} \quad 3x - 1 = 0$$
$$x = -\frac{1}{2} \quad \text{or} \quad x = \frac{1}{3}.$$

These numbers divide the number line into three regions.

A B C
−1/2 1/3

Test a number from each region in the inequality

$$6x^2 + x \geq 1.$$

Region A: Let $x = -1$.

$$6(-1)^2 + (-1) \geq 1 \quad ?$$
$$6 - 1 \geq 1 \quad ?$$
$$5 \geq 1 \quad \textit{True}$$

Region B: Let $x = 0$.

$$6(0)^2 + 0 \geq 1 \quad ?$$
$$0 \geq 1 \quad \textit{False}$$

Region C: Let $x = 1$.

$$6(1)^2 + 1 \geq 1 \quad ?$$
$$6 + 1 \geq 1 \quad ?$$
$$7 \geq 1 \quad \textit{True}$$

The numbers in Regions A and C, including $-\frac{1}{2}$ and $\frac{1}{3}$, are solutions.

Solution set: $\left(-\infty, -\frac{1}{2}\right] \cup \left[\frac{1}{3}, \infty\right)$

19. $(p - 1)(p - 2)(p - 4) < 0$

Solve the equation

$$(p - 1)(p - 2)(p - 4) = 0$$
$$p - 1 = 0 \quad \text{or} \quad p - 2 = 0 \quad \text{or} \quad p - 4 = 0$$
$$p = 1 \quad \text{or} \quad p = 2 \quad \text{or} \quad p = 4.$$

These numbers divide the number line into four regions.

A B C D
1 2 4

Test a number from each region in the inequality

$$(p - 1)(p - 2)(p - 4) < 0.$$

Region A: Let $p = 0$.

$$(0 - 1)(0 - 2)(0 - 4) < 0 \quad ?$$
$$-1(-2)(-4) < 0 \quad ?$$
$$-8 < 0 \quad \textit{True}$$

Region B: Let $p = 1.5$.

$$(1.5 - 1)(1.5 - 2)(1.5 - 4) < 0 \quad ?$$
$$.5(-.5)(-2.5) < 0 \quad ?$$
$$.625 < 0 \quad \textit{False}$$

Region C: Let $p = 3$.

$$\begin{aligned}(3-1)(3-2)(3-4) &< 0 \quad ? \\ 2(1)(-1) &< 0 \quad ? \\ -2 &< 0 \quad \textit{True}\end{aligned}$$

Region D: Let $p = 5$.

$$\begin{aligned}(5-1)(5-2)(5-4) &< 0 \quad ? \\ 4(3)(1) &< 0 \quad ? \\ 12 &< 0 \quad \textit{False}\end{aligned}$$

The numbers in Regions A and C, not including 1, 2, or 4, are solutions.

Solution set: $(-\infty, 1) \cup (2, 4)$

23. $\frac{x-1}{x-4} > 0$

Solve the equation

$$\begin{aligned}\frac{x-1}{x-4} &= 0 \\ x-1 &= 0 \\ x &= 1.\end{aligned}$$

Set the denominator equal to zero and solve.

$$\begin{aligned}x-4 &= 0 \\ x &= 4\end{aligned}$$

These numbers divide the number line into three regions.

A		B		C
	1		4	

Test a number from each region in the inequality

$$\frac{x-1}{x-4} > 0.$$

Region A: Let $x = 0$.

$$\begin{aligned}\frac{0-1}{0-4} &> 0 \quad ? \\ \frac{1}{4} &> 0 \quad \textit{True}\end{aligned}$$

Region B: Let $x = 2$.

$$\begin{aligned}\frac{2-1}{2-4} &> 0 \quad ? \\ \frac{1}{-2} &> 0 \quad \textit{False}\end{aligned}$$

Region C: Let $x = 5$.

$$\begin{aligned}\frac{5-1}{5-4} &> 0 \quad ? \\ 4 &> 0 \quad \textit{True}\end{aligned}$$

The numbers in Regions A and C, not including 1 or 4, are solutions.

Solution set: $(-\infty, 1) \cup (4, \infty)$

27. $\frac{8}{x-2} \geq 2$

Solve the equation

$$\begin{aligned}\frac{8}{x-2} &= 2 \\ 8 &= 2(x-2) \\ 8 &= 2x-4 \\ 12 &= 2x \\ 6 &= x.\end{aligned}$$

Set the denominator equal to zero and solve.

$$\begin{aligned}x-2 &= 0 \\ x &= 2\end{aligned}$$

These numbers divide the number line into three regions.

A		B		C
	2		6	

Test a number from each region in the inequality.

$$\frac{8}{x-2} \geq 2.$$

Region A: Let $x = 0$.

$$\begin{aligned}\frac{8}{0-2} &\geq 2 \quad ? \\ -4 &\geq 2 \quad \textit{False}\end{aligned}$$

Region B: Let $x = 3$.

$$\begin{aligned}\frac{8}{3-2} &\geq 2 \quad ? \\ 8 &\geq 2 \quad \textit{True}\end{aligned}$$

Region C: Let $x = 7$.

$$\begin{aligned}\frac{8}{7-2} &\geq 2 \quad ? \\ \frac{8}{5} &\geq 2 \quad \textit{False}\end{aligned}$$

The numbers in Region B, including 6 (because of "$\geq$") but not 2 (because it makes the denominator 0), are solutions.

Solution set: $(2, 6]$

31. $\frac{a}{a+2} \geq 2$

Solve the equation

$$\frac{a}{a+2} = 2$$

$$\begin{aligned} a &= 2(a+2) \\ a &= 2a+4 \\ -a &= 4 \\ a &= -4. \end{aligned}$$

Set the denominator equal to zero and solve.

$$\begin{aligned} a+2 &= 0 \\ a &= -2 \end{aligned}$$

These numbers divide the number line into three regions.

A B C
-4 -2

Test a number from each region in the inequality

$$\frac{a}{a+2} \geq 2.$$

Region A: Let $a = -5$.

$$\frac{-5}{-5+2} \geq 2 \quad ?$$

$$\frac{5}{3} \geq 2 \qquad \textit{False}$$

Region B: Let $a = -3$.

$$\frac{-3}{-3+2} \geq 2 \quad ?$$

$$3 \geq 2 \qquad \textit{True}$$

Region C: Let $a = 0$.

$$\frac{0}{0+2} \geq 2 \quad ?$$

$$0 \geq 2 \qquad \textit{False}$$

The numbers in Region B, including -4 but not -2 (because it makes the denominator 0), are solutions.

Solution set: $[-4, -2)$

35. $(4-3x)^2 \geq -2$

The square of any real number is always greater than or equal to 0, so any real number satisfies this inequality.

Solution set: $(-\infty, \infty)$

In Exercises 39-42, use the quadratic function

$$s(t) = -16t^2 + 256t.$$

39. Let $s(t) = 624$.

$$\begin{aligned} 624 &= -16t^2 + 256t \\ -16t^2 + 256t - 624 &= 0 \\ 16t^2 - 256t + 624 &= 0 \end{aligned}$$

Divide by 16.

$$\begin{aligned} t^2 - 16t + 39 &= 0 \\ (t-3)(t-13) &= 0 \\ t-3 = 0 \quad \text{or} \quad t-13 &= 0 \\ t = 3 \quad \text{or} \quad t &= 13 \end{aligned}$$

The rock will be 624 ft above the ground at 3 sec and again at 13 sec.

40. If $s(t) > 624$, then

$$\begin{aligned} -16t^2 + 256t &> 624 \\ -16t^2 + 256t - 624 &> 0. \end{aligned}$$

Divide by -16.

$$t^2 - 16t + 39 < 0$$

Solve the equation

$$t^2 - 16t + 39 = 0$$

as in Exercise 39 to find that $t = 3$ or $t = 13$. These numbers divide the number line into three regions.

A B C
3 13

Check a point from each region in the inequality

$$t^2 - 16t + 39 < 0.$$

Region A: Let $t = 0$.

$$\begin{aligned} 0^2 - 16(0) + 39 &< 0 \quad ? \\ 39 &< 0 \qquad \textit{False} \end{aligned}$$

Region B: Let $t = 4$.

$$\begin{aligned} 4^2 - 16(4) + 39 &< 0 \quad ? \\ 16 - 64 + 39 &< 0 \quad ? \\ -9 &< 0 \qquad \textit{True} \end{aligned}$$

Region C: Let $t = 14$.

$$\begin{aligned} 14^2 - 16(14) + 39 &< 0 \quad ? \\ 196 - 224 + 39 &< 0 \quad ? \\ 11 &< 0 \qquad \textit{False} \end{aligned}$$

The solution set includes the points in Region B, not including the endpoints, that is $(3, 13)$. Thus, the rock will be more than 624 ft above the ground between 3 sec and 13 sec.

41. Let $s(t) = 0$.

$$-16t^2 + 256t = 0$$
$$-16t(t - 16) = 0$$
$$-16t = 0 \quad \text{or} \quad t - 16 = 0$$
$$t = 0 \quad \text{or} \quad t = 16$$

The rock will be at ground level at 0 sec (the time when it is initially projected) and at 16 sec (the time when it hits the ground).

42. If $s(t) < 624$, then

$$-16t^2 + 256t < 624$$
$$-16t^2 + 256t - 624 < 0.$$

Divide by -16.

$$t^2 - 16t + 39 > 0$$

As in Exercise 40, the equation

$$t^2 - 16t + 39 = 0$$

has solutions $t = 3$ or $t = 13$. The same regions are indicated on a number line, although this time we must check a point from each of the regions in the inequality

$$t^2 - 16t + 39 > 0.$$

From Exercise 40, points from regions A and C had results greater than zero. The solution set includes the points in these regions, not including the endpoints, and also limited by 0 and 16 from Exercise 41, the times when the rock is on the ground, that is, $(0, 3) \cup (13, 16)$. The rock will be less than 624 ft above the ground between 0 and 3 sec and between 13 and 16 sec.

Chapter 8 Review Exercises

1. $t^2 = 121$

$t = 11 \quad \text{or} \quad t = -11$

Solution set: $\{-11, 11\}$

2. $p^2 = 3$

$p = \sqrt{3} \quad \text{or} \quad p = -\sqrt{3}$

Solution set: $\{-\sqrt{3}, \sqrt{3}\}$

3. $(2x + 5)^2 = 100$

$$2x + 5 = 10 \quad \text{or} \quad 2x + 5 = -10$$
$$2x = 5 \qquad\qquad 2x = -15$$
$$x = \frac{5}{2} \quad \text{or} \quad x = -\frac{15}{2}$$

Solution set: $\{-\frac{15}{2}, \frac{5}{2}\}$

4. $(3k - 2)^2 = -25$

$$3k - 2 = \sqrt{-25} \quad \text{or} \quad 3k - 2 = -\sqrt{-25}$$
$$3k - 2 = 5i \qquad\qquad 3k - 2 = -5i$$
$$3k = 2 + 5i \qquad\qquad 3k = 2 - 5i$$
$$k = \frac{2 + 5i}{3} \quad \text{or} \quad k = \frac{2 - 5i}{3}$$

Solution set: $\{\frac{2+5i}{3}, \frac{2-5i}{3}\}$

5. $x^2 + 4x = 15$

Complete the square.

$$\left(\frac{1}{2} \cdot 4\right)^2 = 2^2 = 4$$

Add 4 to both sides.

$$x^2 + 4x + 4 = 15 + 4$$
$$(x + 2)^2 = 19$$
$$x + 2 = \sqrt{19} \quad \text{or} \quad x + 2 = -\sqrt{19}$$
$$x = -2 + \sqrt{19} \quad \text{or} \quad x = -2 - \sqrt{19}$$

Solution set: $\{-2 + \sqrt{19}, -2 - \sqrt{19}\}$

6. $2m^2 - 3m = -1$

Divide both sides by 2.

$$m^2 - \frac{3}{2}m = -\frac{1}{2}$$

Complete the square.

$$\left[\frac{1}{2}\left(-\frac{3}{2}\right)\right]^2 = \left(-\frac{3}{4}\right)^2 = \frac{9}{16}$$

Add $\frac{9}{16}$ to both sides.

$$m^2 - \frac{3}{2}m + \frac{9}{16} = -\frac{1}{2} + \frac{9}{16}$$
$$\left(m - \frac{3}{4}\right)^2 = -\frac{8}{16} + \frac{9}{16}$$
$$\left(m - \frac{3}{4}\right)^2 = \frac{1}{16}$$

$$m - \frac{3}{4} = \sqrt{\frac{1}{16}} \quad \text{or} \quad m - \frac{3}{4} = -\sqrt{\frac{1}{16}}$$

$$m - \frac{3}{4} = \frac{1}{4} \qquad m - \frac{3}{4} = -\frac{1}{4}$$

$$m = \frac{3}{4} + \frac{1}{4} \qquad m = \frac{3}{4} - \frac{1}{4}$$

$$m = 1 \quad \text{or} \quad m = \frac{1}{2}$$

Solution set: $\{\frac{1}{2}, 1\}$

7. $2y^2 + y - 21 = 0$

$a = 2, b = 1, c = -21$

$$y = \frac{-b \pm \sqrt{b^2 - 4ac}}{2a}$$

$$y = \frac{-1 \pm \sqrt{1^2 - 4(2)(-21)}}{2(2)}$$

$$= \frac{-1 \pm \sqrt{1 + 168}}{4}$$

$$= \frac{-1 \pm \sqrt{169}}{4}$$

$$= \frac{-1 \pm 13}{4}$$

$$y = \frac{-1 + 13}{4} = \frac{12}{4} = 3 \quad \text{or}$$

$$y = \frac{-1 - 13}{4} = -\frac{14}{4} = -\frac{7}{2}$$

Solution set: $\{-\frac{7}{2}, 3\}$

8. $k^2 + 5k = 7$

$k^2 + 5k - 7 = 0$

$a = 1, b = 5, c = -7$

$$k = \frac{-b \pm \sqrt{b^2 - 4ac}}{2a}$$

$$k = \frac{-5 \pm \sqrt{5^2 - 4(1)(-7)}}{2(1)}$$

$$= \frac{-5 \pm \sqrt{25 + 28}}{2}$$

$$k = \frac{-5 \pm \sqrt{53}}{2}$$

Solution set: $\left\{\frac{-5+\sqrt{53}}{2}, \frac{-5-\sqrt{53}}{2}\right\}$

9. $(t + 3)(t - 4) = -2$

$t^2 - t - 12 = -2$

$t^2 - t - 10 = 0$

$a = 1, b = -1, c = -10$

$$t = \frac{-b \pm \sqrt{b^2 - 4ac}}{2a}$$

$$t = \frac{-(-1) \pm \sqrt{(-1)^2 - 4(1)(-10)}}{2(1)}$$

$$= \frac{1 \pm \sqrt{1 + 40}}{2}$$

$$t = \frac{1 \pm \sqrt{41}}{2}$$

Solution set: $\left\{\frac{1+\sqrt{41}}{2}, \frac{1-\sqrt{41}}{2}\right\}$

10. $9p^2 = 42p - 49$

$9p^2 - 42p + 49 = 0$

$a = 9, b = -42, c = 49$

$$p = \frac{-b \pm \sqrt{b^2 - 4ac}}{2a}$$

$$p = \frac{-(-42) \pm \sqrt{(-42)^2 - 4(9)(49)}}{2(9)}$$

$$= \frac{42 \pm \sqrt{1764 - 1764}}{18}$$

$$= \frac{42 \pm 0}{18}$$

$$p = \frac{7}{3}$$

Solution set: $\{\frac{7}{3}\}$

11. $3p^2 = 2(2p - 1)$

$3p^2 = 4p - 2$

$3p^2 - 4p + 2 = 0$

$a = 3, b = -4, c = 2$

$$p = \frac{-b \pm \sqrt{b^2 - 4ac}}{2a}$$

$$p = \frac{-(-4) \pm \sqrt{(-4)^2 - 4(3)(2)}}{2(3)}$$

$$= \frac{4 \pm \sqrt{16 - 24}}{6}$$

$$= \frac{4 \pm \sqrt{-8}}{6}$$

$$= \frac{4 \pm 2i\sqrt{2}}{6}$$

$$= \frac{2(2 \pm i\sqrt{2})}{6}$$

$$p = \frac{2 \pm i\sqrt{2}}{3}$$

Solution set: $\left\{\frac{2+i\sqrt{2}}{3}, \frac{2-i\sqrt{2}}{3}\right\}$

12.
$$\begin{aligned} m(2m-7) &= 3m^2+3 \\ 2m^2-7m &= 3m^2+3 \\ -m^2-7m-3 &= 0 \\ m^2+7m+3 &= 0 \end{aligned}$$

$a=1, b=7, c=3$

$$m = \frac{-b \pm \sqrt{b^2-4ac}}{2a}$$

$$m = \frac{-7 \pm \sqrt{7^2-4(1)(3)}}{2(1)}$$

$$= \frac{-7 \pm \sqrt{49-12}}{2}$$

$$m = \frac{-7 \pm \sqrt{37}}{2}$$

Solution set: $\left\{\frac{-7+\sqrt{37}}{2}, \frac{-7-\sqrt{37}}{2}\right\}$

13. $2x^2+3x+4=0$

$a=2, b=3, c=4$

$$x = \frac{-b \pm \sqrt{b^2-4ac}}{2a}$$

$$x = \frac{-3 \pm \sqrt{3^2-4(2)(4)}}{2(2)}$$

$$= \frac{-3 \pm \sqrt{9-32}}{4}$$

$$= \frac{-3 \pm \sqrt{-23}}{4}$$

$$x = \frac{-3 \pm i\sqrt{23}}{4}$$

Solution set: $\left\{\frac{-3+i\sqrt{23}}{4}, \frac{-3-i\sqrt{23}}{4}\right\}$

14. The student was not correct. The quadratic formula is

$$x = \frac{-b \pm \sqrt{b^2-4ac}}{2a},$$

$$\text{not} \quad x = -b \pm \frac{\sqrt{b^2-4ac}}{2a}.$$

The term $-b$ must be in the numerator.

15. $s(t) = -16t^2+256t$

Let $s(t)=768$ and solve for t.

$$\begin{aligned} 768 &= -16t^2+256t \\ 16t^2-256+768 &= 0 \\ t^2-16t+48 &= 0 \\ (t-4)(t-12) &= 0 \\ t-4=0 \quad &\text{or} \quad t-12=0 \\ t=4 \quad &\text{or} \quad t=12 \end{aligned}$$

The rock will be 768 ft from the ground at 4 sec and at 12 sec.

16. The problem in Exercise 15 has two answers because the rock reaches the height of 768 ft on its way up and again on its way down.

In Exercises 17 and 18,

$$x^2+2x+k=0.$$

17. In order for the given equation to have exactly one real (in this case rational) solution, the discriminant b^2-4ac must equal 0. From the equation $a=1, b=2$, and $c=k$, so

$$\begin{aligned} b^2-4ac = 2^2-4(1)(k) &= 0 \\ 4-4k &= 0 \\ -4k &= -4 \\ k &= 1. \end{aligned}$$

Therefore, $k=1$ will give one real solution.

18. In order for the given equation to have two real (in this case irrational) solutions, the discriminant b^2-4ac must be greater than 0. Using the same values of a, b, and c as in Exercise 17,

$$\begin{aligned} b^2-4ac = 2^2-4(1)(k) &> 0 \\ 4-4k &> 0 \\ -4k &> -4 \\ k &< 1. \end{aligned}$$

Therefore, all k such that $k<1$ will give two real solutions.

19. $a^2+5a+2=0$
$a=1, b=5, c=2$

$$\begin{aligned} b^2-4ac &= 5^2-4(1)(2) \\ &= 25-8 \\ &= 17 \end{aligned}$$

Since the discriminant is positive, but not a perfect square, there are two distinct irrational number solutions. The answer is (c).

20.
$$\begin{aligned} 4c^2 &= 3-4c \\ 4c^2+4c-3 &= 0 \end{aligned}$$

$a=4, b=4, c=-3$

$$\begin{aligned} b^2-4ac &= 4^2-4(4)(-3) \\ &= 16+48 \\ &= 64 \quad \text{or} \quad 8^2 \end{aligned}$$

Since the discriminant is positive and a perfect square, there are two distinct rational number solutions. The answer is (a).

21. $4x^2 = 6x - 8$

$4x^2 - 6x + 8 = 0$

$a = 4, b = -6, c = 8$

$$\begin{aligned} b^2 - 4ac &= (-6)^2 - 4(4)(8) \\ &= 36 - 128 \\ &= -92 \end{aligned}$$

Since the discriminant is negative, there are two distinct imaginary number solutions. The answer is (d).

22. $9z^2 + 30z + 25 = 0$

$a = 9, b = 30, c = 25$

$$\begin{aligned} b^2 - 4ac &= 30^2 - 4(9)(25) \\ &= 900 - 900 \\ &= 0 \end{aligned}$$

Since the discriminant is zero, there is exactly one rational number solution. The answer is (b).

23. $24x^2 - 74x + 45$

$a = 24, b = -74, c = 45$

$$\begin{aligned} b^2 - 4ac &= (-74)^2 - 4(24)(45) \\ &= 5476 - 4320 \\ &= 1156 \quad \text{or} \quad 34^2 \end{aligned}$$

Since the discriminant is positive and a perfect square, the polynomial can be factored.

$$24x^2 - 74x + 45 = (6x - 5)(4x - 9)$$

24. $36x^2 + 69x - 34$

$a = 36, b = 69, c = -34$

$$\begin{aligned} b^2 - 4ac &= 69^2 - 4(36)(-34) \\ &= 4761 + 4896 \\ &= 9657 \end{aligned}$$

Since the discriminant is not a perfect square, the polynomial cannot be factored.

25. $\dfrac{15}{x} = 2x - 1$

Multiply both sides by x.

$$x\left(\frac{15}{x}\right) = x(2x - 1)$$

$$\begin{aligned} 15 &= 2x^2 - x \\ 0 &= 2x^2 - x - 15 \\ 0 &= (2x + 5)(x - 3) \end{aligned}$$

$$2x + 5 = 0 \quad \text{or} \quad x - 3 = 0$$

$$x = -\frac{5}{2} \quad \text{or} \quad x = 3$$

Both potential solutions check.

Solution set: $\{-\frac{5}{2}, 3\}$

26. $\dfrac{1}{y} + \dfrac{2}{y+1} = 2$

Multiply both sides by $y(y + 1)$.

$$\begin{aligned} y(y+1)\left(\frac{1}{y} + \frac{2}{y+1}\right) &= y(y+1) \cdot 2 \\ y + 1 + 2y &= 2y^2 + 2y \\ 0 &= 2y^2 - y - 1 \\ 0 &= (2y + 1)(y - 1) \end{aligned}$$

$$2y + 1 = 0 \quad \text{or} \quad y - 1 = 0$$

$$y = -\frac{1}{2} \quad \text{or} \quad y = 1$$

Both potential solutions check.

Solution set: $\{-\frac{1}{2}, 1\}$

27. $8(3x + 5)^2 + 2(3x + 5) - 1 = 0$

Let $u = 3x + 5$. The equation becomes

$$\begin{aligned} 8u^2 + 2u - 1 &= 0 \\ (4u - 1)(2u + 1) &= 0 \end{aligned}$$

$$4u - 1 = 0 \quad \text{or} \quad 2u + 1 = 0$$

$$u = \frac{1}{4} \quad \text{or} \quad u = -\frac{1}{2}.$$

To find x, substitute $3x + 5$ for u.

$$3x + 5 = \frac{1}{4} \quad \text{or} \quad 3x + 5 = -\frac{1}{2}$$

$$12x + 20 = 1 \qquad\qquad 6x + 10 = -1$$

$$12x = -19 \qquad\qquad 6x = -11$$

$$x = -\frac{19}{12} \quad \text{or} \quad x = -\frac{11}{6}$$

Both potential solutions check.

Solution set: $\{-\frac{11}{6}, -\frac{19}{12}\}$

28. $-2r = \sqrt{\frac{48-20r}{2}}$

Square both sides.

$$(-2r)^2 = \left(\sqrt{\frac{48-20r}{2}}\right)^2$$
$$4r^2 = \frac{48-20r}{2}$$
$$4r^2 = 24 - 10r$$
$$4r^2 + 10r - 24 = 0$$
$$2r^2 + 5r - 12 = 0$$
$$(r+4)(2r-3) = 0$$
$$r + 4 = 0 \quad \text{or} \quad 2r - 3 = 0$$
$$r = -4 \quad \text{or} \quad r = \frac{3}{2}$$

Check -4 in the original equation.

$$-2(-4) = \sqrt{\frac{48-20(-4)}{2}} \quad ?$$
$$8 = \sqrt{64} \quad ?$$
$$8 = 8 \qquad \textit{True}$$

-4 is a solution. Since the square root must be nonnegative, $\frac{3}{2}$ is not a solution.

Solution set: $\{-4\}$

29. $2x^{2/3} - x^{1/3} - 28 = 0$

Let $u = x^{1/3}$, so $u^2 = (x^{1/3})^2 = x^{2/3}$. The equation becomes

$$2u^2 - u - 28 = 0$$
$$(2u+7)(u-4) = 0$$
$$2u + 7 = 0 \quad \text{or} \quad u - 4 = 0$$
$$u = -\frac{7}{2} \quad \text{or} \quad u = 4.$$

To find x, substitute $x^{1/3}$ for u.

$$x^{1/3} = -\frac{7}{2} \quad \text{or} \quad x^{1/3} = 4$$
$$(x^{1/3})^3 = \left(-\frac{7}{2}\right)^3 \qquad (x^{1/3})^3 = 4^3$$
$$x = -\frac{343}{8} \quad \text{or} \quad x = 64$$

Both potential solutions check.

Solution set: $\{-\frac{343}{8}, 64\}$

30. $(x^2+x)^2 = 8(x^2+x) - 12$

Let $u = x^2 + x$. The equation becomes

$$u^2 = 8u - 12$$
$$u^2 - 8u + 12 = 0$$
$$(u-6)(u-2) = 0$$
$$u - 6 = 0 \quad \text{or} \quad u - 2 = 0$$
$$u = 6 \quad \text{or} \quad u = 2.$$

To find x, substitute $x^2 + x$ for u.

$$x^2 + x = 6$$
$$x^2 + x - 6 = 0$$
$$(x-2)(x+3) = 0$$
$$x - 2 = 0 \quad \text{or} \quad x + 3 = 0$$
$$x = 2 \quad \text{or} \quad x = -3$$

or

$$x^2 + x = 2$$
$$x^2 + x - 2 = 0$$
$$(x+2)(x-1) = 0$$
$$x + 2 = 0 \quad \text{or} \quad x - 1 = 0$$
$$x = -2 \quad \text{or} \quad x = 1$$

The potential solutions check.

Solution set: $\{-3, -2, 1, 2\}$

31. Let $x =$ Lisa's speed on the trip to pick up Laurie.

Make a table. Use $d = rt$, or $t = \frac{d}{r}$.

	Distance	Rate	Time
To Laurie	8	x	$\frac{8}{x}$
To the mall	11	$x+15$	$\frac{11}{x+15}$

Time to pick up Laurie	+	time to mall	=	24 min (or .4 hr).
$\frac{8}{x}$	+	$\frac{11}{x+15}$	=	.4

Multiply both sides by $x(x+15)$.

$$x(x+15)\left(\frac{8}{x} + \frac{11}{x+15}\right) = x(x+15) \cdot .4$$
$$8(x+15) + 11x = .4x(x+15)$$
$$8x + 120 + 11x = .4x^2 + 6x$$
$$0 = .4x^2 - 13x - 120$$

Multiply by 10 to clear the decimal.

$$0 = 4x^2 - 130x - 1200$$
$$0 = 2x^2 - 65x - 600$$
$$0 = (x - 40)(2x + 15)$$
$$x - 40 = 0 \quad \text{or} \quad 2x + 15 = 0$$
$$x = 40 \quad \text{or} \quad x = -\frac{15}{2}$$

Speed cannot be negative, so $-\frac{15}{2}$ is not a solution. Lisa's speed on the trip to pick up Laurie was 40 mph.

32. Let $x =$ the time for Linda alone and $x - 2 =$ the time for Ed alone.

Worker	Rate	Time together	Part of job done
Linda	$\frac{1}{x}$	3	$\frac{3}{x}$
Ed	$\frac{1}{x-2}$	3	$\frac{3}{x-2}$

Part by Linda + part by Ed = 1 whole job.

$$\frac{3}{x} + \frac{3}{x-2} = 1$$

Multiply both sides by $x(x - 2)$.

$$x(x-2)\left(\frac{3}{x} + \frac{3}{x-2}\right) = x(x-2) \cdot 1$$
$$3x - 6 + 3x = x^2 - 2x$$
$$0 = x^2 - 8x + 6$$

Use the quadratic formula.
$a = 1, b = -8, c = 6$

$$x = \frac{-b \pm \sqrt{b^2 - 4ac}}{2a}$$
$$x = \frac{8 \pm \sqrt{(-8)^2 - 4(1)(6)}}{2}$$
$$= \frac{8 \pm \sqrt{64 - 24}}{2}$$
$$= \frac{8 \pm \sqrt{40}}{2}$$
$$= \frac{8 \pm 2\sqrt{10}}{2}$$
$$= \frac{2(4 \pm \sqrt{10})}{2}$$
$$= 4 \pm \sqrt{10}$$

$$x = 4 + \sqrt{10} \approx 7.2 \quad \text{or}$$
$$x = 4 - \sqrt{10} \approx .8$$

Reject about .8 as Linda's time, because that would make Ed's time negative. So, Linda's time alone is about 7.2 hr and Ed's time is about $x - 2 = 5.2$ hr.

33. The equation $x = \sqrt{2x + 4}$ can't have a negative solution, because the principal square root can't be negative.

34. The quadratic formula could be used to solve for x^2 in the equation $x^4 - 5x^2 + 6 = 0$. After you solve for x^2, you would have to remember to use the square root property to solve for x.

35. $S = \frac{Id^2}{k}$

Solve for d. Multiply both sides by k, then divide by I.

$$\frac{Sk}{I} = d^2$$
$$d = \pm\sqrt{\frac{Sk}{I}}$$
$$d = \frac{\pm\sqrt{Sk}}{\sqrt{I}}$$

Rationalize the denominator.

$$d = \frac{\pm\sqrt{Sk}}{\sqrt{I}} \cdot \frac{\sqrt{I}}{\sqrt{I}}$$
$$d = \frac{\pm\sqrt{SkI}}{I}$$

36. $k = \frac{rF}{wv^2}$

Solve for v. Multiply both sides by v^2, then divide by k.

$$v^2 = \frac{rF}{kw}$$
$$v = \pm\sqrt{\frac{rF}{kw}}$$
$$v = \frac{\pm\sqrt{rF}}{\sqrt{kw}}$$
$$v = \frac{\pm\sqrt{rF}}{\sqrt{kw}} \cdot \frac{\sqrt{kw}}{\sqrt{kw}}$$
$$v = \frac{\pm\sqrt{rFkw}}{kw}$$

37. $S = 2\pi rh + 2\pi r^2$

Solve for r.

$$2\pi r^2 + 2\pi rh - S = 0$$

Use the quadratic formula with $a = 2\pi, b = 2\pi h$, and $c = -S$.

$$r = \frac{-b \pm \sqrt{b^2 - 4ac}}{2a}$$

$$r = \frac{-2\pi h \pm \sqrt{(2\pi h)^2 - 4(2\pi)(-S)}}{2(2\pi)}$$

$$r = \frac{-2\pi h \pm \sqrt{4\pi^2 h^2 + 8\pi S}}{4\pi}$$

$$r = \frac{-2\pi h \pm 2\sqrt{\pi^2 h^2 + 2\pi S}}{4\pi}$$

$$r = \frac{2(-\pi h \pm \sqrt{\pi^2 h^2 + 2\pi S})}{4\pi}$$

$$r = \frac{-\pi h \pm \sqrt{\pi^2 h^2 + 2\pi S}}{2\pi}$$

38. $mt^2 = 3mt + 6$

Solve for t.

$$mt^2 - 3mt - 6 = 0$$

Use the quadratic formula.
$a = m, b = -3m, c = -6$

$$t = \frac{-b \pm \sqrt{b^2 - 4ac}}{2a}$$

$$t = \frac{3m \pm \sqrt{(-3m)^2 - 4(m)(-6)}}{2m}$$

$$t = \frac{3m \pm \sqrt{9m^2 + 24m}}{2m}$$

39. $f(t) = -16t^2 + 45t + 400$

Let $f(t) = 200$.

$$200 = -16t^2 + 45t + 400$$
$$0 = -16t^2 + 45t + 200$$
$$0 = 16t^2 - 45t - 200$$

$a = 16, b = -45, c = -200$

$$t = \frac{-b \pm \sqrt{b^2 - 4ac}}{2a}$$

$$t = \frac{-(-45) \pm \sqrt{(-45)^2 - 4(16)(-200)}}{2(16)}$$

$$= \frac{45 \pm \sqrt{2025 + 12,800}}{32}$$

$$= \frac{45 \pm \sqrt{14,825}}{32}$$

$$t = \frac{45 + \sqrt{14,825}}{32} \approx 5.2 \quad \text{or}$$

$$t = \frac{45 - \sqrt{14,825}}{32} \approx -2.4$$

Reject the negative solution since time cannot be negative. The ball will reach a height of 200 ft above the ground after 5.2 sec.

40. $s(t) = -16t^2 + 75t + 407$

Let $s(t) = 450$.

$$450 = -16t^2 + 75t + 407$$
$$0 = -16t^2 + 75t - 43$$

$a = -16, b = 75, c = -43$

$$t = \frac{-75 \pm \sqrt{75^2 - 4(-16)(-43)}}{2(-16)}$$

$$= \frac{-75 \pm \sqrt{5625 - 2752}}{-32}$$

$$= \frac{-75 \pm \sqrt{2873}}{-32}$$

$$t = \frac{-75 + \sqrt{2873}}{-32} \approx .7 \quad \text{or}$$

$$t = \frac{-75 - \sqrt{2873}}{-32} \approx 4.0$$

The ball will be 450 ft above the ground after .7 sec (going up) and again after 4.0 sec (coming down).

41. Let $x =$ the width of the original rectangle and $x + 2 =$ the length of the original rectangle.

Adding 1 m to the width and subtracting 1 m from the length of the rectangle gives $x+1$ in each case, which is the length of a side of the square.

$$(\text{length of one side})^2 = \text{Area of square}$$
$$(x+1)^2 = 121$$
$$x + 1 = 11 \quad \text{or} \quad x + 1 = -11$$
$$x = 10 \quad \text{or} \quad x = -12$$

Since width cannot be negative, the original rectangle had a width of 10 m and a length of $x+2 =$ 12 m.

42. Let $x =$ the length of the longer leg,
$2x - 9 =$ the length of the hypotenuse, and
$x - 3 =$ the length of the shorter leg.

Use the Pythagorean theorem.

$$\begin{aligned} a^2 + b^2 &= c^2 \\ x^2 + (x-3)^2 &= (2x-9)^2 \\ x^2 + x^2 - 6x + 9 &= 4x^2 - 36x + 81 \\ -2x^2 + 30x - 72 &= 0 \\ x^2 - 15x + 36 &= 0 \\ (x-12)(x-3) &= 0 \end{aligned}$$

$$x - 12 = 0 \quad \text{or} \quad x - 3 = 0$$
$$x = 12 \quad \text{or} \quad x = 3$$

Reject 3 ft as the length of the longer leg, since it causes the length of the hypotenuse to be negative. The longer leg is 12 ft long, the hypotenuse is $2x - 9 = 15$ ft long, and the shorter leg is $x - 3 = 9$ ft long.

43. Let $x =$ one of the page numbers and $x + 1 =$ the other page number.

The product is 4692, so

$$\begin{aligned} x(x+1) &= 4692 \\ x^2 + x &= 4692 \\ x^2 + x - 4692 &= 0 \\ (x+69)(x-68) &= 0 \end{aligned}$$

$$x + 69 = 0 \quad \text{or} \quad x - 68 = 0$$
$$x = -69 \quad \text{or} \quad x = 68.$$

Reject the negative answer, because no page number is negative. So, one page number is 68, and the other is $x + 1 = 69$.

44. Let $x =$ the width of the border.

$$\begin{aligned} \text{Area of map} &= \text{length} \cdot \text{width} \\ 352 &= (2x+20)(2x+14) \\ 352 &= 4x^2 + 68x + 280 \\ 0 &= 4x^2 + 68x - 72 \\ 0 &= x^2 + 17x - 18 \\ 0 &= (x+18)(x-1) \end{aligned}$$

$$x + 18 = 0 \quad \text{or} \quad x - 1 = 0$$
$$x = -18 \quad \text{or} \quad x = 1$$

Reject the negative answer for length. The mat is 1 in wide.

45. $f(t) = 100t^2 - 300t$ is the distance of the light from the starting point at t min. When the light returns to the starting point, the value of $f(t)$ will be 0.

$$0 = 100t^2 - 300t$$

Divide by 100.

$$\begin{aligned} 0 &= t^2 - 3t \\ 0 &= t(t-3) \end{aligned}$$

$$t = 0 \quad \text{or} \quad t - 3 = 0$$
$$t = 3$$

Since $t = 0$ represents the starting time, the light will return to the starting point in 3 min.

46. Let $p =$ the price at which supply and demand are equal.

$$\text{demand} = \text{supply}$$
$$\frac{25}{p} = 70p + 15$$

Multiply both sides by p.

$$\begin{aligned} p\left(\frac{25}{p}\right) &= p(70p + 15) \\ 25 &= 70p^2 + 15p \\ 0 &= 70p^2 + 15p - 25 \\ 0 &= 14p^2 + 3p - 5 \\ 0 &= (2p-1)(7p+5) \end{aligned}$$

$$2p - 1 = 0 \quad \text{or} \quad 7p + 5 = 0$$
$$p = \frac{1}{2} \quad \text{or} \quad p = -\frac{5}{7}$$

Reject the negative answer for price. The answer is $\frac{1}{2}$ dollar or \$.50.

47. $A = P(1 + r)^2$

Let $A = 10,920.25$ and $P = 10,000$.
Solve for r.

$$\begin{aligned} 10,920.25 &= 10,000(1+r)^2 \\ 1.092025 &= (1+r)^2 \end{aligned}$$

$$1 + r = 1.045 \quad \text{or} \quad 1 + r = -1.045$$
$$r = .045 \quad \text{or} \quad r = -2.045$$

Reject the negative rate. The required interest rate is 4.5%.

48. Let $x =$ time for Jake working alone.

Make a table.

Worker	Rate	Time together	Part of job done
Jake	$\frac{1}{x}$	4	$\frac{4}{x}$
Jim	$\frac{1}{x-1}$	4	$\frac{4}{x-1}$

Part done by Jake	+	part done by Jim	=	1 whole job.
$\frac{4}{x}$	+	$\frac{4}{x-1}$	=	1

Multiply both sides by $x(x-1)$.

$$x(x-1)\left(\frac{4}{x}+\frac{4}{x-1}\right)=x(x-1)\cdot 1$$
$$4(x-1)+4x=x^2-x$$
$$4x-4+4x=x^2-x$$
$$0=x^2-9x+4$$

Use the quadratic formula.
$a=1, b=-9, c=4$

$$x=\frac{-b\pm\sqrt{b^2-4ac}}{2a}$$
$$x=\frac{9\pm\sqrt{(-9)^2-4(1)(4)}}{2(1)}$$
$$=\frac{9\pm\sqrt{65}}{2}$$
$$x=\frac{9+\sqrt{65}}{2}\approx 8.5 \quad \text{or}$$
$$x=\frac{9-\sqrt{65}}{2}\approx .5$$

Reject .5 for Jake's time because that would make Jim's time negative. So, Jake's time is about 8.5 hr and Jim's is about 7.5 hr.

For Exercises 49-53, see the answer graphs in the back of the textbook.

49. $(x-4)(2x+3)>0$

Solve the equation

$$(x-4)(2x+3)=0$$
$$x-4=0 \quad \text{or} \quad 2x+3=0$$
$$x=4 \quad \text{or} \quad x=-\frac{3}{2}.$$

These numbers divide the number line into three regions.

A B C
-3/2 4

Test a number from each region in the inequality

$$(x-4)(2x+3)>0.$$

Region A: Let $x=-2$.

$$(-2-4)[2(-2)+3]>0 \quad ?$$
$$-6(-1)>0 \quad ?$$
$$6>0 \quad \textit{True}$$

Region B: Let $x=2$.

$$(2-4)[2(2)+3]>0 \quad ?$$
$$-2(7)>0 \quad ?$$
$$-14>0 \quad \textit{False}$$

Region C: Let $x=5$.

$$(5-4)[2(5)+3]>0 \quad ?$$
$$13>0 \quad \textit{True}$$

The numbers in Regions A and C, not including $-\frac{3}{2}$ or 4, are solutions.

Solution set: $\left(-\infty,-\frac{3}{2}\right)\cup(4,\infty)$

50. $x^2+x\le 12$

Solve the equation

$$x^2+x=12$$
$$x^2+x-12=0$$
$$(x+4)(x-3)=0$$
$$x+4=0 \quad \text{or} \quad x-3=0$$
$$x=-4 \quad \text{or} \quad x=3.$$

These numbers divide the number line into three regions.

A B C
-4 3

Test a number from each region in the inequality

$$x^2+x\le 12.$$

Region A: Let $x=-5$.

$$(-5)^2+(-5)\le 12 \quad ?$$
$$25-5\le 12 \quad ?$$
$$20\le 12 \quad \textit{False}$$

Region B: Let $x=0$.

$$0^2+0\le 12 \quad ?$$
$$0\le 12 \quad \textit{True}$$

Region C: Let $x=4$.

$$4^2+4\le 12 \quad ?$$
$$16+4\le 12 \quad ?$$
$$20\le 12 \quad \textit{False}$$

The numbers in Region B, including -4 and 3, are solutions.

Solution set: $[-4, 3]$

51. $2k^2 > 5k + 3$

Solve the equation

$$2k^2 = 5k + 3$$
$$2k^2 - 5k - 3 = 0$$
$$(k - 3)(2k + 1) = 0$$
$$k - 3 = 0 \quad \text{or} \quad 2k + 1 = 0$$
$$k = 3 \quad \text{or} \quad k = -\frac{1}{2}.$$

These numbers divide the number line into three regions.

A B C
-1/2 3

Test a number from each region in the inequality

$$2k^2 > 5k + 3.$$

Region A: Let $k = -1$.

$$2(-1)^2 > 5(-1) + 3 \quad ?$$
$$2 > -5 + 3 \quad ?$$
$$2 > -2 \quad \textit{True}$$

Region B: Let $k = 0$.

$$2(0)^2 > 5(0) + 3 \quad ?$$
$$0 > 3 \quad \textit{False}$$

Region C: Let $k = 4$.

$$2(4)^2 > 5(4) + 3 \quad ?$$
$$32 > 20 + 3 \quad ?$$
$$32 > 23 \quad \textit{True}$$

The numbers in Regions A and C, not including $-\frac{1}{2}$ or 3, are solutions.

Solution set: $\left(-\infty, -\frac{1}{2}\right) \cup (3, \infty)$

52. $\dfrac{3y + 4}{y - 2} \leq 1$

Solve the equation

$$\frac{3y + 4}{y - 2} = 1$$
$$3y + 4 = y - 2$$
$$2y = -6$$
$$y = -3.$$

Set the denominator equal to zero and solve.

$$y - 2 = 0$$
$$y = 2$$

These numbers divide the number line into three regions.

A B C
-3 2

Test a number from each region in the inequality

$$\frac{3y + 4}{y - 2} \leq 1.$$

Region A: Let $y = -4$.

$$\frac{3(-4) + 4}{-4 - 2} \leq 1 \quad ?$$
$$\frac{-12 + 4}{-6} \leq 1 \quad ?$$
$$\frac{-8}{-6} \leq 1 \quad ?$$
$$\frac{4}{3} \leq 1 \quad \textit{False}$$

Region B: Let $y = 0$.

$$\frac{3(0) + 4}{0 - 2} \leq 1 \quad ?$$
$$-2 \leq 1 \quad \textit{True}$$

Region C: Let $y = 3$.

$$\frac{3(3) + 4}{3 - 2} \leq 1 \quad ?$$
$$13 \leq 1 \quad \textit{False}$$

The numbers in Region B, including -3 but not 2 (because it makes the denominator 0), are solutions.

Solution set: $[-3, 2)$

53. $(x + 2)(x - 3)(x + 5) \leq 0$

Solve the equation

$$(x + 2)(x - 3)(x + 5) = 0$$
$$x + 2 = 0 \quad \text{or} \quad x - 3 = 0 \quad \text{or} \quad x + 5 = 0$$
$$x = -2 \quad \text{or} \quad x = 3 \quad \text{or} \quad x = -5.$$

These numbers divide the number line into four regions.

A B C D
-5 -2 3

Test a number from each region in the inequality

$$(x+2)(x-3)(x+5) \leq 0.$$

Region A: Let $x = -6$.

$$\begin{aligned}(-6+2)(-6-3)(-6+5) &\leq 0 \quad ?\\ -4(-9)(-1) &\leq 0 \quad ?\\ -36 &\leq 0 \qquad \textit{True}\end{aligned}$$

Region B: Let $x = -3$.

$$\begin{aligned}(-3+2)(-3-3)(-3+5) &\leq 0 \quad ?\\ -1(-6)(2) &\leq 0 \quad ?\\ 12 &\leq 0 \qquad \textit{False}\end{aligned}$$

Region C: Let $x = 0$.

$$\begin{aligned}(0+2)(0-3)(0+5) &\leq 0 \quad ?\\ 2(-3)(5) &\leq 0 \quad ?\\ -30 &\leq 0 \qquad \textit{True}\end{aligned}$$

Region D: Let $x = 4$.

$$\begin{aligned}(4+2)(4-3)(4+5) &\leq 0 \quad ?\\ 6(1)(9) &\leq 0 \quad ?\\ 54 &\leq 0 \qquad \textit{False}\end{aligned}$$

The numbers in Regions A and C, including $-5, -2$, and 3, are solutions.

Solution set: $(-\infty, -5] \cup [-2, 3]$

54. $(4m+3)^2 \leq -4$

The square of a real number is never negative.

Solution set: $\emptyset$

55. $V = r^2 + R^2h$

Solve for R.

$$\begin{aligned}V - r^2 &= R^2h\\ R^2h &= V - r^2\\ R^2 &= \frac{V-r^2}{h}\\ R &= \pm\sqrt{\frac{V-r^2}{h}}\\ R &= \frac{\pm\sqrt{V-r^2}}{\sqrt{h}}\\ R &= \frac{\pm\sqrt{V-r^2}}{\sqrt{h}} \cdot \frac{\sqrt{h}}{\sqrt{h}}\\ R &= \frac{\pm\sqrt{Vh-r^2h}}{h}\end{aligned}$$

56. $$\begin{aligned}3t^2 - 6t &= -4\\ 3t^2 - 6t + 4 &= 0\end{aligned}$$

Use the quadratic formula.
$a = 3, b = -6, c = 4$

$$\begin{aligned}t &= \frac{-b \pm \sqrt{b^2-4ac}}{2a}\\ t &= \frac{6 \pm \sqrt{(-6)^2 - 4(3)(4)}}{2(3)}\\ &= \frac{6 \pm \sqrt{-12}}{6}\\ &= \frac{6 \pm 2i\sqrt{3}}{6}\\ &= \frac{2(3 \pm i\sqrt{3})}{6}\\ t &= \frac{3 \pm i\sqrt{3}}{3}\end{aligned}$$

Solution set: $\left\{\frac{3+i\sqrt{3}}{3}, \frac{3-i\sqrt{3}}{3}\right\}$

57. $(b^2-2b)^2 = 11(b^2-2b) - 24$

Let $u = b^2 - 2b$. The equation becomes

$$\begin{aligned}u^2 &= 11u - 24\\ u^2 - 11u + 24 &= 0\\ (u-8)(u-3) &= 0\\ u - 8 = 0 \quad &\text{or} \quad u - 3 = 0\\ u = 8 \quad &\text{or} \quad u = 3.\end{aligned}$$

To find b, substitute $b^2 - 2b$ for u.

$$\begin{aligned}b^2 - 2b &= 8\\ b^2 - 2b - 8 &= 0\\ (b-4)(b+2) &= 0\\ b - 4 = 0 \quad &\text{or} \quad b + 2 = 0\\ b = 4 \quad &\text{or} \quad b = -2\end{aligned}$$

or

$$\begin{aligned}b^2 - 2b &= 3\\ b^2 - 2b - 3 &= 0\\ (b-3)(b+1) &= 0\\ b - 3 = 0 \quad &\text{or} \quad b + 1 = 0\\ b = 3 \quad &\text{or} \quad b = -1\end{aligned}$$

The potential solutions all check.

Solution set: $\{-2, -1, 3, 4\}$

58. $(r-1)(2r+3)(r+6)<0$

Solve the equation

$$(r-1)(2r+3)(r+6)=0$$

$$r-1=0 \quad \text{or} \quad 2r+3=0 \quad \text{or} \quad r+6=0$$

$$r=1 \quad \text{or} \quad r=-\frac{3}{2} \quad \text{or} \quad r=-6.$$

These numbers divide the number line into four regions.

A B C D
–6 –3/2 1

Test a number from each region in the inequality

$$(r-1)(2r+3)(r+6)<0.$$

Region A: Let $r=-7$.

$$(-7-1)[2(-7)+3](-7+6)<0 \quad ?$$
$$-8(-11)(-1)<0 \quad ?$$
$$-88<0 \quad \textit{True}$$

Region B: Let $r=-2$.

$$(-2-1)[2(-2)+3](-2+6)<0 \quad ?$$
$$-3(-1)(4)<0 \quad ?$$
$$12<0 \quad \textit{False}$$

Region C: Let $r=0$.

$$(0-1)[2(0)+3](0+6)<0 \quad ?$$
$$-1(3)(6)<0 \quad ?$$
$$-18<0 \quad \textit{True}$$

Region D: Let $r=2$.

$$(2-1)[2(2)+3](2+6)<0 \quad ?$$
$$1(7)(8)<0 \quad ?$$
$$56<0 \quad \textit{False}$$

The numbers in Regions A and C, not including $-6, -\frac{3}{2}$, or 1, are solutions.

Solution set: $(-\infty,-6)\cup\left(-\frac{3}{2},1\right)$

59. $(3k+11)^2=7$

$$3k+11=\sqrt{7} \quad \text{or} \quad 3k+11=-\sqrt{7}$$

$$3k=-11+\sqrt{7} \qquad 3k=-11-\sqrt{7}$$

$$k=\frac{-11+\sqrt{7}}{3} \quad \text{or} \quad k=\frac{-11-\sqrt{7}}{3}$$

Solution set: $\left\{\frac{-11+\sqrt{7}}{3}, \frac{-11-\sqrt{7}}{3}\right\}$

60. $p=\sqrt{\frac{yz}{6}}$

Solve for y. Square both sides.

$$p^2=\left(\sqrt{\frac{yz}{6}}\right)^2$$

$$p^2=\frac{yz}{6}$$

$$\frac{6p^2}{z}=y$$

$$y=\frac{6p^2}{z}$$

61.
$$-5x^2=-8x+3$$
$$-5x^2+8x-3=0$$
$$5x^2-8x+3=0$$
$$(5x-3)(x-1)=0$$
$$5x-3=0 \quad \text{or} \quad x-1=0$$
$$x=\frac{3}{5} \quad \text{or} \quad x=1$$

Solution set: $\left\{\frac{3}{5},1\right\}$

62. $6+\frac{15}{s^2}=-\frac{19}{s}$

Multiply both sides by s^2.

$$s^2\left(6+\frac{15}{s^2}\right)=s^2\left(-\frac{19}{s}\right)$$

$$6s^2+15=-19s$$
$$6s^2+19s+15=0$$
$$(3s+5)(2s+3)=0$$
$$3s+5=0 \quad \text{or} \quad 2s+3=0$$
$$s=-\frac{5}{3} \quad \text{or} \quad s=-\frac{3}{2}$$

Both potential solutions check.

Solution set: $\left\{-\frac{5}{3},-\frac{3}{2}\right\}$

63. $\frac{-2}{x+5}\le -5$

Solve the equation

$$\frac{-2}{x+5}=-5$$

$$-2=-5(x+5)$$
$$-2=-5x-25$$
$$5x=-23$$
$$x=-\frac{23}{5}.$$

Set the denominator equal to zero and solve.

$$x+5=0$$
$$x=-5$$

These numbers divide the number line into three regions.

A B C

-5 -23/5

Test a number from each region in the inequality

$$\frac{-2}{x+5} \leq -5.$$

Region A: Let $x = -6$.

$$\frac{-2}{-6+5} \leq -5 \quad ?$$

$$2 \leq -5 \qquad \textit{False}$$

Region B: Let $x = -\frac{24}{5}$.

$$\frac{-2}{-\frac{24}{5}+5} \leq -5 \quad ?$$

$$\frac{-2}{\frac{1}{5}} \leq -5 \quad ?$$

$$-10 \leq -5 \qquad \textit{True}$$

Region C: Let $x = 0$.

$$\frac{-2}{0+5} \leq -5 \quad ?$$

$$\frac{-2}{5} \leq -5 \qquad \textit{False}$$

The numbers in Region B, including $-\frac{23}{5}$ but not -5, (because it makes the denominator 0), are solutions.

Solution set: $\left(-5, -\frac{23}{5}\right]$

64. Let $x =$ the time for the slower pipe to fill the tank alone and
$x - 3 =$ the time for the faster pipe to fill the tank alone.

Make a table.

Pipe	Rate	Time together	Part of job done
Slower pipe	$\frac{1}{x}$	2	$\frac{2}{x}$
Faster pipe	$\frac{1}{x-3}$	2	$\frac{2}{x-3}$

Part done by slower pipe	+	part done by faster pipe	=	1 whole job.
$\frac{2}{x}$	+	$\frac{2}{x-3}$	=	1

Multiply both sides by $x(x-3)$.

$$x(x-3)\left(\frac{2}{x}+\frac{2}{x-3}\right) = x(x-3)\cdot 1$$
$$2x - 6 + 2x = x^2 - 3x$$
$$0 = x^2 - 7x + 6$$
$$0 = (x-6)(x-1)$$
$$x - 6 = 0 \quad \text{or} \quad x - 1 = 0$$
$$x = 6 \quad \text{or} \quad x = 1$$

Reject 1 as the time for the slower pipe, because that would yield a negative time for the faster pipe. So, the slower pipe would take **6 hr**, and the faster pipe would take $x - 3 = 3$ hr.

65. Let $x =$ Phong's speed.

Make a table. Use $d = rt$, or $t = \frac{d}{r}$.

	Distance	Rate	Time
Upstream	20	$x-3$	$\frac{20}{x-3}$
Downstream	20	$x+3$	$\frac{20}{x+3}$

Time upstream	+	time downstream	=	7 hr.
$\frac{20}{x-3}$	+	$\frac{20}{x+3}$	=	7

Multiply both sides by $(x+3)(x-3)$.

$$(x+3)(x-3)\left(\frac{20}{x-3}+\frac{20}{x+3}\right)$$
$$= (x+3)(x-3)(7)$$
$$20(x+3) + 20(x-3) = 7(x^2 - 9)$$
$$20x + 60 + 20x - 60 = 7x^2 - 63$$
$$0 = 7x^2 - 40x - 63$$
$$0 = (x-7)(7x+9)$$
$$x - 7 = 0 \quad \text{or} \quad 7x + 9 = 0$$
$$x = 7 \quad \text{or} \quad x = -\frac{9}{7}$$

Reject $-\frac{9}{7}$ since speed can't be negative. Phong's speed was 7 mph.

66. $\frac{2}{x-4}+\frac{1}{x}=\frac{11}{5}$

Multiply both sides by $5x(x-4)$.

$$5x(x-4)\left(\frac{2}{x-4}+\frac{1}{x}\right)=5x(x-4)\left(\frac{11}{5}\right)$$
$$10x+5(x-4)=11x(x-4)$$
$$10x+5x-20=11x^2-44x$$
$$0=11x^2-59x+20$$
$$0=(x-5)(11x-4)$$
$$x-5=0 \quad \text{or} \quad 11x-4=0$$
$$x=5 \quad \text{or} \quad x=\frac{4}{11}$$

Both potential solutions check.

Solution set: $\{\frac{4}{11},5\}$

67. $y^2=-242$

$$y=\sqrt{-242} \quad \text{or} \quad y=-\sqrt{-242}$$
$$y=\sqrt{-121\cdot 2} \qquad y=-\sqrt{-121\cdot 2}$$
$$y=11i\sqrt{2} \quad \text{or} \quad y=-11i\sqrt{2}$$

Solution set: $\{-11i\sqrt{2},11i\sqrt{2}\}$

68. $(8k-7)^2\geq -1$

The square of any real number is always greater than or equal to 0, so any real number satisfies this inequality.

Solution set: $(-\infty,\infty)$

Chapter 8 Test

1. $t^2=54$

$$t=\sqrt{54} \quad \text{or} \quad t=-\sqrt{54}$$
$$t=3\sqrt{6} \quad \text{or} \quad t=-3\sqrt{6}$$

Solution set: $\{-3\sqrt{6},3\sqrt{6}\}$

2. $(7x+3)^2=25$

$$7x+3=5 \quad \text{or} \quad 7x+3=-5$$
$$7x=2 \qquad 7x=-8$$
$$x=\frac{2}{7} \quad \text{or} \quad x=-\frac{8}{7}$$

Solution set: $\{-\frac{8}{7},\frac{2}{7}\}$

3. $x^2+2x=1$

Complete the square.

$$\left(\frac{1}{2}\cdot 2\right)^2=1^2=1$$

Add 1 to both sides.

$$x^2+2x+1=1+1$$
$$(x+1)^2=2$$
$$x+1=\sqrt{2} \quad \text{or} \quad x+1=-\sqrt{2}$$
$$x=-1+\sqrt{2} \quad \text{or} \quad x=-1-\sqrt{2}$$

Solution set: $\{-1+\sqrt{2},-1-\sqrt{2}\}$

4. $2x^2-3x-1=0$

$a=2, b=-3, c=-1$

$$x=\frac{-b\pm\sqrt{b^2-4ac}}{2a}$$
$$x=\frac{-(-3)\pm\sqrt{(-3)^2-4(2)(-1)}}{2(2)}$$
$$x=\frac{3\pm\sqrt{17}}{4}$$

Solution set: $\left\{\frac{3+\sqrt{17}}{4},\frac{3-\sqrt{17}}{4}\right\}$

5. $3t^2-4t=-5$

$$3t^2-4t+5=0$$

$a=3, b=-4, c=5$

$$t=\frac{-b\pm\sqrt{b^2-4ac}}{2a}$$
$$t=\frac{-(-4)\pm\sqrt{(-4)^2-4(3)(5)}}{2(3)}$$
$$=\frac{4\pm\sqrt{-44}}{6}$$
$$=\frac{4\pm 2i\sqrt{11}}{6}$$
$$=\frac{2(2\pm i\sqrt{11})}{6}$$
$$t=\frac{2\pm i\sqrt{11}}{3}$$

Solution set: $\left\{\frac{2+i\sqrt{11}}{3},\frac{2-i\sqrt{11}}{3}\right\}$

6. $3x=\sqrt{\frac{9x+2}{2}}$

Square both sides.

$$9x^2=\frac{9x+2}{2}$$
$$18x^2=9x+2$$
$$18x^2-9x-2=0$$

As directed, use the quadratic formula.
$a = 18, b = -9, c = -2$

$$x = \frac{-b \pm \sqrt{b^2 - 4ac}}{2a}$$
$$x = \frac{-(-9) \pm \sqrt{(-9)^2 - 4(18)(-2)}}{2(18)}$$
$$= \frac{9 \pm \sqrt{225}}{36}$$
$$= \frac{9 \pm 15}{36}$$
$$x = \frac{9+15}{36} = \frac{24}{36} = \frac{2}{3} \quad \text{or}$$
$$x = \frac{9-15}{36} = \frac{-6}{36} = -\frac{1}{6}$$

The potential solution, $\frac{2}{3}$, checks. Since the square root must be nonnegative, $-\frac{1}{6}$ is not a solution.

Solution set: $\left\{\frac{2}{3}\right\}$

7. Let x = Maretha's time alone and
$x - 2$ = Lillaana's time alone.

Make a table.

	Rate	Time together	Part of job done
Maretha	$\frac{1}{x}$	5	$\frac{5}{x}$
Lillaana	$\frac{1}{x-2}$	5	$\frac{5}{x-2}$

Part done by Maretha + part done by Lillaana = 1 whole job.

$$\frac{5}{x} + \frac{5}{x-2} = 1$$

Multiply both sides by $x(x-2)$.

$$x(x-2)\left(\frac{5}{x} + \frac{5}{x-2}\right) = x(x-2) \cdot 1$$
$$5x - 10 + 5x = x^2 - 2x$$
$$0 = x^2 - 12x + 10$$

Use the quadratic formula.
$a = 1, b = -12, c = 10$

$$x = \frac{-b \pm \sqrt{b^2 - 4ac}}{2a}$$
$$x = \frac{-(-12) \pm \sqrt{(-12)^2 - 4(1)(10)}}{2}$$
$$= \frac{12 \pm \sqrt{104}}{2}$$
$$= \frac{12 \pm 2\sqrt{26}}{2}$$
$$= \frac{2(6 \pm \sqrt{26})}{2}$$
$$x = 6 + \sqrt{26} \approx 11.1 \quad \text{or}$$
$$x = 6 - \sqrt{26} \approx .9$$

Reject .9 for Maretha's time, because that would yield a negative time for Lillaana. So, Maretha's time is 11.1 hr and Lillaana's time is $x - 2 = 9.1$ hr.

8. If k is a negative number, then $4k$ is also negative, so the equation $x^2 = 4k$ will have two imaginary solutions. The answer is (a).

9. $2x^2 - 8x - 3 = 0$
$a = 2, b = -8, c = -3$

$$b^2 - 4ac = (-8)^2 - 4(2)(-3)$$
$$= 64 + 24$$
$$= 88$$

The discriminant, 88, is positive but not a perfect square, so there will be two distinct irrational number solutions.

10. $3 - \frac{16}{x} - \frac{12}{x^2} = 0$

Multiply both sides by x^2.

$$x^2\left(3 - \frac{16}{x} - \frac{12}{x^2}\right) = x^2 \cdot 0$$
$$3x^2 - 16x - 12 = 0$$
$$(3x + 2)(x - 6) = 0$$
$$3x + 2 = 0 \quad \text{or} \quad x - 6 = 0$$
$$x = -\frac{2}{3} \quad \text{or} \quad x = 6$$

Both potential solutions check.

Solution set: $\left\{-\frac{2}{3}, 6\right\}$

11. $4x^2 + 7x - 3 = 0$

Use the quadratic formula.

$a = 4, b = 7, c = -3$

$$x = \frac{-b \pm \sqrt{b^2 - 4ac}}{2a}$$

$$x = \frac{-7 \pm \sqrt{7^2 - 4(4)(-3)}}{2(4)}$$

$$x = \frac{-7 \pm \sqrt{97}}{8}$$

Solution set: $\left\{\frac{-7+\sqrt{97}}{8}, \frac{-7-\sqrt{97}}{8}\right\}$

12. $9x^4 + 4 = 37x^2$

$9x^4 - 37x^2 + 4 = 0$

Let $u = x^2$ and $u^2 = (x^2)^2 = x^4$. The equation becomes

$$9u^2 - 37u + 4 = 0$$
$$(9u - 1)(u - 4) = 0$$
$$9u - 1 = 0 \quad \text{or} \quad u - 4 = 0$$
$$u = \frac{1}{9} \quad \text{or} \quad u = 4.$$

To find x, substitute x^2 for u.

$$x^2 = \frac{1}{9} \quad \text{or } x^2 = 4$$
$$x = \frac{1}{3} \text{ or } x = -\frac{1}{3} \text{ or } x = 2 \text{ or } x = -2$$

The potential solutions check.

Solution set: $\{-2, -\frac{1}{3}, \frac{1}{3}, 2\}$

13. $12 = (2d + 1)^2 + (2d + 1)$

Let $u = 2d + 1$. The equation becomes

$$12 = u^2 + u$$
$$0 = u^2 + u - 12$$
$$0 = (u + 4)(u - 3)$$
$$u + 4 = 0 \quad \text{or} \quad u - 3 = 0$$
$$u = -4 \quad \text{or} \quad u = 3.$$

To find d, substitute $2d + 1$ for u.

$$2d + 1 = -4 \quad \text{or} \quad 2d + 1 = 3$$
$$2d = -5 \qquad 2d = 2$$
$$d = -\frac{5}{2} \quad \text{or} \quad d = 1$$

Both potential solutions check.

Solution set: $\{-\frac{5}{2}, 1\}$

14. $S = 4\pi r^2$

Solve for r.

$$\frac{S}{4\pi} = r^2$$
$$r = \pm\sqrt{\frac{S}{4\pi}}$$
$$r = \frac{\pm\sqrt{S}}{2\sqrt{\pi}}$$
$$r = \frac{\pm\sqrt{S}}{2\sqrt{\pi}} \cdot \frac{\sqrt{\pi}}{\sqrt{\pi}}$$
$$r = \frac{\pm\sqrt{\pi S}}{2\pi}$$

15. Let $f(x) = 24.64$ in the quadratic function. Solve for x.

$$f(x) = 3.23x^2 - 1.89x + 1.06$$
$$24.64 = 3.23x^2 - 1.89x + 1.06$$
$$0 = 3.23x^2 - 1.89x - 23.58$$

Use the quadratic formula.

$a = 3.23, b = -1.89, c = -23.58$

$$x = \frac{-b \pm \sqrt{b^2 - 4ac}}{2a}$$
$$x = \frac{1.89 \pm \sqrt{(-1.89)^2 - 4(3.23)(-23.58)}}{2(3.23)}$$
$$= \frac{1.89 \pm \sqrt{308.2257}}{6.46}$$
$$\approx \frac{1.89 \pm 17.56}{6.46}$$
$$x \approx \frac{1.89 + 17.56}{6.46} \approx 3 \text{ or}$$
$$x \approx \frac{1.89 - 17.56}{6.46} \approx -2.4$$

Reject the negative solution. If $x = 0$ corresponds to 1982, then $x = 3$ corresponds to 1985 (1982 + 3 = 1985). The number of cases reached 24.64 thousand in 1985.

16. Let x = Sandi's rate.

Make a table. Use $d = rt$, or $t = \frac{d}{r}$.

	Distance	Rate	Time
Upstream	10	$x - 3$	$\frac{10}{x-3}$
Downstream	10	$x + 3$	$\frac{10}{x+3}$

Time upstream	+	time downstream	=	3.5 hr.
$\frac{10}{x-3}$	+	$\frac{10}{x+3}$	=	3.5

Multiply both sides by $2(x+3)(x-3)$.

$$2(x+3)(x-3)\left(\frac{10}{x-3}+\frac{10}{x+3}\right)$$
$$= 2(x+3)(x-3)3.5$$
$$20(x+3)+20(x-3)=7(x^2-9)$$
$$20x+60+20x-60=7x^2-63$$
$$0=7x^2-40x-63$$
$$0=(x-7)(7x+9)$$
$$x-7=0 \quad \text{or} \quad 7x+9=0$$
$$x=7 \quad \text{or} \quad x=-\frac{9}{7}$$

Reject $-\frac{9}{7}$ since rate can't be negative. Sandi's rate was 7 mph.

17. Let $x =$ the width of the walk.

The area of the walk is equal to the area of the outer figure minus the area of the pool. So,

$$152=(10+2x)(24+2x)-24(10)$$
$$152=240+68x+4x^2-240$$
$$0=4x^2+68x-152$$
$$0=x^2+17x-38$$
$$0=(x+19)(x-2)$$
$$x+19=0 \quad \text{or} \quad x-2=0$$
$$x=-19 \quad \text{or} \quad x=2.$$

Reject -19 since width can't be negative. The walk is 2 ft wide.

18. Let $x =$ the height of the tower, and $2x+2 =$ the distance from the point to the top.

The distance from the base to the point is 30 m. These three segments form a right triangle, so the Pythagorean theorem applies.

$$a^2+b^2=c^2$$
$$x^2+30^2=(2x+2)^2$$
$$x^2+900=4x^2+8x+4$$
$$0=3x^2+8x-896$$
$$0=(x-16)(3x+56)$$
$$x-16=0 \quad \text{or} \quad 3x+56=0$$
$$x=16 \quad \text{or} \quad x=-\frac{56}{3}$$

Reject $-\frac{56}{3}$ since height can't be negative. The tower is 16 m high.

For Exercises 19 and 20, see the answer graphs in the back of the textbook.

19. $$2x^2+7x>15$$
$$2x^2+7x-15>0$$

Solve the equation

$$2x^2+7x-15=0$$
$$(2x-3)(x+5)=0$$
$$2x-3=0 \quad \text{or} \quad x+5=0$$
$$x=\frac{3}{2} \quad \text{or} \quad x=-5.$$

These numbers divide the number line into three regions.

A B C
-5 3/2

Test a number from each region in the inequality

$$2x^2+7x>15.$$

Region A: Let $x=-6$.

$$2(-6)^2+7(-6)>15 \quad ?$$
$$72-42>15 \quad ?$$
$$30>15 \quad \textit{True}$$

Region B: Let $x=0$.

$$2(0)^2+7(0)>15 \quad ?$$
$$0>15 \quad \textit{False}$$

Region C: Let $x=2$.

$$2(2)^2+7(2)>15 \quad ?$$
$$8+14>15 \quad ?$$
$$22>15 \quad \textit{True}$$

The numbers in Regions A and C, not including -5 or $\frac{3}{2}$, are solutions.

Solution set: $(-\infty,-5)\cup\left(\frac{3}{2},\infty\right)$

20. $\frac{5}{t-4}\le 1$

Solve the equation

$$\frac{5}{t-4}=1$$
$$5=t-4$$
$$t=9.$$

Set the denominator equal to zero and solve.

$$t-4=0$$
$$t=4$$

These numbers divide the number line into three regions.

A B C

4 9

Test a number from each region in the inequality

$$\frac{5}{t-4} \leq 1.$$

Region A: Let $t = 0$.

$$\frac{5}{0-4} \leq 1 \quad ?$$
$$\frac{5}{-4} \leq 1 \qquad \textit{True}$$

Region B: Let $t = 5$.

$$\frac{5}{5-4} \leq 1 \quad ?$$
$$5 \leq 1 \qquad \textit{False}$$

Region C: Let $t = 10$.

$$\frac{5}{10-4} \leq 1 \quad ?$$
$$\frac{5}{6} \leq 1 \qquad \textit{True}$$

The numbers in Regions A and C, including 9 but not 4 (because it makes the denominator 0), are solutions.

Solution set: $(-\infty, 4) \cup [9, \infty)$

Cumulative Review Exercises (Chapters 1-8)

In Exercises 1-4,

$$S = \left\{-\frac{7}{3}, -2, -\sqrt{3}, 0, .7, \sqrt{12}, \sqrt{-8}, 7, \frac{32}{3}\right\}.$$

1. The elements of S that are integers are $-2, 0$, and 7.

2. The elements of S that are rational numbers are $-\frac{7}{3}, -2, 0, .7, 7$, and $\frac{32}{3}$.

3. The elements of S that are real numbers are $-\frac{7}{3}, -2, -\sqrt{3}, 0, .7, \sqrt{12}, 7$, and $\frac{32}{3}$. This is all the elements except $\sqrt{-8}$.

4. All of the elements of S are complex numbers, since the complex number system includes the set of real numbers.

5. $|-3| + 8 - |-9| - (-7+3)$
$= 3 + 8 - 9 - (-4)$
$= 3 + 8 - 9 + 4$
$= 6$

6. $2(-3)^2 + (-8)(-5) + (-17)$
$= 2(9) + 40 - 17$
$= 18 + 40 - 17$
$= 41$

7. $-2x + 4 = 5(x-4) + 17$
$-2x + 4 = 5x - 20 + 17$
$-2x + 4 = 5x - 3$
$7 = 7x$
$1 = x$

Solution set: $\{1\}$

8. $|3y - 7| \leq 1$
$-1 \leq 3y - 7 \leq 1$
$6 \leq 3y \leq 8$
$2 \leq y \leq \frac{8}{3}$

Solution set: $\left[2, \frac{8}{3}\right]$

9. $|4z + 2| > 7$
$4z + 2 > 7$ or $4z + 2 < -7$
$4z > 5$ $\quad 4z < -9$
$z > \frac{5}{4}$ or $z < -\frac{9}{4}$

Solution set: $\left(-\infty, -\frac{9}{4}\right) \cup \left(\frac{5}{4}, \infty\right)$

10. $2x - 4y = 7$

Solve the equation for y to get the slope-intercept form, $y = mx + b$.

$$2x - 4y = 7$$
$$-4y = -2x + 7$$
$$y = \frac{1}{2}x - \frac{7}{4}$$

The slope m is $\frac{1}{2}$ and the y-intercept, since $b = -\frac{7}{4}$, is $\left(0, -\frac{7}{4}\right)$.

11. Through $(2, -1)$; perpendicular to $-3x + y = 5$

Find the slope of the given line.

$$-3x + y = 5$$
$$y = 3x + 5$$

The slope is 3. The negative reciprocal of 3 is $-\frac{1}{3}$, so the slope of the line through $(2, -1)$ is $-\frac{1}{3}$. Use the point-slope form with $m = -\frac{1}{3}$ and $(x_1, y_1) = (2, -1)$ to find the equation of this line.

$$y - y_1 = m(x - x_1)$$
$$y - (-1) = -\frac{1}{3}(x - 2)$$
$$y + 1 = -\frac{1}{3}(x - 2)$$

Multiply by 3 to clear the fraction.

$$3y + 3 = -1(x - 2)$$
$$3y + 3 = -x + 2$$
$$x + 3y = -1$$

12. $$\left(\frac{x^{-3}y^2}{x^5y^{-2}}\right)^{-1} = \left(x^{-3-5}y^{2-(-2)}\right)^{-1} = (x^{-8}y^4)^{-1} = x^8y^{-4} = \frac{x^8}{y^4}$$

13. $$\frac{(4x^{-2})^2(2y^3)}{8x^{-3}y^5} = \frac{16x^{-4}(2y^3)}{8x^{-3}y^5} = \frac{4x^{-4}y^3}{x^{-3}y^5} = 4x^{-4-(-3)}y^{3-5} = 4x^{-1}y^{-2} = \frac{4}{xy^2}$$

14. $$f(x) = \sqrt{2x - 5}$$
$$f(10) = \sqrt{2(10) - 5} = \sqrt{20 - 5} = \sqrt{15}.$$

Since the quantity under the radical sign must be greater than or equal to 0,

$$2x - 5 \geq 0$$
$$2x \geq 5$$
$$x \geq \frac{5}{2}.$$

The domain of the function $\left[\frac{5}{2}, \infty\right)$.

15. $$5x - 3y = 17 \quad (1)$$
$$2y = 4x - 12 \quad (2)$$

Solve equation (2) for y.

$$2y = 4x - 12 \quad (2)$$
$$y = 2x - 6$$

Substitute $2x - 6$ for y in equation (1) to find x.

$$5x - 3y = 17 \quad (1)$$
$$5x - 3(2x - 6) = 17$$
$$5x - 6x + 18 = 17$$
$$-x = -1$$
$$x = 1$$

Since $x = 1$,

$$y = 2x - 6 = 2(1) - 6 = -4.$$

Solution set: $\{(1, -4)\}$

16. If $D_x = 4, D_y = 3, D_z = -6$, and $D = 12$, then

$$\frac{D_x}{D} = \frac{4}{12} = \frac{1}{3},$$
$$\frac{D_y}{D} = \frac{3}{12} = \frac{1}{4}, \text{ and}$$
$$\frac{D_z}{D} = \frac{-6}{12} = -\frac{1}{2}.$$

Solution set: $\left\{\left(\frac{1}{3}, \frac{1}{4}, -\frac{1}{2}\right)\right\}$

17. $$\left(\frac{2}{3}t + 9\right)^2 = \left(\frac{2}{3}t\right)^2 + 2\left(\frac{2}{3}t\right)(9) + 9^2 = \frac{4}{9}t^2 + 12t + 81$$

18. $$(3t^3 + 5t^2 - 8t + 7) - (6t^3 + 4t - 8)$$
$$= 3t^3 - 6t^3 + 5t^2 - 8t - 4t + 7 + 8$$
$$= -3t^3 + 5t^2 - 12t + 15$$

19. Divide $4x^3 + 2x^2 - x + 26$ by $x + 2$.

Use synthetic division.

−2)	4	2	−1	26
		−8	12	−22
	4	−6	11	4

The answer is

$$4x^2 - 6x + 11 + \frac{4}{x + 2}.$$

20. $$16x - x^3 = x(16 - x^2) = x(4 + x)(4 - x)$$

21. $24m^2 + 2m - 15 = (4m - 3)(6m + 5)$

22. $9x^2 - 30xy + 25y^2$

Use the perfect square formula

$$(a - b)^2 = a^2 - 2ab + b^2,$$

where $a = 3x$ and $b = 5y$.
So,

$$9x^2 - 30xy + 25y^2 = (3x - 5y)^2.$$

23. $8x^3 + 27y^3$

Use the sum of two cubes formula

$$a^3 + b^3 = (a+b)(a^2 - ab + b^2),$$

where $a = 2x$ and $b = 3y$.
So,

$$8x^3 + 27y^3 = (2x + 3y)(4x^2 - 6xy + 9y^2).$$

24. $\dfrac{x^2 - 3x - 10}{x^2 + 3x + 2} \cdot \dfrac{x^2 - 2x - 3}{x^2 + 2x - 15}$

$$= \frac{(x-5)(x+2)}{(x+2)(x+1)} \cdot \frac{(x-3)(x+1)}{(x+5)(x-3)}$$

$$= \frac{x-5}{x+5}$$

25. $\dfrac{5t+2}{-6} \div \dfrac{15t+6}{5}$

Multiply by the reciprocal.

$$= \frac{5t+2}{-6} \cdot \frac{5}{15t+6}$$

$$= \frac{5t+2}{-6} \cdot \frac{5}{3(5t+2)}$$

$$= \frac{5}{-18} \text{ or } -\frac{5}{18}$$

26. $\dfrac{3}{2-k} - \dfrac{5}{k} + \dfrac{6}{k^2 - 2k}$

$$= \frac{3}{2-k} - \frac{5}{k} + \frac{6}{k(k-2)}$$

$$= \frac{-3}{k-2} - \frac{5}{k} + \frac{6}{k(k-2)}$$

The LCD is $k(k-2)$.

$$= \frac{-3k}{(k-2)k} - \frac{5(k-2)}{k(k-2)} + \frac{6}{k(k-2)}$$

$$= \frac{-3k - 5(k-2) + 6}{k(k-2)}$$

$$= \frac{-3k - 5k + 10 + 6}{k(k-2)}$$

$$= \frac{-8k + 16}{k(k-2)}$$

$$= \frac{-8(k-2)}{k(k-2)}$$

$$= -\frac{8}{k}$$

27. $\dfrac{\frac{r}{s} - \frac{s}{r}}{\frac{r}{s} + 1}$

Multiply the numerator and denominator by rs, the LCD of all the fractions.

$$= \frac{\left(\frac{r}{s} - \frac{s}{r}\right) rs}{\left(\frac{r}{s} + 1\right) rs}$$

$$= \frac{r^2 - s^2}{r^2 + rs}$$

$$= \frac{(r-s)(r+s)}{r(r+s)}$$

$$= \frac{r-s}{r}$$

28. $\sqrt[3]{\dfrac{27}{16}} = \dfrac{\sqrt[3]{27}}{\sqrt[3]{16}} = \dfrac{3}{2\sqrt[3]{2}} = \dfrac{3 \cdot \sqrt[3]{4}}{2\sqrt[3]{2} \cdot \sqrt[3]{4}}$

$$= \frac{3\sqrt[3]{4}}{2\sqrt[3]{8}} = \frac{3\sqrt[3]{4}}{2 \cdot 2} = \frac{3\sqrt[3]{4}}{4}$$

29. $\dfrac{2}{\sqrt{7} - \sqrt{5}} = \dfrac{2(\sqrt{7} + \sqrt{5})}{(\sqrt{7} - \sqrt{5})(\sqrt{7} + \sqrt{5})}$

$$= \frac{2(\sqrt{7} + \sqrt{5})}{7-5}$$

$$= \frac{2(\sqrt{7} + \sqrt{5})}{2}$$

$$= \sqrt{7} + \sqrt{5}$$

30. $2x = \sqrt{\dfrac{5x+2}{3}}$

Square both sides.

$$(2x)^2 = \left(\sqrt{\frac{5x+2}{3}}\right)^2$$

$$4x^2 = \frac{5x+2}{3}$$

$$12x^2 = 5x + 2$$

$$12x^2 - 5x - 2 = 0$$

$$(3x-2)(4x+1) = 0$$

$$3x - 2 = 0 \quad \text{or} \quad 4x + 1 = 0$$

$$x = \frac{2}{3} \quad \text{or} \quad x = -\frac{1}{4}$$

The square root of a real number cannot be negative, so $-\frac{1}{4}$ is not a solution. The potential solution $\frac{2}{3}$ checks.

Solution set: $\{\frac{2}{3}\}$

31. $\frac{3}{x-3} - \frac{2}{x-2} = \frac{3}{x^2-5x+6}$

$\frac{3}{x-3} - \frac{2}{x-2} = \frac{3}{(x-3)(x-2)}$

$x-3 \neq 0$ or $x-2 \neq 0$, so $x \neq 3$ or $x \neq 2$.

Multiply by $(x-3)(x-2)$.

$$(x-3)(x-2)\left(\frac{3}{x-3} - \frac{2}{x-2}\right)$$
$$= (x-3)(x-2)\left(\frac{3}{(x-3)(x-2)}\right)$$
$$3(x-2) - 2(x-3) = 3$$
$$3x - 6 - 2x + 6 = 3$$
$$x = 3$$

But $x \neq 3$ since it makes the denominator 0.

Solution set: $\emptyset$

32. $$(r-5)(2r+3) = 1$$
$$2r^2 - 7r - 15 = 1$$
$$2r^2 - 7r - 16 = 0$$

Use the quadratic formula.
$a = 2, b = -7, c = -16$

$$r = \frac{-b \pm \sqrt{b^2-4ac}}{2a}$$
$$r = \frac{-(-7) \pm \sqrt{(-7)^2 - 4(2)(-16)}}{2(2)}$$
$$= \frac{7 \pm \sqrt{49+128}}{4}$$
$$r = \frac{7 \pm \sqrt{177}}{4}$$

Solution set: $\left\{\frac{7+\sqrt{177}}{4}, \frac{7-\sqrt{177}}{4}\right\}$

33. $b^4 - 5b^2 + 4 = 0$

Let $u = b^2$ and $u^2 = (b^2)^2 = b^4$. The equation becomes

$$u^2 - 5u + 4 = 0$$
$$(u-4)(u-1) = 0$$
$$u - 4 = 0 \quad \text{or} \quad u - 1 = 0$$
$$u = 4 \quad \text{or} \quad u = 1.$$

To find b, substitute b^2 for u.

$$b^2 = 4 \quad \text{or} \quad b^2 = 1$$
$$b = 2 \quad \text{or} \quad b = -2 \quad \text{or} \quad b = 1 \quad \text{or} \quad b = -1$$

The potential solutions check.

Solution set: $\{-2, -1, 1, 2\}$

34. Let x = the width of the rectangle.

The perimeter of the rectangle is 20 in. Since the formula for perimeter is $P = 2L + 2W$,

$$20 = 2L + 2x.$$

Solve the equation for L.

$$20 - 2x = 2L$$
$$10 - x = L$$

The area of the rectangle is 21 in^2. Since the formula for area is $A = LW$,

$$21 = (10-x)x$$
$$21 = 10x - x^2$$
$$x^2 - 10x + 21 = 0$$
$$(x-7)(x-3) = 0$$
$$x - 7 = 0 \quad \text{or} \quad x - 3 = 0$$
$$x = 7 \quad \text{or} \quad x = 3.$$

Reject 7 for the width since width is not longer than length. So, the width is 3 in, and the length is $10 - x = 10 - 3 = 7$ in.

35. Let $x =$ the distance traveled by the southbound car and
$2x - 38 =$ the distance traveled by the eastbound car.

Since the cars are traveling at right angles with one another, the Pythagorean theorem can be applied.

$$a^2 + b^2 = c^2$$
$$x^2 + (2x-38)^2 = 95^2$$
$$x^2 + 4x^2 - 152x + 1444 = 9025$$
$$5x^2 - 152x - 7581 = 0$$
$$(x-57)(5x+133) = 0$$
$$x - 57 = 0 \quad \text{or} \quad 5x + 133 = 0$$
$$x = 57 \quad \text{or} \quad x = -\frac{133}{5}$$

Reject $-\frac{133}{5}$ since distance can't be negative. The southbound car traveled 57 mi, and the eastbound car traveled $2x - 38 = 2(57) - 38 = 76$ mi.

36. Let $x =$ the number of units from Supplier I,
$y =$ the number of units from Supplier II, and
$z =$ the number of units from Supplier III.

The system of equations is

$$\begin{aligned} x + y + z &= 100 \quad (1) \\ 80x + 50y + 65z &= 5990 \quad (2) \\ x &= z. \quad (3) \end{aligned}$$

To eliminate y, multiply equation (1) by -50 and add the result to equation (2).

$$\begin{aligned} -50x - 50y - 50z &= -5000 \\ 80x + 50y + 65z &= 5990 \quad (2) \\ \hline 30x \qquad + 15z &= 990 \quad (4) \end{aligned}$$

Since $x = z$ from equation (3), substitute z for x in equation (4) and solve for z.

$$\begin{aligned} 30x + 15z &= 990 \quad (4) \\ 30z + 15z &= 990 \\ 45z &= 990 \\ z &= 22 \end{aligned}$$

So, $x = z = 22$. Substitute 22 for x and z in equation (1) to find y.

$$\begin{aligned} x + y + z &= 100 \quad (1) \\ 22 + y + 22 &= 100 \\ y + 44 &= 100 \\ y &= 56 \end{aligned}$$

There were 22 units ordered from Suppliers I and III and 56 units ordered from Supplier II.

37. Let $f(t) = 70$ in the function. Solve for t.

$$\begin{aligned} f(t) &= -16t^2 + 80t + 50 \\ 70 &= -16t^2 + 80t + 50 \\ 0 &= -16t^2 + 80t - 20 \end{aligned}$$

Divide by -4.

$$0 = 4t^2 - 20t + 5$$

Use the quadratic formula.
$a = 4, b = -20, c = 5$

$$t = \frac{-b \pm \sqrt{b^2 - 4ac}}{2a}$$

$$t = \frac{20 \pm \sqrt{(-20)^2 - 4(4)(5)}}{2(4)}$$

$$= \frac{20 \pm \sqrt{320}}{8}$$

$$\approx \frac{20 \pm 17.9}{8}$$

$$t = \frac{20 + 17.9}{8} \approx 4.7 \text{ or}$$

$$t = \frac{20 - 17.9}{8} \approx .3$$

The object reaches a height of 70 ft at about .3 sec and about 4.7 sec.

38. $z^2 - 2z = 15$

To complete the square, take half the coefficient of z and square it.

$$\left[\frac{1}{2}(-2)\right]^2 = 1$$

So,

$$\begin{aligned} z^2 - 2z + 1 &= 15 + 1 \\ (z - 1)^2 &= 16 \\ z - 1 = 4 \quad &\text{or} \quad z - 1 = -4 \\ z = 5 \quad &\text{or} \quad z = -3. \end{aligned}$$

Solution set: $\{-3, 5\}$

39. $2x^2 - 4x - 3 = 0$

$a = 2, b = -4, c = -3$

$$x = \frac{-b \pm \sqrt{b^2 - 4ac}}{2a}$$

$$x = \frac{-(-4) \pm \sqrt{(-4)^2 - 4(2)(-3)}}{2(2)}$$

$$= \frac{4 \pm \sqrt{16 + 24}}{4}$$

$$= \frac{4 \pm \sqrt{40}}{4}$$

$$= \frac{4 \pm 2\sqrt{10}}{4}$$

$$= \frac{2(2 + \sqrt{10})}{4}$$

$$x = \frac{2 \pm \sqrt{10}}{2}$$

Solution set: $\left\{\frac{2+\sqrt{10}}{2}, \frac{2-\sqrt{10}}{2}\right\}$

40.

$$\frac{x + 3}{x - 5} > 2$$

$$\frac{x + 3}{x - 5} - 2 > 0$$

$$\frac{x + 3}{x - 5} - \frac{2(x - 5)}{x - 5} > 0$$

$$\frac{x + 3 - 2x + 10}{x - 5} > 0$$

$$\frac{-x + 13}{x - 5} > 0$$

The number 13 makes the numerator 0, and 5 makes the denominator 0. Use these numbers to divide a number line into regions.

A B C

5 13

Test a number from each region in the inequality

$$\frac{x+3}{x-5} > 2.$$

Region A: Let $x = 0$.

$$\frac{0+3}{0-5} > 2 \quad ?$$

$$-\frac{3}{5} > 2 \qquad \textit{False}$$

Region B: Let $x = 6$.

$$\frac{6+3}{6-5} > 2 \quad ?$$

$$9 > 2 \qquad \textit{True}$$

Region C: Let $x = 14$.

$$\frac{14+3}{14-5} > 2 \quad ?$$

$$\frac{17}{9} > 2 \qquad \textit{False}$$

The numbers in Region B, not including 5 or 13, are solutions.

Solution set: $(5, 13)$

Chapter 9

GRAPHS OF NONLINEAR FUNCTIONS AND CONIC SECTIONS

9.1 Graphs of Quadratic Functions; Vertical Parabolas

9.1 Margin Exercises

For Exercises 1-3, see the answer graphs for the margin exercises in the textbook.

1. **(a)** The graph of

$$f(x) = x^2 + 3$$

has the same shape as the graph of $f(x) = x^2$. Here $k = 3$, so the graph is shifted up 3 units and has vertex $(0, 3)$.

(b) The graph of

$$f(x) = x^2 - 1$$

also has the same shape as the graph of $f(x) = x^2$. Here $k = -1$, so the graph is shifted 1 unit down and has vertex $(0, -1)$.

2. **(a)** The parabola

$$f(x) = (x - 3)^2$$

has the same shape as $f(x) = x^2$ but shifted 3 units to the right, since 3 would cause $x - 3$ to equal 0. The vertex is $(3, 0)$.

(b) The parabola

$$f(x) = (x + 2)^2$$

is like $f(x) = x^2$ but shifted 2 units to the left, since -2 would cause $x + 2$ to equal 0. The vertex is $(-2, 0)$.

3. **(a)** $f(x) = (x+2)^2 - 1$ has a graph like $f(x) = x^2$ but shifted 2 units to the left (since $x + 2 = 0$ if $x = -2$) and 1 unit down (because of the -1). The vertex is $(-2, -1)$. The points $(-3, 0)$ and $(-1, 0)$ are also on the graph.

(b) $f(x) = (x - 2)^2 + 5$ is shifted 2 units to the right (since $x - 2 = 0$ if $x = 2$) and 5 units up (because of the $+5$), so its vertex is $(2, 5)$.

4. An equation of the form $f(x) = a(x - h)^2 + k$ has a parabola for its graph. If $a > 0$, the parabola opens upward; if $a < 0$, the parabola open downward.

(a) $f(x) = -\frac{2}{3}x^2$

Here, $a = -\frac{2}{3} < 0$, so the parabola opens downward.

(b) $f(x) = \frac{3}{4}x^2 + 1$

Here, $a = \frac{3}{4} > 0$, so the parabola opens upward.

(c) $f(x) = -2x^2 - 3$

Here, $a = -2 < 0$, so the parabola opens downward.

(d) $f(x) = 3x^2 + 2$

Here, $a = 3 > 0$, so the parabola opens upward.

5. If, in $f(x) = a(x - h)^2 + k$, $0 < |a| < 1$, then the graph is wider than $f(x) = x^2$; if $|a| > 1$, the graph is narrower.

(a) $f(x) = -\frac{2}{3}x^2$

Here, $|a| = \left|-\frac{2}{3}\right| = \frac{2}{3} < 1$, so the graph is wider than $f(x) = x^2$.

(b) $f(x) = \frac{3}{4}x^2 + 1$

Here, $|a| = \left|\frac{3}{4}\right| = \frac{3}{4} < 1$, so the parabola is wider than $f(x) = x^2$.

(c) $f(x) = -2x^2 - 3$

Here, $|a| = |-2| = 2 > 1$, so the parabola is narrower than $f(x) = x^2$.

(d) $f(x) = 3x^2 + 2$

Here, $|a| = |3| = 3 > 1$, so the parabola is narrower than $f(x) = x^2$.

6. $f(x) = \frac{1}{2}(x - 2)^2 + 1$

This equation has a graph like $f(x) = x^2$ but shifted 2 units to the right and 1 unit up. It has vertex $(2, 1)$. Since $a = \frac{1}{2} > 0$, the parabola opens upward. Since $|a| = \left|\frac{1}{2}\right| = \frac{1}{2} < 1$, the parabola is wider than the graph of $f(x) = x^2$. If $x = 0$, then $f(x) = 3$, so $(0, 3)$ is on the graph. If $x = 4$, then $f(x) = 3$, so $(4, 3)$ is also on the graph. See the answer graph for the margin exercises in the textbook.

9.1 Section Exercises

3. The graph of $f(x) = -3x^2$ has no vertical or horizontal shift, so its vertex is at the origin, $(0, 0)$.

7. The graph of $f(x) = (x-1)^2$ has a horizontal shift of 1 unit to the right (since $x-1=0$ when $x=1$). Its vertex is at $(1, 0)$.

11. The graph of $f(x) = -.4x^2$ opens downward because the coefficient of x^2 is negative. The .4 makes the parabola wider than the graph of $f(x) = x^2$.

15. The graph of $f(x) = (x+3)^2 - 4$ is shifted to the left 3 units (since $x+3=0$ if $x=-3$) and down 4 units (because of the -4).

    The graph of $f(x) = (x-5)^2 - 8$ is shifted to the right 5 units (since $x-5=0$ if $x=5$) and down 8 units (because of the -8).

For Exercises 19-27, see the answer graphs in the back of the textbook.

19. $f(x) = -2x^2$

    Since $a = -2 < 0$, the graph opens downward. Since $|a| = |-2| = 2 > 1$, the graph is narrower than the graph of $f(x) = x^2$. The graph has no horizontal or vertical shift, so the vertex is $(0, 0)$. The points $(2, -8)$ and $(-2, -8)$ are on the graph.

23. $f(x) = -x^2 + 2$

    Since $a = -1 < 0$, the graph opens downward. Since $|a| = |-1| = 1$, the graph has the same shape as $f(x) = x^2$. The graph is shifted 2 units upward (because of the $+2$), so the vertex is $(0, 2)$. The points $(2, -2)$ and $(-2, -2)$ are on the graph.

27. $f(x) = (x+2)^2 - 1$

    Here, $a = 1$, so the graph opens upward and has the same shape as $f(x) = x^2$. The graph is shifted 2 units to the left (since $x+2=0$ if $x=-2$) and 1 unit down (because of the -1), so the vertex is $(-2, -1)$. The points $(-1, 0)$ and $(-3, 0)$ are on the graph.

31. The graph of $f(x) = x^2 + 6$ would be shifted 6 units upward from the graph of $g(x) = x^2$.

32. To graph $f(x) = x + 6$, plot the intercepts $(-6, 0)$ and $(0, 6)$, and draw the line through them. See the answer graph in the back of the textbook.

33. When considering the graph of $f(x) = x + 6$, the y-intercept is 6. The graph of $g(x) = x$ has y-intercept 0. Therefore, the graph of $f(x) = x + 6$ is shifted 6 units upward compared to the graph of $g(x) = x$.

34. The graph of $f(x) = (x-6)^2$ is shifted 6 units to the right compared to the graph of $g(x) = x^2$.

35. To graph $f(x) = x - 6$, plot the intercepts $(6, 0)$ and $(0, -6)$, and draw the line through them. See the answer graph in the back of the textbook.

36. When considering the graph of $f(x) = x - 6$, its x-intercept is 6 as compared to the graph of $g(x) = x$ with x-intercept 0. The graph of $f(x) = x - 6$ is shifted 6 units to the right compared to the graph of $g(x) = x$.

9.2 More About Quadratic Functions; Horizontal Parabolas

9.2 Margin Exercises

1. **(a)** $f(x) = x^2 - 6x + 7$

   To find the vertex, complete the square on $x^2 - 6x$.

$$\left[\frac{1}{2}(-6)\right]^2 = (-3)^2 = 9$$

   Add and subtract 9 on the right.

$$\begin{aligned} f(x) &= x^2 - 6x + 9 - 9 + 7 \\ f(x) &= (x^2 - 6x + 9) - 9 + 7 \\ f(x) &= (x^2 - 6x + 9) - 2 \\ f(x) &= (x-3)^2 - 2 \end{aligned}$$

   The vertex is $(3, -2)$.

   (b) $f(x) = x^2 + 4x - 9$

   To find the vertex, complete the square on $x^2 + 4x$.

$$\left[\frac{1}{2}(4)\right]^2 = 2^2 = 4$$

   Add and subtract 4 on the right.

$$\begin{aligned} f(x) &= x^2 + 4x + 4 - 4 - 9 \\ f(x) &= (x^2 + 4x + 4) - 4 - 9 \\ f(x) &= (x^2 + 4x + 4) - 13 \\ f(x) &= (x+2)^2 - 13 \end{aligned}$$

   The vertex is $(-2, -13)$.

2. **(a)** $f(x) = 2x^2 - 4x + 1$

Factor out 2 from the first two terms.

$$f(x) = 2(x^2 - 2x) + 1$$

Complete the square on $x^2 - 2x$.

$$\left[\frac{1}{2}(-2)\right]^2 = (-1)^2 = 1$$

Add and subtract 1.

$$\begin{aligned} f(x) &= 2(x^2 - 2x + 1 - 1) + 1 \\ f(x) &= 2(x^2 - 2x + 1) + 2(-1) + 1 \\ f(x) &= 2(x^2 - 2x + 1) - 2 + 1 \\ f(x) &= 2(x - 1)^2 - 1 \end{aligned}$$

The vertex is $(1, -1)$.

(b) $f(x) = -\frac{1}{2}x^2 + 2x - 3$

Factor out $-\frac{1}{2}$ from the first two terms.

$$f(x) = -\frac{1}{2}(x^2 - 4x) - 3$$

Complete the square on $x^2 - 4x$.

$$\left[\frac{1}{2}(-4)\right]^2 = (-2)^2 = 4$$

Add and subtract 4.

$$\begin{aligned} f(x) &= -\frac{1}{2}(x^2 - 4x + 4 - 4) - 3 \\ f(x) &= -\frac{1}{2}(x^2 - 4x + 4) - \frac{1}{2}(-4) - 3 \\ f(x) &= -\frac{1}{2}(x^2 - 4x + 4) + 2 - 3 \\ f(x) &= -\frac{1}{2}(x - 2)^2 - 1 \end{aligned}$$

The vertex is $(2, -1)$.

3. **(a)** In $f(x) = -2x^2 + 3x - 1, a = -2, b = 3$, and $c = -1$.

The vertex is $\left(\frac{-b}{2a}, f\left(\frac{-b}{2a}\right)\right)$, so

$$\frac{-b}{2a} = \frac{-3}{2(-2)} = \frac{-3}{-4} = \frac{3}{4},$$

and

$$\begin{aligned} f\left(\frac{3}{4}\right) &= -2\left(\frac{3}{4}\right)^2 + 3\left(\frac{3}{4}\right) - 1 \\ &= -2\left(\frac{9}{16}\right) + \frac{9}{4} - 1 \\ &= -\frac{9}{8} + \frac{18}{8} - \frac{8}{8} \\ &= \frac{1}{8}. \end{aligned}$$

The vertex is $\left(\frac{3}{4}, \frac{1}{8}\right)$.

(b) In $f(x) = 4x^2 - x + 5, a = 4, b = -1$, and $c = 5$.

The vertex is $\left(\frac{-b}{2a}, f\left(\frac{-b}{2a}\right)\right)$, so

$$\frac{-b}{2a} = \frac{-(-1)}{2(4)} = \frac{1}{8},$$

and

$$\begin{aligned} f\left(\frac{1}{8}\right) &= 4\left(\frac{1}{8}\right)^2 - \frac{1}{8} + 5 \\ &= 4\left(\frac{1}{64}\right) - \frac{1}{8} + 5 \\ &= \frac{1}{16} - \frac{2}{16} + \frac{80}{16} \\ &= \frac{79}{16}. \end{aligned}$$

The vertex is $\left(\frac{1}{8}, \frac{79}{16}\right)$.

4. $f(x) = x^2 - 6x + 5$

Find the y-intercept. Let $x = 0$.

$$f(0) = 0^2 - 6(0) + 5 = 5$$

The y-intercept is $(0, 5)$. Find any x-intercepts. Let $f(x) = 0$.

$$\begin{aligned} f(x) &= x^2 - 6x + 5 \\ 0 &= x^2 - 6x + 5 \\ 0 &= (x - 5)(x - 1) \\ x - 5 = 0 \quad &\text{or} \quad x - 1 = 0 \\ x = 5 \quad &\text{or} \quad x = 1 \end{aligned}$$

The x-intercepts are $(5, 0)$ and $(1, 0)$. To find the vertex, complete the square on $x^2 - 6x$.

$$\left[\frac{1}{2}(-6)\right]^2 = (-3)^2 = 9$$

Add and subtract 9.

$$\begin{aligned} f(x) &= x^2 - 6x + 9 - 9 + 5 \\ f(x) &= (x - 3)^2 - 4 \end{aligned}$$

The vertex is at $(3, -4)$.

By symmetry, another point on the graph is $(6, 5)$. The graph opens upward because $a = 1 > 0$. See the answer graph for the margin exercises in the textbook. Since x can be any real number, the domain is $(-\infty, \infty)$. Since the lowest point of the parabola is $(3, -4)$, the range is $[-4, \infty)$.

5. **(a)** $f(x) = 4x^2 - 20x + 25$

Here, $a = 4, b = -20$, and $c = 25$, so

$$\begin{aligned} b^2 - 4ac &= (-20)^2 - 4(4)(25) \\ &= 400 - 400 \\ &= 0. \end{aligned}$$

Since the discriminant is 0, the graph has only one x-intercept.

(b) $f(x) = 2x^2 + 3x + 5$

Here, $a = 2, b = 3$, and $c = 5$, so

$$\begin{aligned} b^2 - 4ac &= 3^2 - 4(2)(5) \\ &= 9 - 40 \\ &= -31. \end{aligned}$$

Since the discriminant is negative, the graph has no x-intercepts.

(c) $f(x) = -3x^2 - x + 2$

Here, $a = -3, b = -1$, and $c = 2$, so

$$\begin{aligned} b^2 - 4ac &= (-1)^2 - 4(-3)(2) \\ &= 1 + 24 \\ &= 25. \end{aligned}$$

Since the discriminant is positive, the graph has two x-intercepts.

6. Let $x =$ the width of the field.
(See the figure in the textbook.)

Then, $x + x +$ length $= 100$
length $= 100 - 2x$.

The area is the product of the length and the width.

$$A = (100 - 2x)x$$
$$A = 100x - 2x^2$$
$$A = -2x^2 + 100x$$

Here, $a = -2, b = 100$, and $c = 0$, so

$$h = \frac{-b}{2a} = \frac{-100}{2(-2)} = 25 \text{ and}$$

$$\begin{aligned} f(h) &= -2(25)^2 + 100(25) \\ &= -2(625) + 2500 \\ &= -1250 + 2500 \\ &= 1250. \end{aligned}$$

The graph is a parabola that opens downward with vertex at $(25, 1250)$. The vertex of the graph shows that the maximum area will be 1250 ft^2. This area will occur if the width is 25 ft and the length is $100 - 2x$, or 50 ft.

7. $x = (y + 1)^2 - 4$

Since the roles of x and y are reversed, this graph has its vertex at $(-4, -1)$. It opens to the right, the positive x-direction, and has the same shape as $y = x^2$. The points $(0, 1)$ and $(0, -3)$ are on the graph. See the answer graph for the margin exercises in the textbook.

8. **(a)** $x = 2y^2 - 6y + 5$
$x = 2(y^2 - 3y) + 5$

Complete the square on $y^2 - 3y$.

$$\left[\frac{1}{2}(-3)\right]^2 = \left(-\frac{3}{2}\right)^2 = \frac{9}{4}$$

Add and subtract $\frac{9}{4}$.

$$\begin{aligned} x &= 2\left(y^2 - 3y + \frac{9}{4} - \frac{9}{4}\right) + 5 \\ &= 2\left(y^2 - 3y + \frac{9}{4}\right) - 2\left(\frac{9}{4}\right) + 5 \\ &= 2\left(y^2 - 3y + \frac{9}{4}\right) - \frac{9}{2} + 5 \\ x &= 2\left(y - \frac{3}{2}\right)^2 + \frac{1}{2} \end{aligned}$$

The vertex is $\left(\frac{1}{2}, \frac{3}{2}\right)$. Since $a = 2 > 0$, the graph opens to the right. Since y can be any real number, the range is $(-\infty, \infty)$. The domain is $\left[\frac{1}{2}, \infty\right)$.

(b) $x = -y^2 + 2y + 5$
$x = -(y^2 - 2y) + 5$

Complete the square on $y^2 - 2y$.

$$\left[\frac{1}{2}(-2)\right]^2 = (-1)^2 = 1$$

Add and subtract 1.

$$\begin{aligned} x &= -(y^2 - 2y + 1 - 1) + 5 \\ &= -(y^2 - 2y + 1) + 1 + 5 \\ x &= -(y - 1)^2 + 6 \end{aligned}$$

The vertex is $(6, 1)$. Since $a = -1 < 0$, the graph opens to the left. The domain is $(-\infty, 6]$, and the range is $(-\infty, \infty)$.

9.2 Section Exercises

3. If the equation is in the form $ax^2+bx+c=0$, the value of b^2-4ac, the discriminant, can be used to tell the number of x-intercepts of the graph.
If $b^2-4ac>0$, the parabola has two x-intercepts.
If $b^2-4ac=0$, the parabola has one x-intercept (the vertex).
If $b^2-4ac<0$, the parabola has no x-intercepts.

7. $y=-x^2+5x+3$

Use the vertex formula.
$a=-1, b=5, c=3$

$$\frac{-b}{2a}=\frac{-5}{2(-1)}=\frac{5}{2}$$

$y=f(x)$, so

$$\begin{aligned} f\left(\frac{-b}{2a}\right) &= f\left(\frac{5}{2}\right) \\ &= -\left(\frac{5}{2}\right)^2+5\left(\frac{5}{2}\right)+3 \\ &= -\frac{25}{4}+\frac{25}{2}+3 \\ &= \frac{-25+50+12}{4} \\ &= \frac{37}{4}. \end{aligned}$$

The vertex is at

$$\left(\frac{-b}{2a}, f\left(\frac{-b}{2a}\right)\right)=\left(\frac{5}{2},\frac{37}{4}\right).$$

Because $a=-1$, the parabola opens downward and has the same shape as the graph of $f(x)=x^2$.

$$\begin{aligned} b^2-4ac &= 5^2-4(-1)(3) \\ &= 25+12 \\ &= 37 \end{aligned}$$

The discriminant is positive, so the parabola has two x-intercepts.

For Exercises 11 and 15, see the answer graphs in the back of the textbook.

11. $f(x)=x^2+4x+3$

Evaluate $f(0)$ to find the y-intercept.

$$f(0)=0^2+4(0)+3=3$$

The y-intercept is $(0,3)$.

Solve $f(x)=0$ to find any x-intercepts.

$$\begin{aligned} 0 &= x^2+4x+3 \\ 0 &= (x+3)(x+1) \\ x+3 &= 0 \quad \text{or} \quad x+1=0 \\ x &= -3 \quad \text{or} \quad x=-1 \end{aligned}$$

The x-intercepts are $(-3,0)$ and $(-1,0)$.

$a=1, b=4$, and $c=3$. Use the vertex formula with

$$\frac{-b}{2a}=\frac{-4}{2(1)}=-2$$

$$f(-2)=(-2)^2+4(-2)+3=-1.$$

The vertex is at $(-2,-1)$.

Because $a=1$, the graph opens upward and has the same shape as the graph of $f(x)=x^2$. By symmetry, another point on the graph is $(-4,3)$. Since x can be any real number, the domain is $(-\infty,\infty)$. Since $(-2,-1)$ is the lowest point of the parabola, the range is $[-1,\infty)$.

15. $x=-\frac{1}{5}y^2+2y-4$

The roles of x and y are reversed, so this is a horizontal parabola.

Let $x=0$ to find the y-intercepts.

$$0=-\frac{1}{5}y^2+2y-4$$

$a=-\frac{1}{5}$ (or $-.2$), $b=2, c=-4$

$$\begin{aligned} y &= \frac{-2\pm\sqrt{2^2-4(-.2)(-4)}}{2(-.2)} \\ &= \frac{-2\pm\sqrt{4-3.2}}{-.4}=\frac{-2\pm\sqrt{.8}}{-.4} \end{aligned}$$

$y\approx 2.8$ or $y\approx 7.2$

The y-intercepts are $(0,2.8)$ and $(0,7.2)$. Let $y=0$ to find any x-intercepts.

$$x=-\frac{1}{5}(0)^2+2(0)-4=-4$$

The x-intercept is $(-4,0)$.

Factor out $-\frac{1}{5}$, then complete the square to find the vertex.

$$x=-\frac{1}{5}(y^2-10y)-4$$

$$\left[\frac{1}{2}(-10)\right]^2=(-5)^2=25$$

Add and subtract 25.

$$x = -\frac{1}{5}(y^2 - 10y + 25 - 25) - 4$$
$$x = -\frac{1}{5}(y^2 - 10y + 25) - \frac{1}{5}(-25) - 4$$
$$x = -\frac{1}{5}(y^2 - 10y + 25) + 5 - 4$$
$$x = -\frac{1}{5}(y - 5)^2 + 1$$

The vertex is $(1, 5)$. Since $a = -\frac{1}{5} < 0$, the graph opens to the left. Also, $|a| = \left|-\frac{1}{5}\right| = \frac{1}{5} < 1$, so the graph is wider than the graph of $y = x^2$. By symmetry, the point $(-4, 10)$ is also on the graph.

19. $f(x) = x^2 - 8x + 18$

Find the vertex.
$a = 1, b = -8, c = 18$

$$\frac{-b}{2a} = \frac{-(-8)}{2(1)} = 4$$
$$f(4) = 4^2 - 8(4) + 18 = 2$$

The vertex is at $(4, 2)$. This matches the calculator-generated graph in choice B.

23. Let $x =$ the width of the lot.
$640 - 2x =$ the length of the lot.

Area is length times width so

$$\begin{aligned} A &= x(640 - 2x) \\ &= 640x - 2x^2 \\ &= -2x^2 + 640x. \end{aligned}$$

Use the vertex formula.
$a = -2, b = 640, c = 0$

$$\frac{-b}{2a} = \frac{-640}{2(-2)} = 160$$
$$\begin{aligned} f(160) &= -2(160)^2 + 640(160) \\ &= -51,200 + 102,400 \\ &= 51,200 \end{aligned}$$

The graph is a parabola that opens downward so the maximum occurs at the vertex (160, 51,200). The maximum area is 51,200 ft^2 if the width x is 160 ft and the length is

$$640 - 2x = 640 - 2(160) = 320 \text{ ft}.$$

27. Let $x =$ one number,
$60 - x =$ the other number,
and $P =$ the product.

$$\begin{aligned} P &= x(60 - x) \\ &= 60x - x^2 \\ &= -x^2 + 60x \end{aligned}$$

This parabola opens downward so the maximum occurs at the vertex.
$a = -1, b = 60, c = 0$

$$\frac{-b}{2a} = \frac{-60}{2(-1)} = 30$$

$x = 30$ when the product is the maximum.

$$60 - x = 30.$$

The two numbers are 30 and 30.

31. From Exercise 30, the vertex is at $(2.1, 12.341)$. Since the parabola opens downward, the vertex is the point with the largest y-coordinate. This means that 2.1 yr after 1988, which is 1990, there were 12.341 million jobs, the largest number of jobs during the time covered by the graph.

33. **(a)** $x^2 - 4x + 3 = 0$ is solved by finding the x-intercepts of the graph.

Solution set: $\{1, 3\}$

(b) $x^2 - 4x + 3 > 0$ is solved by finding the x-values for which y is greater than zero (above the x-axis).

Solution set: $(-\infty, 1) \cup (3, \infty)$

(c) $x^2 - 4x + 3 < 0$ is solved by finding the x-values for which y is less than zero (below the x-axis).

Solution set: $(1, 3)$

34. **(a)** $3x^2 + 10x - 8 = 0$ is solved by finding the x-intercepts of the graph.

Solution set: $\{-4, \frac{2}{3}\}$

(b) $3x^2 + 10x - 8 \geq 0$ is solved by finding the x-values for which y is greater than or equal to zero (on or above the x-axis).

Solution set: $(-\infty, -4] \cup [\frac{2}{3}, \infty)$

(c) $3x^2 + 10x - 8 < 0$ is solved by finding the x-values for which y is less than zero (below the x-axis).

Solution set: $(-4, \frac{2}{3})$

35. **(a)** $-x^2 + 3x + 10 = 0$ is solved by finding the x-intercepts of the graph.

Solution set: $\{-2, 5\}$

(b) $-x^2+3x+10 \geq 0$ is solved by finding the x-values for which y is greater than or equal to zero (on or above the x-axis.).

Solution set: $[-2, 5]$

(c) $-x^2+3x+10 \leq 0$ is solved by finding the x-values for which y is less than or equal to zero (on or below the x-axis).

Solution set: $(-\infty, -2] \cup [5, \infty)$

36. **(a)** $-2x^2-x+15=0$ is solved by finding the x-intercepts of the graph.

Solution set: $\{-3, \frac{5}{2}\}$

(b) $-2x^2-x+15 \geq 0$ is solved by finding the x-values for which y is greater than or equal to zero (on or above the x-axis).

Solution set: $[-3, \frac{5}{2}]$

(c) $-2x^2-x+15 \leq 0$ is solved by finding the x-values for which y is less than or equal to zero (on or below the x-axis).

Solution set: $(-\infty, -3] \cup [\frac{5}{2}, \infty)$

9.3 Graphs of Elementary Functions and Circles

9.3 Margin Exercises

For Exercises 1-4, see the answer graphs for the margin exercises in the textbook.

1. **(a)** $f(x)=\sqrt{x+4}$

The graph of this function is found by shifting the graph of the square root function $f(x)=\sqrt{x}$ 4 units to the left.

x	-4	-3	0
y	0	1	2

(b) $f(x)=\dfrac{1}{x}-2$

This is the graph of the reciprocal function $f(x)=\frac{1}{x}$ shifted 2 units downward. Since $x \neq 0$ (or a denominator of 0 results), the line $x=0$ is an asymptote. Since $\frac{1}{x} \neq 0$, the line $y=-2$ is also an asymptote.

x	$\frac{1}{2}$	1	2
y	0	-1	$-\frac{3}{2}$

x	$-\frac{1}{3}$	$-\frac{1}{2}$	-1	-2
y	-5	-4	-3	$-\frac{5}{2}$

(c) $f(x)=|x+2|+1$

This is the graph of the absolute value function $f(x)=|x|$ shifted 2 units to the left (since $x+2=0$ if $x=-2$) and 1 unit upward (because of the $+1$).

x	-4	-3	-2	-1	0
y	3	2	1	2	3

2. If the point (x, y) is on the circle, the distance from (x, y) to the center $(0, 0)$ is 4. By the distance formula,

$$\sqrt{(x_2-x_1)^2+(y_2-y_1)^2}=d$$
$$\sqrt{(x-0)^2+(y-0)^2}=4$$
$$\sqrt{x^2+y^2}=4.$$

Square both sides.

$$x^2+y^2=16$$

The equation of the circle is $x^2+y^2=16$.

3. Center at $(3, -2)$; radius 4

$$\sqrt{(x_2-x_1)^2+(y_2-y_1)^2}=d$$
$$\sqrt{(x-3)^2+[y-(-2)]^2}=4$$
$$\sqrt{(x-3)^2+(y+2)^2}=4$$

Square both sides.

$$(x-3)^2+(y+2)^2=16$$

4. $(x-5)^2+(y+2)^2=9$

The equation of the circle is an equation of the form

$$(x-h)^2+(y-k)^2=r^2$$
$$(x-5)^2+[y-(-2)]^2=3^2.$$

The center (h, k) is $(5, -2)$. The radius r is 3.

5. To find the center and radius, add the appropriate constants to complete both squares on the left.

$$x^2+y^2-6x+8y-11=0$$
$$(x^2-6x \quad)+(y^2+8y \quad)=11$$
$$(x^2-6x+9-9)+(y^2+8y+16-16)=11$$
$$(x^2-6x+9)+(y^2+8y+16)-9-16=11$$
$$(x^2-6x+9)+(y^2+8y+16)=11+25$$
$$(x-3)^2+(y+4)^2=36$$
$$(x-3)^2+(y+4)^2=6^2$$

The circle has center at $(3, -4)$ and radius 6.

6. $f(x) = \sqrt{36 - x^2}$

Replace $f(x)$ with y.

$$y = \sqrt{36 - x^2}$$

Square both sides.

$$y^2 = 36 - x^2$$
$$x^2 + y^2 = 36$$

The graph of this equation is a circle with center $(0, 0)$ and radius 6. Since $f(x)$ represents the principal square root in the original equation, $f(x)$ must be nonnegative. The graph of $f(x)$ is the upper half of the circle. See the answer graph for the margin exercises in the textbook.

9.3 Section Exercises

3. The lowest point on the graph of $f(x) = |x|$ has coordinates $(0, 0)$.

7. $f(x) = |x - 2| - 2$

The graph of this function is found by shifting the graph of $f(x) = |x|$ 2 units to the right (since $x - 2 = 0$ if $x = 2$) and 2 units downward (because of the -2). The correct graph is choice A.

For Exercises 11 and 15, see the answer graphs in the back of the textbook.

11. $f(x) = \frac{1}{x} + 1$

This is the graph of the reciprocal function $f(x) = \frac{1}{x}$ shifted 1 unit upward. Since $x \neq 0$ (or a denominator of 0 results), the line $x = 0$ is an asymptote. Also since $\frac{1}{x} \neq 0$, the line $y = 1$ is an asymptote.

x	$\frac{1}{3}$	$\frac{1}{2}$	1	2	3
y	4	3	2	$\frac{3}{2}$	$\frac{4}{3}$

x	-3	-2	-1	$-\frac{1}{2}$	$-\frac{1}{3}$
y	$\frac{2}{3}$	$\frac{1}{2}$	0	-1	-2

15. $f(x) = \frac{1}{x - 2}$

This is the graph of the reciprocal function $f(x) = \frac{1}{x}$ shifted 2 units to the right. Since $x \neq 2$ (or a denominator of 0 results), the line $x = 2$ is an asymptote. Since $\frac{1}{x-2} \neq 0$, the line $y = 0$ is also an asymptote.

x	$\frac{5}{2}$	3	4	5
y	2	1	$\frac{1}{2}$	$\frac{1}{3}$

x	-1	0	1	$\frac{3}{2}$
y	$-\frac{1}{3}$	$-\frac{1}{2}$	-1	-2

19. $(x-3)^2 + (y-2)^2 = 25$ is a circle with center $(3, 2)$ and radius 5. The answer is graph B.

23. In a circle, there are some x-values that yield two y-values. In a function, every x-value yields one and only one y-value. (A circle also fails the vertical line test.)

27. $x^2 + y^2 - 2x + 4y - 4 = 0$

To find the center and radius, complete the square on x and the square on y.

$$(x^2 - 2x \quad) + (y^2 + 4y \quad) = 4$$
$$(x^2 - 2x + 1 - 1) + (y^2 + 4y + 4 - 4) = 4$$
$$(x^2 - 2x + 1) + (y^2 + 4y + 4) - 1 - 4 = 4$$
$$(x^2 - 2x + 1) + (y^2 + 4y + 4) = 1 + 4 + 4$$
$$(x - 1)^2 + (y + 2)^2 = 9$$
$$(x - 1)^2 + (y + 2)^2 = 3^2$$

The center is $(1, -2)$, and the radius is 3.

For Exercises 31-39, see the answer graphs in the back of the textbook.

31.
$$x^2 + y^2 = 9$$
$$(x - 0)^2 + (y - 0)^2 = 3^2$$

Here, $h = 0, k = 0$, and $r = 3$, so the graph is a circle with center (h, k) at $(0, 0)$ and radius 3.

35. $(x + 3)^2 + (y - 2)^2 = 9$
$(x + 3)^2 + (y - 2)^2 = 3^2$

Here, $h = -3, k = 2$, and $r = 3$. The graph is a circle with center (h, k) at $(-3, 2)$ and radius 3.

39. $f(x) = \sqrt{16 - x^2}$

Replace $f(x)$ with y.

$$y = \sqrt{16 - x^2}$$

Square both sides.

$$y^2 = 16 - x^2$$
$$x^2 + y^2 = 16$$

This equation is the graph of a circle with center $(0, 0)$ and radius 4. Since $f(x)$ represents a principal square root in the original equation, $f(x)$ is nonnegative and its graph is the upper half of the circle.

43. **(a)** $|x-(-p)| = |x+p|$

(b) The point should have coordinates $(p,0)$ because the distance from the point to the origin should equal the distance from the line to the origin.

(c) The distance from (x,y) to $(p,0)$ is

$$\sqrt{(x-p)^2+(y-0)^2} = \sqrt{(x-p)^2+y^2}.$$

(d) Using the results from parts (a) and (c), these distances should be equal.

$$\sqrt{(x-p)^2+y^2} = |x+p|$$

Square both sides.

$$\begin{aligned}(x-p)^2+y^2 &= (x+p)^2\\ x^2-2xp+p^2+y^2 &= x^2+2xp+p^2\\ y^2 &= 4xp\\ y^2 &= 4px\end{aligned}$$

9.4 Ellipses and Hyperbolas

9.4 Margin Exercises

For Exercises 1 and 2, see the answer graphs for the margin exercises in the textbook.

1. **(a)** $\frac{x^2}{4}+\frac{y^2}{25}=1$

This equation is the graph of an ellipse in the form

$$\frac{x^2}{a^2}+\frac{y^2}{b^2}=1.$$

Here $a=2$, so the x-intercepts are $(2,0)$ and $(-2,0)$. Also $b=5$, so the y-intercepts are $(0,5)$ and $(0,-5)$. Plot the intercepts, and draw the ellipse through them.

(b) $\frac{x^2}{64}+\frac{y^2}{49}=1$

The x-intercepts for this ellipse are $(8,0)$ and $(-8,0)$. The y-intercepts are $(0,7)$ and $(0,-7)$. Plot the intercepts, and draw the ellipse through them.

2. **(a)** $\frac{x^2}{4}-\frac{y^2}{25}=1$

The graph is a hyperbola with $a=2$ and $b=5$. The x-intercepts are $(2,0)$ and $(-2,0)$. There are no y-intercepts. The points $(2,5), (2,-5), (-2,-5)$, and $(-2,5)$ are the corners of the fundamental rectangle that determines the asymptotes. Graph a branch of the hyperbola through each intercept and approaching the asymptotes.

(b) $\frac{y^2}{81}-\frac{x^2}{64}=1$

The graph is a hyperbola with $a=8$ and $b=9$. The y-intercepts are $(0,9)$ and $(0,-9)$. There are no x-intercepts. The corners of the fundamental rectangle are $(8,9), (8,-9), (-8,-9)$, and $(-8,9)$. Extend the diagonals of the rectangle through these points to get the asymptotes. Graph a branch of the hyperbola through each intercept and approaching the asymptotes.

3. **(a)** $3x^2 = 27-4y^2$

Both variables are squared, so the graph is either an ellipse or a hyperbola. Get the variables on one side of the equation.

$$\begin{aligned}3x^2+4y^2 &= 27\\ \frac{x^2}{9}+\frac{4y^2}{27} &= 1\\ \frac{x^2}{9}+\frac{y^2}{\frac{27}{4}} &= 1\end{aligned}$$

Since the equation is in the form

$$\frac{x^2}{a^2}+\frac{y^2}{b^2}=1,$$

the graph is an ellipse.

(b)
$$\begin{aligned}6x^2 &= 100+2y^2\\ 6x^2-2y^2 &= 100\\ \frac{x^2}{\frac{50}{3}}-\frac{y^2}{50} &= 1\end{aligned}$$

Because of the minus sign and since both variables are squared, the graph of the equation is a hyperbola.

(c) $3x^2 = 27-4y$
$4y = -3x^2+27$

This is an equation of a vertical parabola since only one variable, x, is squared.

(d)
$$\begin{aligned}3x^2 &= 27-3y^2\\ 3x^2+3y^2 &= 27\\ x^2+y^2 &= 9\end{aligned}$$

This is an equation of the form $(x-h)^2+(y-k)^2 = r^2$, where $(h,k)=(0,0)$ and $r=3$, so the graph of the equation is a circle.

4. $\frac{y}{3} = -\sqrt{1 - \frac{x^2}{4}}$

Square both sides.

$$\frac{y^2}{9} = 1 - \frac{x^2}{4}$$
$$\frac{x^2}{4} + \frac{y^2}{9} = 1$$

This is the equation of an ellipse with intercepts $(2,0), (-2,0), (0,3)$, and $(0,-3)$. Since $\frac{y}{3}$ equals a negative square root in the original equation, y must be negative. The graph is the lower half of the ellipse. See the answer graph for the margin exercises in the textbook.

9.4 Section Exercises

3. $\frac{x^2}{9} - \frac{y^2}{25} = 1$

This is a hyperbola that opens to the left and right. It goes through $(3,0)$ and $(-3,0)$. This is graph D.

For Exercises 7-31, see the answer graphs in the back of the textbook.

7. $\frac{x^2}{9} + \frac{y^2}{25} = 1$

The equation is in the form $\frac{x^2}{a^2} + \frac{y^2}{b^2} = 1$. The graph is an ellipse with $a = 3$ and $b = 5$. The x-intercepts are $(3,0)$ and $(-3,0)$. The y-intercepts are $(0,5)$ and $(0,-5)$. Plot the intercepts, and draw the ellipse through them.

11. $\frac{x^2}{49} + \frac{y^2}{25} = 1$

The x-intercepts are $(7,0)$ and $(-7,0)$. The y-intercepts are $(0,5)$ and $(0,-5)$. Plot the intercepts, and draw the ellipse through them.

15. $\frac{x^2}{16} - \frac{y^2}{9} = 1$

The equation is in the form $\frac{x^2}{a^2} - \frac{y^2}{b^2} = 1$. The graph is a hyperbola with $a = 4$ and $b = 3$. The x-intercepts are $(4,0)$ and $(-4,0)$. There are no y-intercepts. The corners of the fundamental rectangle are $(4,3), (4,-3), (-4,-3)$, and $(-4,3)$. Extend the diagonals of the rectangle through these points to get the asymptotes. Graph a branch of the hyperbola through each intercept approaching the asymptotes.

19. $\frac{x^2}{25} - \frac{y^2}{36} = 1$

This is a hyperbola with $a = 5$ and $b = 6$. The x-intercepts are $(5,0)$ and $(-5,0)$. There are no y-intercepts. To sketch the graph, draw the diagonals of the fundamental rectangle with corners $(5,6), (5,-6), (-5,-6)$, and $(-5,6)$. Graph a branch of the hyperbola through each intercept approaching the asymptotes.

23. $4x^2 + y^2 = 16$

$$\frac{x^2}{4} + \frac{y^2}{16} = 1$$

$a = 2, b = 4$

The graph is an ellipse. The x-intercepts are $(2,0)$ and $(-2,0)$; the y-intercepts are $(0,4)$ and $(0,-4)$. Plot the intercepts, and draw the ellipse through them.

27.
$$9x^2 = 144 + 16y^2$$
$$9x^2 - 16y^2 = 144$$
$$\frac{x^2}{16} - \frac{y^2}{9} = 1$$

The equation is a hyperbola with $a = 4$ and $b = 3$. The x-intercepts are $(4,0)$ and $(-4,0)$. There are no y-intercepts. To sketch the graph, draw the diagonals of the fundamental rectangle with corners $(4,3), (4,-3), (-4,-3)$, and $(-4,3)$. These are the asymptotes. Graph a branch of the hyperbola through each intercept approaching the asymptotes.

31. $y = \sqrt{\frac{x+4}{2}}$

Square both sides.

$$y^2 = \frac{x+4}{2}$$
$$2y^2 = x + 4$$
$$x = 2y^2 - 4$$
$$x = 2(y-0)^2 - 4$$

This is the equation of a horizontal parabola with $h = -4$ and $k = 0$. The vertex is $(-4,0)$. Since $a = 2 > 0$, the graph opens to the right. Also, $|a| = |2| = 2 > 0$, so the graph is narrower than the graph of $y = x^2$. Since y represents a principal square root in the original equation, y is nonnegative and its graph is the upper half of the parabola.

35. $\dfrac{x^2}{5013}+\dfrac{y^2}{4970}=1$

The equation represents an ellipse with $a=\sqrt{5013}$ and $b=\sqrt{4970}$. The x-intercepts are $(\sqrt{5013},0)$ and $(-\sqrt{5013},0)$, and the y-intercepts are $(0,\sqrt{4970})$ and $(0,-\sqrt{4970})$. The sun is one of the foci, $(c,0)$ or $(-c,0)$.

$$\begin{aligned}\text{Since } c^2&=a^2-b^2,\\ c^2&=5013-4970\\ c^2&=43\\ c&=\sqrt{43}.\end{aligned}$$

(a) If the sun is the focus $(-c,0)$, then Venus is at its greatest distance at $(a,0)$, so the greatest distance is

$$\begin{aligned}|-c|+|a|&=\sqrt{43}+\sqrt{5013}\\ &\approx 77.4\text{ million mi.}\end{aligned}$$

(b) If the sun is the focus $(-c,0)$, then Venus is at its smallest distance at $(-a,0)$, so the smallest distance is

$$\begin{aligned}|-a|-|-c|&=\sqrt{5013}-\sqrt{43}\\ &\approx 64.2\text{ million mi.}\end{aligned}$$

9.5 Nonlinear Systems of Equations

9.5 Margin Exercises

1. **(a)** $x^2+y^2=10 \quad (1)$
$x=y+2 \quad (2)$

Substitute $y+2$ for x in equation (1).

$$\begin{aligned}x^2+y^2&=10 \quad (1)\\ (y+2)^2+y^2&=10\\ y^2+4y+4+y^2&=10\\ 2y^2+4y-6&=0\\ y^2+2y-3&=0\\ (y+3)(y-1)&=0\end{aligned}$$

$$y+3=0 \quad\text{or}\quad y-1=0$$
$$y=-3 \quad\text{or}\quad y=1$$

Substitute these values for y in equation (2) and solve for x.

If $y=-3$, then

$$\begin{aligned}x&=y+2 \quad (2)\\ x&=-3+2\\ x&=-1.\end{aligned}$$

If $y=1$, then

$$\begin{aligned}x&=y+2 \quad (2)\\ x&=1+2\\ x&=3.\end{aligned}$$

Solution set: $\{(-1,-3),(3,1)\}$

(b) $x^2-2y^2=8 \quad (1)$
$y+x=6 \quad (2)$

Solve equation (2) for y.

$$y=6-x$$

Substitute $6-x$ for y in equation (1).

$$\begin{aligned}x^2-2y^2&=8 \quad (1)\\ x^2-2(6-x)^2&=8\\ x^2-2(x^2-12x+36)&=8\\ x^2-2x^2+24x-72&=8\\ -x^2+24x-80&=0\\ x^2-24x+80&=0\\ (x-4)(x-20)&=0\end{aligned}$$

$$x-4=0 \quad\text{or}\quad x-20=0$$
$$x=4 \quad\text{or}\quad x=20$$

Since $y=6-x$, if $x=4$, then

$$y=6-4=2.$$

If $x=20$, then

$$y=6-20=-14.$$

Solution set: $\{(4,2),(20,-14)\}$

2. **(a)** $xy=8 \quad (1)$
$x+y=6 \quad (2)$

Solve equation (1) for x.

$$x=\frac{8}{y}$$

Substitute $\frac{8}{y}$ for x in equation (2).

$$\begin{aligned}x+y&=6 \quad (2)\\ \frac{8}{y}+y&=6\end{aligned}$$

Multiply by y.

$$\begin{aligned}8+y^2&=6y\\ y^2-6y+8&=0\\ (y-2)(y-4)&=0\end{aligned}$$

$$y-2=0 \quad \text{or} \quad y-4=0$$
$$y=2 \quad \text{or} \quad y=4$$

Since $x=\frac{8}{y}$, if $y=2$, then

$$x=\frac{8}{2}=4.$$

If $y=4$, then

$$x=\frac{8}{4}=2.$$

Solution set: $\{(4,2),(2,4)\}$

(b) $xy+10=0 \quad (1)$
$4x+9y=-2 \quad (2)$

Solve for y in equation (1).

$$y=-\frac{10}{x}$$

Substitute $-\frac{10}{x}$ for y in equation (2).

$$4x+9y=-2 \quad (2)$$
$$4x+9\left(-\frac{10}{x}\right)=-2$$
$$4x-\frac{90}{x}=-2$$

Multiply by x.

$$4x^2-90=-2x$$
$$4x^2+2x-90=0$$
$$2x^2+x-45=0$$
$$(2x-9)(x+5)=0$$
$$2x-9=0 \quad \text{or} \quad x+5=0$$
$$x=\frac{9}{2} \quad \text{or} \quad x=-5$$

Since $y=-\frac{10}{x}$, if $x=\frac{9}{2}$, then

$$y=\frac{-10}{\frac{9}{2}}=-\frac{20}{9}.$$

If $x=-5$, then

$$y=\frac{-10}{-5}=2.$$

Solution set: $\left\{(-5,2),\left(\frac{9}{2},-\frac{20}{9}\right)\right\}$

3. **(a)**
$$\begin{aligned} x^2+y^2 &= 41 \quad (1)\\ x^2-y^2 &= 9 \quad (2)\\ \hline 2x^2 &= 50 \quad \textit{Add}\\ x^2 &= 25 \\ x=5 \quad &\text{or} \quad x=-5 \end{aligned}$$

Substitute each value of x in equation (1) to find y. First substitute 5 for x.

$$x^2+y^2=41 \quad (1)$$
$$5^2+y^2=41$$
$$25+y^2=41$$
$$y^2=16$$
$$y=4 \quad \text{or} \quad y=-4$$

Now substitute -5 for x in equation (1).

$$x^2+y^2=41 \quad (1)$$
$$(-5)^2+y^2=41$$
$$25+y^2=41$$
$$y^2=16$$
$$y=4 \quad \text{or} \quad y=-4$$

Solution set: $\{(5,4),(5,-4),(-5,4),(-5,-4)\}$

(b) $x^2+3y^2=40 \quad (1)$
$4x^2-y^2=4 \quad (2)$

Multiply equation (2) by 3 and add the result to equation (1).

$$\begin{aligned} 12x^2-3y^2 &= 12\\ x^2+3y^2 &= 40 \quad (1)\\ \hline 13x^2 \quad &= 52\\ x^2 &= 4\\ x=2 \quad &\text{or} \quad x=-2 \end{aligned}$$

Substitute 2 for x in equation (1) to find y.

$$x^2+3y^2=40 \quad (1)$$
$$2^2+3y^2=40$$
$$4+3y^2=40$$
$$3y^2=36$$
$$y^2=12$$
$$y=2\sqrt{3} \quad \text{or} \quad y=-2\sqrt{3}$$

Since x is squared in equation (1), the values of y, $2\sqrt{3}$ or $-2\sqrt{3}$, are the same for $x=2$ and $x=-2$.

Solution set:

$\{(2,2\sqrt{3}),(2,-2\sqrt{3}),(-2,2\sqrt{3}),(-2,-2\sqrt{3})\}$

4. **(a)** $x^2+xy+y^2=3 \quad (1)$
$x^2 \qquad +y^2=5 \quad (2)$

Multiply equation (2) by -1 and add the result to equation (1).

$$\begin{aligned} -x^2 \qquad -y^2 &= -5\\ x^2+xy+y^2 &= 3 \quad (1)\\ \hline xy \qquad &= -2 \quad (3) \end{aligned}$$

Solve equation (3) for y to get $y = -\frac{2}{x}$. Substitute $-\frac{2}{x}$ for y in equation (2).

$$x^2 + y^2 = 5 \quad (2)$$
$$x^2 + \left(-\frac{2}{x}\right)^2 = 5$$
$$x^2 + \frac{4}{x^2} = 5$$

Multiply by x^2.

$$x^4 + 4 = 5x^2$$
$$x^4 - 5x^2 + 4 = 0$$
$$(x^2 - 4)(x^2 - 1) = 0$$
$$x^2 = 4 \quad \text{or} \quad x^2 = 1$$
$$x = 2 \quad \text{or} \quad x = -2 \quad \text{or} \quad x = 1 \quad \text{or} \quad x = -1$$

Since $y = -\frac{2}{x}$, substitute these values for x to find the values of y.

If $x = 1$, then $y = -2$.
If $x = -1$, then $y = 2$.
If $x = 2$, then $y = -1$.
If $x = -2$, then $y = 1$.

Solution set: $\{(1, -2), (-1, 2), (2, -1), (-2, 1)\}$

(b)
$$x^2 + 7xy - 2y^2 = -8 \quad (1)$$
$$-2x^2 + 4y^2 = 16 \quad (2)$$

Multiply equation (1) by 2 and add the result to equation (2).

$$\begin{array}{rl} 2x^2 + 14xy - 4y^2 &= -16 \\ -2x^2 \qquad + 4y^2 &= \;\;16 \quad (2) \\ \hline 14xy &= 0 \\ xy &= 0 \end{array}$$

Either $x = 0$ or $y = 0$. If $x = 0$, then substitute 0 for x in equation (1) to find y.

$$x^2 + 7xy - 2y^2 = -8 \quad (1)$$
$$0 + 0 - 2y^2 = -8$$
$$y^2 = 4$$
$$y = 2 \quad \text{or} \quad y = -2$$

If $y = 0$, then substitute 0 for y in equation (1) to find x.

$$x^2 + 7xy - 2y^2 = -8 \quad (1)$$
$$x^2 + 0 + 0 = -8$$
$$x = \sqrt{-8} \quad \text{or} \quad x = -\sqrt{-8}$$
$$x = 2i\sqrt{2} \quad \text{or} \quad x = -2i\sqrt{2}$$

Solution set: $\{(0, 2), (0, -2), (2i\sqrt{2}, 0), (-2i\sqrt{2}, 0)\}$

9.5 Section Exercises

3. If the graphs are a line and a circle, the graphs can intersect at one point, two points, or no points. As a result there are zero, one, or two solutions, not three.

In Exercises 7-27, check each solution in the equations of the original system.

7.
$$y = x^2 + 6x + 9 \quad (1)$$
$$x + y = 3 \quad (2)$$

Substitute $x^2 + 6x + 9$ for y in equation (2).

$$x + y = 3 \quad (2)$$
$$x + (x^2 + 6x + 9) = 3$$
$$x^2 + 7x + 9 = 3$$
$$x^2 + 7x + 6 = 0$$
$$(x + 6)(x + 1) = 0$$
$$x + 6 = 0 \quad \text{or} \quad x + 1 = 0$$
$$x = -6 \quad \text{or} \quad x = -1$$

Substitute these values for x in equation (2) and solve for y.

If $x = -6$, then

$$x + y = 3 \quad (2)$$
$$-6 + y = 3$$
$$y = 9.$$

If $x = -1$, then

$$x + y = 3 \quad (2)$$
$$-1 + y = 3$$
$$y = 4.$$

Solution set: $\{(-6, 9), (-1, 4)\}$

11.
$$xy = 4 \quad (1)$$
$$3x + 2y = -10 \quad (2)$$

Solve equation (1) for y to get $y = \frac{4}{x}$. Substitute $\frac{4}{x}$ for y in equation (2) to find x.

$$3x + 2y = -10 \quad (2)$$
$$3x + 2\left(\frac{4}{x}\right) = -10$$

Multiply by x.

$$3x^2 + 8 = -10x$$
$$3x^2 + 10x + 8 = 0$$
$$(3x + 4)(x + 2) = 0$$

$$3x+4=0 \quad \text{or} \quad x+2=0$$
$$3x=-4$$
$$x=-\frac{4}{3} \quad \text{or} \quad x=-2$$

Since $y=\frac{4}{x}$, if $x=-\frac{4}{3}$, then

$$y=\frac{4}{-\frac{4}{3}}=-3.$$

If $x=-2$, then

$$y=\frac{4}{-2}=-2.$$

Solution set: $\{(-2,-2),(-\frac{4}{3},-3)\}$

15. $y=3x^2+6x \quad (1)$
$y=x^2-x-6 \quad (2)$

Substitute x^2-x-6 for y in equation (1) to find x.

$$y=3x^2+6x \quad (1)$$
$$x^2-x-6=3x^2+6x$$
$$2x^2+7x+6=0$$
$$(2x+3)(x+2)=0$$
$$2x+3=0 \quad \text{or} \quad x+2=0$$
$$2x=-3$$
$$x=-\frac{3}{2} \quad \text{or} \quad x=-2$$

Substitute $-\frac{3}{2}$ for x in equation (1) to find y.

$$y=3x^2+6x \quad (1)$$
$$y=3\left(-\frac{3}{2}\right)^2+6\left(-\frac{3}{2}\right)$$
$$y=3\left(\frac{9}{4}\right)+6\left(-\frac{6}{4}\right)$$
$$y=\frac{27}{4}-\frac{36}{4}$$
$$y=-\frac{9}{4}$$

Substitute -2 for x in equation (1) to find y.

$$y=3x^2+6x \quad (1)$$
$$y=3(-2)^2+6(-2)$$
$$y=12-12$$
$$y=0$$

Solution set: $\{(-\frac{3}{2},-\frac{9}{4}),(-2,0)\}$

19. $3x^2+2y^2=12 \quad (1)$
$x^2+2y^2=4 \quad (2)$

Multiply equation (2) by -1 and add the result to equation (1).

$$-x^2-2y^2=-4$$
$$3x^2+2y^2=12 \quad (1)$$
$$2x^2 \quad = 8$$
$$x^2=4$$
$$x=2 \quad \text{or} \quad x=-2$$

Substitute 2 for x in equation (2) to find y.

$$x^2+2y^2=4 \quad (2)$$
$$2^2+2y^2=4$$
$$4+2y^2=4$$
$$2y^2=0$$
$$y^2=0$$
$$y=0$$

Substitute -2 for x in equation (2) to find y.

$$x^2+2y^2=4 \quad (2)$$
$$(-2)^2+2y^2=4$$
$$4+2y^2=4$$
$$2y^2=0$$
$$y^2=0$$
$$y=0$$

Solution set: $\{(-2,0),(2,0)\}$

23. $2x^2+2y^2=8 \quad (1)$
$3x^2+4y^2=24 \quad (2)$

Multiply equation (1) by -2 and add the result to equation (2).

$$-4x^2-4y^2=-16$$
$$3x^2+4y^2=24 \quad (2)$$
$$-x^2 \quad = 8$$
$$x^2=-8$$
$$x=\sqrt{-8} \quad \text{or} \quad x=-\sqrt{-8}$$
$$x=i\sqrt{4\cdot 2} \qquad x=-i\sqrt{4\cdot 2}$$
$$x=2i\sqrt{2} \quad \text{or} \quad x=-2i\sqrt{2}$$

Simplify equation (1) by dividing both sides by 2.

$$2x^2+2y^2=8 \quad (1)$$
$$x^2+y^2=4$$

Substitute $2i\sqrt{2}$ for x to find y.

$$\begin{aligned} x^2+y^2&=4\\ (2i\sqrt{2})^2+y^2&=4\\ -8+y^2&=4 \quad i^2=-1\\ y^2&=12\\ y=\sqrt{12} \quad &\text{or} \quad y=-\sqrt{12}\\ y=\sqrt{4\cdot 3} \quad & \quad y=-\sqrt{4\cdot 3}\\ y=2\sqrt{3} \quad &\text{or} \quad y=-2\sqrt{3}\end{aligned}$$

Since x is squared in the equation $x^2+y^2=4$, the values of y, $2\sqrt{3}$ or $-2\sqrt{3}$, are the same for $x=2i\sqrt{2}$ and $x=-2i\sqrt{2}$.

Solution set: $\{(2i\sqrt{2},2\sqrt{3}),(2i\sqrt{2},-2\sqrt{3}),$
$(-2i\sqrt{2},2\sqrt{3}),(-2i\sqrt{2},-2\sqrt{3})\}$

27. $3x^2+2xy-3y^2=5 \quad (1)$
$-x^2-3xy+\ y^2=3 \quad (2)$

Multiply equation (2) by 3 and add the result to equation (1).

$$\begin{aligned} -3x^2-9xy+3y^2&=\ 9\\ 3x^2+2xy-3y^2&=\ 5 \quad (1)\\ \hline -7xy \qquad &=14\\ x&=\frac{14}{-7y}\\ x&=-\frac{2}{y}\end{aligned}$$

Substitute $-\frac{2}{y}$ for x in equation (2).

$$\begin{aligned} -x^2-3xy+y^2&=3 \quad (2)\\ -\left(-\frac{2}{y}\right)^2-3\left(-\frac{2}{y}\right)y+y^2&=3\\ -\left(\frac{4}{y^2}\right)+6+y^2&=3\end{aligned}$$

Multiply by y^2.

$$\begin{aligned} -4+6y^2+y^4&=3y^2\\ y^4+3y^2-4&=0\\ (y^2+4)(y^2-1)&=0\\ y^2+4=0 \quad &\text{or} \quad y^2-1=0\\ y^2=-4 \quad & \quad y^2=1\\ y=2i \quad \text{or} \quad y=-2i \quad &\text{or} \quad y=1 \quad \text{or} \quad y=-1\end{aligned}$$

Since $x=-\frac{2}{y}$, substitute these values for y to find the values of x.

If $y=2i$, then

$$x=-\frac{2}{2i}=-\frac{1}{i}=-\frac{1}{i}\cdot\frac{i}{i}=\frac{-i}{-1}=i.$$

If $y=-2i$, then

$$x=-\frac{2}{-2i}=\frac{1}{i}=\frac{1}{i}\cdot\frac{i}{i}=\frac{i}{-1}=-i.$$

If $y=1$, then

$$x=-\frac{2}{1}=-2.$$

If $y=-1$, then

$$x=-\frac{2}{-1}=2.$$

Solution set: $\{(i,2i),(-i,-2i),(2,-1),(-2,1)\}$

31. $y=\frac{1}{2}x^2 \quad (1)$
$x+y=4 \quad (2)$

Replace y with $\frac{1}{2}x^2$ in equation (2).

$$x+\frac{1}{2}x^2=4$$

Multiply by 2.

$$\begin{aligned} 2x+x^2&=8\\ x^2+2x-8&=0\\ (x+4)(x-2)&=0\\ x+4=0 \quad &\text{or} \quad x-2=0\\ x=-4 \quad &\text{or} \quad x=2\end{aligned}$$

The graphing calculator screen shows the solution $(2,2)$. If $x=-4$ in equation (1), then

$$y=\frac{1}{2}(-4)^2=\frac{1}{2}(16)=8,$$

and the other solution is $(-4,8)$.

35. $px=16 \quad (1)$
$p=10x+12 \quad (2)$

Substitute $10x+12$ for p in equation (1).

$$\begin{aligned} px&=16 \quad (1)\\ (10x+12)x&=16\\ 10x^2+12x&=16\\ 10x^2+12x-16&=0\\ 5x^2+6x-8&=0\\ (5x-4)(x+2)&=0\\ 5x-4=0 \quad &\text{or} \quad x+2=0\\ 5x=4 \quad &\\ x=\frac{4}{5} \quad &\text{or} \quad x=-2\end{aligned}$$

Since x cannot be negative, eliminate -2 as a value of x. Substitute $\frac{4}{5}$ for x in equation (1) to find y.

$$px = 16 \quad (1)$$
$$p\left(\frac{4}{5}\right) = 16$$
$$p = 16\left(\frac{5}{4}\right)$$
$$p = 20$$

Since the demand x is in thousands,

$$x = \frac{4}{5}(1000) = 800.$$

The equilibrium price is \$20; the supply/demand at that price is 800 calculators.

9.6 Second-Degree Inequalities; Systems of Inequalities

9.6 Margin Exercises

See the answer graphs for the margin exercises in the textbook.

1. $y \geq (x+1)^2 - 5$

   The boundary, $y = (x+1)^2 - 5$, is a parabola opening upward with vertex at $(-1, -5)$, drawn as a solid curve because of " $\geq$." Use the point $(0,0)$ as a test point.

   $$0 \geq (0+1)^2 - 5$$
   $$0 \geq 1 - 5$$
   $$0 \geq -4 \quad \textit{True}$$

   Shade the region on the side of the parabola that contains $(0,0)$. This is the region inside the parabola.

2. $x^2 + 4y^2 > 36$

   $$\frac{x^2}{36} + \frac{y^2}{9} > 1$$

   The boundary is the ellipse

   $$\frac{x^2}{36} + \frac{y^2}{9} = 1$$

   with intercepts $(6,0), (-6,0), (0,3)$, and $(0,-3)$, drawn as a dashed curve because of " $>$." Test $(0,0)$.

   $$0^2 + 4(0)^2 > 36$$
   $$0 > 36 \quad \textit{False}$$

   Shade the side of the ellipse that does not contain $(0,0)$. This is the region outside the ellipse.

3. $3x - 4y \geq 12$
   $x + 3y \geq 6$
   $y \leq 2$

   The boundary, $3x - 4y = 12$, is a solid line with intercepts $(4,0)$ and $(0,-3)$. Test $(0,0)$.

   $$3(0) - 4(0) \geq 12$$
   $$0 \geq 12 \quad \textit{False}$$

   Shade the side of the line that does not contain $(0,0)$.

   The boundary, $x + 3y = 6$, is a solid line with intercepts $(6,0)$ and $(0,2)$. Test $(0,0)$.

   $$0 + 3(0) \geq 6$$
   $$0 \geq 6 \quad \textit{False}$$

   Shade the side of the line that does not contain $(0,0)$.

   The boundary, $y = 2$, is a solid horizontal line through $(0,2)$. Shade the region below the line since $y \leq 2$.

   The intersection of the three shaded regions is the graph of the solution set of the system.

4. $y \geq x^2 + 1$

   $$\frac{x^2}{9} + \frac{y^2}{4} \geq 1$$

   $y \leq 5$

   The boundary, $y = x^2 + 1$, is a parabola opening upward with vertex at $(0,1)$, drawn as a solid curve because of " $\geq$." Test $(0,0)$.

   $$0 \geq 0^2 + 1$$
   $$0 \geq 1 \quad \textit{False}$$

   Shade the region on the side of the parabola that does not contain $(0,0)$. This is the region inside the parabola.

   The boundary, $\frac{x^2}{9} + \frac{y^2}{4} = 1$, is the ellipse with intercepts $(3,0), (-3,0), (0,2)$, and $(0,-2)$, drawn as a solid curve because of " $\geq$." Test $(0,0)$.

   $$\frac{0^2}{9} + \frac{0^2}{4} \geq 1$$
   $$0 \geq 1 \quad \textit{False}$$

Shade the side of the ellipse that does not contain $(0,0)$. This is the region outside the ellipse.

The boundary $y = 5$ is a solid horizontal line through $(0,5)$. Shade the region below the line since $y \leq 5$.

The graph of the solution set of the system includes all points common to the three graphs, that is, the intersection of the three shaded regions.

9.6 Section Exercises

For Exercises 3-15, see the answer graphs in the back of the textbook.

3.
$$\begin{aligned} y^2 &> 4 + x^2 \\ y^2 - x^2 &> 4 \\ \frac{y^2}{4} - \frac{x^2}{4} &> 1 \end{aligned}$$

The boundary, $\frac{y^2}{4} - \frac{x^2}{4} = 1$, is a hyperbola with y-intercepts $(0,2)$ and $(0,-2)$ and asymptotes formed by the diagonals of the fundamental rectangle with corners at $(2,2), (2,-2), (-2,-2)$, and $(-2,2)$. The hyperbola has dashed branches because of " $>$." Test $(0,0)$.

$$\begin{aligned} 0^2 &> 4 + 0^2 \\ 0 &> 4 \quad \textit{False} \end{aligned}$$

Shade the sides of the hyperbola that do not contain $(0,0)$. These are the regions inside the branches of the hyperbola.

7.
$$\begin{aligned} 2y^2 &\geq 8 - x^2 \\ x^2 + 2y^2 &\geq 8 \\ \frac{x^2}{8} + \frac{y^2}{4} &\geq 1 \end{aligned}$$

The boundary, $\frac{x^2}{8} + \frac{y^2}{4} = 1$, is the ellipse with intercepts $(2\sqrt{2},0), (-2\sqrt{2},0), (0,2)$, and $(0,-2)$, drawn as a solid curve because of " $\geq$." Test $(0,0)$.

$$\begin{aligned} 2(0^2) &\geq 8 - 0^2 \\ 2 &\geq 8 \quad \textit{False} \end{aligned}$$

Shade the side of the ellipse that does not contain $(0,0)$. This is the region outside the ellipse.

11.
$$\begin{aligned} 9x^2 &> 16y^2 + 144 \\ 9x^2 - 16y^2 &> 144 \\ \frac{x^2}{16} - \frac{y^2}{9} &> 1 \end{aligned}$$

The boundary, $\frac{x^2}{16} - \frac{y^2}{9} = 1$, is a hyperbola with x-intercepts $(4,0)$ and $(-4,0)$ and asymptotes formed by the diagonals of the fundamental rectangle with corners at $(4,3), (4,-3), (-4,-3)$, and $(-4,3)$. The hyperbola has dashed branches because of " $>$." Test $(0,0)$.

$$\begin{aligned} 9(0)^2 &> 16(0)^2 + 144 \\ 0 &> 144 \quad \textit{False} \end{aligned}$$

Shade the sides of the hyperbola that do not contain $(0,0)$. These are the regions inside the branches of the hyperbola.

15. $x \leq -y^2 + 6y - 7$

Factor out -1, then complete the square to find the vertex.

$$\begin{aligned} x &\leq -(y^2 - 6y \quad) - 7 \\ x &\leq -(y^2 - 6y + 9 - 9) - 7 \\ x &\leq -(y^2 - 6y + 9) + 9 - 7 \\ x &\leq -(y-3)^2 + 2 \end{aligned}$$

The boundary, $x = -(y-3)^2 + 2$, is a solid horizontal parabola with vertex at $(2,3)$ that opens to the left. $(-7,0)$ and $(-7,6)$ are points on the graph. Test $(0,0)$.

$$\begin{aligned} 0 &\leq -0^2 + 6 \cdot 0 - 7 \\ 0 &\leq -7 \quad \textit{False} \end{aligned}$$

Shade the side of parabola that does not contain $(0,0)$. This is the region inside the parabola.

19.
$$\begin{aligned} x^2 + y^2 &< 25 \\ y &> -2 \end{aligned}$$

The boundary, $x^2 + y^2 = 25$, is a dashed circle with center $(0,0)$ and radius 5. When $(0,0)$ is tested, a true statement, $0 < 25$, results, so the inside of the circle is shaded. The graph of $y = -2$ is a dashed horizontal line through $(0,-2)$ with shading above the line, since $y > -2$. The correct answer is (c).

For Exercises 23-31, see the answer graphs in the back of the textbook.

23.
$$\begin{aligned} 5x - 3y &\leq 15 \\ 4x + y &\geq 4 \end{aligned}$$

The boundary, $5x - 3y = 15$, is a solid line with intercepts $(3,0)$ and $(0,-5)$. Test $(0,0)$.

$$\begin{aligned} 5(0) - 3(0) &\leq 15 \\ 0 &\leq 15 \quad \textit{True} \end{aligned}$$

Shade the side of the line that contains $(0,0)$.

The boundary, $4x+y=4$, is a solid line with intercepts $(1,0)$ and $(0,4)$. Test $(0,0)$.

$$\begin{aligned} 4(0)+0 &\geq 4 \\ 0 &\geq 4 \quad \textit{False} \end{aligned}$$

Shade the side of the line that does not contain $(0,0)$.

The graph of the system is the intersection of the two shaded regions.

27. $y > x^2 - 4$
$y < -x^2 + 3$

The boundary, $y = x^2 - 4$, is a dashed parabola with vertex $(0,-4)$ that opens upward. Test $(0,0)$.

$$\begin{aligned} 0 &> 0^2 - 4 \\ 0 &> -4 \quad \textit{True} \end{aligned}$$

Shade the side of the parabola that contains $(0,0)$. This is the region inside the parabola.

The boundary, $y = -x^2 + 3$, is a dashed parabola with vertex $(0,3)$ that opens downward. Test $(0,0)$.

$$\begin{aligned} 0 &< -0^2 + 3 \\ 0 &< 3 \quad \textit{True} \end{aligned}$$

Shade the side of the parabola that contains $(0,0)$. This is the region inside the parabola.

The graph of the system is the intersection of the two shaded regions.

31. $y \leq -x^2$
$y \geq x - 3$
$y \leq -1$
$x < 1$

The boundary, $y = -x^2$, is a solid parabola with vertex at $(0,0)$ that opens downward. Test $(0,-1)$.

$$\begin{aligned} -1 &\leq -0^2 \\ -1 &\leq 0 \quad \textit{True} \end{aligned}$$

Shade the side of the parabola that contains $(0,-1)$. This is the region inside the parabola.

The boundary, $y = x - 3$, is a solid line with intercepts $(3,0)$ and $(0,-3)$. Test $(0,0)$.

$$\begin{aligned} 0 &\geq 0 - 3 \\ 0 &\geq -3 \quad \textit{True} \end{aligned}$$

Shade the side of the line that contains $(0,0)$.

For $y \leq -1$, shade below the solid horizontal line $y = -1$.

For $x < 1$, shade to the left of the dashed vertical line $x = 1$.

The intersection of the four shaded regions is the graph of the system.

35. $y > x^2 + 2$
$y < 5$

The graph of $y > x^2 + 2$ will be the region inside a parabola that opens upward and has been shifted two units up. The graph of $y < 5$ will be the region below the line $y = 5$.

The graph of the system is choice B.

Chapter 9 Review Exercises

1. The graph of $f(x) = 3x^2 - 2$ has a vertical shift of 2 units downward (because of the -2). Its vertex is at $(0,-2)$.

2. The graph of $f(x) = 6 - 2x^2$ or $f(x) = -2x^2 + 6$ has a vertical shift of 6 units upward (because of the $+6$). Its vertex is at $(0,6)$.

3. The graph of $f(x) = (x+2)^2$ has a horizontal shift of -2, that is, 2 units to the left (since $x + 2 = 0$ if $x = -2$). Its vertex is at $(-2,0)$.

4. The graph of $f(x) = -(x-1)^2$ has a horizontal shift of 1 unit to the right (since $x - 1 = 0$ if $x = 1$). Its vertex is at $(1,0)$.

5. The graph of $f(x) = (x-3)^2 + 7$ has a horizontal shift of 3 units to the right (since $x-3 = 0$ if $x = 3$) and a vertical shift of 7 units upward (because of the $+7$). Its vertex is at $(3,7)$.

6. The graph of $f(x) = \frac{4}{3}(x-2)^2 + 1$ has a horizontal shift of 2 units to the right (since $x-2 = 0$ if $x = 2$) and a vertical shift of 1 unit upward (because of the $+1$). Its vertex is at $(2,1)$.

7. The graph of $x = (y+2)^2 + 3$ has the roles of x and y reversed. It has a horizontal shift of 3 units to the right (because of the $+3$) and a vertical shift of -2 units, that is, 2 units downward (since $y + 2 = 0$ if $y = -2$). Its vertex is at $(3,-2)$.

8. The graph of $x = (y-3)^2 - 4$ has a horizontal shift of -4, that is, 4 units to the left (because of the -4) and a vertical shift of 3 units upward (since $y - 3 = 0$ if $y = 3$). Its vertex is at $(-4, 3)$.

For Exercises 9-12, see the answer graphs in the back of the textbook.

9. $f(x) = 4x^2 + 4x - 2$

Use the vertex formula with $a = 4$, $b = 4$, $c = -2$.

$$\frac{-b}{2a} = \frac{-4}{2(4)} = -\frac{1}{2}$$

$$f\left(-\frac{1}{2}\right) = 4\left(-\frac{1}{2}\right)^2 + 4\left(-\frac{1}{2}\right) - 2$$
$$= 1 - 2 - 2 = -3$$

The vertex is $\left(-\frac{1}{2}, -3\right)$.

Since $a = 4 > 0$, the parabola opens upward. Also, $|a| = |4| = 4 > 1$, so the graph is narrower than the graph of $y = x^2$. The points $(-2, 6)$, $(0, -2)$, and $(1, 6)$ are on the graph. Since x can be any real number, the domain is $(-\infty, \infty)$. The vertex $\left(-\frac{1}{2}, -3\right)$ is the lowest point on the graph, so the range is $[-3, \infty)$.

10. $f(x) = -2x^2 + 8x - 5$

This time factor out -2, then complete the square to find the vertex.

$$f(x) = -2(x^2 - 4x \qquad) - 5$$
$$f(x) = -2(x^2 - 4x + 4 - 4) - 5$$
$$f(x) = -2(x^2 - 4x + 4) + 8 - 5$$
$$f(x) = -2(x - 2)^2 + 3$$

The vertex is $(2, 3)$.

Here, $a = -2 < 0$, so the parabola opens downward. Also, $|a| = |-2| = 2 > 1$, so the graph is narrower than the graph of $y = x^2$. The points $(0, -5)$, $(1, 1)$, and $(3, 1)$ are on the graph. The domain is $(-\infty, \infty)$. Since the vertex $(2, 3)$ is the highest point on the graph, the range is $(-\infty, 3]$.

11. $x = -\frac{1}{2}y^2 + 6y - 14$

Factor out $-\frac{1}{2}$, then complete the square to find the vertex. Since the roles of x and y are reversed, this is a horizontal parabola.

$$x = -\frac{1}{2}(y^2 - 12y \qquad) - 14$$
$$x = -\frac{1}{2}(y^2 - 12y + 36 - 36) - 14$$
$$x = -\frac{1}{2}(y^2 - 12y + 36) + 18 - 14$$
$$x = -\frac{1}{2}(y - 6)^2 + 4$$

The vertex is $(4, 6)$.

Here, $a = -\frac{1}{2} < 0$, so the parabola opens to the left. Also, $|a| = \left|-\frac{1}{2}\right| = \frac{1}{2} < 1$, so the graph is wider than the graph of $y = x^2$. The points $(-14, 0)$, $(2, 4)$, and $(2, 8)$ are on the graph.

12. $x = 2y^2 + 8y + 3$

Factor out 2, then complete the square to find the vertex. Since the roles of x and y are reversed, this is a horizontal parabola.

$$x = 2(y^2 + 4y \qquad) + 3$$
$$x = 2(y^2 + 4y + 4 - 4) + 3$$
$$x = 2(y^2 + 4y + 4) - 8 + 3$$
$$x = 2(y + 2)^2 - 5$$

The vertex is $(-5, -2)$.

Here, $a = 2 > 0$, so the parabola opens to the right. Also, $|a| = |2| = 2 > 1$, so the graph is narrower than the graph of $y = x^2$. The points $(3, 0)$ and $(3, -4)$ are on the graph.

13. A parabola can intersect the x-axis at zero, one, or two points. If the equation is in the form $ax^2 + bx + c = 0$, the value of $b^2 - 4ac$ is the discriminant.
If $b^2 - 4ac > 0$, the parabola has two x-intercepts.
If $b^2 - 4ac = 0$, the parabola has one x-intercept (the vertex).
If $b^2 - 4ac < 0$, the parabola has no x-intercepts.

14. $f(x) = a(x - h)^2 + k$

Since $a < 0$, the parabola opens downward. Since $h > 0$ and $k < 0$, the x-coordinate is positive and the y-coordinate is negative. Therefore, the vertex is in quadrant IV. The correct graph is (a).

15. $f(x) = .2x^2 - 1.6x + 3.3$

Factor out .2, then complete the square to find the vertex.

$$\begin{aligned} f(x) &= .2(x^2 - 8x \qquad) + 3.3 \\ f(x) &= .2(x^2 - 8x + 16 - 16) + 3.3 \\ f(x) &= .2(x^2 - 8x + 16) - 3.2 + 3.3 \\ f(x) &= .2(x - 4)^2 + .1 \end{aligned}$$

The vertex is $(4, .1)$. Since $a = .2 > 0$, the parabola opens upward and the minimum sales occurs at the vertex. Here, $x = 4$ is the number of years since 1982, so $1982 + 4 = 1986$. Minimum sales occurred in 1986 and were \$.1 billion.

16. $f(t) = -16t^2 + 160t$

The equation represents a parabola. Since $a = -16 < 0$, the parabola opens downward. The maximum time and height occur at the vertex of the parabola,

$$\left(\frac{-b}{2a}, f\left(\frac{-b}{2a}\right)\right).$$

Using the standard form of the equation, $a = -16, b = 160$, and $c = 0$, so

$$\frac{-b}{2a} = \frac{-160}{2(-16)} = 5,$$

and

$$f(5) = -16(5)^2 + 160(5) = -400 + 800 = 400.$$

The vertex is $(5, 400)$, so the time at which the maximum height is reached is 5 sec. The maximum height is 400 ft.

17. Let $x =$ the length of the rectangle and $y =$ the width of the rectangle.

The perimeter is 200 m, so

$$\begin{aligned} 200 &= 2x + 2y \\ 100 &= x + y. \quad (1) \end{aligned}$$

The area is

$$A = xy. \quad (2)$$

Solve equation (1) for y.

$$\begin{aligned} 100 &= x + y \quad (1) \\ 100 - x &= y \end{aligned}$$

Substitute $100 - x$ for y in equation (2).

$$\begin{aligned} A &= xy \quad (2) \\ A &= x(100 - x) \\ A &= -x^2 + 100x \end{aligned}$$

To find the vertex, complete the square.

$$\begin{aligned} A &= -x^2 + 100x \\ A &= -1(x^2 - 100x + 2500 - 2500) \\ A &= -1(x^2 - 100x + 2500) + (-1)(-2500) \\ A &= -(x - 50)^2 + 2500 \end{aligned}$$

This is a parabola with vertex $(50, 2500)$, and since $a = -1 < 0$, the graph opens downward. Thus, its maximum area occurs at the vertex where $x = 50$ and

$$y = 100 - x = 100 - 50 = 50.$$

Thus, length and width are both 50 m, and the rectangle is a square.

18. Let $x =$ one number.
Then $10 - x =$ the other number.
Let $p =$ the product of the numbers.

$$\begin{aligned} p &= x(10 - x) \\ p &= 10x - x^2 \\ p &= -x^2 + 10x \end{aligned}$$

Complete the square to find the vertex of the parabola.

$$\begin{aligned} p &= -1(x^2 - 10x + 25 - 25) \\ p &= -1(x^2 - 10x + 25) + 25 \\ p &= -(x - 5)^2 + 25 \end{aligned}$$

The vertex is $(5, 25)$. Since $a = -1 < 0$, the parabola opens downward. Thus, the maximum product is at the vertex $(5, 25)$ where $x = 5$ and $10 - x = 10 - 5 = 5$. The product is a maximum when the two numbers are 5 and 5.

For Exercises 19-21, see the answer graphs in the back of the textbook.

19. $f(x) = |x + 4|$

This is the graph of the absolute value function $f(x) = |x|$ shifted 4 units to the left (since $x + 4 = 0$ if $x = -4$).

x	-6	-5	-4	-3	-2
y	2	1	0	1	2

20. $f(x) = \dfrac{1}{x-4}$

This is the graph of the reciprocal function $f(x) = \frac{1}{x}$ shifted 4 units to the right. Since $x \neq 4$ (or a denominator of 0 results), the line $x = 4$ is an asymptote. Since $\frac{1}{x-4} \neq 0$, the line $y = 0$ is also an asymptote.

x	$4\frac{1}{2}$	5	6
y	2	1	$\frac{1}{2}$

x	2	3	$3\frac{1}{2}$
y	$-\frac{1}{2}$	-1	-2

21. $f(x) = \sqrt{x} + 3$

This is the graph of the square root function $f(x) = \sqrt{x}$ shifted 3 units upward.

x	0	1	4
y	3	4	5

22. Center $(-2, 4)$, $r = 3$
A circle has equation

$$\begin{aligned} (x-h)^2 + (y-k)^2 &= r^2, \\ \text{so} \quad [x-(-2)]^2 + (y-4)^2 &= 3^2 \\ (x+2)^2 + (y-4)^2 &= 9. \end{aligned}$$

23. Center $(-1, -3), r = 5$
A circle has equation

$$\begin{aligned} (x-h)^2 + (y-k)^2 &= r^2, \\ \text{so} \quad [x-(-1)]^2 + [y-(-3)]^2 &= 5^2 \\ (x+1)^2 + (y+3)^2 &= 25. \end{aligned}$$

24. Center $(4, 2), r = 6$
A circle has equation

$$\begin{aligned} (x-h)^2 + (y-k)^2 &= r^2, \\ \text{so} \quad (x-4)^2 + (y-2)^2 &= 6^2 \\ (x-4)^2 + (y-2)^2 &= 36. \end{aligned}$$

25. $x^2 + y^2 + 6x - 4y - 3 = 0$

Complete the squares on x and y.

$$\begin{aligned} (x^2 + 6x \quad\;) + (y^2 - 4y \quad\;) &= 3 \\ (x^2 + 6x + 9 - 9) + (y^2 - 4y + 4 - 4) &= 3 \\ (x^2 + 6x + 9) + (y^2 - 4y + 4) - 9 - 4 &= 3 \\ (x+3)^2 + (y-2)^2 &= 16 \\ (x+3)^2 + (y-2)^2 &= 4^2 \end{aligned}$$

The circle has center at $(-3, 2)$ and radius 4.

26. $x^2 + y^2 - 8x - 2y + 13 = 0$

Complete the squares on x and y.

$$\begin{aligned} (x^2 - 8x \quad\;) + (y^2 - 2y \quad\;) &= -13 \\ (x^2 - 8x + 16 - 16) + (y^2 - 2y + 1 - 1) &= -13 \\ (x^2 - 8x + 16) + (y^2 - 2y + 1) - 16 - 1 &= -13 \\ (x-4)^2 + (y-1)^2 &= 4 \\ (x-4)^2 + (y-1)^2 &= 2^2 \end{aligned}$$

The circle has center at $(4, 1)$ and radius 2.

27.
$$\begin{aligned} 2x^2 + 2y^2 + 4x + 20y &= -34 \\ x^2 + y^2 + 2x + 10y &= -17 \\ (x^2 + 2x \quad\;) + (y^2 + 10y \quad\;) &= -17 \\ (x^2 + 2x + 1 - 1) + (y^2 + 10y + 25 - 25) &= -17 \\ (x^2 + 2x + 1) + (y^2 + 10y + 25) - 1 - 25 &= -17 \\ (x+1)^2 + (y+5)^2 &= 9 \\ (x+1)^2 + (y+5)^2 &= 3^2 \end{aligned}$$

The circle has center at $(-1, -5)$ and radius 3.

28.
$$\begin{aligned} 4x^2 + 4y^2 - 24x + 16y &= 48 \\ x^2 + y^2 - 6x + 4y &= 12 \\ (x^2 - 6x \quad\;) + (y^2 + 4y \quad\;) &= 12 \\ (x^2 - 6x + 9 - 9) + (y^2 + 4y + 4 - 4) &= 12 \\ (x^2 - 6x + 9) + (y^2 + 4y + 4) - 9 - 4 &= 12 \\ (x-3)^2 + (y+2)^2 &= 25 \\ (x-3)^2 + (y+2)^2 &= 5^2 \end{aligned}$$

The circle has center at $(3, -2)$ and radius 5.

For Exercises 29-34, see the answer graphs in the back of the textbook.

29. $\dfrac{x^2}{16} + \dfrac{y^2}{9} = 1$

$a = 4,\ b = 3$
The graph is an ellipse with x-intercepts $(4, 0)$ and $(-4, 0)$ and y-intercepts $(0, 3)$ and $(0, -3)$. Plot the intercepts, and draw the ellipse through them.

30. $\dfrac{x^2}{49} + \dfrac{y^2}{25} = 1$

$a = 7, b = 5$

The graph is an ellipse with x-intercepts $(7, 0)$ and $(-7, 0)$ and y-intercepts $(0, 5)$ and $(0, -5)$. Plot the intercepts, and draw the ellipse through them.

31. $\dfrac{x^2}{16} - \dfrac{y^2}{25} = 1$

$a = 4,\ b = 5$

The graph is a hyperbola with x-intercepts $(4, 0)$ and $(-4, 0)$ and asymptotes that are the extended diagonals of the rectangle with corners $(4,5), (4,-5)$, $(-4,-5)$, and $(-4,5)$. Graph a branch of the hyperbola through each intercept approaching the asymptotes.

32. $\frac{y^2}{25} - \frac{x^2}{4} = 1$

$a = 2, b = 5$

The graph is a hyperbola with y-intercepts $(0,5)$ and $(0,-5)$ and asymptotes that are the extended diagonals of the rectangle with corners $(2,5), (2,-5)$, $(-2,-5)$, and $(-2,5)$. Graph a branch of the hyperbola through each intercept approaching the asymptotes.

33. $x^2 + 9y^2 = 9$

$\frac{x^2}{9} + \frac{y^2}{1} = 1$

$a = 3, b = 1$

The graph is an ellipse with x-intercepts $(3,0)$ and $(-3,0)$ and y-intercepts $(0,1)$ and $(0,-1)$. Plot the intercepts and draw the ellipse through them.

34. $f(x) = -\sqrt{16 - x^2}$

Replace $f(x)$ with y.

$$y = -\sqrt{16 - x^2}$$

Square both sides.

$$y^2 = 16 - x^2$$
$$x^2 + y^2 = 16$$

This equation is the graph of a circle with center $(0,0)$ and radius 4. Since $f(x)$ represents a negative square root, $f(x)$ is negative and its graph is the lower half of the circle.

35. $$x^2 + y^2 = 64$$
$$(x-0)^2 + (y-0)^2 = 8^2$$

The graph is a circle.

36. $y = 2x^2 - 3$

The equation has one squared variable, x, so the graph of the equation is a parabola.

37. $$y^2 = 2x^2 - 8$$
$$2x^2 - y^2 = 8$$
$$\frac{x^2}{4} - \frac{y^2}{8} = 1$$

Since both variables are squared and because of the minus sign, the graph of this equation is a hyperbola.

38. $$y^2 = 8 - 2x^2$$
$$2x^2 + y^2 = 8$$
$$\frac{x^2}{4} + \frac{y^2}{8} = 1$$

Since both variables are squared and because of the plus sign, the graph of this equation is an ellipse.

39. $x = y^2 + 4$

$x = (y-0)^2 + 4$

The equation has one squared variable, y, so the graph of the equation is a horizontal parabola.

40. $$x^2 - y^2 = 64$$
$$\frac{x^2}{64} - \frac{y^2}{64} = 1$$

Since both variables are squared and because of the minus sign, the graph of this equation is a hyperbola.

41. $$2y = 3x - x^2 \quad (1)$$
$$x + 2y = -12 \quad (2)$$

Substitute equation (1) for $2y$ in equation (2).

$$x + 2y = -12 \quad (2)$$
$$x + (3x - x^2) = -12$$
$$-x^2 + 4x + 12 = 0$$
$$x^2 - 4x - 12 = 0$$
$$(x-6)(x+2) = 0$$
$$x - 6 = 0 \quad \text{or} \quad x + 2 = 0$$
$$x = 6 \quad \text{or} \quad x = -2$$

Substitute these values for x in equation (2) to find y.

If $x = 6$, then

$$x + 2y = -12 \quad (2)$$
$$6 + 2y = -12$$
$$2y = -18$$
$$y = -9.$$

If $x = -2$, then

$$x + 2y = -12 \quad (2)$$
$$-2 + 2y = -12$$
$$2y = -10$$
$$y = -5.$$

Solution set: $\{(6,-9), (-2,-5)\}$

42. $y+1=x^2+2x \quad (1)$
$y+2x=4 \quad (2)$

Rewrite equation (2) as

$$y-4=-2x. \quad (3)$$

Multiply equation (3) by -1 and add the result to equation (1).

$$\begin{aligned} -y+4 &= 2x \\ y+1 &= x^2+2x \quad (1) \\ \hline 5 &= x^2+4x \end{aligned}$$

Rewrite the resulting equation, then factor it.

$$\begin{aligned} x^2+4x-5&=0 \\ (x+5)(x-1)&=0 \\ x+5=0 \quad &\text{or} \quad x-1=0 \\ x=-5 \quad &\text{or} \quad x=1 \end{aligned}$$

Substitute these values for x in equation (2) to find y.

If $x=-5$, then

$$\begin{aligned} y+2x&=4 \quad (2) \\ y+2(-5)&=4 \\ y-10&=4 \\ y&=14. \end{aligned}$$

If $x=1$, then

$$\begin{aligned} y+2x&=4 \quad (2) \\ y+2(1)&=4 \\ y&=2. \end{aligned}$$

Solution set: $\{(1,2),(-5,14)\}$

43. $x^2+3y^2=28 \quad (1)$
$y-x=-2 \quad (2)$

Solve equation (2) for y.

$$y=x-2$$

Substitute $x-2$ for y in equation (1).

$$\begin{aligned} x^2+3y^2&=28 \quad (1) \\ x^2+3(x-2)^2&=28 \\ x^2+3(x^2-4x+4)-28&=0 \\ x^2+3x^2-12x+12-28&=0 \\ 4x^2-12x-16&=0 \\ x^2-3x-4&=0 \\ (x-4)(x+1)&=0 \\ x-4=0 \quad &\text{or} \quad x+1=0 \\ x=4 \quad &\text{or} \quad x=-1 \end{aligned}$$

Since $y=x-2$, if $x=4$, then

$$y=4-2=2.$$

If $x=-1$, then

$$y=-1-2=-3.$$

Solution set: $\{(4,2),(-1,-3)\}$

44. $xy=8 \quad (1)$
$x-2y=6 \quad (2)$

Solve equation (2) for x.

$$x=2y+6$$

Substitute $2y+6$ for x in equation (1) to find y.

$$\begin{aligned} xy&=8 \quad (1) \\ (2y+6)y&=8 \\ 2y^2+6y-8&=0 \\ y^2+3y-4&=0 \\ (y+4)(y-1)&=0 \\ y+4=0 \quad &\text{or} \quad y-1=0 \\ y=-4 \quad &\text{or} \quad y=1 \end{aligned}$$

Substitute these values for y in equation (1) to find x.

If $y=-4$, then

$$\begin{aligned} xy&=8 \quad (1) \\ -4x&=8 \\ x&=-2. \end{aligned}$$

If $y=1$, then

$$\begin{aligned} xy&=8 \quad (1) \\ x(1)&=8 \\ x&=8. \end{aligned}$$

Solution set: $\{(-2,-4),(8,1)\}$

45. $x^2+y^2=6 \quad (1)$
$x^2-2y^2=-6 \quad (2)$

Multiply equation (2) by -1 and add the result to equation (1).

$$\begin{aligned} -x^2+2y^2&=6 \\ x^2+y^2&=6 \quad (1) \\ \hline 3y^2&=12 \\ y^2&=4 \\ y=2 \quad &\text{or} \quad y=-2 \end{aligned}$$

Substitute these values for y in equation (1) to find x.

If $y = 2$, then

$$\begin{aligned} x^2 + y^2 &= 6 \quad (1) \\ x^2 + 2^2 &= 6 \\ x^2 + 4 &= 6 \\ x^2 &= 2 \\ x = \sqrt{2} \quad &\text{or} \quad x = -\sqrt{2}. \end{aligned}$$

If $y = -2$, then

$$\begin{aligned} x^2 + y^2 &= 6 \quad (1) \\ x^2 + (-2)^2 &= 6 \\ x^2 + 4 &= 6 \\ x^2 &= 2 \\ x = \sqrt{2} \quad &\text{or} \quad x = -\sqrt{2}. \end{aligned}$$

Solution set:
$\{(\sqrt{2}, 2), (-\sqrt{2}, 2), (\sqrt{2}, -2), (-\sqrt{2}, -2)\}$

46. $3x^2 - 2y^2 = 12 \quad (1)$
$x^2 + 4y^2 = 18 \quad (2)$

Multiply equation (1) by 2 and add the result to equation (2).

$$\begin{aligned} 6x^2 - 4y^2 &= 24 \\ x^2 + 4y^2 &= 18 \quad (2) \\ \hline 7x^2 \qquad &= 42 \\ x^2 &= 6 \\ x = \sqrt{6} \quad &\text{or} \quad x = -\sqrt{6} \end{aligned}$$

Substitute these values for x in equation (2) to find y.

If $x = \sqrt{6}$, then

$$\begin{aligned} x^2 + 4y^2 &= 18 \quad (2) \\ (\sqrt{6})^2 + 4y^2 &= 18 \\ 6 + 4y^2 &= 18 \\ 4y^2 &= 12 \\ y^2 &= 3 \\ y = \sqrt{3} \quad &\text{or} \quad y = -\sqrt{3}. \end{aligned}$$

If $x = -\sqrt{6}$, then

$$\begin{aligned} x^2 + 4y^2 &= 18 \quad (2) \\ (-\sqrt{6})^2 + 4y^2 &= 18 \\ 6 + 4y^2 &= 18 \\ 4y^2 &= 12 \\ y^2 &= 3 \\ y = \sqrt{3} \quad &\text{or} \quad y = -\sqrt{3}. \end{aligned}$$

Solution set:
$\{(\sqrt{6}, \sqrt{3}), (\sqrt{6}, -\sqrt{3}), (-\sqrt{6}, \sqrt{3}), (-\sqrt{6}, -\sqrt{3})\}$

47. A circle and a line can intersect in zero, one, or two points, so zero, one, or two solutions are possible.

48. A parabola and a hyperbola can intersect in zero, one, two, three, or four points, so zero, one, two, three, or four solutions are possible.

For Exercises 49-62, see the answer graphs in the back of the textbook.

49.
$$\begin{aligned} 9x^2 &\geq 16y^2 + 144 \\ 9x^2 - 16y^2 &\geq 144 \\ \frac{x^2}{16} - \frac{y^2}{9} &\geq 1 \end{aligned}$$

The boundary, $\frac{x^2}{16} - \frac{y^2}{9} = 1$, is a solid hyperbola with x-intercepts $(4, 0)$ and $(-4, 0)$. The asymptotes are the extended diagonals of the rectangle with corners $(4, 3), (4, -3), (-4, -3)$, and $(-4, 3)$. Test $(0, 0)$.

$$\begin{aligned} 9(0)^2 &\geq 16(0)^2 + 144 \\ 0 &\geq 144 \quad \textit{False} \end{aligned}$$

Shade the sides of the hyperbola that do not contain $(0, 0)$. These are the regions inside the branches of the hyperbola.

50. $4x^2 + y^2 \geq 16$

$$\frac{x^2}{4} + \frac{y^2}{16} \geq 1$$

The boundary, $\frac{x^2}{4} + \frac{y^2}{16} = 1$, is a solid ellipse with intercepts $(2, 0), (-2, 0), (0, 4)$, and $(0, -4)$. Test $(0, 0)$.

$$\begin{aligned} 4(0)^2 + 0^2 &\geq 16 \\ 0 &\geq 16 \quad \textit{False} \end{aligned}$$

Shade the side of the ellipse that does not contain $(0, 0)$. This is the region outside the ellipse.

51. $y < -(x + 2)^2 + 1$

The boundary, $y = -(x + 2)^2 + 1$, is a dashed vertical parabola with vertex $(-2, 1)$. Since $a = -1 < 0$, the parabola opens downward. Also, $|a| = |-1| = 1$, so the graph has the same shape as the graph of $y = x^2$. Test $(0, 0)$.

$$\begin{aligned} 0 &< -(0 + 2)^2 + 1 \\ 0 &< -(4) + 1 \\ 0 &< -3 \quad \textit{False} \end{aligned}$$

Shade the side of the parabola that does not contain $(0, 0)$. This is the region inside the parabola.

52. $2x + 5y \leq 10$
$3x - \ y \leq \ 6$

The boundary, $2x + 5y = 10$, is a solid line with intercepts $(5,0)$ and $(0,2)$. Test $(0,0)$.

$$2(0) + 5(0) \leq 10$$
$$0 \leq 10 \quad \textit{True}$$

Shade the side of the line that contains $(0,0)$.

The boundary, $3x - y = 6$, is a solid line with intercepts $(2,0)$ and $(0,-6)$. Test $(0,0)$.

$$3(0) - 0 \leq 6$$
$$0 \leq 6 \quad \textit{True}$$

Shade the side of the line that contains $(0,0)$.

The graph of the system is the intersection of the two shaded regions.

53. $|x| \leq 2$
$|y| > 1$
$4x^2 + 9y^2 \leq 36$

The equation of the boundary, $|x| = 2$, can be written as

$$x = -2 \quad \text{or} \quad x = 2.$$

The graph is these two solid vertical lines. Since $0 \leq 2$ is true, the region between the lines, containing $(0,0)$, is shaded.

The boundary, $|y| = 1$, consists of the two dashed horizontal lines $y = 1$ and $y = -1$. Since $0 > 1$ is false, the regions above and below the lines, not containing $(0,0)$, are shaded.

The boundary given by

$$4x^2 + 9y^2 = 36$$
$$\text{or} \quad \frac{x^2}{9} + \frac{y^2}{4} = 1$$

is graphed as a solid ellipse with intercepts $(3,0)$, $(-3,0)$, $(0,2)$, and $(0,-2)$. Test $(0,0)$.

$$4(0)^2 + 9(0)^2 \leq 36$$
$$0 \leq 36 \quad \textit{True}$$

The region inside the ellipse, containing $(0,0)$, is shaded.

The graph of the system consists of the regions that include the common points of the three shaded regions.

54. $9x^2 \leq 4y^2 + 36$
$x^2 + y^2 \leq 16$

The equation of the first boundary is

$$9x^2 = 4y^2 + 36$$
$$9x^2 - 4y^2 = 36$$
$$\frac{x^2}{4} - \frac{y^2}{9} = 1.$$

The graph is a solid hyperbola with x-intercepts $(2,0)$ and $(-2,0)$. The asymptotes are the extended diagonals of the rectangle with corners $(2,3)$, $(2,-3)$, $(-2,-3)$, and $(-2,3)$. Test $(0,0)$.

$$9(0)^2 \leq 4(0)^2 + 36$$
$$0 \leq 36 \quad \textit{True}$$

Shade the region between the branches of the hyperbola that contains $(0,0)$.

The equation of the second boundary is $x^2 + y^2 = 16$. This is a solid circle with center $(0,0)$ and radius 4. Test $(0,0)$.

$$0^2 + 0^2 \leq 16$$
$$0 \leq 16 \quad \textit{True}$$

Shade the region inside the circle.

The graph of the system is the intersection of the shaded regions which is between the two branches of the hyperbola and inside the circle.

55. $\dfrac{x^2}{64} + \dfrac{y^2}{25} = 1$

$a = 8, b = 5$

The graph is an ellipse with intercepts $(8,0)$, $(-8,0)$, $(0,5)$, and $(0,-5)$. Plot the intercepts, and draw the ellipse through them.

56. $\dfrac{y^2}{4} - 1 = \dfrac{x^2}{9}$

$\dfrac{y^2}{4} - \dfrac{x^2}{9} = 1$

$a = 3, b = 2$

The graph is a hyperbola with y-intercepts $(0,2)$ and $(0,-2)$ and asymptotes that are the extended diagonals of the rectangle with corners $(3,2)$, $(3,-2)$, $(-3,-2)$, and $(-3,2)$. Draw a branch of the hyperbola through each intercept and approaching the asymptotes.

57. $x^2 + y^2 = 25$

This equation may be written $(x-0)^2+(y-0)^2 = 5^2$. The graph is a circle with center at $(0,0)$ and radius 5.

58. $y = 2(x-2)^2 - 3$

The graph is a parabola with vertex $(2,-3)$. Since $a = 2 > 0$, the parabola opens upward. Also, $|a| = |2| = 2 > 1$, so the graph is narrower than the graph of $y = x^2$. The points $(0,5)$ and $(4,5)$ are on the graph.

59. $x^2 - 9y^2 = 9$

$$\frac{x^2}{9} - \frac{y^2}{1} = 1$$

$a = 3, b = 1$

The graph is a hyperbola with x-intercepts (3,0) and $(-3,0)$ and asymptotes that are the extended diagonals of the rectangle with corners $(3,1)$, $(3,-1)$, $(-3,-1)$, and $(-3,1)$. Graph a branch of the hyperbola through each intercept and approaching the asymptotes.

60. $f(x) = \sqrt{4-x}$

Replace $f(x)$ with y.

$$y = \sqrt{4-x}$$

Square both sides.

$$y^2 = 4 - x$$
$$x = -y^2 + 4$$
$$x = -1(y-0)^2 + 4$$

This equation is the graph of a horizontal parabola with vertex at $(4,0)$. Since $a = -1 < 0$, the graph opens to the left. Also, $|a| = |-1| = 1$, so the graph has the same shape as the graph of $y = x^2$. The points $(0,2)$ and $(3,1)$ are on the graph. Since $f(x)$ represents a principal square root, $f(x)$ is nonnegative and its graph is the upper half of the parabola.

61. $3x + 2y \geq 0$
$y \leq 4$
$x \leq 4$

The boundary, $3x+2y = 0$, is a solid line through $(0,0)$ and $(2,-3)$. Test $(0,1)$.

$$3(0) + 2(1) \geq 0$$
$$2 \geq 0 \quad \textit{True}$$

Shade the side of the line that contains $(0,1)$.

The boundary, $y = 4$, is a solid horizontal line through $(0,4)$. Since $y \leq 4$, shade below the line.

The boundary, $x = 4$, is a solid vertical line through $(4,0)$. Since $x \leq 4$, shade the region to the left of the line.

The graph of the system is the intersection of the three shaded regions.

62. $4y > 3x - 12$
$x^2 < 16 - y^2$

The boundary, $4y = 3x - 12$, is a dashed line with intercepts $(0,-3)$ and $(4,0)$. Test $(0,0)$.

$$4(0) > 3(0) - 12$$
$$0 > -12 \quad \textit{True}$$

Shade the side of the line that contains $(0,0)$.

The boundary, $x^2 = 16 - y^2$ or $x^2 + y^2 = 16$, is a dashed circle with center $(0,0)$ and radius 4. Test $(0,0)$.

$$0^2 < 16 - 0^2$$
$$0 < 16 \quad \textit{True}$$

Shade the region inside the circle.

The graph of the system is the intersection of the two shaded regions, that is, the region inside the circle and above the line.

Chapter 9 Test

For Exercises 1 and 2, see the answer graphs in the back of the textbook.

1. $f(x) = \frac{1}{2}x^2 - 2$

This is the graph of the parabola $f(x) = x^2$ shifted vertically -2 units, that is, 2 units down. Its vertex is at $(0,-2)$. Since $a = \frac{1}{2} > 0$, the parabola opens upward. Also, $|a| = \left|\frac{1}{2}\right| = \frac{1}{2} < 1$, so the graph of the parabola is wider than the graph of $f(x) = x^2$. The points $(2,0)$ and $(-2,0)$ are on the graph.

2. $f(x) = -x^2 + 4x - 1$

Use the vertex formula with $a = -1$, $b = 4$, $c = -1$.

$$\frac{-b}{2a} = \frac{-4}{2(-1)} = 2$$

$$f(2) = -2^2 + 4(2) - 1 = -4 + 8 - 1 = 3$$

The graph is a parabola with vertex at $(2, 3)$. Since $a = -1 < 0$, the parabola opens downward. Also, $|a| = |-1| = 1$, so the graph has the same shape as the graph of $f(x) = x^2$. The points $(0, -1)$ and $(4, -1)$ are on the graph. The domain is $(-\infty, \infty)$, and the range, since the vertex $(2, 3)$ is the highest point on the graph, is $(-\infty, 3]$.

3. **(a)** Since $x = 1$ corresponds to 1983, substitute 1 for x in the quadratic function.

$$\begin{aligned} f(x) &= -2.6x^2 + 11.7x + 22.5 \\ f(1) &= -2.6(1)^2 + 11.7(1) + 22.5 \\ &= -2.6 + 11.7 + 22.5 \\ &= 31.6. \end{aligned}$$

Then 31.6 thousand or 31,600 salmon returned in 1983.

(b) The graph of the quadratic function

$f(x) = -2.6x^2 + 11.7x + 22.5$

is a parabola that opens downward. The vertex will provide maximum values. Use the vertex formula.

$a = -2.6, b = 11.7, c = 22.5$

$$\frac{-b}{2a} = \frac{-11.7}{2(-2.6)} = 2.25$$

$$\begin{aligned} f(2.25) &= -2.6(2.25)^2 + 11.7(2.25) + 22.5 \\ &= 35.6625 \end{aligned}$$

The vertex is at $(2.25, 35.6625)$. The x-value gives the year. Since $x = 0$ corresponds to 1982, $x = 2.25$ corresponds to sometime during 1984 (1982 + 2 = 1984). To the nearest thousand, the number of salmon that returned that year is about 36 thousand or 36,000.

4. $f(x) = (x - 2)^2 + 2$

This is the graph of a parabola $f(x) = x^2$ shifted horizontally 2 units to the right (since $x - 2 = 0$ if $x = 2$) and 2 units upward (because of the $+2$). The correct graph is choice B.

For Exercises 5 and 6, see the answer graphs in the back of the textbook.

5. $f(x) = |x - 3| + 4$

This is the graph of the absolute value function $f(x) = |x|$ shifted 3 units to the right (since $x-3 = 0$ if $x = 3$) and 4 units upward (because of the $+4$).

Its vertex is at $(3, 4)$.

x	0	1	2	3	4	5	6
y	7	6	5	4	5	6	7

6. $(x - 2)^2 + (y + 3)^2 = 16$
$(x - 2)^2 + (y + 3)^2 = 4^2$

The graph is a circle with center at $(2, -3)$ and radius 4.

7. $x^2 + y^2 + 8x - 2y = 8$

To find the center and radius, complete the squares on x and y.

$$\begin{aligned} (x^2 + 8x \quad) + (y^2 - 2y \quad) &= 8 \\ (x^2 + 8x + 16 - 16) + (y^2 - 2y + 1 - 1) &= 8 \\ (x^2 + 8x + 16) + (y^2 - 2y + 1) - 16 - 1 &= 8 \\ (x + 4)^2 + (y - 1)^2 &= 25 \\ (x + 4)^2 + (y - 1)^2 &= 5^2 \end{aligned}$$

The graph is a circle with center at $(-4, 1)$ and radius 5.

8. The equation of a parabola can be written as

$$y = a(x - h)^2 + k$$
$$\text{or} \quad x = a(y - k)^2 + h.$$

The equation will have a term containing a variable squared and a term which has the other variable to the first power.

For Exercises 9-12, see the answer graphs in the back of the textbook.

9. $f(x) = \sqrt{9 - x^2}$

Replace $f(x)$ with y.

$$y = \sqrt{9 - x^2}$$

Square both sides.

$$\begin{aligned} y^2 &= 9 - x^2 \\ x^2 + y^2 &= 9 \end{aligned}$$

The graph is a circle with center at $(0, 0)$ and radius 3. Since $f(x)$ represents a principal square root, $f(x)$ is nonnegative and its graph is the upper half of the circle.

10. $4x^2 + 9y^2 = 36$

$$\frac{x^2}{9} + \frac{y^2}{4} = 1$$

The equation is in $\frac{x^2}{a^2} + \frac{y^2}{b^2} = 1$ form with $a = 3$ and $b = 2$. The graph is an ellipse with intercepts $(3, 0), (-3, 0), (0, 2)$, and $(0, -2)$. Plot the intercepts, and draw the ellipse through them.

11. $16y^2 - 4x^2 = 64$

$$\frac{y^2}{4} - \frac{x^2}{16} = 1$$

The equation is in $\frac{y^2}{b^2} - \frac{x^2}{a^2} = 1$ form with $a = 4$ and $b = 2$. The graph is a hyperbola with y–intercepts $(0, 2)$ and $(0, -2)$ and asymptotes that are extended diagonals of the rectangle with corners $(4, 2), (4, -2), (-4, -2)$, and $(-4, 2)$. Draw a branch of the hyperbola through each intercept and approaching the asymptotes.

12. $x = -(y - 2)^2 + 2$

Since the roles of x and y are reversed, this graph is a horizontal parabola that has been shifted 2 units to the right (because of the $+2$) and 2 units upward (since $y - 2 = 0$ if $y = 2$). Its vertex is at $(2, 2)$. Since $a = -1 < 0$, the graph opens to the left. Also, $|a| = |-1| = 1$, so the graph has the same shape as the graph of $y = x^2$. The points $(-2, 0)$ and $(-2, 4)$ are on the graph.

13. $6x^2 + 4y^2 = 12$

$$\frac{x^2}{2} + \frac{y^2}{3} = 1$$

This is the graph of an ellipse or a hyperbola since both variables are squared. Because of the plus sign, the graph of the equation is an ellipse.

14.
$$\begin{aligned} 16x^2 &= 144 + 9y^2 \\ 16x^2 - 9y^2 &= 144 \\ \frac{x^2}{9} - \frac{y^2}{16} &= 1 \end{aligned}$$

Since both variables are squared and because of the minus sign, the graph of the equation is a hyperbola.

15.
$$\begin{aligned} 4y^2 + 4x &= 9 \\ 4x &= -4y^2 + 9 \\ x &= -y^2 + \frac{9}{4} \end{aligned}$$

This is the graph of a horizontal parabola since only one variable, y, is squared.

16.
$$\begin{aligned} 2x - y &= 9 \quad (1) \\ xy &= 5 \quad (2) \end{aligned}$$

Solve equation (1) for y.

$$y = 2x - 9$$

Substitute $2x - 9$ for y in equation (2).

$$\begin{aligned} xy &= 5 \quad (2) \\ x(2x - 9) &= 5 \\ 2x^2 - 9x &= 5 \\ 2x^2 - 9x - 5 &= 0 \\ (2x + 1)(x - 5) &= 0 \\ 2x + 1 = 0 \quad &\text{or} \quad x - 5 = 0 \\ 2x = -1 \quad & \\ x = -\frac{1}{2} \quad &\text{or} \quad x = 5 \end{aligned}$$

Substitute these values for x in equation (2) to find y.

If $x = -\frac{1}{2}$, then

$$\begin{aligned} xy &= 5 \quad (2) \\ -\frac{1}{2}y &= 5 \\ y &= -10. \end{aligned}$$

If $x = 5$, then

$$\begin{aligned} xy &= 5 \quad (2) \\ 5y &= 5 \\ y &= 1. \end{aligned}$$

Solution set: $\left\{\left(-\frac{1}{2}, -10\right), (5, 1)\right\}$

17.
$$\begin{aligned} x - 4 &= 3y \quad (1) \\ x^2 + y^2 &= 8 \quad (2) \end{aligned}$$

Solve equation (1) for x.

$$x = 3y + 4$$

Substitute $3y + 4$ for x in equation (2).

$$\begin{aligned} x^2 + y^2 &= 8 \quad (2) \\ (3y + 4)^2 + y^2 &= 8 \\ 9y^2 + 24y + 16 + y^2 &= 8 \\ 10y^2 + 24y + 8 &= 0 \\ 5y^2 + 12y + 4 &= 0 \\ (5y + 2)(y + 2) &= 0 \\ 5y + 2 = 0 \quad &\text{or} \quad y + 2 = 0 \\ 5y = -2 \quad & \\ y = -\frac{2}{5} \quad &\text{or} \quad y = -2 \end{aligned}$$

Since $x = 3y + 4$, substitute these values for y to find x.

If $y = -\frac{2}{5}$, then

$$x = 3\left(-\frac{2}{5}\right) + 4 = -\frac{6}{5} + 4 = \frac{14}{5}.$$

If $y = -2$, then

$$x = 3(-2) + 4 = -6 + 4 = -2.$$

Solution set: $\{(-2,-2), (\frac{14}{5}, -\frac{2}{5})\}$

18. $x^2 + y^2 = 25$ (1)
$x^2 - 2y^2 = 16$ (2)

Multiply equation (1) by 2 and add the result to equation (2).

$$\begin{aligned} 2x^2 + 2y^2 &= 50 \\ x^2 - 2y^2 &= 16 \quad (2) \\ \hline 3x^2 \qquad &= 66 \\ x^2 &= 22 \\ x = \sqrt{22} \quad &\text{or} \quad x = -\sqrt{22} \end{aligned}$$

Substitute these values for x in equation (1).

If $x = \sqrt{22}$, then

$$\begin{aligned} x^2 + y^2 &= 25 \quad (1) \\ (\sqrt{22})^2 + y^2 &= 25 \\ 22 + y^2 &= 25 \\ y^2 &= 3 \\ y = \sqrt{3} \quad &\text{or} \quad y = -\sqrt{3}. \end{aligned}$$

If $x = -\sqrt{22}$, then

$$\begin{aligned} x^2 + y^2 &= 25 \quad (1) \\ (-\sqrt{22})^2 + y^2 &= 25 \\ 22 + y^2 &= 25 \\ y^2 &= 3 \\ y = \sqrt{3} \quad &\text{or} \quad y = -\sqrt{3}. \end{aligned}$$

Solution set:
$\{(\sqrt{22}, \sqrt{3}), (\sqrt{22}, -\sqrt{3}), (-\sqrt{22}, \sqrt{3}), (-\sqrt{22}, -\sqrt{3})\}$

For Exercises 19 and 20, see the answer graphs in the back of the textbook.

19. $y \le x^2 - 2$

The boundary, $y = x^2 - 2$, is a solid parabola with vertex $(0,-2)$. Since $a = 1 > 0$, the parabola opens upward. It also has the same shape as $y = x^2$. The points $(2,2)$ and $(-2,2)$ are on the graph. Test $(0,0)$.

$$\begin{aligned} 0 &\le 0^2 - 2 \\ 0 &\le -2 \quad \textit{False} \end{aligned}$$

Shade the side of the parabola that does not contain $(0,0)$. This is the region outside the parabola.

20. $x^2 + 25y^2 \le 25$
$x^2 + \quad y^2 \le 9$

The first boundary, $\frac{x^2}{25} + \frac{y^2}{1} = 1$, is a solid ellipse with intercepts $(5,0), (-5,0), (0,1)$, and $(0,-1)$. Test $(0,0)$.

$$\begin{aligned} 0^2 + 25 \cdot 0^2 &\le 25 \\ 0 &\le 25 \quad \textit{True} \end{aligned}$$

Shade the region inside the ellipse.

The boundary, $x^2 + y^2 = 9$, is a solid circle with center $(0,0)$ and radius 3. Test $(0,0)$.

$$\begin{aligned} 0^2 + 0^2 &\le 9 \\ 0 &\le 9 \quad \textit{True} \end{aligned}$$

Shade the region inside the circle.

The solution of the system is the intersection of the two shaded regions, that is, the region inside the ellipse and inside the circle.

Cumulative Review Exercises (Chapters 1-9)

1. $-10 + |-5| - |3| + 4 = -10 + 5 - 3 + 4$
$= -4$

2.
$$\begin{aligned} 4 - (2x + 3) + x &= 5x - 3 \\ 4 - 2x - 3 + x &= 5x - 3 \\ -x + 1 &= 5x - 3 \\ -6x &= -4 \\ x &= \frac{2}{3} \end{aligned}$$

Solution set: $\{\frac{2}{3}\}$

3.
$$\begin{aligned} -4k + 7 &\ge 6k + 1 \\ -10k &\ge -6 \end{aligned}$$

Divide by -10; reverse the direction of the inequality sign.

$$k \le \frac{-6}{-10}$$

$$k \le \frac{3}{5}$$

Solution set: $(-\infty, \frac{3}{5}]$

4. $|5m| - 6 = 14$

$$\begin{aligned} |5m| &= 20 \\ 5m = 20 \quad &\text{or} \quad -5m = -20 \\ m = 4 \quad &\text{or} \quad m = -4 \end{aligned}$$

Solution set: $\{-4, 4\}$

5. $|2p - 5| > 15$

$$\begin{aligned} 2p - 5 > 15 \quad &\text{or} \quad 2p - 5 < -15 \\ 2p > 20 \quad &\quad\quad 2p < -10 \\ p > 10 \quad &\text{or} \quad p < -5 \end{aligned}$$

Solution set: $(-\infty, -5) \cup (10, \infty)$

6. Let $(x_1, y_1) = (2, 5)$ and $(x_2, y_2) = (-4, 1)$.

$$m = \frac{(y_2 - y_1)}{(x_2 - x_1)} = \frac{(1-5)}{(-4-2)} = \frac{-4}{-6} = \frac{2}{3}$$

7. Through $(-3, -2)$; perpendicular to $2x - 3y = 7$

Write $2x - 3y = 7$ in slope-intercept form.

$$\begin{aligned} -3y &= -2x + 7 \\ y &= \frac{2}{3}x - \frac{7}{3} \end{aligned}$$

The slope is $\frac{2}{3}$. Perpendicular lines have slopes that are negative reciprocals of each other. Since $\frac{2}{3}\left(-\frac{3}{2}\right) = -1$, a line perpendicular to the given line will have slope $-\frac{3}{2}$. Let $m = -\frac{3}{2}$ and $(x_1, y_1) = (-3, -2)$ in the point-slope form.

$$\begin{aligned} y - y_1 &= m(x - x_1) \\ y - (-2) &= -\frac{3}{2}[x - (-3)] \\ y + 2 &= -\frac{3}{2}(x + 3) \end{aligned}$$

Multiply by 2 to clear the fraction.

$$\begin{aligned} 2y + 4 &= -3(x + 3) \\ 2y + 4 &= -3x - 9 \\ 3x + 2y &= -13 \end{aligned}$$

8. $$\begin{aligned} (5y - 3)^2 &= (5y)^2 - 2(5y)3 + 3^2 \\ &= 25y^2 - 30y + 9 \end{aligned}$$

9. $$\begin{aligned} (2r + 7)(6r - 1) &= 12r^2 - 2r + 42r - 7 \\ &= 12r^2 + 40r - 7 \end{aligned}$$

10. $(8x^4 - 4x^3 + 2x^2 + 13x + 8) \div (2x + 1)$

$$\begin{array}{r} 4x^3 - 4x^2 + 3x + 5 \\ 2x + 1 \overline{)8x^4 - 4x^3 + 2x^2 + 13x + 8} \\ \underline{8x^4 + 4x^3} \\ -8x^3 + 2x^2 \\ \underline{-8x^3 - 4x^2} \\ 6x^2 + 13x \\ \underline{6x^2 + 3x} \\ 10x + 8 \\ \underline{10x + 5} \\ 3 \end{array}$$

Answer: $4x^3 - 4x^2 + 3x + 5 + \dfrac{3}{2x + 1}$

11. $12x^2 - 7x - 10 = (4x - 5)(3x + 2)$

12. $2y^4 + 5y^2 - 3$

Let $y^2 = p$. Then $(y^2)^2$ or $y^4 = p^2$.

$$\begin{aligned} 2y^4 + 5y^2 - 3 &= 2p^2 + 5p - 3 \\ &= (2p - 1)(p + 3) \end{aligned}$$

Now substitute y^2 for p.

$$= (2y^2 - 1)(y^2 + 3)$$

13. $$\begin{aligned} z^4 - 1 &= (z^2 + 1)(z^2 - 1) \\ &= (z^2 + 1)(z + 1)(z - 1) \end{aligned}$$

14. $$\begin{aligned} a^3 - 27b^3 &= a^3 - (3b)^3 \\ &= (a - 3b)(a^2 + 3ab + 9b^2) \end{aligned}$$

15. $$\begin{aligned} \frac{5x - 15}{24} \cdot \frac{64}{3x - 9} &= \frac{5(x-3)}{24} \cdot \frac{64}{3(x-3)} \\ &= \frac{40}{9} \end{aligned}$$

16. $\dfrac{y^2 - 4}{y^2 - y - 6} \div \dfrac{y^2 - 2y}{y - 1}$

Multiply by the reciprocal.

$$= \frac{y^2 - 4}{y^2 - y - 6} \cdot \frac{y - 1}{y^2 - 2y}$$

Factor and simplify.

$$= \frac{(y + 2)(y - 2)}{(y - 3)(y + 2)} \cdot \frac{(y - 1)}{y(y - 2)}$$

$$= \frac{y - 1}{y(y - 3)}$$

17. $\dfrac{5}{c+5} - \dfrac{2}{c+3}$

The LCD is $(c+5)(c+3)$.

$$= \frac{5(c+3)}{(c+5)(c+3)} - \frac{2(c+5)}{(c+3)(c+5)}$$
$$= \frac{5c+15-2c-10}{(c+5)(c+3)}$$
$$= \frac{3c+5}{(c+5)(c+3)}$$

18. $$\frac{p}{p^2+p} + \frac{1}{p^2+p} = \frac{p+1}{p^2+p}$$
$$= \frac{p+1}{p(p+1)} = \frac{1}{p}$$

19. Let $x =$ the time to do the job working together.

Make a table.

Worker	Rate	Time together	Part of job done
Kareem	$\frac{1}{3}$	x	$\frac{x}{3}$
Jamal	$\frac{1}{2}$	x	$\frac{x}{2}$

Part done by Kareem	plus	part done by Jamal	equals	1 whole job.
$\frac{x}{3}$	$+$	$\frac{x}{2}$	$=$	1

Multiply by the LCD, 6.

$$6\left(\frac{x}{3}+\frac{x}{2}\right) = 6\cdot 1$$
$$2x+3x = 6$$
$$5x = 6$$
$$x = \frac{6}{5} \text{ or } 1\frac{1}{5}$$

It takes $\frac{6}{5}$ or $1\frac{1}{5}$ hr to do the job together.

20. $$\left(\frac{4}{3}\right)^{-1} = \frac{1}{\frac{4}{3}} = 1\cdot\frac{3}{4} = \frac{3}{4}$$

21. $$\frac{(2a)^{-2}a^4}{a^{-3}} = \frac{2^{-2}a^{-2}a^4}{a^{-3}} = \frac{2^{-2}a^2}{a^{-3}}$$
$$= \frac{a^2a^3}{2^2} = \frac{a^5}{4}$$

22. $$4\sqrt[3]{16} - 2\sqrt[3]{54} = 4\sqrt[3]{8\cdot 2} - 2\sqrt[3]{27\cdot 2}$$
$$= 4\cdot 2\sqrt[3]{2} - 2\cdot 3\sqrt[3]{2}$$
$$= 8\sqrt[3]{2} - 6\sqrt[3]{2} = 2\sqrt[3]{2}$$

23. $$\frac{3\sqrt{5x}}{\sqrt{2x}} = \frac{3\sqrt{5x}\cdot\sqrt{2x}}{\sqrt{2x}\cdot\sqrt{2x}} = \frac{3\sqrt{10x^2}}{2x}$$
$$= \frac{3x\sqrt{10}}{2x} = \frac{3\sqrt{10}}{2}$$

24. $\dfrac{5+3i}{2-i}$

Multiply by the conjugate of the denominator.

$$= \frac{(5+3i)(2+i)}{(2-i)(2+i)}$$
$$= \frac{10+5i+6i+3i^2}{4-i^2}$$
$$= \frac{10+11i+3(-1)}{4-(-1)}$$
$$= \frac{7+11i}{5} = \frac{7}{5} + \frac{11}{5}i$$

25. $2\sqrt{k} = \sqrt{5k+3}$

Square both sides.

$$4k = 5k+3$$
$$-k = 3$$
$$k = -3$$

Since k must be nonnegative so that $\sqrt{k}$ is a real number, -3 cannot be a solution.

Solution set: $\emptyset$

26. $$10q^2 + 13q = 3$$
$$10q^2 + 13q - 3 = 0$$
$$(5q-1)(2q+3) = 0$$
$$5q-1 = 0 \quad \text{or} \quad 2q+3 = 0$$
$$5q = 1 \qquad\qquad 2q = -3$$
$$q = \frac{1}{5} \quad \text{or} \quad q = -\frac{3}{2}$$

Solution set: $\left\{\frac{1}{5}, -\frac{3}{2}\right\}$

27. $$(4x-1)^2 = 8$$
$$16x^2 - 8x + 1 = 8$$
$$16x^2 - 8x - 7 = 0$$

Use the quadratic formula with $a = 16, b = -8$, and $c = -7$.

$$x = \frac{-b \pm \sqrt{b^2 - 4ac}}{2a}$$
$$x = \frac{-(-8) \pm \sqrt{(-8)^2 - 4(16)(-7)}}{2(16)}$$
$$= \frac{8 \pm \sqrt{64 + 448}}{32}$$
$$= \frac{8 \pm \sqrt{512}}{32}$$
$$= \frac{8 \pm \sqrt{256 \cdot 2}}{32}$$
$$= \frac{8 \pm 16\sqrt{2}}{32}$$
$$x = \frac{8(1 \pm 2\sqrt{2})}{32} = \frac{1 \pm 2\sqrt{2}}{4}$$

Solution set: $\left\{\frac{1+2\sqrt{2}}{4}, \frac{1-2\sqrt{2}}{4}\right\}$

28. $3k^2 - 3k - 2 = 0$

Let $a = 3, b = -3$, and $c = -2$ in the quadratic formula.

$$k = \frac{-b \pm \sqrt{b^2 - 4ac}}{2a}$$
$$k = \frac{-(-3) \pm \sqrt{(-3)^2 - 4(3)(-2)}}{2(3)}$$
$$k = \frac{3 \pm \sqrt{9 + 24}}{6}$$
$$k = \frac{3 \pm \sqrt{33}}{6}$$

Solution set: $\left\{\frac{3+\sqrt{33}}{6}, \frac{3-\sqrt{33}}{6}\right\}$

29. $2(x^2 - 3)^2 - 5(x^2 - 3) = 12$

Let $u = x^2 - 3$.

$$2u^2 - 5u = 12$$
$$2u^2 - 5u - 12 = 0$$
$$(2u + 3)(u - 4) = 0$$
$$2u + 3 = 0 \quad \text{or} \quad u - 4 = 0$$
$$2u = -3$$
$$u = -\frac{3}{2} \quad \text{or} \quad u = 4$$

Substitute $x^2 - 3$ for u to find x.

If $u = -\frac{3}{2}$, then

$$x^2 - 3 = -\frac{3}{2}$$
$$x^2 = \frac{6}{2} - \frac{3}{2}$$
$$x^2 = \frac{3}{2}$$
$$x = \pm\sqrt{\frac{3}{2}}$$
$$x = \pm\frac{\sqrt{3}}{\sqrt{2}} \cdot \frac{\sqrt{2}}{\sqrt{2}} = \pm\frac{\sqrt{6}}{2}.$$

If $u = 4$, then

$$x^2 - 3 = 4$$
$$x^2 = 7$$
$$x = \pm\sqrt{7}.$$

Solution set: $\left\{-\frac{\sqrt{6}}{2}, \frac{\sqrt{6}}{2}, -\sqrt{7}, \sqrt{7}\right\}$

30. Solve $F = \frac{kwv^2}{r}$ for v.

$$F = \frac{kwv^2}{r}$$
$$F \cdot r = \frac{kwv^2}{r} \cdot r$$
$$Fr = kwv^2$$
$$\frac{Fr}{kw} = \frac{kwv^2}{kw}$$
$$\frac{Fr}{kw} = v^2$$

Take the square root of each side.

$$\pm\sqrt{\frac{Fr}{kw}} = v$$
$$\frac{\pm\sqrt{Fr}}{\sqrt{kw}} \cdot \frac{\sqrt{kw}}{\sqrt{kw}} = v$$
$$\frac{\pm\sqrt{Frkw}}{kw} = v$$

For Exercises 31-35, see the answer graphs in the back of the textbook.

31. $f(x) = -3x + 5$

Replace $f(x)$ with y to get the equation

$$y = -3x + 5.$$

The y-intercept is $(0, 5)$ and $m = -3$ or $\frac{-3}{1}$. Plot $(0, 5)$. From $(0, 5)$, move down 3 units and right 1 unit. Draw the line through these two points.

32. $f(x) = -2(x-1)^2 + 3$

The graph is a parabola that has been shifted 1 unit to the right and 3 units upward, so its vertex is at $(1,3)$. Since $a = -2 < 0$, the parabola opens downward. Also $|a| = |-2| = 2 > 1$, so the graph is narrower than the graph of $f(x) = x^2$. The points $(0,1)$ and $(2,1)$ are on the graph.

33. $f(x) = \sqrt{x-2}$

Replace $f(x)$ with y.

$$y = \sqrt{x-2}$$

Square both sides.

$$\begin{aligned} y^2 &= x - 2 \\ x &= y^2 + 2 \\ x &= (y-0)^2 + 2 \end{aligned}$$

The graph is a horizontal parabola with vertex at $(2,0)$. Since $a = 1 > 0$, the parabola opens to the right. The points $(3,1)$ and $(6,2)$ are on the graph. Since y represents a principal square root, y must be nonnegative and its graph is the top half of the parabola.

34. $\dfrac{x^2}{4} - \dfrac{y^2}{16} = 1$

$a = 2, b = 4$

The graph is a hyperbola with x-intercepts $(2,0)$ and $(-2,0)$ and asymptotes that are the extended diagonals of the rectangle with corners $(2,4), (2,-4), (-2,-4)$, and $(-2,4)$. Draw a branch of the hyperbola through each intercept and approaching the asymptotes.

35. $\dfrac{x^2}{25} + \dfrac{y^2}{16} \le 1$

The boundary, $\frac{x^2}{25} + \frac{y^2}{16} = 1$, is a solid ellipse with $a = 5$ and $b = 4$. The intercepts are $(5,0)$, $(-5,0)$, $(0,4)$, and $(0,-4)$. Test $(0,0)$.

$$\begin{aligned} \frac{0^2}{25} + \frac{0^2}{16} &\le 1 \\ 0 &\le 1 \quad \textit{True} \end{aligned}$$

Shade the region inside the ellipse.

36. $\begin{aligned} 3x - y &= 12 \quad (1) \\ 2x + 3y &= -3 \quad (2) \end{aligned}$

Multiply equation (1) by 3 and add the result to equation (2).

$$\begin{aligned} 9x - 3y &= 36 \\ 2x + 3y &= -3 \quad (2) \\ \hline 11x \qquad &= 33 \\ x &= 3 \end{aligned}$$

Substitute 3 for x in equation (1) to find y.

$$\begin{aligned} 3x - y &= 12 \quad (1) \\ 3(3) - y &= 12 \\ 9 - y &= 12 \\ -y &= 3 \\ y &= -3 \end{aligned}$$

Solution set: $\{(3,-3)\}$

37. $\begin{aligned} x + y - 2z &= 9 \quad (1) \\ 2x + y + z &= 7 \quad (2) \\ 3x - y - z &= 13 \quad (3) \end{aligned}$

Add equation (2) and equation (3).

$$\begin{aligned} 2x + y + z &= 7 \quad (2) \\ 3x - y - z &= 13 \quad (3) \\ \hline 5x \qquad &= 20 \\ x &= 4 \end{aligned}$$

Multiply equation (1) by -1 and add the result to equation (2).

$$\begin{aligned} -x - y + 2z &= -9 \\ 2x + y + z &= 7 \quad (2) \\ \hline x \qquad + 3z &= -2 \quad (4) \end{aligned}$$

Substitute 4 for x in equation (4) to find z.

$$\begin{aligned} x + 3z &= -2 \quad (4) \\ 4 + 3z &= -2 \\ 3z &= -6 \\ z &= -2 \end{aligned}$$

Substitute 4 for x and -2 for z in equation (2) to find y.

$$\begin{aligned} 2x + y + z &= 7 \quad (2) \\ 2(4) + y - 2 &= 7 \\ 8 + y - 2 &= 7 \\ y + 6 &= 7 \\ y &= 1 \end{aligned}$$

Solution set: $\{(4,1,-2)\}$

38. $$xy = -5 \quad (1)$$
$$2x + y = 3 \quad (2)$$

Solve equation (2) for y.

$$y = -2x + 3$$

Substitute $-2x + 3$ for y in equation (1).

$$\begin{aligned} xy &= -5 \quad (1) \\ x(-2x+3) &= -5 \\ -2x^2 + 3x &= -5 \\ -2x^2 + 3x + 5 &= 0 \\ 2x^2 - 3x - 5 &= 0 \\ (2x-5)(x+1) &= 0 \end{aligned}$$

$$2x - 5 = 0 \quad \text{or} \quad x + 1 = 0$$
$$2x = 5$$
$$x = \frac{5}{2} \quad \text{or} \quad x = -1$$

Substitute these values for x in equation (1) to find y.

If $x = \frac{5}{2}$, then

$$\begin{aligned} xy &= -5 \quad (1) \\ \frac{5}{2}y &= -5 \\ y &= -2. \end{aligned}$$

If $x = -1$, then

$$\begin{aligned} xy &= -5 \quad (1) \\ -1 \cdot y &= -5 \\ y &= 5. \end{aligned}$$

Solution set: $\{(-1,5), (\frac{5}{2}, -2)\}$

39. Let $s =$ Al's speed,
$2s =$ Bev's speed,
$t =$ Bev's time, and
$t + \frac{1}{2} =$ Al's time.

	Distance	Rate	Time
Al	20	s	$t + \frac{1}{2}$
Bev	20	$2s$	t

Since $d = rt$, the system of equations is

$$s\left(t + \frac{1}{2}\right) = 20 \quad (1)$$
$$2st = 20. \quad (2)$$

Solve equation (2) for s.

$$\begin{aligned} 2st &= 20 \quad (2) \\ s &= \frac{20}{2t} \\ s &= \frac{10}{t} \end{aligned}$$

Substitute $\frac{10}{t}$ for s in equation (1) to find t.

$$\begin{aligned} s\left(t + \frac{1}{2}\right) &= 20 \quad (1) \\ \frac{10}{t}\left(t + \frac{1}{2}\right) &= 20 \\ 10 + \frac{5}{t} &= 20 \end{aligned}$$

Multiply by t.

$$\begin{aligned} 10t + 5 &= 20t \\ 5 &= 10t \\ \frac{5}{10} &= t \\ \frac{1}{2} &= t \end{aligned}$$

Since $s = \frac{10}{t}$ and $t = \frac{1}{2}$,

$$s = \frac{10}{\frac{1}{2}} = 20.$$

So, Al's speed was 20 mph. Then Bev's speed was $2 \cdot 20$ or 40 mph.

40. Let $t =$ the number of twenties and
$t + 8 =$ the number of fives.

$$\begin{aligned} 5(t+8) + 20t &= 215 \\ 5t + 40 + 20t &= 215 \\ 25t + 40 &= 215 \\ 25t &= 175 \\ t &= 7 \\ t + 8 = 7 + 8 &= 15 \end{aligned}$$

There were 15 fives and 7 twenties.

Chapter 10

EXPONENTIAL AND LOGARITHMIC FUNCTIONS

10.1 Inverse Functions

10.1 Margin Exercises

1. **(a)** $\{(1,2),(2,4),(3,3),(4,5)\}$

   Every x-value in the function corresponds to only one y-value, and every y-value corresponds to only one x-value, so this is a one-to-one function. Reverse the order of each ordered pair to get the inverse. The inverse is

   $$\{(2,1),(4,2),(3,3),(5,4)\}$$

   (b) $\{(0,3),(-1,2),(1,3)\}$

   Since the y-value 3 corresponds to x-values 0 and 1, the function is not one-to-one. Therefore, it does not have an inverse.

   (c)

Height	Miles
5000	87
15,000	149
20,000	172
35,000	228

   Every value in the height column corresponds to only one value in the miles column, and every value in the miles column corresponds to only one value in the height column, so this is a one-to-one function. Reverse the table columns to obtain the inverse.

Height	Miles
87	5000
149	15,000
172	20,000
228	35,000

2. **(a)** Since a horizontal line will intersect the graph in no more than one point, the function is one-to-one.

   (b) Since a horizontal line will intersect the graph in two points, the function is not one-to-one.

3. To find the equation of the inverse of a function, (1) determine if the function is one-to-one; (2) replace $f(x)$ with y; (3) exchange x and y; (4) solve for y; (5) replace y with $f^{-1}(x)$.

(a) The graph of $f(x) = 3x - 4$ is a line. By the horizontal line test, it is a one-to-one function. To find the inverse, replace $f(x)$ with y.

$$y = 3x - 4$$

Exchange x and y.

$$x = 3y - 4$$

Solve for y.

$$x + 4 = 3y$$
$$\frac{x+4}{3} = y$$

Replace y with $f^{-1}(x)$.

$$f^{-1}(x) = \frac{x+4}{3}$$

(b) $f(x) = x^3 + 1$ is one-to-one because of the exponent, which produces one value of $f(x)$ for each value of x. To find the inverse, replace $f(x)$ with y.

$$y = x^3 + 1$$

Exchange x and y.

$$x = y^3 + 1$$

Solve for y.

$$x - 1 = y^3$$

Take the cube root of each side.

$$\sqrt[3]{x-1} = y$$

Replace y with $f^{-1}(x)$.

$$f^{-1}(x) = \sqrt[3]{x-1}$$

(c) $f(x) = (x-3)^2$ is not a one-to-one function because two x-values, such as 1 and 5, give the same function value, in this case 4. Thus, the function does not have an inverse.

4. $f(x) = \sqrt[3]{1-x}$

(a) To find $f(2)$, substitute 2 for x.

$$f(2) = \sqrt[3]{1-2} = \sqrt[3]{-1} = -1$$

(b) $f(9) = \sqrt[3]{1-9} = \sqrt[3]{-8} = -2$

(c) To find $f^{-1}(-1)$, first find $f^{-1}(x)$. Replace $f(x)$ with y.

$$y = \sqrt[3]{1-x}$$

Exchange x and y.

$$x = \sqrt[3]{1-y}$$

Solve for y; cube each side.

$$x^3 = 1-y$$
$$y = 1-x^3$$

Replace y with $f^{-1}(x)$.

$$f^{-1}(x) = 1-x^3$$

Now replace x with -1.

$$\begin{aligned} f^{-1}(-1) &= 1-(-1)^3 \\ &= 1-(-1) \\ &= 2 \end{aligned}$$

Alternately, from part (a),

$$f(2) = -1 \text{ so } f^{-1}(-1) = 2.$$

(d) Since $f(9) = -2$ from part (b), $f^{-1}(-2) = 9$.

5. See the answer graphs for the margin exercises in the textbook.

(a) The given graph goes through $(-5,1)$ and $(1,-1)$. Then the inverse will go through $(1,-5)$ and $(-1,1)$. Use these points and symmetry about $y = x$ to complete the graph of the inverse.

(b) Points on the given graph are $(-3,0), (-2,2)$, and $(0,3)$. Then the inverse will contain $(0,-3)$, $(2,-2)$, and $(3,0)$. Use these points and symmetry about $y = x$ to complete the graph of the inverse.

(c) If the given graph goes through $\left(-1,\frac{1}{2}\right), (0,1)$, $(1,2)$, and $(2,4)$, then the inverse will go through $\left(\frac{1}{2},-1\right), (1,0), (2,1)$, and $(4,2)$. Use these points and symmetry about $y = x$ to complete the graph of the inverse.

10.1 Section Exercises

3. f is a one-to-one function with the point (a,b) on its graph. To form the inverse, the x- and y-coordinates are interchanged. Then (b,a) must be on the graph of the inverse of f. The correct answer is (a).

7. $\{(3,6),(2,10),(5,12)\}$ is a one-to-one function, since each x-value corresponds to only one y-value and each y-value corresponds to only one x-value. To find the inverse, interchange x and y in each ordered pair. The inverse is

$$\{(6,3),(10,2),(12,5)\}$$

11. The graph of $f(x) = 2x+4$ is a line, so by the horizontal line test $f(x)$ is a one-to-one function. To find the inverse, replace $f(x)$ with y.

$$y = 2x+4$$

Interchange x and y.

$$x = 2y+4$$

Solve for y.

$$x-4 = 2y$$
$$\frac{x-4}{2} = y$$

Replace y with $f^{-1}(x)$.

$$f^{-1}(x) = \frac{x-4}{2}$$

15. $f(x) = 3x^2+2$ is not a one-to-one function because two x-values, such as 1 and -1, both have the same y-value, in this case 5. The graph of this function is a vertical parabola which does not pass the horizontal line test.

19. The graph of $f(x) = mx+b$, a linear equation, passes the horizontal line test (as long as $m \neq 0$) so it is one-to-one and has an inverse.

23. **(a)** $f(x) = 2^x$

To find $f(0)$, substitute 0 for x.

$$f(0) = 2^0 = 1$$

(b) From part (a), we know that $(0,1)$ belongs to f. Since f is a one-to-one function, it follows that $(1,0)$ belongs to f^{-1}. Therefore, $f^{-1}(1) = 0$.

27. **(a)** This graph does not pass the horizontal line test, so the function is not one-to-one.

For Exercises 31 and 35, see the answer graphs in the back of the textbook.

31. $f(x) = 2x - 1$
$y = 2x - 1$

The graph is a line through $(-2, -5)$, $(0, -1)$, and $(3, 5)$. Plot these points, and draw the line through them. Then the inverse will be a line through $(-5, -2)$, $(-1, 0)$, and $(5, 3)$. Plot these points, and draw the dashed line through them.

35. $f(x) = \sqrt{x}, x \geq 0$

Complete the table of values.

x	$f(x)$
0	0
1	1
4	2

Plot these points, and connect them with a smooth curve. Since $f(x)$ is one-to-one, make a table of values for $f^{-1}(x)$ by interchanging x and y.

x	$f^{-1}(x)$
0	0
1	1
2	4

Plot these points, and connect them with a dashed smooth curve.

39. The graph of $y_1 = x^3 + 5$ passes the horizontal line test, so it is a one-to-one function and has an inverse y_2.

$$y_1 = x^3 + 5$$

Exchange x and y_1.

$$x = {y_1}^3 + 5$$
$$x - 5 = {y_1}^3$$

Take the cube root of each side.

$$\sqrt[3]{x - 5} = y_1$$

Replace y_1 with y_2.

$$y_2 = \sqrt[3]{x - 5}$$

10.2 Exponential Functions

10.2 Margin Exercises

In Exercises 1 and 2, choose values of x and find the corresponding values of y. Then plot the points, and draw a smooth curve through them. See the answer graphs for the margin exercises in the textbook.

1. **(a)** $f(x) = 10^x$

x	-2	-1	0	1	2
y	.01	.1	1	10	100

(b) $g(x) = \left(\frac{1}{4}\right)^x$

x	-2	-1	0	1	2
y	16	4	1	$\frac{1}{4}$	$\frac{1}{16}$

2. $y = 2^{4x-3}$

x	$4x - 3$	y
0	-3	$\frac{1}{8}$
$\frac{3}{4}$	0	1
1	1	2
-1	-7	$\frac{1}{128}$

3. **(a)** $25^x = 125$

Rewrite each side with the same base.

$$(5^2)^x = 5^3$$
$$5^{2x} = 5^3$$

If $a^x = a^y$, then $x = y$. Set the exponents equal.

$$2x = 3$$
$$x = \frac{3}{2}$$

Check $\frac{3}{2}$ in the original equation.

$$25^{3/2} = 125 \quad ?$$
$$(25^{1/2})^3 = 125 \quad ?$$
$$5^3 = 125 \quad ?$$
$$125 = 125 \quad \textit{True}$$

Solution set: $\{\frac{3}{2}\}$

(b) $4^x = 32$

Write each side with the same base.

$$(2^2)^x = 2^5$$
$$2^{2x} = 2^5$$

Set the exponents equal.

$$2x = 5$$
$$x = \frac{5}{2}$$

Check.

$$4^{5/2} = 32 \quad ?$$
$$(4^{1/2})^5 = 32 \quad ?$$
$$2^5 = 32 \quad ?$$
$$32 = 32 \qquad \textit{True}$$

Solution set: $\{\frac{5}{2}\}$

(c) $81^p = 27$

Write each side with the same base.

$$(3^4)^p = 3^3$$
$$3^{4p} = 3^3$$

Set the exponents equal.

$$4p = 3$$
$$p = \frac{3}{4}$$

Check.

$$81^{3/4} = 27 \quad ?$$
$$(81^{1/4})^3 = 27 \quad ?$$
$$3^3 = 27 \quad ?$$
$$27 = 27 \qquad \textit{True}$$

Solution set: $\{\frac{3}{4}\}$

4. $A(t) = 100(3)^{-.5t}$

(a) Let $t = 0$, and find $A(0)$.

$$A(0) = 100(3)^{-.5(0)} = 100(3)^0 = 100$$

The amount is 100 g.

(b) Let $t = 2$, and find $A(2)$.

$$A(2) = 100(3)^{-.5(2)} = 100(3)^{-1} = 100\left(\frac{1}{3}\right) = 33\frac{1}{3}$$

The amount is $33\frac{1}{3}$ g.

(c) Let $t = 10$, and find $A(10)$.

$$A(10) = 100(3)^{-.5(10)} = 100(3)^{-5} = \frac{100}{3^5} \approx 41$$

The amount is about .41 g.

(d) To graph $A(t) = 100(3)^{-.5t}$, use the values for t and $A(t)$ from parts (a)-(c).

t	0	2	10
$A(t)$	100	$33\frac{1}{3}$	.41

Plot the points, and draw a smooth curve through them. See the answer graph for the margin exercises in the textbook.

10.2 Section Exercises

3. The graph of the exponential function $y = a^x$ *does not* have an x-intercept because the graph will approach the x-axis, but never touch it. To have an x-intercept, y must be zero. If a is not zero, then a^x can never be zero.

7. $g(x) = \left(\frac{1}{3}\right)^x$

Make a table of values.

x	-2	-1	0	1	2
y	9	3	1	$\frac{1}{3}$	$\frac{1}{9}$

Plot points from the table, and draw a smooth curve through them. See the answer graph in the back of the textbook.

11. $f(x) = a^x$ is one-to-one and thus has an inverse. Each value of $f(x)$ corresponds to one and only one value of x.

In Exercises 15-23, check each solution in the original equation.

15. $100^x = 1000$

Write each side with the same base.

$$(10^2)^x = 10^3$$
$$10^{2x} = 10^3$$

If $a^x = a^y$, then $x = y$. Set the exponents equal.

$$2x = 3$$
$$x = \frac{3}{2}$$

Solution set: $\{\frac{3}{2}\}$

19. $5^x = \dfrac{1}{125}$

$5^x = \left(\dfrac{1}{5}\right)^3$

Write each side with the same base.

$$5^x = 5^{-3}$$

Set the exponents equal.

$$x = -3$$

Solution set: $\{-3\}$

23. $\left(\dfrac{3}{2}\right)^x = \dfrac{8}{27}$

$\left(\dfrac{3}{2}\right)^x = \left(\dfrac{2}{3}\right)^3$

Write each side with the same base.

$$\left(\frac{3}{2}\right)^x = \left(\frac{3}{2}\right)^{-3}$$

Set the exponents equal.

$$x = -3$$

Solution set: $\{-3\}$

27. $S(x) = 74,741(1.17)^x$

Since the year 1976 corresponds to $x = 0$, the year 1986 corresponds to $x = 10$ $(1986 - 1976 = 10)$.

$$S(10) = 74,741(1.17)^{10} \approx 360,000$$

The average salary in 1986 was about $360,000.

10.3 Logarithmic Functions

10.3 Margin Exercises

1.

Exponential Form	Logarithmic Form
$2^5 = 32$	$\log_2 32 = 5$
$100^{1/2} = 10$	$\log_{100} 10 = \dfrac{1}{2}$
$8^{2/3} = 4$	$\log_8 4 = \dfrac{2}{3}$
$6^{-4} = \dfrac{1}{1296}$	$\log_6 \dfrac{1}{1296} = -4$

2. **(a)** $\log_3 27 = x$

Change to exponential form.

$$3^x = 27$$

Write with the same base.

$$3^x = 3^3$$

Set the exponents equal.

$$x = 3$$

Check this solution.

$$\log_3 27 = 3 \quad ?$$
$$3^3 = 27 \quad ?$$
$$27 = 27 \qquad \textit{True}$$

Solution set: $\{3\}$

(b) $\log_5 p = 2$

Change to exponential form.

$$5^2 = p$$
$$25 = p$$

The solution checks.

Solution set: $\{25\}$

(c) $\log_m \dfrac{1}{16} = -4$

Change to exponential form.

$$m^{-4} = \frac{1}{16}$$
$$m^{-4} = \frac{1}{2^4}$$
$$m^{-4} = 2^{-4}$$

Set the bases equal.

$$m = 2$$

This solution checks.

Solution set: $\{2\}$

3. **(a)** $\log_{2/5} \frac{2}{5} = 1$ since $\left(\frac{2}{5}\right)^1 = \frac{2}{5}$.

(b) $\log_{.4} 1 = 0$ since $(.4)^0 = 1$.

4. See the answer graphs for the margin exercises in the textbook.

(a) $y = \log_3 x$ is equivalent to $x = 3^y$. Choose values for y and find x.

x	$\frac{1}{9}$	$\frac{1}{3}$	1	3	9
y	-2	-1	0	1	2

Plot the points, and draw a smooth curve through them.

(b) $y = \log_{1/10} x$ is equivalent to $x = \left(\frac{1}{10}\right)^y$. Choose values for y and find x.

x	10	1	$\frac{1}{10}$	$\frac{1}{100}$
y	-1	0	1	2

Plot the points, and draw a smooth curve through them.

5. $P(t) = 80\log_{10}(t+10)$

(a) Let $t = 0$, and find $P(0)$.

$$\begin{aligned} P(0) &= 80\log_{10}(0+10) \\ &= 80\log_{10} 10 \end{aligned}$$

Note that $\log_{10} 10 = 1$.

$$\begin{aligned} &= 80(1) \\ P(0) &= 80 \end{aligned}$$

The population after 0 days is 80 mites.

(b) Let $t = 90$, and find $P(90)$.

$$\begin{aligned} P(90) &= 80\log_{10}(90+10) \\ &= 80\log_{10} 100 \end{aligned}$$

Note that $\log_{10} 100 = 2$.

$$\begin{aligned} &= 80(2) \\ P(90) &= 160 \end{aligned}$$

The population after 90 days is 160 mites.

(c) Let $t = 990$, and find $P(990)$.

$$\begin{aligned} P(990) &= 80\log_{10}(990+10) \\ &= 80\log_{10} 1000 \end{aligned}$$

Note that $\log_{10} 1000 = 3$.

$$\begin{aligned} &= 80(3) \\ P(990) &= 240 \end{aligned}$$

The population after 990 days is 240 mites.

(d) To graph $P(t) = 80\log_{10}(t+10)$, use the values for t and $P(t)$ from parts (a)-(c).

t	0	90	990
$P(t)$	80	160	240

Plot the points, and draw a smooth curve through them. See the answer graph for the margin exercises in the textbook.

10.3 Section Exercises

3. The base is 4, the exponent (logarithm) is 5, and the number is 1024, so $4^5 = 1024$ becomes $\log_4 1024 = 5$ in logarithmic form.

7. The base is 10, the exponent (logarithm) is -3, and the number is .001, so $10^{-3} = .001$ becomes $\log_{10} .001 = -3$ in logarithmic form.

11. The base is 10, the logarithm (exponent) is -4, and the number is $\frac{1}{10{,}000}$, so $\log_{10} \frac{1}{10{,}000} = -4$ becomes $10^{-4} = \frac{1}{10{,}000}$ in exponential form.

15. The value of the expression $\log_9 3$ is the power of 9 that yields 3. The student must determine how to obtain 3 from 9 with a radical. Since $\sqrt{9} = 3$ or $9^{1/2} = 3$, $\log_9 3 = \frac{1}{2}$.

In Exercises 19-31, check each solution in the original equation.

19. $\log_x 9 = \dfrac{1}{2}$

Change to exponential form.

$$x^{1/2} = 9$$

Square both sides.

$$\begin{aligned} (x^{1/2})^2 &= 9^2 \\ x^1 &= 81 \\ x &= 81 \end{aligned}$$

Solution set: $\{81\}$

23. $\log_{12} x = 0$

Change to exponential form.

$$\begin{aligned} 12^0 &= x \\ 1 &= x \end{aligned}$$

Solution set: $\{1\}$

27. $\log_x \frac{1}{25} = -2$

Change to exponential form.

$$x^{-2} = \frac{1}{25}$$

$$x^{-2} = \left(\frac{1}{5}\right)^2$$

Write each side with the same exponent.

$$x^{-2} = 5^{-2}$$

Set the bases equal.

$$x = 5$$

Solution set: $\{5\}$

31. $\log_\pi \pi^4 = x$

Change to exponential form.

$$\pi^x = \pi^4$$

Set the exponents equal.

$$x = 4$$

Solution set: $\{4\}$

For Exercises 35 and 39, see the answer graphs in the back of the textbook.

35. $y = \log_3 x$

Change to exponential form.

$$3^y = x$$

Make a table of values.

x	$\frac{1}{9}$	$\frac{1}{3}$	1	3	9
y	-2	-1	0	1	2

(Note that the values in this table are the reverse of those in the table for Exercise 5 in Section 10.2.) Plot the points, and draw a smooth curve through them.

39. $y = \log_{2.718} x$

Change to exponential form.

$$2.718^y = x$$

Make a table of values.

x	.4	1	2.7
y	-1	0	1

Plot the points, and draw a smooth curve through them.

43. The values of t are on the horizontal axis, and the values of $f(t)$ are on the vertical axis. Read the value of $f(t)$ from the graph for the given value of t. At $t = 0, f(0) = 8$.

47. The number 1 is not used as a base for a logarithmic function since the function would look like $x = 1^y$ in exponential form. Then, for any real value of y, the statement $1 = 1$ would always be the result.

51. $R = \log_{10} \frac{x}{x_0}$

Change to exponential form.

$$10^R = \frac{x}{x_0} \text{ or } x_0 10^R = x$$

Let $R = 6.7$ for the Northridge earthquake, x_1.

$$x_0 10^{6.7} = x_1$$
$$5{,}000{,}000 x_0 \approx x_1$$

Let $R = 7.3$ for the Landers earthquake, x_2.

$$x_0 10^{7.3} = x_2$$
$$20{,}000{,}000 x_0 \approx x_2$$

The ratio of x_2 to x_1, is

$$\frac{20{,}000{,}000 x_0}{5{,}000{,}000 x_0} = 4.$$

The Landers quake was about 4 times more powerful.

10.4 Properties of Logarithms

10.4 Margin Exercises

1. Use the multiplication property for logarithms.

$$\log_b xy = \log_b x + \log_b y$$

(a) $\log_6(5 \cdot 8) = \log_6 5 + \log_6 8$

(b) $\log_4 3 + \log_4 7 = \log_4(3 \cdot 7)$
$= \log_4 21$

(c) $\log_8 8k = \log_8 8 + \log_8 k$
$= 1 + \log_8 k \ (k > 0)$

(d) $\log_5 m^2 = \log_5(m \cdot m)$
$= \log_5 m + \log_5 m$
$= 2 \log_5 m \ (m > 0)$

2. Use the division property for logarithms.

$$\log_b \frac{x}{y} = \log_b x - \log_b y$$

(a) $\log_7 \frac{9}{4} = \log_7 9 - \log_7 4$

(b) $\log_3 p - \log_3 q = \log_3 \frac{p}{q}$

(c) $\log_4 \frac{3}{16} = \log_4 3 - \log_4 16$
$= \log_4 3 - 2$

3. Use the power property for logarithms.

$$\log_b x^r = r \log_b x$$

(a) $\log_3 5^2 = 2 \log_3 5$

(b) $\log_a x^4 = 4 \log_a x \quad (x > 0)$

(c) $\log_b \sqrt{8} = \log_b 8^{1/2}$
$= \frac{1}{2} \log_b 8$

(d) $\log_2 \sqrt[3]{2} = \log_2 2^{1/3}$
$= \frac{1}{3} \log_2 2$
$= \frac{1}{3}(1) = \frac{1}{3}$

4. Use the property

$$\log_b b^x = x \text{ or } b^{\log_b x} = x.$$

(a) By the first property,

$$\log_{10} 10^3 = 3.$$

(b) By the first property,

$$\log_2 8 = \log_2 2^3 = 3.$$

(c) By the second property,

$$5^{\log_5 3} = 3.$$

5. Use the properties of logarithms.

(a) $\log_6 36m^5$
$= \log_6(6^2 m^5)$

Use the multiplication property.

$= \log_6 6^2 + \log_6 m^5$

Note that $\log_6 6^2 = 2$; use the power property.

$= 2 + 5 \log_6 m$

(b) $\log_2 \sqrt{9z}$
$= \log_2 3\sqrt{z}$
$= \log_2 3z^{1/2}$

Use the multiplication property.

$= \log_2 3 + \log_2 z^{1/2}$

Use the power property.

$= \log_2 3 + \frac{1}{2} \log_2 z$

(c) $\log_q \frac{8r}{m-1} \quad (m \neq 1, q \neq 1)$

Use the division property.

$= \log_q(8r) - \log_q(m-1)$

Use the multiplication property.

$= \log_q 8 + \log_q r - \log_q(m-1)$

(d) $\log_4(3x + y)$ cannot be written as a sum of logarithms since

$$\log_b(M + N) \neq \log_b M + \log_b N.$$

There is no property of logarithms to rewrite the logarithm of a sum.

10.4 Section Exercises

3. Use the power property.

$$\log_a b^k = k \log_a b$$
$$(a > 0, a \neq 1, b > 0)$$

7. $\log_2 8^{1/4}$

Write 8 as a power of 2.

$= \log_2(2^3)^{1/4}$

Multiply exponents.

$= \log_2 2^{3/4}$

Use the power property.

$= \frac{3}{4} \log_2 2$

Note that $\log_2 2 = 1$.

$= \frac{3}{4}(1) = \frac{3}{4}$

11. $\log_3 \frac{\sqrt[3]{4}}{x^2y}$

$= \log_3 \frac{4^{1/3}}{x^2y}$

Use the division property.

$= \log_3 4^{1/3} - \log_3(x^2y)$

Use the multiplication property.

$= \log_3 4^{1/3} - (\log_3 x^2 + \log_3 y)$
$= \log_3 4^{1/3} - \log_3 x^2 - \log_3 y$

Use the power property.

$= \frac{1}{3}\log_3 4 - 2\log_3 x - \log_3 y$

15. $\log_2 \frac{\sqrt[3]{x}\cdot\sqrt[5]{y}}{r^2}$

$= \log_2 \frac{x^{1/3}y^{1/5}}{r^2}$

Use the multiplication and division properties.

$= \log_2 x^{1/3} + \log_2 y^{1/5} - \log_2 r^2$

Use the power property.

$= \frac{1}{3}\log_2 x + \frac{1}{5}\log_2 y - 2\log_2 r$

19. By the multiplication property,

$$\log_b x + \log_b y = \log_b xy.$$

23. $(\log_a r - \log_a s) + 3\log_a t$

Use the division and power properties.

$= \log_a \frac{r}{s} + \log_a t^3$

Use the multiplication property.

$= \log_a \frac{rt^3}{s}$

27. By the multiplication property,

$\log_{10}(x+3) + \log_{10}(x-3)$
$= \log_{10}(x+3)(x-3)$
$= \log_{10}(x^2-9).$

31. For the power rule

$$\log_b x^r = r\log_b x$$

to be true, x must be in the domain of $g(x) = \log_b x$, that is, $(0, \infty)$. You cannot take the logarithm of a negative number.

33. Let $\log_3 81 = x$.

Write in exponential form.

$$3^x = 81$$

Write each side with the same base.

$$3^x = 3^4$$

Set the exponents equal.

$$x = 4$$

Therefore,

$$\log_3 81 = 4.$$

34. $\log_3 81$ means the exponent to which 3 must be raised in order to obtain 81.

35. Using the result from Exercise 33,

$$3^{\log_3 81} = 3^4 = 81.$$

36. $\log_2 19$ means the exponent to which 2 must be raised in order to obtain 19.

37. Keeping in mind the result from Exercise 35,

$$2^{\log_2 19} = 19.$$

38. To find $k^{\log_k m}$, first assume $\log_k m = y$. This means changing to an exponential equation, $k^y = m$. Therefore,

$$k^{\log_k m} = k^y = m.$$

10.5 Evaluating Logarithms

10.5 Margin Exercises

1. Enter the given number. Then use the log key on a calculator to find the logarithm to the nearest ten thousandth.

(a) $\log 41,600 \approx 4.6191$

(b) $\log 43.5 \approx 1.6385$

(c) $\log .442 \approx -.3546$

2. Use $\text{pH} = -\log[H_3O^+]$.

Find the pH of each solution with the given hydronium ion concentration.

(a) 3.7×10^{-8}

$$\begin{aligned} pH &= -\log(3.7 \times 10^{-8}) \\ &= -(\log 3.7 + \log 10^{-8}) \\ &\approx -(.5682 - 8) \\ &= -(-7.4318) \\ pH &\approx 7.4 \end{aligned}$$

(b) 1.2×10^{-3}

$$\begin{aligned} pH &= -\log(1.2 \times 10^{-3}) \\ &= -(\log 1.2 + \log 10^{-3}) \\ &\approx -(.0792 - 3) \\ &= -(-2.9208) \\ pH &\approx 2.9 \end{aligned}$$

3. Use $pH = -\log[H_3O^+]$.

Find the hydronium ion concentration of solutions with the given pHs.

(a) 4.6

$$\begin{aligned} 4.6 &= -\log[H_3O^+] \\ -4.6 &= \log[H_3O^+] \\ 10^{-4.6} &= [H_3O^+] \\ [H_3O^+] &\approx 2.5 \times 10^{-5} \end{aligned}$$

(b) 7.5

$$\begin{aligned} 7.5 &= -\log[H_3O^+] \\ -7.5 &= \log[H_3O^+] \\ 10^{-7.5} &= [H_3O^+] \\ [H_3O^+] &\approx 3.2 \times 10^{-8} \end{aligned}$$

4. Enter the given number. Then use the $\ln x$ key on a calculator to find the natural logarithm.

(a) $\ln .01 \approx -4.6052$

(b) $\ln 27 \approx 3.2958$

(c) $\ln 529 \approx 6.2710$

5. Since 1995 is represented by $t = 0$, 1998 is represented by $t = 3$ $(1998 - 1995 = 3)$. Let $t = 3$ in the function.

$$\begin{aligned} f(t) &= 29.5e^{.052t} \\ f(3) &= 29.5e^{.052(3)} \\ &= 29.5e^{.156} \\ &\approx 34.5 \end{aligned}$$

In 1998, the population will be about 34.5 million.

6. Use the function

$$N(r) = -5000 \ln r.$$

Let $r = .2$.

$$\begin{aligned} N(.2) &= -5000 \ln .2 \\ &\approx -5000(-1.6094) \\ &= 8047 \end{aligned}$$

About 8000 yr have passed.

7. Use the change-of-base rule.

$$\log_a x = \frac{\log_b x}{\log_b a}$$

(a) $\log_3 17 = \dfrac{\log 17}{\log 3} \approx 2.5789$

(b) $\log_9 121 = \dfrac{\ln 121}{\ln 9} \approx 2.1827$

10.5 Section Exercises

3. $10^0 = 1$ and $10^1 = 10$, so $\log 1 = 0$ and $\log 10 = 1$. Thus, the value of $\log 5$ must lie between 0 and 1.

7. Enter 43, and use the log key on your calculator.

$$\log 43 \approx 1.6335$$

11. $\log .0326 \approx -1.4868$

15. Enter 7.84, and use the $\ln x$ key on your calculator.

$$\ln 7.84 \approx 2.0592$$

19. $\ln 388.1 \approx 5.9613$

23. $\ln 10 \approx 2.3026$

27. Use the change-of-base rule.

$$\log_a x = \frac{\log_b x}{\log_b a}$$

$$\log_{\sqrt{2}} \pi = \frac{\ln \pi}{\ln \sqrt{2}} \approx 3.3030$$

31. Let $m =$ the number of letters in your first name and $n =$ the number of letters in your last name.

Answers will vary, but suppose the name is Paul Bunyan, with $m = 4$ and $n = 6$.

(a) $\log_m n = \log_4 6$

This is the exponent to which 4 must be raised to get 6.

(b) Use the change-of-base rule to find $\log_4 6$.

$$\log_4 6 = \frac{\log 6}{\log 4} \approx 1.29248125$$

(c) Here, $m = 4$. Use the y^x (or x^y) key on your calculator.

$$4^{1.29248125} \approx 6$$

The result is 6, the value of n.

35. Grapes have a hydronium ion concentration of 5.0×10^{-5}.

$$\begin{aligned} pH &= -\log[H_3O^+] \\ &= -\log[5.0 \times 10^{-5}] \end{aligned}$$

Use the multiplication property.

$$\begin{aligned} &= -(\log 5.0 + \log 10^{-5}) \\ &\approx -(.6990 - 5) \\ &= -.6990 + 5 \\ pH &\approx 4.3 \end{aligned}$$

39. Spinach has a pH value of 5.4.

$$\begin{aligned} pH &= -\log[H_3O^+] \\ 5.4 &= -\log[H_3O^+] \\ -5.4 &= \log[H_3O^+] \\ 10^{-5.4} &= [H_3O^+] \\ [H_3O^+] &\approx 4.0 \times 10^{-6} \end{aligned}$$

43. In 1985, $t = 0$, and in 1994, $t = 9$ $(1994 - 1985 = 9)$.

$$\begin{aligned} N(t) &= 1757e^{.0264t} \\ N(9) &= 1757e^{(.0264)(9)} \\ &= 1757e^{.2376} \\ &\approx 1757(1.2682) \\ &\approx 2228 \end{aligned}$$

About 2228 million books (or about 2.2 billion) were sold in 1994.

47. At the bottom of the display, we see that $x = 10$ and $y = 730.04885$. This means that in the year equivalent to $x = 10$, which is $1980 + 10 - 1 = 1989$, there were approximately 730 thousand (or 730,000) Cesarean births.

51.
$$\begin{aligned} N(r) &= -5000 \ln r \\ N(.10) &= -5000 \ln .10 \\ &= -5000(-2.3026) \\ &\approx 11,500 \end{aligned}$$

About 11,500 yr have elapsed.

10.6 Exponential and Logarithmic Equations and Their Applications

10.6 Margin Exercises

In Exercises 1-5, check each solution in the original equation.

1. **(a)**
$$\begin{aligned} 2^p &= 9 \\ \log 2^p &= \log 9 \\ p \log 2 &= \log 9 \\ p &= \frac{\log 9}{\log 2} \approx 3.170 \end{aligned}$$

Solution set: $\{3.170\}$

(b)
$$\begin{aligned} 10^k &= 4 \\ \log 10^k &= \log 4 \\ k \log 10 &= \log 4 \\ k &= \frac{\log 4}{\log 10} \approx \frac{.6021}{1} \\ &\approx .602 \end{aligned}$$

Solution set: $\{.602\}$

2. $e^{-.01t} = .38$

Take base e logarithms on both sides.

$$\begin{aligned} \ln e^{-.01t} &= \ln .38 \\ -.01t \ln e &= \ln .38 \\ -.01t &= \ln .38 \quad \textit{ln e = 1} \\ t &= \frac{\ln .38}{-.01} \\ t &\approx 96.8 \end{aligned}$$

Solution set: $\{96.8\}$

3. $\log_5 \sqrt{x - 7} = 1$

Change to exponential form.

$$5^1 = \sqrt{x - 7}$$

Square both sides.

$$25 = x - 7$$
$$32 = x$$

Solution set: $\{32\}$

4. $\log_8(2x+5) + \log_8 3 = \log_8 33$

Use the multiplication property.

$$\log_8[3(2x+5)] = \log_8 33$$

If $\log_b x = \log_b y$, then $x = y$.

$$3(2x+5) = 33$$
$$2x + 5 = 11$$
$$2x = 6$$
$$x = 3$$

Solution set: $\{3\}$

5. $\log_3 2x - \log_3(3x+15) = -2$

Use the division property.

$$\log_3 \frac{2x}{3x+15} = -2$$

Change to exponential form.

$$3^{-2} = \frac{2x}{3x+15}$$
$$\frac{1}{9} = \frac{2x}{3x+15}$$

Multiply by $9(3x+15)$.

$$3x + 15 = 18x$$
$$15 = 15x$$
$$1 = x$$

Solution set: $\{1\}$

6. Use $A = P\left(1+\frac{r}{n}\right)^{nt}$ with $P = 2000, r = .05, n = 1$, and $t = 10$.

$$A = 2000\left(1 + \frac{.05}{1}\right)^{1(10)} = 2000(1.05)^{10} \approx 3258$$

The value is about \$3260.

7. $y = 50e^{-.002t}$

(a) Let $t = 180$.

$$y = 50e^{-.002(180)} = 50e^{-.36} \approx 34.9$$

About 34.9 watts will be available.

(b) The original amount of power is found when $t = 0$.

$$y = 50e^{-.002(0)} = 50e^0 = 50$$

Half the original power is 25 watts. Let $y = 25$, and solve for t.

$$25 = 50e^{-.002t}$$

Divide both sides by 50.

$$.5 = e^{-.002t}$$

Take base e logarithms on both sides.

$$\ln .5 = \ln e^{-.002t}$$

Use the power property.

$$\ln .5 = -.002t \ln e$$
$$\ln .5 = -.002t \qquad \textit{ln } e = 1$$
$$\frac{\ln .5}{-.002} = t$$
$$346.57 \approx t$$

After about 347 days, the power will be half the original amount.

10.6 Section Exercises

3. $2^x = 64$

After trying several powers of 2, we find that $2^6 = 64$, and so $x = 6$. The solution set is $\{6\}$.

To find the exact value of $\frac{\log 64}{\log 2}$, use this process.

$$2^x = 64$$
$$\log 2^x = \log 64$$
$$x \log 2 = \log 64$$
$$x = \frac{\log 64}{\log 2}$$

But, we know that $x = 6$ from our earlier work. Therefore,

$$\frac{\log 64}{\log 2} = 6.$$

In Exercises 7-19, and 27-35, check each solution in the original equation.

7.
$$9^{-x+2} = 13$$
$$\log 9^{-x+2} = \log 13$$
$$(-x+2)\log 9 = \log 13$$
$$-x\log 9 + 2\log 9 = \log 13$$
$$-x\log 9 = \log 13 - 2\log 9$$
$$x\log 9 = 2\log 9 - \log 13$$
$$x = \frac{2\log 9 - \log 13}{\log 9}$$
$$x \approx .833$$

Solution set: $\{.833\}$

11. $e^{.006x} = 30$

Take base e logarithms on both sides.

$$\ln e^{.006x} = \ln 30$$
$$.006x \ln e = \ln 30$$
$$.006x = \ln 30 \qquad \textit{ln } e = 1$$
$$x = \frac{\ln 30}{.006}$$
$$x \approx 566.866$$

Solution set: $\{566.866\}$

15.
$$\ln e^{.04x} = \sqrt{3}$$
$$.04x \ln e = \sqrt{3}$$
$$.04x = \sqrt{3} \qquad \textit{ln } e = 1$$
$$x = \frac{\sqrt{3}}{.04}$$
$$x \approx 43.301$$

Solution set: $\{43.301\}$

19. $\log_3(6x+5) = 2$

Change to exponential form.

$$3^2 = 6x + 5$$
$$9 = 6x + 5$$
$$4 = 6x$$
$$\frac{4}{6} = x$$
$$\frac{2}{3} = x$$

Solution set: $\{\frac{2}{3}\}$

23. If $x = 2$ in $\log_4(x-3)$, we have $\log_4(2-3)$ or $\log_4(-1)$. We are asked to find the logarithm of a negative number. This process is not defined.

27. $\log_5(3t+2) - \log_5 t = \log_5 4$
$$\log_5 \frac{3t+2}{t} = \log_5 4$$
$$\frac{3t+2}{t} = 4$$
$$3t + 2 = 4t$$
$$2 = t$$

Solution set: $\{2\}$

31. $\log_2 x + \log_2(x-7) = 3$
$$\log_2[x(x-7)] = 3$$

Change to exponential form.

$$2^3 = x(x-7)$$
$$8 = x^2 - 7x$$
$$x^2 - 7x - 8 = 0$$
$$(x-8)(x+1) = 0$$
$$x - 8 = 0 \quad \text{or} \quad x + 1 = 0$$
$$x = 8 \quad \text{or} \quad x = -1$$

Check each potential solution. The number 8 makes the equation true. The number -1 must be rejected because it yields an equation in which the logarithm of a negative number must be found.

Solution set: $\{8\}$

35. $\log_2 x + \log_2(x-6) = 4$
$$\log_2[x(x-6)] = 4$$

Change to exponential form.

$$2^4 = x(x-6)$$
$$16 = x^2 - 6x$$
$$x^2 - 6x - 16 = 0$$
$$(x-8)(x+2) = 0$$
$$x - 8 = 0 \quad \text{or} \quad x + 2 = 0$$
$$x = 8 \quad \text{or} \quad x = -2$$

Check each potential solution. The number 8 makes the equation true. The number -2 must be rejected because it yields an equation in which the logarithm of a negative number must be found.

Solution set: $\{8\}$

39. Find $A(t) = 400e^{-.032t}$ when $t = 25$.

$$A(25) = 400e^{-.032(25)} = 400e^{-.8} \approx 179.73$$

About 180 g of lead will be left.

43. $A = Pe^{rt}$

Here $r = 4.5\%$ (.045), and if the principal doubles, then $A = 2P$.

$$2P = Pe^{.045t}$$

Divide by P.

$$2 = e^{.045t}$$

Take base e logarithms on both sides.

$$\ln 2 = \ln e^{.045t}$$
$$\ln 2 = .045t \ln e$$
$$\ln 2 = .045t \qquad \textit{ln } e = 1$$
$$\frac{\ln 2}{.045} = t$$
$$15.4 \approx t$$

The money will double in about 15.4 yr.

Chapter 10 Review Exercises

1. Since a horizontal line intersects the graph in two points, the function is not one-to-one.

2. Since every horizontal line intersects the graph in no more than one point, the function is one-to-one.

3. The function $f(x) = -3x + 7$ is a linear function. By the horizontal line test, it is a one-to-one function. To find the inverse, replace $f(x)$ with y.

$$y = -3x + 7$$

Interchange x and y.

$$x = -3y + 7$$

Solve for y.

$$x - 7 = -3y$$
$$\frac{x-7}{-3} = y$$
$$\text{or } \frac{7-x}{3} = y$$

Replace y with $f^{-1}(x)$.

$$f^{-1}(x) = \frac{x-7}{-3} \text{ or } \frac{7-x}{3}$$

4. $f(x) = \sqrt[3]{6x-4}$

The cube root causes each value of x to be matched with only one value of $f(x)$. The function is one-to-one. To find the inverse, replace $f(x)$ with y.

$$y = \sqrt[3]{6x-4}$$

Cube both sides.

$$y^3 = 6x - 4$$

Interchange x and y.

$$x^3 = 6y - 4$$

Solve for y.

$$x^3 + 4 = 6y$$
$$\frac{x^3+4}{6} = y$$

Replace y with $f^{-1}(x)$.

$$\frac{x^3+4}{6} = f^{-1}(x)$$

5. $f(x) = -x^2 + 3$

This is an equation of a vertical parabola which opens downward. Since a horizontal line will intersect the graph in two points, the function is not one-to-one.

6. $f(x) = x$

This is a linear function. It is a one-to-one function. To find the inverse, replace $f(x)$ with y.

$$y = x$$

Interchange x and y.

$$x = y$$

Replace y with $f^{-1}(x)$.

$$x = f^{-1}(x)$$

For Exercises 7-12, see the answer graphs in the back of the textbook.

7. The graph is a linear function through $(0, 1)$ and $(3, 0)$. The graph of $f^{-1}(x)$ will include the points $(1, 0)$ and $(0, 3)$, found by interchanging x and y. Plot these points, and draw a straight line through them.

8. The graph is a curve through $(1, 2), (0, 1)$, and $\left(-1, \frac{1}{2}\right)$. Interchange x and y to get $(2, 1), (1, 0)$, and $\left(\frac{1}{2}, -1\right)$ which are on $f^{-1}(x)$. Plot these points, and draw a smooth curve through them.

9. $f(x) = 3^x$

Make a table of values.

x	-2	-1	0	1	2
y	$\frac{1}{9}$	$\frac{1}{3}$	1	3	9

Plot the points from the table, and draw a smooth curve through them.

10. $f(x) = \left(\frac{1}{3}\right)^x$

Make a table of values.

x	-2	-1	0	1	2
y	9	3	1	$\frac{1}{3}$	$\frac{1}{9}$

Plot the points from the table, and draw a smooth curve through them.

11. $y = 3^{x+1}$

Make a table of values.

x	-3	-2	-1	0	1
$x+1$	-2	-1	0	1	2
y	$\frac{1}{9}$	$\frac{1}{3}$	1	3	9

Plot the points from the table, and draw a smooth curve through them.

12. $y = 2^{2x+3}$

Make a table of values.

x	-2	$-\frac{3}{2}$	-1	0	$\frac{1}{2}$
$2x+3$	-1	0	1	3	4
y	$\frac{1}{2}$	1	2	8	16

Plot the points from the table, and draw a smooth curve through them.

In Exercises 13-15, check each solution in the original equation.

13. $4^{3x} = 8^{x+4}$

Write each side with the same base.

$$(2^2)^{3x} = 2^{3(x+4)}$$
$$2^{6x} = 2^{3x+12}$$

If $a^x = a^y$, then $x = y$. Set the exponents equal.

$$6x = 3x + 12$$
$$3x = 12$$
$$x = 4$$

Solution set: $\{4\}$

14. $$\left(\frac{1}{27}\right)^{x-1} = 9^{2x}$$

$$\left[\left(\frac{1}{3}\right)^3\right]^{x-1} = (3^2)^{2x}$$

Write each side with the same base.

$$(3^{-3})^{x-1} = (3^2)^{2x}$$
$$3^{-3x+3} = 3^{4x}$$

Set the exponents equal.

$$-3x + 3 = 4x$$
$$3 = 7x$$
$$\frac{3}{7} = x$$

Solution set: $\{\frac{3}{7}\}$

15. $5^x = 1$

Write each side with the same base.

$$5^x = 5^0$$

Set the exponents equal.

$$x = 0$$

Solution set: $\{0\}$

16. $y = a^x \ (a > 0, a \neq 1)$

To find the y-intercept, let $x = 0$.

$$y = a^0$$
$$y = 1$$

The point $(0, 1)$ is the y-intercept.

17. The outcome of Exercise 16 suggests $f(0) = 1$, so that any base $a > 0, a \neq 1$, has the property that $a^0 = 1$. In other words, any positive number not equal to 1, raised to the zero power, is 1.

For Exercises 18 and 19, see the answer graphs in the back of the textbook.

18. $g(x) = \log_3 x$

This is the inverse of the function $f(x) = 3^x$ in Exercise 9. Reverse the values in the table there to get the table for $g(x)$.

x	$\frac{1}{9}$	$\frac{1}{3}$	1	3	9
y	-2	-1	0	1	2

Plot the points from the table, and draw a smooth curve through them.

19. $g(x) = \log_{1/3} x$

This is the inverse of the function $f(x) = \left(\frac{1}{3}\right)^x$ in Exercise 10. Reverse the values in the table there to get the table for $g(x)$.

x	9	3	1	$\frac{1}{3}$	$\frac{1}{9}$
y	-2	-1	0	1	2

Plot the points from the table, and draw a smooth curve through them.

20. **(a)** The base is 5, the logarithm (exponent) is 4, and the number is 625, so $\log_5 625 = 4$ becomes $5^4 = 625$ in exponential form.

(b) The base is 5, the exponent (logarithm) is -2, and the number is .04, so $5^{-2} = .04$ becomes $\log_5 .04 = -2$ in logarithmic form.

21. **(a)** $\log_b a$ is the exponent to which b must be raised to obtain a.

(b) Let $\log_b a = y$.
Then in exponential form

$$b^y = a.$$

Replace y with $\log_b a$, so

$$b^{\log_b a} = a.$$

22. $\log_4 3x^2$

Use the multiplication property.

$$= \log_4 3 + \log_4 x^2$$

Use the power property.

$$= \log_4 3 + 2\log_4 x$$

23. $\log_2 \dfrac{p^2 r}{\sqrt{z}} = \log_2 \dfrac{p^2 r}{z^{1/2}}$

Use the multiplication and division properties.

$$= \log_2 p^2 + \log_2 r - \log_2 z^{1/2}$$

Use the power property.

$$= 2\log_2 p + \log_2 r - \frac{1}{2}\log_2 z$$

24. $\log_b 3 + \log_b x - 2\log_b y$

Use the multiplication and power properties.

$$= \log_b 3x - \log_b y^2$$

Use the division property.

$$= \log_b \frac{3x}{y^2}$$

25. $\log_3(x+7) - \log_3(4x+6)$

Use the division property.

$$= \log_3 \frac{x+7}{4x+6}$$

26. Enter 28.9, and use the log key on your calculator.

$$\log 28.9 \approx 1.4609$$

27. $\log .257 \approx -.5901$

28. $\log 10^{4.8613} = x$
$\log_{10} 10^{4.8613} = x$

Change to exponential form.

$$10^x = 10^{4.8613}$$

Set the exponents equal.

$$x = 4.8613$$

So,

$$\log 10^{4.8613} = 4.8613.$$

29. Enter 28.9, and use the $\ln x$ key on your calculator.

$$\ln 28.9 \approx 3.3638$$

30. $\ln .257 \approx -1.3587$

31. $\ln e^{4.8613} = x$

The base of a natural logarithm is e, so in exponential form

$$e^x = e^{4.8613}.$$

Set the exponents equal.

$$x = 4.8613$$

So,

$$\ln e^{4.8613} = 4.8613.$$

32. **(a)** $\log 356.8 \approx 2.552424846$

(b) $\log 35.68 \approx 1.552424846$

(c) $\log 3.568 \approx 0.552424846$

(d) The whole number part of the answers (2, 1, or 0) varies, whereas the decimal part (.552424846) remains the same.

In Exercises 33-35, use the change-of-base rule.

$$\log_a x = \frac{\log_b x}{\log_b a}$$

33. $\log_{16} 13 = \frac{\log 13}{\log 16} \approx .9251$

34. $\log_4 12 = \frac{\log 12}{\log 4} \approx 1.7925$

35. $\log_{\sqrt{6}} \sqrt{13} = \frac{\log \sqrt{13}}{\log \sqrt{6}} \approx 1.4315$

36. $H(t) = 500 \log_3 (2t + 3)$

(a) Let $t = 0$.

$$\begin{aligned} H(0) &= 500 \log_3 (2 \cdot 0 + 3) \\ &= 500 \log_3 3 \\ &= 500 \cdot 1 \\ H(0) &= 500 \end{aligned}$$

Initially there were 500 hares.

(b) Let $t = 3$.

$$\begin{aligned} H(3) &= 500 \log_3 (2 \cdot 3 + 3) \\ &= 500 \log_3 9 \\ &= 500 \cdot 2 \\ &= 1000 \end{aligned}$$

After 3 yr there were 1000 hares.

(c) Let $t = 12$.

$$\begin{aligned} H(12) &= 500 \log_3 (2 \cdot 12 + 3) \\ &= 500 \log_3 27 \\ &= 500 \cdot 3 \\ &= 1500 \end{aligned}$$

After 12 yr there were 1500 hares.

(d) No, the change-of-base rule was not needed.

37. Milk has a hydronium ion concentration of 4.0×10^{-7}.

$$\begin{aligned} \text{pH} &= -\log[H_3O^+] \\ &= -\log[4.0 \times 10^{-7}] \end{aligned}$$

Use the multiplication property.

$$\begin{aligned} &= -(\log 4.0 + \log 10^{-7}) \\ &\approx -(.6021 - 7) \\ &= -.6021 + 7 \\ \text{pH} &\approx 6.4 \end{aligned}$$

38. Crackers have a hydronium ion concentration of 3.8×10^{-9}.

$$\begin{aligned} \text{pH} &= -\log[H_3O^+] \\ &= -\log[3.8 \times 10^{-9}] \end{aligned}$$

Use the multiplication property.

$$\begin{aligned} &= -(\log 3.8 + \log 10^{-9}) \\ &\approx -(.5798 - 9) \\ &= -.5798 + 9 \\ \text{pH} &\approx 8.4 \end{aligned}$$

39. $C(t) = 9e^{.026t}$

(a) If $t = 0$ represents 1990, then $t = 10$ represents 2000 $(2000 - 1990 = 10)$. Find $C(10)$.

$$C(10) = 9e^{.026(10)} = 9e^{.26} \approx 11.7$$

The population will be about 11.7 million in 2000.

(b) If $t = 0$ represents 1990, $t = -20$ represents 1970 $(1970 - 1990 = -20)$. Find $C(-20)$.

$$C(-20) = 9e^{.026(-20)} = 9e^{-.52} \approx 5.4$$

The population was about 5.4 million in 1970.

40. $Q(t) = 500e^{-.05t}$

(a) Let $t = 0$.

$$\begin{aligned} Q(0) &= 500e^{-.05(0)} \\ &= 500e^0 \\ &= 500(1) \\ &= 500 \end{aligned}$$

There are 500 g.

(b) Let $t = 4$.

$$\begin{aligned} Q(4) &= 500e^{-.05(4)} \\ &= 500e^{-.2} \\ &\approx 500(.8187) \\ &= 409.35 \end{aligned}$$

There will be about 409 g in 4 days.

In Exercises 41-49, check each solution in the original equation.

41.
$$\begin{aligned} 3^x &= 9.42 \\ \log 3^x &= \log 9.42 \\ x \log 3 &= \log 9.42 \\ x &= \frac{\log 9.42}{\log 3} \\ x &\approx 2.042 \end{aligned}$$

Solution set: $\{2.042\}$

42.
$$\begin{aligned} 2^{x-1} &= 15 \\ \log 2^{x-1} &= \log 15 \\ (x-1)\log 2 &= \log 15 \\ x-1 &= \frac{\log 15}{\log 2} \\ x &= \frac{\log 15}{\log 2}+1 \\ x &\approx 4.907 \end{aligned}$$

Solution set: $\{4.907\}$

43. $e^{.06x} = 3$

Take base e logarithms on both sides.

$$\begin{aligned} \ln e^{.06x} &= \ln 3 \\ .06x \ln e &= \ln 3 \\ .06x &= \ln 3 \qquad \textit{ln } e = 1 \\ x &= \frac{\ln 3}{.06} \\ x &\approx 18.310 \end{aligned}$$

Solution set: $\{18.310\}$

44. $\log_3(9x+8) = 2$

Change to exponential form.

$$\begin{aligned} 3^2 &= 9x+8 \\ 9 &= 9x+8 \\ 1 &= 9x \\ \frac{1}{9} &= x \end{aligned}$$

Solution set: $\{\frac{1}{9}\}$

45. $\log_5(y+6)^3 = 2$

Change to exponential form.

$$\begin{aligned} 5^2 &= (y+6)^3 \\ 25 &= (y+6)^3 \end{aligned}$$

Take the cube root of each side.

$$\begin{aligned} \sqrt[3]{25} &= y+6 \\ \sqrt[3]{25}-6 &= y \end{aligned}$$

Solution set: $\{-6+\sqrt[3]{25}\}$

46. $\log_3(p+2) - \log_3 p = \log_3 2$

Use the division property.

$$\begin{aligned} \log_3 \frac{p+2}{p} &= \log_3 2 \\ \frac{p+2}{p} &= 2 \\ p+2 &= 2p \\ 2 &= p \end{aligned}$$

Solution set: $\{2\}$

47.
$$\begin{aligned} \log(2x+3) &= \log x + 1 \\ \log_{10}(2x+3) - \log_{10} x &= 1 \end{aligned}$$

Use the division property.

$$\log_{10} \frac{2x+3}{x} = 1$$

Change to exponential form.

$$\begin{aligned} 10^1 &= \frac{2x+3}{x} \\ 10x &= 2x+3 \\ 8x &= 3 \\ x &= \frac{3}{8} \end{aligned}$$

Solution set: $\{\frac{3}{8}\}$

48. $\log_4 x + \log_4(8-x) = 2$

Use the multiplication property.

$$\log_4[x(8-x)] = 2$$

Change to exponential form.

$$\begin{aligned} 4^2 &= x(8-x) \\ 16 &= 8x - x^2 \\ x^2 - 8x + 16 &= 0 \\ (x-4)(x-4) &= 0 \\ x-4 &= 0 \\ x &= 4 \end{aligned}$$

Solution set: $\{4\}$

49. $\log_2 x + \log_2(x+15) = 4$

Use the multiplication property.

$$\log_2[x(x+15)] = 4$$

Change to exponential form.

$$\begin{aligned} 2^4 &= x(x+15) \\ 16 &= x^2 + 15x \\ x^2 + 15x - 16 &= 0 \\ (x+16)(x-1) &= 0 \\ x+16 = 0 \quad &\text{or} \quad x-1 = 0 \\ x = -16 \quad &\text{or} \quad x = 1 \end{aligned}$$

Check each potential solution. The number 1 makes the equation true. The number -16 must be rejected because it yields an equation in which the logarithm of a negative number must be found.

Solution set: $\{1\}$

50. $C(t) = 9e^{.026t}$

Let $C(t) = 10$.

$$10 = 9e^{.026t}$$
$$\frac{10}{9} = e^{.026t}$$

Take base e logarithms on both sides.

$$\ln \frac{10}{9} = \ln e^{.026t}$$

Use the power property.

$$\ln \frac{10}{9} = .026t \ln e$$
$$\ln \frac{10}{9} = .026t \qquad \textit{ln e = 1}$$
$$\frac{\ln \frac{10}{9}}{.026} = t$$
$$4 \approx t$$

Since $t = 0$ represents 1990, $t = 4$ represents 1990 + 4 = 1994.

51. $Q(t) = 500e^{-.05t}$

The original amount is found when $t = 0$. From Exercise 40(a), $Q(0) = 500$. Half the original substance is then $\frac{500}{2} = 250$. Let $Q(t) = 250$, and solve for t.

$$250 = 500e^{-.05t}$$
$$.5 = e^{-.05t}$$

Take base e logarithms on both sides.

$$\ln .5 = \ln e^{-.05t}$$

Use the power property.

$$\ln .5 = -.05t \ \ln e$$
$$\ln .5 = -.05t \qquad \textit{ln e = 1}$$
$$\frac{\ln .5}{-.05} = t$$
$$13.86 \approx t$$

The half life is nearly 14 yr.

52. Use $A = P\left(1 + \frac{r}{n}\right)^{nt}$ with $P = 6500, r = .03, n = 365$, and $t = 3$.

$$A = 6500\left(1 + \frac{.03}{365}\right)^{365(3)} \approx 7112.11$$

In 3 yr the value would be about \$7112.11.

53. Use $A = P\left(1 + \frac{r}{n}\right)^{nt}$.

Plan A: Let $P = 1000, r = .04, n = 4$, and $t = 3$.

$$A = 1000\left(1 + \frac{.04}{4}\right)^{4(3)} \approx 1126.83$$

Plan B: Let $P = 1000, r = .039, n = 12$, and $t = 3$.

$$A = 1000\left(1 + \frac{.039}{12}\right)^{12(3)} \approx 1123.91$$

Plan A is the better plan by \$2.92.

In Exercises 54-60, check each solution in the original equation.

54. $\log_3(x + 9) = 4$

Change to exponential form.

$$3^4 = x + 9$$
$$81 = x + 9$$
$$72 = x$$

Solution set: $\{72\}$

55. $\log_2 32 = x$

Change to exponential form.

$$2^x = 32$$

Write each side with the same base.

$$2^x = 2^5$$

If $a^x = a^y$, then $x = y$. Set the exponents equal.

$$x = 5$$

Solution set: $\{5\}$

56. $\log_x \frac{1}{81} = 2$

Change to exponential form.

$$x^2 = \frac{1}{81}$$
$$x^2 = \left(\frac{1}{9}\right)^2$$

Take the square root of both sides.

$$x = \frac{1}{9} \quad \text{or} \quad x = -\frac{1}{9}$$

Reject $x = -\frac{1}{9}$ since the base of a logarithm cannot be negative.

Solution set: $\left\{\frac{1}{9}\right\}$

57. $27^x = 81$

Write each side with the same base.

$$(3^3)^x = 3^4$$
$$3^{3x} = 3^4$$

Set the exponents equal.

$$3x = 4$$
$$x = \frac{4}{3}$$

Solution set: $\{\frac{4}{3}\}$

58. $2^{2x-3} = 8$

Write each side with the same base.

$$2^{2x-3} = 2^3$$

Set the exponents equal.

$$2x - 3 = 3$$
$$2x = 6$$
$$x = 3$$

Solution set: $\{3\}$

59. $\log_3(x+1) - \log_3 x = 2$

Use the division property.

$$\log_3 \frac{x+1}{x} = 2$$

Change to exponential form.

$$3^2 = \frac{x+1}{x}$$
$$9x = x + 1$$
$$8x = 1$$
$$x = \frac{1}{8}$$

Solution set: $\{\frac{1}{8}\}$

60. $\log(3x - 1) = \log 10$

$$3x - 1 = 10$$
$$3x = 11$$
$$x = \frac{11}{3}$$

Solution set: $\{\frac{11}{3}\}$

Chapter 10 Test

1. **(a)** $f(x) = x^2 + 9$

This function is not one-to-one. The graph of $f(x)$ is a vertical parabola. A horizontal line will intersect the graph more than once.

(b) This function is one-to-one. A horizontal line will not intersect the graph in more than one point.

2. $f(x) = \sqrt[3]{x+7}$

Replace $f(x)$ with y.

$$y = \sqrt[3]{x+7}$$

Interchange x and y.

$$x = \sqrt[3]{y+7}$$

Solve for y. Cube both sides.

$$x^3 = y + 7$$
$$x^3 - 7 = y$$

Replace y with $f^{-1}(x)$.

$$f^{-1}(x) = x^3 - 7$$

For Exercises 3-5, see the answer graphs in the back of the textbook.

3. By the horizontal line test, $f(x)$ is a one-to-one function and has an inverse. Choose some points on the graph of $f(x)$, such as $(4,0)$, $(3,-1)$, and $(0,-2)$. To graph the inverse, interchange the x- and y-values to get $(0,4), (-1,3)$, and $(-2,0)$. Plot these points, and draw a smooth curve through them.

4. $y = 6^x$

Make a table of values.

x	-2	-1	0	1
y	$\frac{1}{36}$	$\frac{1}{6}$	1	6

Plot these points, and draw a smooth curve through them.

5. $y = \log_6 x$

Change to exponential form.

$$x = 6^y$$

Make a table of values.

x	$\frac{1}{36}$	$\frac{1}{6}$	1	6
y	-2	-1	0	1

Plot these points, and draw a smooth curve through them.

6. $y = 6^x$ and $y = \log_6 x$ are inverse functions. To use the graph of the function from Exercise 4 to obtain the graph of the function in Exercise 5, interchange the x- and y-coordinates of the ordered pairs $\left(-2, \frac{1}{36}\right), \left(-1, \frac{1}{6}\right), (0,1)$, and $(1,6)$ to get $\left(\frac{1}{36}, -2\right), \left(\frac{1}{6}, -1\right), (1,0)$, and $(6,1)$. Plot these points, and draw the curve through them.

7. $5^x = \dfrac{1}{625}$

$$5^x = \left(\frac{1}{5}\right)^4$$

Write each side with the same base.

$$5^x = 5^{-4}$$

If $a^x = a^y$, then $x = y$. Set the exponents equal.

$$x = -4$$

This solution checks.

Solution set: $\{-4\}$

8. $2^{3x-7} = 8^{2x+2}$

Write each side with the same base.

$$2^{3x-7} = 2^{3(2x+2)}$$

Set the exponents equal.

$$\begin{aligned} 3x - 7 &= 3(2x+2) \\ 3x - 7 &= 6x + 6 \\ -13 &= 3x \\ -\frac{13}{3} &= x \end{aligned}$$

This solution checks.

Solution set: $\left\{-\frac{13}{3}\right\}$

9. $y = 5000(2)^{-.15t}$

(a) The original value is when $t = 0$.

$$\begin{aligned} y &= 5000(2)^{-.15(0)} \\ &= 5000(2)^0 \\ &= 5000(1) \\ y &= 5000 \end{aligned}$$

The original value is \$5000.

(b) The value after 5 yr is when $t = 5$.

$$\begin{aligned} y &= 5000(2)^{-.15(5)} \\ &= 5000(2)^{-.75} \\ y &\approx 2973.018 \end{aligned}$$

The value after 5 yr is about \$2973.

(c) The value after 10 yr is when $t = 10$.

$$\begin{aligned} y &= 5000(2)^{-.15(10)} \\ &= 5000(2)^{-1.5} \\ y &= 1767.767 \end{aligned}$$

The value after 10 yr is about \$1768.

(d) Use the results of parts (a)-(c) to make a table of values.

t	0	5	10
y	5000	2973	1768

On a coordinate system, with the horizontal axis labeled t and the vertical axis labeled y, plot the points $(0, 5000), (5, 2973)$, and $(10, 1768)$. Draw a smooth curve through them. See the answer graph in the back of the textbook.

10. The base is 4, the exponent (logarithm) is -2, and the number is .0625, so $4^2 = .0625$ becomes $\log_4 .0625 = -2$ in logarithmic form.

11. The base is 7, the logarithm (exponent) is 2, and the number is 49, so $\log_7 49 = 2$ becomes $7^2 = 49$ in exponential form.

In Exercises 12-14, check each solution in the original equation.

12. $\log_{1/2} x = -5$

Change to exponential form.

$$\begin{aligned} \left(\frac{1}{2}\right)^{-5} &= x \\ 2^5 &= x \\ 32 &= x \end{aligned}$$

Solution set: $\{32\}$

13. $x = \log_9 3$

Change to exponential form.

$$9^x = 3$$

Write each side with the same base.

$$(3^2)^x = 3$$
$$3^{2x} = 3^1$$

If $a^x = a^y$, then $x = y$. Set the exponents equal.

$$2x = 1$$
$$x = \frac{1}{2}$$

Solution set: $\{\frac{1}{2}\}$

14. $\log_x 16 = 4$

Change to exponential form.

$$x^4 = 16$$

Write each side with the same exponent.

$$x^4 = 2^4$$

Set the bases equal.

$$x = 2$$

Solution set: $\{2\}$

15. The value of $\log_2 32$ is *5*. This means that if we raise *2* to the *fifth* power, the result is *32*.

16. $\log_3 x^2 y$

Use the multiplication property.

$$= \log_3 x^2 + \log_3 y$$

Use the power property.

$$= 2\log_3 x + \log_3 y$$

17. $\log_5 \left(\frac{\sqrt{x}}{yz}\right)$

Use the division property.

$$= \log_5 \sqrt{x} - \log_5 yz$$

Use the multiplication property.

$$= \log_5 x^{1/2} - (\log_5 y + \log_5 z)$$

Use the power property.

$$= \frac{1}{2}\log_5 x - \log_5 y - \log_5 z$$

18. $3\log_b s - \log_b t$

Use the power property.

$$= \log_b s^3 - \log_b t$$

Use the division property.

$$= \log_b \frac{s^3}{t}$$

19. $\frac{1}{4}\log_b r + 2\log_b s - \frac{2}{3}\log_b t$

Use the power property.

$$= \log_b r^{1/4} + \log_b s^2 - \log_b t^{2/3}$$

Use the multiplication and division properties.

$$= \log_b \frac{r^{1/4}s^2}{t^{2/3}}$$

20. $\log 21.3 \approx 1.3284$

21. $\ln .43 \approx -.8440$

22. **(a)** $\log_6 45 = x$ means $6^x = 45$. Because $6^2 = 36$ and $6^3 = 216$, the value of $\log_6 45 = x$ must be between 2 and 3.

(b) Use the change-of-base rule.

$$\log_a x = \frac{\log_b x}{\log_b a}$$
$$\log_6 45 = \frac{\log 45}{\log 6} \approx 2.1245$$

23. $$3^x = 78$$
$$\ln 3^x = \ln 78$$

Use the power property.

$$x \ln 3 = \ln 78$$
$$x = \frac{\ln 78}{\ln 3}$$
$$x \approx 3.9656$$

Solution set: $\{3.9656\}$

24. $\log_8(x+5) + \log_8(x-2) = \log_8 8$

Use the multiplication property.

$$\log_8(x+5)(x-2) = \log_8 8$$
$$(x+5)(x-2) = 8$$
$$x^2 + 3x - 10 = 8$$
$$x^2 + 3x - 18 = 0$$
$$(x+6)(x-3) = 0$$
$$x + 6 = 0 \quad \text{or} \quad x - 3 = 0$$
$$x = -6 \quad \text{or} \quad x = 3$$

Check both potential solutions. The number 3 makes the equation true. The number -6 must be rejected because it yields an equation in which the logarithm of a negative number must be found.

Solution set: $\{3\}$

25. $y = 5000e^{-.104t}$

(a) After 15 yr, the value of t is 15.

$$y = 5000e^{-.104(15)}$$
$$y = 5000e^{-1.56}$$
$$y \approx 1050.68$$

After 15 yr, the value of the copier will be about \$1051.

(b) To find the original value, let $t = 0$.

$$y = 5000e^{-.104t}$$
$$y = 5000e^{-.104(0)}$$
$$y = 5000e^{0}$$
$$y = 5000$$

The original value is \$5000, so half the original value is \$2500. Substitute 2500 for y to find t.

$$2500 = 5000e^{-.104t}$$
$$.5 = e^{-.104t}$$

Take base e logarithms on both sides.

$$\ln .5 = \ln e^{-.104t}$$

Use the power property.

$$\ln .5 = -.104t \ln e$$
$$\ln .5 = -.104t \qquad \textit{ln e = 1}$$
$$\frac{\ln .5}{-.104} = t$$
$$6.66 \approx t$$

It will take about 7 yr for the copier to reach half of the original value.

Cumulative Review Exercises (Chapters 1-10)

For Exercises 1-4,

$$S = \left\{-\frac{9}{4}, -2, -\sqrt{2}, 0, .6, \sqrt{11}, \sqrt{-8}, 6, \frac{30}{3}\right\}.$$

1. The integers are $-2, 0, 6$, and $\frac{30}{3}$ (or 10).

2. The rational numbers are $-\frac{9}{4}, -2, 0, .6, 6$, and $\frac{30}{3}$ (or 10). Each can be expressed as a quotient of two integers.

3. The irrational numbers are $-\sqrt{2}$ and $\sqrt{11}$.

4. All are real numbers except $\sqrt{-8}$.

5. $|-8| + 6 - |-2| - (-6 + 2)$
$= 8 + 6 - 2 - (-4)$
$= 8 + 6 - 2 + 4 = 16$

6. $-12 - |-3| - 7 - |-5|$
$= -12 - 3 - 7 - 5 = -27$

7. $2(-5) + (-8)(4) - (-3)$
$= -10 - 32 + 3 = -39$

8.
$$7 - (3 + 4a) + 2a = -5(a - 1) - 3$$
$$7 - 3 - 4a + 2a = -5a + 5 - 3$$
$$4 - 2a = -5a + 2$$
$$3a = -2$$
$$a = -\frac{2}{3}$$

Solution set: $\{-\frac{2}{3}\}$

9.
$$2m + 2 \le 5m - 1$$
$$-3m \le -3$$

Divide by -3; reverse the inequality.

$$m \ge 1$$

Solution set: $[1, \infty)$

10. $|2x - 5| = 9$
$$2x - 5 = 9 \quad \text{or} \quad 2x - 5 = -9$$
$$2x = 14 \qquad\qquad 2x = -4$$
$$x = 7 \quad \text{or} \quad x = -2$$

Solution set: $\{-2, 7\}$

11.
$$|3p| - 4 = 12$$
$$|3p| = 16$$
$$3p = 16 \quad \text{or} \quad 3p = -16$$
$$p = \frac{16}{3} \quad \text{or} \quad p = -\frac{16}{3}$$

Solution set: $\{-\frac{16}{3}, \frac{16}{3}\}$

12.
$$|3k - 8| \le 1$$
$$-1 \le 3k - 8 \le 1$$
$$7 \le 3k \le 9$$
$$\frac{7}{3} \le k \le 3$$

Solution set: $[\frac{7}{3}, 3]$

13. $|4m + 2| > 10$
$$4m + 2 > 10 \quad \text{or} \quad 4m + 2 < -10$$
$$4m > 8 \qquad\qquad 4m < -12$$
$$m > 2 \quad \text{or} \quad m < -3$$

Solution set: $(-\infty, -3) \cup (2, \infty)$

14. $(2p+3)(3p-1) = 6p^2 - 2p + 9p - 3$
$= 6p^2 + 7p - 3$

15. $(4k-3)^2 = (4k)^2 - 2(4k)(3) + 3^2$
$= 16k^2 - 24k + 9$

16. $(3m^3 + 2m^2 - 5m) - (8m^3 + 2m - 4)$
$= 3m^3 + 2m^2 - 5m - 8m^3 - 2m + 4$
$= 3m^3 - 8m^3 + 2m^2 - 5m - 2m + 4$
$= -5m^3 + 2m^2 - 7m + 4$

17.
$$\begin{array}{r} 2t^3 + 5t^2 - 3t + 4 \\ 3t+1\overline{)6t^4 + 17t^3 - 4t^2 + 9t + 4} \\ \underline{6t^4 + 2t^3} \qquad\qquad\qquad \\ 15t^3 - 4t^2 \qquad\qquad \\ \underline{15t^3 + 5t^2} \qquad\qquad \\ -9t^2 + 9t \qquad \\ \underline{-9t^2 - 3t} \qquad \\ 12t + 4 \\ \underline{12t + 4} \\ 0 \end{array}$$

Answer: $2t^3 + 5t^2 - 3t + 4$

18. $8x + x^3 = x(8 + x^2)$

19. $24y^2 - 7y - 6 = (8y+3)(3y-2)$

20. $5z^3 - 19z^2 - 4z = z(5z^2 - 19z - 4)$
$= z(5z+1)(z-4)$

21. $16a^2 - 25b^4$

Use the difference of two squares formula,

$$x^2 - y^2 = (x+y)(x-y),$$

where $x = 4a$ and $y = 5b^2$.

$$16a^2 - 25b^2 = (4a + 5b^2)(4a - 5b^2)$$

22. $8c^3 + d^3$

Use the sum of two cubes formula,

$$x^3 + y^3 = (x+y)(x^2 - xy + y^2),$$

where $x = 2c$ and $y = d$.

$$8c^3 + d^3 = (2c + d)(4c^2 - 2cd + d^2)$$

23. $16r^2 + 56rq + 49q^2$
$= (4r)^2 + 2(4r)(7q) + (7q)^2$

Use the perfect square formula,

$$x^2 + 2xy + y^2 = (x+y)^2,$$

where $x = 4r$ and $y = 7q$.

$$16r^2 + 56rq + 49q^2 = (4r + 7q)^2$$

24. $\dfrac{(5p^3)^4(-3p^7)}{2p^2(4p^4)} = \dfrac{(5^4p^{12})(-3p^7)}{8p^6}$
$= \dfrac{(625)(-3)p^{19}}{8p^6}$
$= -\dfrac{1875p^{13}}{8}$

25. $\dfrac{x^2-9}{x^2+7x+12} \div \dfrac{x-3}{x+5}$

Multiply by the reciprocal.

$= \dfrac{x^2-9}{x^2+7x+12} \cdot \dfrac{x+5}{x-3}$

Factor and simplify.

$= \dfrac{(x+3)(x-3)}{(x+3)(x+4)} \cdot \dfrac{(x+5)}{(x-3)}$

$= \dfrac{x+5}{x+4}$

26. $\dfrac{2}{k+3} - \dfrac{5}{k-2}$

The LCD is $(k+3)(k-2)$.

$= \dfrac{2(k-2)}{(k+3)(k-2)} - \dfrac{5(k+3)}{(k-2)(k+3)}$

$= \dfrac{2k - 4 - 5k - 15}{(k+3)(k-2)}$

$= \dfrac{-3k-19}{(k+3)(k-2)}$

27. $\dfrac{3}{p^2-4p} - \dfrac{4}{p^2+2p}$

$= \dfrac{3}{p(p-4)} - \dfrac{4}{p(p+2)}$

The LCD is $p(p-4)(p+2)$.

$= \dfrac{3(p+2)}{p(p-4)(p+2)} - \dfrac{4(p-4)}{p(p+2)(p-4)}$

$= \dfrac{3p + 6 - 4p + 16}{p(p-4)(p+2)}$

$= \dfrac{22-p}{p(p-4)(p+2)}$

28. Let $x =$ the amount of candy at \$1.00 per pound.

	Amount of candy	Price per pound	Total price
First candy	x	1.00	$1.00x$
Second candy	10	1.96	1.96(10)
Mixture	$10 + x$	1.60	$1.60(10 + x)$

Solve the equation.

$$1.00x + 1.96(10) = 1.60(10 + x)$$

Multiply by 10 to clear the decimals.

$$\begin{aligned} 10x + 196 &= 16(10 + x) \\ 10x + 196 &= 160 + 16x \\ 36 &= 6x \\ 6 &= x \end{aligned}$$

Use 6 lb of the $1.00 candy.

29. $\left(\frac{5}{4}\right)^{-2} = \left(\frac{4}{5}\right)^{2} = \frac{16}{25}$

30. $\frac{6^{-3}}{6^2} = \frac{1}{6^3 \cdot 6^2} = \frac{1}{6^5}$

31. $$\begin{aligned} 2\sqrt{32} - 5\sqrt{98} &= 2\sqrt{16 \cdot 2} - 5\sqrt{49 \cdot 2} \\ &= 2 \cdot 4\sqrt{2} - 5 \cdot 7\sqrt{2} \\ &= 8\sqrt{2} - 35\sqrt{2} \\ &= -27\sqrt{2} \end{aligned}$$

32. $$\begin{aligned} (5 + 4i)(5 - 4i) &= 25 - 16i^2 \\ &= 25 - 16(-1) \\ &= 25 + 16 = 41 \end{aligned}$$

33. $$\begin{aligned} 10p^2 + p - 2 &= 0 \\ (5p - 2)(2p + 1) &= 0 \end{aligned}$$

$$5p - 2 = 0 \quad \text{or} \quad 2p + 1 = 0$$
$$5p = 2 \qquad\qquad 2p = -1$$
$$p = \frac{2}{5} \quad \text{or} \quad p = -\frac{1}{2}$$

Solution set: $\{-\frac{1}{2}, \frac{2}{5}\}$

34. $k^2 + 2k - 8 > 0$

Solve the equation

$$\begin{aligned} k^2 + 2k - 8 &= 0 \\ (k + 4)(k - 2) &= 0 \end{aligned}$$
$$k + 4 = 0 \quad \text{or} \quad k - 2 = 0$$
$$k = -4 \quad \text{or} \quad k = 2.$$

These numbers divide the number line into three regions.

A B C
−4 2

Test a number from each region in the inequality

$$k^2 + 2k - 8 > 0.$$

Region A: Let $k = -5$.

$$\begin{aligned} (-5)^2 + 2(-5) - 8 &> 0 \quad ? \\ 25 - 10 - 8 &> 0 \quad ? \\ 7 &> 0 \quad \textit{True} \end{aligned}$$

Region B: Let $k = 0$.

$$\begin{aligned} 0^2 + 2(0) - 8 &> 0 \quad ? \\ -8 &> 0 \quad \textit{False} \end{aligned}$$

Region C: Let $k = 3$.

$$\begin{aligned} 3^2 + 2(3) - 8 &> 0 \quad ? \\ 9 + 6 - 8 &> 0 \quad ? \\ 7 &> 0 \quad \textit{True} \end{aligned}$$

The numbers in Regions A and C, not including −4 or 2 because of " >," are solutions.

Solution set: $(-\infty, -4) \cup (2, \infty)$

35. $y = 1.7x + 230$

(a) Let $x = 1982 - 1980 = 2$.

$$\begin{aligned} y &= 1.7(2) + 230 \\ &= 3.4 + 230 \\ y &= 233.4 \end{aligned}$$

The population in 1982 was 233.4 million.

(b) Let $x = 1985 - 1980 = 5$.

$$\begin{aligned} y &= 1.7(5) + 230 \\ &= 8.5 + 230 \\ y &= 238.5 \end{aligned}$$

The population in 1985 was 238.5 million.

(c) Let $x = 1990 - 1980 = 10$.

$$\begin{aligned} y &= 1.7(10) + 230 \\ &= 17 + 230 \\ y &= 247 \end{aligned}$$

The population in 1990 was 247 million.

(d) Let $y = 315$.

$$\begin{aligned} 315 &= 1.7x + 230 \\ 85 &= 1.7x \\ 50 &= x \end{aligned}$$

Since $x = 50$, this corresponds to the year $1980 + 50 = 2030$. Based on this equation, the population will be 315 million in the year 2030.

36. Use the ordered pairs (1986, 70,000,000) and (1994, 25,300,000). The rate of change is the slope.

$$\begin{aligned} m &= \frac{\text{change in } y}{\text{change in } x} \\ &= \frac{70{,}000{,}000 - 25{,}300{,}000}{1986 - 1994} \\ &= \frac{44{,}700{,}000}{-8} \\ &= -5{,}587{,}500 \end{aligned}$$

The rate of change was $-5,587,500$ dollars/yr.

37. Through $(5,-1)$; parallel to $3x - 4y = 12$
Find the slope of

$$\begin{aligned} 3x - 4y &= 12 \\ -4y &= -3x + 12 \\ y &= \frac{3}{4}x - 3. \end{aligned}$$

The slope is $\frac{3}{4}$, so a line parallel to it also has slope $\frac{3}{4}$.

Let $m = \frac{3}{4}$ and $(x_1, y_1) = (5, -1)$ in the point-slope form.

$$\begin{aligned} y - y_1 &= m(x - x_1) \\ y - (-1) &= \frac{3}{4}(x - 5) \\ y + 1 &= \frac{3}{4}(x - 5) \end{aligned}$$

Multiply by 4 to clear the fraction.

$$\begin{aligned} 4(y + 1) &= 3(x - 5) \\ 4y + 4 &= 3x - 15 \\ 4y - 3x &= -19 \\ 3x - 4y &= 19 \end{aligned}$$

For Exercises 38-42, see the answer graphs in the back of the textbook.

38. $y = -2.5x + 5$

Since the equation is in slope-intercept form, the slope is -2.5 and the y-intercept is $(0, 5)$. Plot $(0, 5)$. To find another point on the line, move down 2.5 units and to the right 1 unit. Draw a line through the two points.

39. $-4x + y \leq 5$

Graph the line $-4x + y = 5$, which has intercepts $(0, 5)$ and $\left(-\frac{5}{4}, 0\right)$, as a solid line because the inequality involves " $\leq$." Test $(0,0)$, which yields $0 \leq 5$, a true statement. Shade the region that includes $(0,0)$.

40. $y = \dfrac{1}{3}(x-1)^2 + 2$

The graph is a vertical parabola with vertex at $(1, 2)$. Since $a = \frac{1}{3} > 0$, the graph opens upward. Also, $|a| = \left|\frac{1}{3}\right| = \frac{1}{3} < 1$, so the graph is wider than the graph of $f(x) = x^2$. The points $\left(0, 2\frac{1}{3}\right)$, $(-2, 5)$, and $(4, 5)$ are also on the graph.

41. $\dfrac{x^2}{9} + \dfrac{y^2}{16} = 1$
$a = 3, b = 4$

The graph is an ellipse centered at $(0,0)$ with x-intercepts $(3,0)$ and $(-3,0)$ and y-intercepts $(0,4)$ and $(0,-4)$. Plot the intercepts, and draw the ellipse through them.

42. $25x^2 - 16y^2 = 400$

In $\dfrac{x^2}{a^2} - \dfrac{y^2}{b^2} = 1$ form, the equation is

$$\frac{x^2}{16} - \frac{y^2}{25} = 1.$$

The graph is a hyperbola with $a = 4$ and $b = 5$. The x-intercepts are $(4,0)$ and $(-4,0)$. To sketch the graph, draw the diagonals of the fundamental rectangle with corners $(4,5), (4,-5), (-4,-5)$, and $(-4,5)$. These are the asymptotes. Graph a branch of the hyperbola through each intercept approaching the asymptotes.

43. **(a)** Only the graphs of the equations in Exercises 38 and 40 pass the vertical line test and are functions.

(b) Only the graph of the equation in Exercise 38 passes the horizontal line test. Thus, this is the graph of a one-to-one function that has an inverse.

44. $5x - 3y = 14 \quad (1)$
$2x + 5y = 18 \quad (2)$

Multiply equation (1) by 5 and equation (2) by 3. Then add the results.

$$\begin{array}{rl} 25x - 15y = & 70 \\ 6x + 15y = & 54 \\ \hline 31x \qquad = & 124 \\ x = & 4 \end{array}$$

Substitute 4 for x in equation (1) to find y.

$$\begin{aligned} 5x - 3y &= 14 \quad (1) \\ 5(4) - 3y &= 14 \\ 20 - 3y &= 14 \\ -3y &= -6 \\ y &= 2 \end{aligned}$$

Solution set: $\{(4, 2)\}$

45. $$\begin{aligned} x + 2y + 3z &= 11 \quad (1) \\ 3x - y + z &= 8 \quad (2) \\ 2x + 2y - 3z &= -12 \quad (3) \end{aligned}$$

To eliminate z, add equations (1) and (3).

$$\begin{aligned} x + 2y + 3z &= 11 \quad (1) \\ 2x + 2y - 3z &= -12 \quad (3) \\ \hline 3x + 4y \quad &= -1 \quad (4) \end{aligned}$$

To eliminate z again, multiply equation (2) by 3 and add the result to equation (3).

$$\begin{aligned} 9x - 3y + 3z &= 24 \\ 2x + 2y - 3z &= -12 \quad (3) \\ \hline 11x - y \quad &= 12 \quad (5) \end{aligned}$$

Multiply equation (5) by 4 and add the result to equation (4).

$$\begin{aligned} 44x - 4y &= 48 \\ 3x + 4y &= -1 \quad (4) \\ \hline 47x \quad &= 47 \\ x &= 1 \end{aligned}$$

Substitute 1 for x in equation (5) to find y.

$$\begin{aligned} 11x - y &= 12 \quad (5) \\ 11(1) - y &= 12 \\ 11 - y &= 12 \\ -y &= 1 \\ y &= -1 \end{aligned}$$

Substitute 1 for x and -1 for y in equation (2) to find z.

$$\begin{aligned} 3x - y + z &= 8 \quad (2) \\ 3(1) - (-1) + z &= 8 \\ 3 + 1 + z &= 8 \\ 4 + z &= 8 \\ z &= 4 \end{aligned}$$

Solution set: $\{(1, -1, 4)\}$

46. $\begin{vmatrix} 2 & 4 & 5 \\ 1 & 3 & 0 \\ 0 & -1 & -2 \end{vmatrix}$ Expand about column 1.

$$\begin{aligned} &= 2\begin{vmatrix} 3 & 0 \\ -1 & -2 \end{vmatrix} - 1\begin{vmatrix} 4 & 5 \\ -1 & -2 \end{vmatrix} + 0\begin{vmatrix} 4 & 5 \\ 3 & 0 \end{vmatrix} \\ &= 2(-6 - 0) - 1[-8 - (-5)] + 0 \\ &= -12 - 1(-3) \\ &= -12 + 3 = -9 \end{aligned}$$

47. $f(x) = 2^x$

Make a table of values.

x	-2	-1	0	1	2
y	$\frac{1}{4}$	$\frac{1}{2}$	1	2	4

Plot the ordered pairs from the table, and draw a smooth curve through the points. See the answer graph in the back of the textbook.

48. $5^{x+3} = \left(\frac{1}{25}\right)^{3x+2}$

$$5^{x+3} = \left[\left(\frac{1}{5}\right)^2\right]^{3x+2}$$

Write each side with the same base.

$$\begin{aligned} 5^{x+3} &= (5^{-2})^{3x+2} \\ 5^{x+3} &= 5^{-2(3x+2)} \end{aligned}$$

If $a^x = a^y$, then $x = y$. Set the exponents equal.

$$\begin{aligned} x + 3 &= -2(3x + 2) \\ x + 3 &= -6x - 4 \\ 7x &= -7 \\ x &= -1 \end{aligned}$$

This solution checks.

Solution set: $\{-1\}$

49. $f(x) = \log_3 x$

Write in exponential form.

$$\begin{aligned} y &= \log_3 x \\ 3^y &= x \end{aligned}$$

Make a table of values.

x	$\frac{1}{9}$	$\frac{1}{3}$	1	3	9
y	-2	-1	0	1	2

Plot the ordered pairs, and draw a smooth curve through the points. See the answer graph in the back of the textbook.

50. $\log_5 x + \log_5(x+4) = 1$

Use the multiplication property.

$$\log_5[x(x+4)] = 1$$

Change to exponential form.

$$\begin{aligned} 5^1 &= x(x+4) \\ 5 &= x^2 + 4x \\ x^2 + 4x - 5 &= 0 \\ (x+5)(x-1) &= 0 \end{aligned}$$

$$x + 5 = 0 \quad \text{or} \quad x - 1 = 0$$
$$x = -5 \quad \text{or} \quad x = 1$$

Check both potential solutions. The number 1 makes the equation true. The number -5 must be rejected because it yields an equation in which the logarithm of a negative number must be found.

Solution set: $\{1\}$